Lecture Notes in Mechanical Engineering

Lecture Notes in Mechanical Engineering (LNME) publishes the latest developments in Mechanical Engineering—quickly, informally and with high quality. Original research reported in proceedings and post-proceedings represents the core of LNME. Volumes published in LNME embrace all aspects, subfields and new challenges of mechanical engineering. Topics in the series include:

- Engineering Design
- Machinery and Machine Elements
- Mechanical Structures and Stress Analysis
- Automotive Engineering
- Engine Technology
- Aerospace Technology and Astronautics
- Nanotechnology and Microengineering
- Control, Robotics, Mechatronics
- MEMS
- Theoretical and Applied Mechanics
- Dynamical Systems, Control
- Fluid Mechanics
- Engineering Thermodynamics, Heat and Mass Transfer
- Manufacturing
- Precision Engineering, Instrumentation, Measurement
- Materials Engineering
- Tribology and Surface Technology

To submit a proposal or request further information, please contact the Springer Editor in your country:

China: Dr. Mengchu Huang at mengchu.huang@springer.com
India: Priya Vyas at priya.vyas@springer.com
Rest of Asia, Australia, New Zealand: Swati Meherishi at swati.meherishi@springer.com
All other countries: Dr. Leontina Di Cecco at Leontina.dicecco@springer.com

To submit a proposal for a monograph, please check our Springer Tracts in Mechanical Engineering at http://www.springer.com/series/11693 or contact Leontina.dicecco@springer.com

Indexed by SCOPUS. The books of the series are submitted for indexing to Web of Science.

More information about this series at http://www.springer.com/series/11236

Lung-Jieh Yang · A. Noorul Haq ·
Lenin Nagarajan
Editors

Proceedings of ICDMC 2019

Design, Materials, Cryogenics, and Constructions

Springer

Editors
Lung-Jieh Yang
Tamkang University
New Taipei City, Taiwan

A. Noorul Haq
National Institute of Technology
Tiruchirappalli
Tiruchirappalli, Tamil Nadu, India

Lenin Nagarajan
Vel Tech Rangarajan Dr. Sagunthala R&D
Institute of Science and Technology
Chennai, Tamil Nadu, India

ISSN 2195-4356 ISSN 2195-4364 (electronic)
Lecture Notes in Mechanical Engineering
ISBN 978-981-15-3630-4 ISBN 978-981-15-3631-1 (eBook)
https://doi.org/10.1007/978-981-15-3631-1

This Springer imprint is published by the registered company Springer Nature Singapore Pte Ltd.
The registered company address is: 152 Beach Road, #21-01/04 Gateway East, Singapore 189721, Singapore

Preface

Indian manufacturing industries need to give more emphasis on matching quality with international competitors after globalization. Countries' economy now depends on experts from Indian manufacturing industries. Unless innovations are made and recent technologies are introduced at this juncture in Indian manufacturing industries, the dream of becoming a superpower by another decade would be a dream only. The international market is now targeted at the Indian population and our manufacturers to implement the emerging new technologies for the same. This International Conference on Design, Materials, Cryogenics and Constructions (ICDMC'19) was the great boon for the Indian industries to get a glimpse of global scenario on sustainable design, newer materials development, applications of low-temperature materials and sustainable constructions. ICDMC'19 made a common platform that helped to have free interaction between industries, researchers and academicians in India as well as in global level.

The conference attracted 125 participants from Germany, France, Bangladesh, South Africa and India. Among them, more than 100 participants were from external premier institutions like National Institute of Technology from different states, Sastra University, Vellore Institute of Technology, etc.; industries like Ashok Leyland, WABCO India, India Pistons, etc.; and research laboratories like Indira Gandhi Centre for Atomic Research, Indian Space Research Organisation, etc. The papers were classified into their major contribution and were discussed in eight technical sessions for two days.

The speakers representing top-notch industries from Tamil Nadu, India, and the premier institutions from Germany and France were invited to deliver the inaugural addresses and keynote addresses and contribute as session chairs to adequately bridge the institution–industry gap.

New Taipei City, Taiwan	Lung-Jieh Yang
Tiruchirappalli, India	A. Noorul Haq
Chennai, India	Lenin Nagarajan

Acknowledgements

The conference was supported by the Vel Tech Rangarajan Dr. Sagunthala R&D Institute of Science and Technology, Chennai, India, and by the knowledge partners given by Mias School of Business Ctu in Prague; Indo German Center for Higher Education; Ecole D'ingenieurs; Gelsenkirchen Bocholt Recklinghausen, University of Applied Sciences; Institut für werkzeuglose Fertigung GmbH; FH Aachen University of Applied Sciences.

Contents

About the Editors

Dr. Lung-Jieh Yang is presently working as Professor in the Department of Mechanical and Electro-Mechanical Engineering, Tamkang University, Taiwan. He obtained his B.S. (Aerospace Engineering) from National Cheng-Kung University, Taiwan, M.S. (Mechanical Engineering) from Tamkang University, Taiwan and Ph.D. (Applied Mechanics) from National Taiwan University, Taiwan. His total teaching and research experience is 21 years. His fields of speciality are Micro Aerial Vehicles & Aerospace Engineering and Polymer MEMS. He has published 60 papers in International Journals and 93 papers in International Conferences. He received 09 patents from US and Taiwan Patent Offices.

Dr. A. Noorul Haq is working as Professor (Higher Administrative Grade) in the Department of Production Engineering, National Institute of Technology, Tiruchirappalli, Tamilnadu, India. He completed his B.E. (Mechanical Engineering) at Annamalai University, M.E. (Mechanical Engineering) at University of Madras, Tamil Nadu, and Ph.D. (Mechanical Engineering) at Indian Institute of Technology, Delhi. His total teaching and research experience is 42 years. His major areas of research interests include Operations Management, Industrial Engineering, Optimization Techniques and Supply Chain Management. He has published 141 research articles out of which 96 are in internationally reputed journals.

Dr. Lenin Nagarajan is currently working as Professor in the Department of Mechanical Engineering, Vel Tech Rangarajan Dr. Sagunthala R&D Institute of Science and Technology, Chennai, India. He obtained his B.E. (Mechanical Engineering) from Madurai Kamaraj University, Tamil Nadu, M.E. (Production Engineering) and Ph.D. (Mechanical Engineering) from Anna University, Tamil Nadu. He is having 13 years of experience in Teaching and Research. His major areas of research interests include Facility Layout Design, Manufacturing Optimization and Composite Materials. He has published 25 papers in respected international journals.

Numerical Investigation on Amphibious UAV Using Turbulent Models for Drag Reduction

P. Gokul Raj, Balasubramanian Esakki, P. Vikram, and Lung-Jieh Yang

Abstract Unmanned Aerial Vehicles (UAVs) have gained significant consideration lately because of its flexibility in arrangement for multiple activities. Especially amphibious UAVs' integration of air cushion vehicle and multirotor has huge demand in military, maritime and seaside protection applications. Steadiness and execution of these sorts of vehicles profoundly rely upon streamlined cooperation of multirotor regarding different wind conditions. The present work focuses on limiting the drag and enhancing the streamlined execution attributes. CFD examination is performed through considering different turbulent models such as k-ω, k-ε and SST k-ω (shear pressure transport) to assess the co-efficient of drag of amphibious UAV. Static investigation is performed through varying the Angle of Attack (AoA) from 0° to 10° under relative velocities of 3, 5, 8 and 10 m/s. The turbulence kinetic energy shapes anticipated the streamline of wind stream around the vehicle.

Keywords Amphibious UAV · CFD · RANS model · Drag · Blunt body

P. Gokul Raj · B. Esakki (✉)
Department of Mechanical Engineering, Vel Tech Rangarajan Dr. Sagunthala R&D Institute of Science and Technology, Avadi, Chennai 600062, India
e-mail: esak.bala@gmail.com

P. Gokul Raj
e-mail: gokulrajprithiv@gmail.com

P. Vikram
Department of Aeronautical Engineering, Vel Tech Rangarajan Dr. Sagunthala R&D Institute of Science and Technology, Avadi, Chennai 600062, India
e-mail: Vikram.veeran@gmail.com

L.-J. Yang
School of Mechanical and Electro Mechanical Engineering, Tamkang University, New Taipei City, Taiwan
e-mail: ljyang@mail.tku.edu.tw

L.-J. Yang et al. (eds.), *Proceedings of ICDMC 2019*,
Lecture Notes in Mechanical Engineering,
https://doi.org/10.1007/978-981-15-3631-1_1

1 Introduction

Unmanned Aerial Vehicles (UAVs) are dominatingly utilized in various applications [1, 2] including accuracy agribusiness, environmental monitoring, aerial imaging, pursuit and protection, observation and surveillance, control line and telecom tower assessments and so on. Be that as it may, the use of UAVs in water quality checking and gathering of water tests in remote water bodies is rare. A UAV which can fly, land and skim along the water surface forcing parcel of difficulties as far as control in flight transition, selection of materials, propulsion, energy consumption and pay load capacity [3]. Also, different factors, for example, as durability, reliability, safety and minimal cost, are utmost important for industrial demand and client necessity. There are few drifting UAVs which have been produced and popularized in the market [4]. These vehicles are planned to cover substantial regions of water bodies in limited ability to focus time. In contrast to other gliding vehicles, because of the guideline of air cushion vehicle [5], the erosion between the vehicle and water surface is kept away, thereby gigantic measure of vitality is spared. The vertical take-off and landing capacity of vehicle can position the vehicle in exact water areas crosswise over waterways, lakes and other water bodies to perform water quality analysis. One of the streamlined parameters affecting continuance of land and/or water capable vehicle is drag. There are few investigations led to ascertain the drag of fixed wing and rotating wing vehicle. Sitaraman and Baeder [6] completed streamlined investigation of quadrotor utilizing Navier–Stokes condition, and wake associations are considered. Steijl et al. [7] contemplated rotor and fuselage collaboration utilizing CFD investigation and sliding plane procedure [8]. The fierce stream qualities amid inviscid and viscid liquids are re-enacted in CFD to dissect the conduct of covers of multirotor framework. Biava et al. [9] inspected the streamlined conduct of helicopter through CFD and exploratory examinations. Kusyumov et al. [10] completed CFD examination of ANSAT helicopter to decide lift and drag powers in viscous flow conditions. Yoon et al. [11] dissected stream appropriation over settled wing aerofoil through shifted AoA and co-effective of drag and lift is obtained. Abudarag et al. [12–15] considered the stream partition among rotor and fuselage utilizing CFD, and fierce stream attributes are analysed. The present work focuses on performing CFD analysis of various tempestuous models, for example, k-ω, k-ε and SST k-ω to determine the co-efficient of drag under different AoA and relative velocity conditions.

2 Modelling

The conceptualization amphibian model is designed through inculcating the principles of multirotor and hover craft as shown in Fig. 1. In order to reduce the computational effort, the model is scaled down to factor of 1:0.25. Also, to maintain the

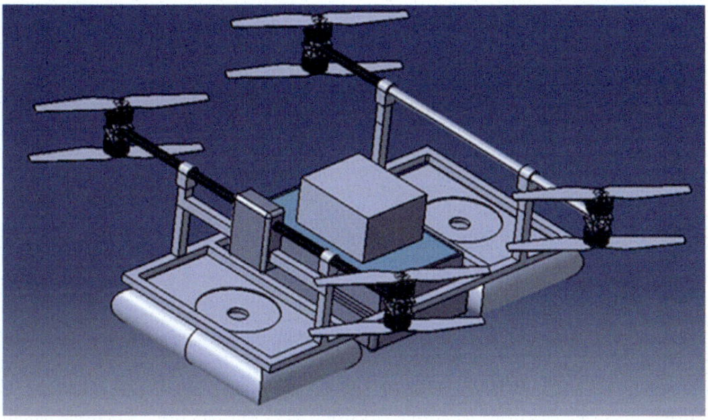

Fig. 1 Conceptualized model of amphibious vehicle

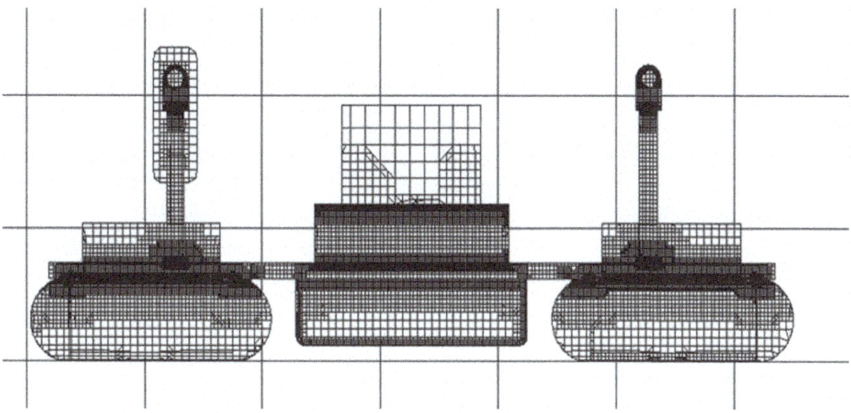

Fig. 2 Structured mesh of the scaled-down amphibious vehicle

Reynolds number with reference to the prototype, velocity is increased four times and various wind speed conditions are accounted for simulation studies.

The scaled-down amphibian structure is meshed with cut-cell element using ICEM tool. Grid quality is verified with skewness and orthogonality checks. The structured meshed image of the scaled-down model is as shown in Fig. 2.

3 Computational Fluid Dynamic Analysis

CFD analysis is performed through varying the AoA (0°–10°) under various relative air velocity conditions. Simulation studies are conducted for the following four cases of velocity conditions.

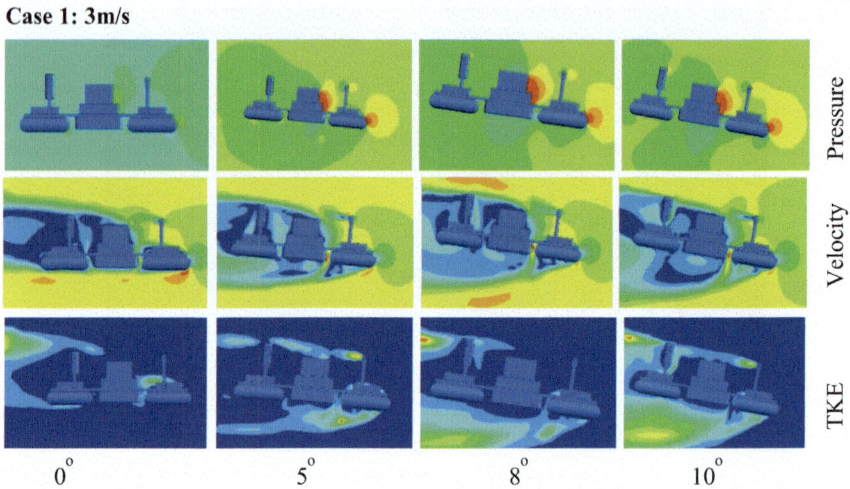

Fig. 3 Velocity, pressure and turbulence kinetic energy contours at 3 m/s for different AoA

Case 1: 3 m/s

At 0° AoA, high turn around stream district is happened behind the water sampler module. While expanding the AoA, the velocity is streamlined and at 8° AoA, reverse flow around amphibious vehicle is streamlined which lessens drag (Fig. 3). Further, increment of AoA prompts arrangement of fierce area at the back of vehicle that may cause increment of the drag. It is clear from pressure contour of differed AoA is that up to 8°, there is a decline in pattern of weight, further increment of AoA causes increment in pressure in the upstream district, and there is a probability of instability of vehicle. Additionally, over 8° of AoA, kinematic energy disturbance is expanded which may prompt vibration and unfit to control the vehicle in the ideal way.

Case 2: 5 m/s

Similar phenomenon is observed as in the case of 3 m/s. However, the intensity of velocity and pressure is quite high as compared to 3 m/s (Fig. 4).

Case 3: 8 m/s

Figure 5 shows the variation velocity, pressure and turbulence kinematic energy contours at 8 m/s for all the four different AoA.

Case 4: 10 m/s

The intensity of turbulence is increased at high relative air speed which can be seen in Fig. 6. For various wind speed conditions and AoA, the co-efficient of drag and lift is estimated which is given in Table 1. It is observed that increase in AoA and wind speed causes increase in drag and decrease in lift. The lift-to-drag ratio for the wind speed of 8 m/s given in Table 2 reveals that at 5° AoA, high amount of lift is generated with minimal drag.

For various turbulent models such as k-ω, k-ε and SST k-ω, CFD analysis is performed and corresponding drag force is determined. Since k-ω is considered as a

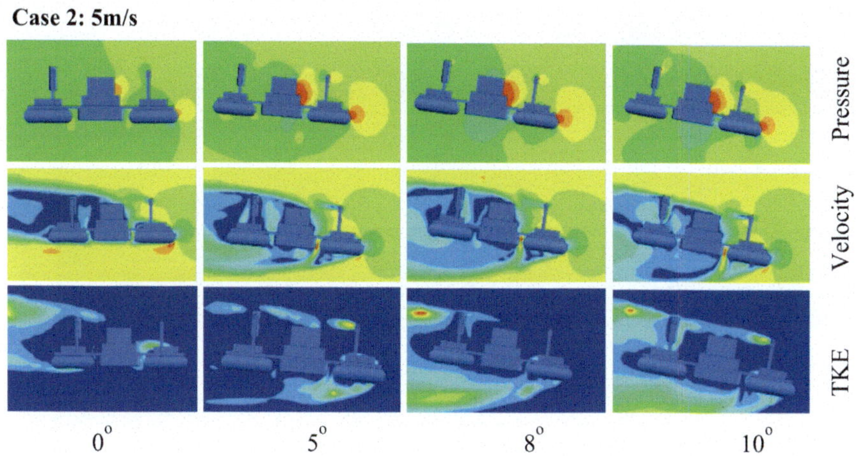

Fig. 4 Velocity, pressure and turbulence kinetic energy contours at 5 m/s for different AoA

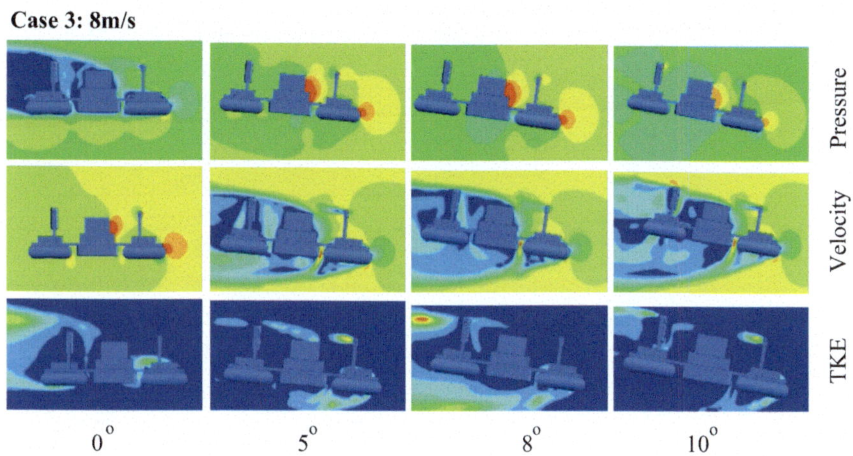

Fig. 5 Velocity, pressure and turbulence kinetic energy contours at 8 m/s for different AoA

standard model to measure the drag force, with reference to that error is calculated. Minimum error is obtained for these models which are given in Table 3, and they can be used to calculate the drag force.

4 Conclusion

CFD analysis is performed for the designed amphibian structure through varying the AoA from 0° to 10° under different wind speed conditions (3, 5, 8 and 10 m/s), and

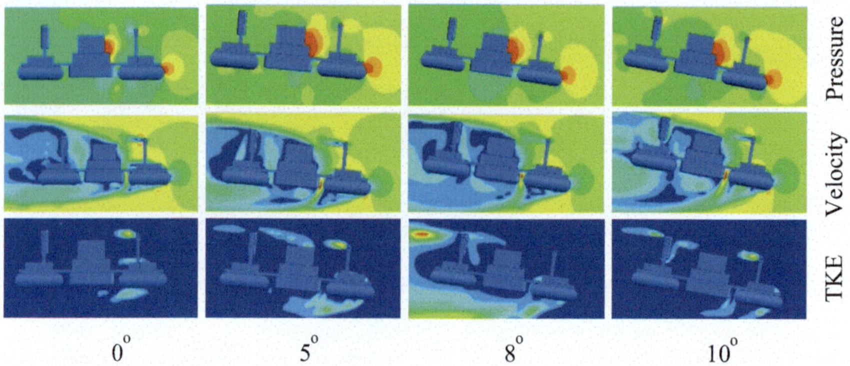

Fig. 6 Velocity, pressure and turbulence kinetic energy contours at 10 m/s for different AoA

Table 1 Estimation of drag for various AoA under different wind speed conditions

Wind speed (m/s)	Parameters	Angle of Attack			
		0°	5°	8°	10°
3	C_D	0.566	0.534	0.489	0.483
	C_L	0.164	0.171	0.083	0.082
	Drag	0.783	0.931	1.009	1.050
5	C_D	0.562	0.532	0.489	0.481
	C_L	0.159	0.172	0.086	0.075
	Drag	2.160	2.577	2.805	2.903
8	C_D	0.573	0.531	0.485	0.463
	C_L	0.152	0.173	0.081	0.023
	Drag	5.631	6.588	7.114	7.159
10	C_D	0.618	0.531	0.486	0.465
	C_L	0.123	0.172	0.081	0.080
	Drag	9.481	10.274	11.133	11.237

Table 2 Lift-to-drag ratio at 8 m/s

AoA (°)	Drag (*N*)	Lift (*N*)	L/D ratio
0	5.631	1.499	0.266
5	6.588	2.151	0.326
8	7.114	1.187	0.166
10	7.159	0.369	0.051

Table 3 Computation of drag using various turbulent models

AoA (°)	Relative velocity (m/s)	Drag (*N*) (k-ω model)	Drag (*N*) (k-ε model)	Drag (*N*) (SST k-ω model)	Error (k-ω/SST k-ω)	Error (k-ω/k-ε)
5	5	2.568	2.610	2.577	−0.009	−0.042
	8	6.572	6.619	6.588	−0.016	−0.047
8	5	2.770	2.784	2.805	−0.035	−0.014
	8	7.107	7.121	7.114	−0.007	−0.014

corresponding co-efficient of drag and lift force are calculated. At 8 m/s and 5° AoA, maximum L/D ratio is obtained in comparison to other operating conditions, and hence, it is well suited for cruise flight. Comparative evaluation of various turbulent models such as k-ω, k-ε and SST k-ω suggested that error between standard k-ω and other models is minimum, and hence, they can be also used for dynamic analysis.

References

1. Valavanis KP, Vachtsevanos GJ (2014) Handbook of unmanned aerial vehicles. Springer
2. Hassanalian M, Abdelkefi A (2017) Classifications, applications, and design challenges of drones: a review. Prog Aerosp Sci 91:99–131
3. Pisanich G, Morris S (2002) Fielding an amphibious UAV—development, results, and lessons learned. In: Digital avionics systems conference, vol 2, pp 8C4–8C4
4. Boxerbaum AS, Werk P, Quinn RD, Vaidyanathan R (2005) Design of an autonomous amphibious robot for surf zone operation: part I mechanical design for multi-mode mobility. In: Advanced intelligent mechatronics proceedings, IEEE/ASME international conference, pp 1459–1464
5. Amyot JR (2013) Hovercraft technology, economics and applications, vol 11. North Holland
6. Sitaraman J, Baeder J (2002). Analysis of quad-tiltrotor blade aerodynamic loads using coupled CFD/free wake analysis. In: 20th AIAA applied aerodynamics conference, p 2813
7. Steijl R, Barakos G, Badcock K (2007) CFD Analysis of rotor-fuselage aerodynamics based on a sliding mesh algorithm
8. Grondin G, Thipyopas C, Moschetta JM (2010) Aerodynamic Analysis of a multi-mission short shrouded coaxial UAV: part III-CFD for hovering flight. In: 28th AIAA applied aerodynamics conference, p 5073
9. Biava M, Khier W, Vigevano L (2012) CFD prediction of air flow past a full helicopter configuration. Aerosp Sci Technol 19(1):3–18
10. Kusyumov A, Mikhailov SA, Garipov AO, Nikolaev EI, Barakos G (2015) CFD simulation of fuselage aerodynamics of the ANSAT helicopter prototype. Trans Control Mech Syst 1(7):318–324
11. Yoon S, Lee HC, Pulliam TH (2016) Computational analysis of multi-rotor flows. In: 54th AIAA aerospace sciences meeting, p 0812
12. Abudarag S, Yagoub R, Elfatih H, Filipovic Z (2017) Computational analysis of unmanned aerial vehicle (UAV). In: AIP conference proceedings, vol 1798, no 1. AIP Publishing, p 020001
13. Lesieur M (1987) Turbulence in fluids: stochastic and numerical modelling. Nijhoff, Boston, MA

14. Versteeg HK, Malalasekera W (2007) An introduction to computational fluid dynamics: the finite volume method. Pearson Education
15. Ferziger JH, Peric M (2012) Computational methods for fluid dynamics. Springer Science & Business Media

Design and Simulation of Self-balancing Robot Using Reaction Sphere

M. Vignesh and A. S. Praveen

Abstract This paper highlights the work for designing a reaction sphere pendulum. First, the equation of motion for the reaction sphere pendulum is developed using Lagrangian's mechanics. 3D model of the experimental setup is designed using SOLIDWORKS, and its constructional methods are discussed in this paper. Based on the system parameters derived from 3D model, equation of motion was simulated using MATLAB and its stabilization behaviour was observed for LQR controller. Simulation results highlighted that the system was able to stabilize to its desired position within 5 s, when disturbed from 18°.

Keywords Reaction sphere · Self-balancing · Inverted pendulum

1 Introduction

Reaction wheel pendulum is a simple pendulum which has a motor-driven fly wheel attached opposite to its pivoted end. Reaction wheel pendulum is two degree-of-freedom robot. Pendulum pivoted at its base constitutes the first degree of freedom, and the controlled rotating wheel is the second degree of freedom. Since the two DOF are controlled by a single actuator, i.e. motor, reaction wheel pendulum comes under category of nonlinear control of under actuated systems.

Reaction wheel system is primarily used in stabilizing satellites when they drift from their intended position. Three reaction wheels are used in satellites to aid stabilization in all the three axes. However, in recent times, reaction sphere actuator has been a keen area of interest among researchers. Single reaction sphere actuator has potential to replace three reaction wheels in a balancing system.

This paper focuses on the design and simulation of a reaction sphere-based balancing pendulum. Initially modified equation of motion for the system is developed,

M. Vignesh (✉) · A. S. Praveen
Department of Mechanical Engineering, Vel Tech Rangarajan Dr. Sagunthala R&D Institute of Science and Technology, Avadi, Chennai 600062, India
e-mail: vigneshmanoharan@yahoo.co.in

L.-J. Yang et al. (eds.), *Proceedings of ICDMC 2019*,
Lecture Notes in Mechanical Engineering,
https://doi.org/10.1007/978-981-15-3631-1_2

and then, the construction of experimental design is discussed. Based on the parameters derived from the experimental design, simulation is performed and its stability is observed.

2 Equation of Motion

Schematic of a reaction sphere pendulum is shown in Fig. 1, and its parameters are described below:

q_1 Angle that pendulum makes with vertical in clockwise direction (°)
q_2 Angle that sphere makes with the vertical in clockwise direction (°)
q_3 Angle that drive wheel makes with the vertical in clockwise direction (°)
D_S Diameter of the sphere (m)
D_W Diameter of the drive wheel (m)
i D_S/D_W, speed ratio of the sphere to drive wheel.

$\dot{q}_1$ Angular velocity of the pendulum shaft and support system (rad/s)
$\dot{q}_2$ Angular velocity of the sphere (rad/s)
$\dot{q}_3$ Angular velocity of the drive wheel (rad/s)
$\ddot{q}_1$ Angular acceleration of the pendulum shaft and support system (rad/s^2)
$\ddot{q}_2$ Angular acceleration of the sphere (rad/s^2)
$\ddot{q}_3$ Angular acceleration of the drive wheel (rad/s^2).

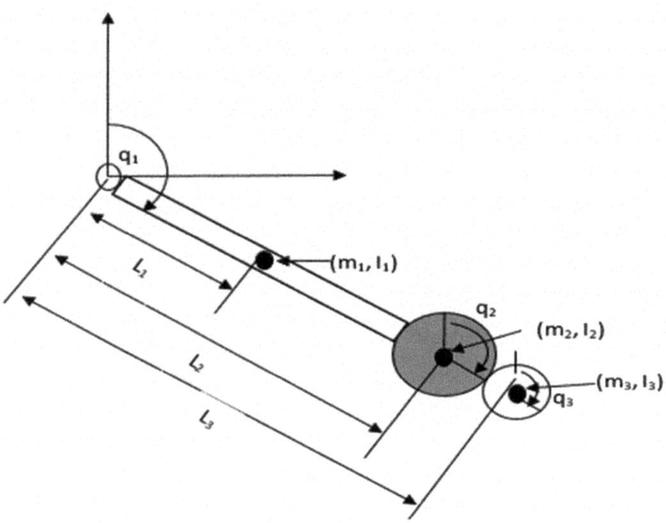

Fig. 1 Schematic of reaction sphere pendulum

$$\ddot{q}_1 = \frac{(m_{22} * \emptyset(q_1))}{(m_{11} * m_{22} - m_{12} * m_{21})} - \frac{(m_{12} * \tau)}{(m_{11} * m_{22} - m_{12} * m_{21})} \tag{1}$$

$$\ddot{q}_2 = \frac{(m_{11} * \tau)}{(m_{11} * m_{22} - m_{12} * m_{21})} - \frac{(m_{21} * \emptyset(q_1))}{(m_{11} * m_{22} - m_{12} * m_{21})} \tag{2}$$

where $m_{11} = \left(m_1 * l_1^2 + I_1 + m_2 * l_2^2 + I_2 + m_3 * l_3^2 + I_3\right)$, $m_{12} = I_2 + i * I_3$, $\emptyset m_{21} = I_2 + i * I_3$, $m_{22} = \left(I_2 + i^2\right)$ and $\emptyset(q_1) = M * g * \sin(q_1)$.

3 Experimental Design

As illustrated in Fig. 2, base of the experimental setup consists of pillow block bearings fixed firmly to a work table. Steel shaft is supported on the bearings provided on the pillow blocks with an overhanging end, and axial movement is constrained with help of set screws provided on the pillow blocks. Shaft support is connected to the overhanging end of the steel shaft and locked with help of screw provided on the shaft support. Pendulum shaft bottom is connected to the shaft support with help of mounting holes provided on the bottom of the pendulum shaft. Pendulum shaft is received on the provision on the pendulum shaft bottom, and its axial movement is arrested with help of a screw. Pendulum shaft top is connected to the other end of the pendulum shaft in a similar manner mentioned above. Ball bearings or ball transfer units are mounted on the provisions provided in the pendulum shaft top. Spherical ball is assembled into the recess provided, and it contacts the bearing ball provided. Mounting bracket is mounted on to the pendulum shaft top using the mounting holes provided. Stepper motor is mounted on the top side, and later, drive wheel is attached to the motor using set screws provided.

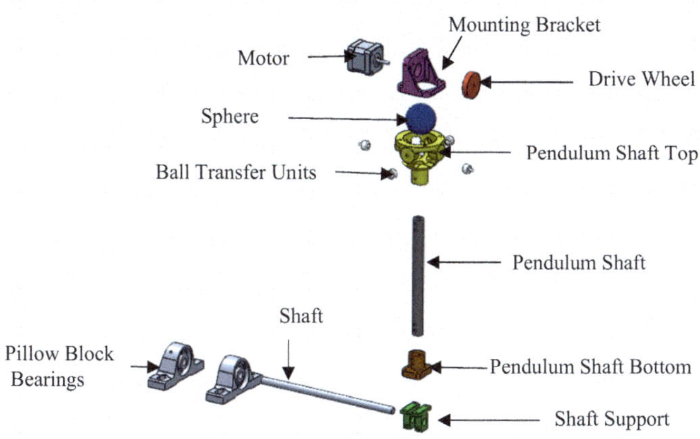

Fig. 2 Experimental setup design

4 Simulation

4.1 Controllability

Equation of motion can be converted into a linearized system as described in [5].

Equation of motion is converted into linearized form as described below:

$$\frac{\mathrm{d}}{\mathrm{d}t}\begin{bmatrix} q_1 \\ \dot{q}_1 \\ q_2 \\ \dot{q}_2 \end{bmatrix} = \begin{bmatrix} 0 & 1\,0\,0 \\ \frac{(m_{22}*M*g)}{(m_{11}*m_{22}-m_{12}*m_{21})} & 0\,0\,0 \\ 0 & 0\,0\,1 \\ -\frac{(m_{21}*M*g)}{(m_{11}*m_{22}-m_{12}*m_{21})} & 0\,0\,0 \end{bmatrix}\begin{bmatrix} q_1 \\ \dot{q}_1 \\ q_2 \\ \dot{q}_2 \end{bmatrix} + \tau\begin{bmatrix} 0 \\ -\frac{(m_{12})}{(m_{11}*m_{22}-m_{12}*m_{21})} \\ 0 \\ \frac{(m_{11})}{(m_{11}*m_{22}-m_{12}*m_{21})} \end{bmatrix} \tag{3}$$

The above Eq. (3) is of the form

$$\frac{\mathbf{d}\boldsymbol{X}}{\mathbf{d}\boldsymbol{t}} = \boldsymbol{AX} + \boldsymbol{B\tau} \tag{4}$$

Then,

$$A = \begin{bmatrix} 0 & 1\,0\,0 \\ \frac{(m_{22}*M*g)}{(m_{11}*m_{22}-m_{12}*m_{21})} & 0\,0\,0 \\ 0 & 0\,0\,1 \\ -\frac{(m_{21}*M*g)}{(m_{11}*m_{22}-m_{12}*m_{21})} & 0\,0\,0 \end{bmatrix}, B = \begin{bmatrix} 0 \\ -\frac{(m_{12})}{(m_{11}*m_{22}-m_{12}*m_{21})} \\ 0 \\ \frac{(m_{11})}{(m_{11}*m_{22}-m_{12}*m_{21})} \end{bmatrix}, X = \begin{bmatrix} q_1 \\ \dot{q}_1 \\ q_2 \\ \dot{q}_2 \end{bmatrix}$$

Controllability of the system shows whether a system at an initially disturbed state can be driven to final state within the stipulated time. Theorem of controllability states that a dynamic system with *A*, *B* is controllable if controllability matrix *C* which is given by

$$C = [B\ AB\ A^2B\ \ldots\ A^{n-1}B] \tag{5}$$

has a full row rank of *n*.

A and *B* matrices are defined into MATLAB, and then, the following command is used to find the controllability.

`Syntax: Rank (ctrb (A, B))`

As shown in Fig. 3, MATLAB output returns rank as 4. Since the rank of

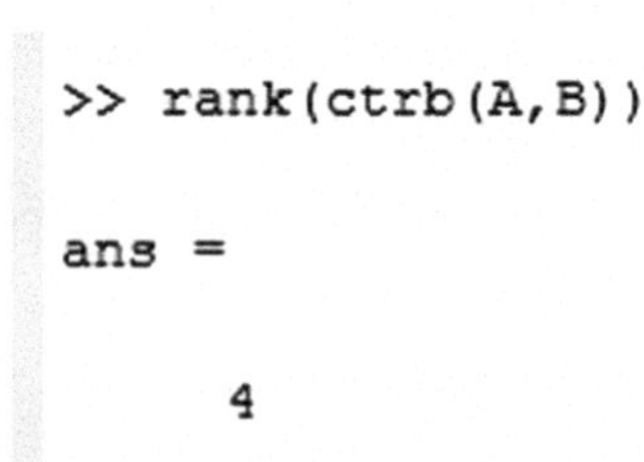

Fig. 3 Output of controllability simulation in MATLAB

Table 1 Parameters for equation of motion

Parameters	Value	Parameters	Value
m_1	0.412 kg	D_W	0.0425 m
m_2	0.510 kg	i	D_S/D_W
m_3	0.014 kg	G	9.81 N m/s^2
l_1	0.11896298 m	I_1	(0.00809528) kg m^2
l_2	0.27279807 m	I_2	(0.00012763) kg m^2
l_3	0.3189 m	I_3	(0.00000343) kg m^2
D_S	0.050 m	T_m	27.4×10^{-3} N m/A

controllability matrix (4 × 4) is a full-rank matrix, i.e. 4, system is controllable.

4.2 *Equation of Motion Simulation*

Equations of motion (1) and (2) are coupled differential equations which are modelled in MATLAB for solving. Parameters used in the simulation are listed in Table 1.

4.3 *Simulation with No Control Torque*

The system was simulated with the following initial conditions: angle of the pendulum $(q_1) = 18°$, velocity of the pendulum $(\dot{q}_1) = 0$ rad/s, angle of the sphere $(q_2) = 0°$, velocity of the sphere $(\dot{q}_1) = 0$ rad/s, time period $(t) = 15$ s.

As shown in Fig. 4. when the pendulum is released from its disturbed state of 18°, it keeps oscillating around 180° and does not converge to the stable equilibrium. Since the friction and air drag parameters are not considered in the governing equation, pendulum does not converge to its neutral position.

4.4 *Simulation with Control Torque*

Linear Quadratic Regulator (LQR) controller estimates the error signal by calculating the difference of measured value against the reference value of the position and then produces a proportional control output. Schematic of the controller is shown in Fig. 6. LQR assumes a linear dynamic system, where $X(t)$ is the state at time t, $u(t)$ is the input at time t, K is a feedback gain matrix, and Q and R are weighing matrixes.

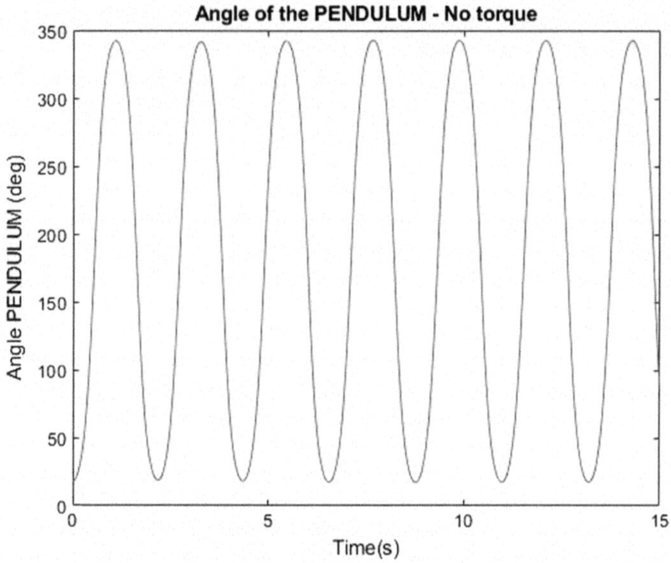

Fig. 4 Simulation output with no control torque

$$X = AX + B\tau \tag{6}$$

$$u(t) = -KX \tag{7}$$

Initial condition for the simulation is as follows:

Angle of the pendulum $(q_1) = 18°$, velocity of the pendulum $(\dot{q}_1) = 0$ rad/s, angle of the sphere $(q_2) = 0°$, velocity of the sphere $(\dot{q}_1) = 0$ rad/s, time period $(t) = 15$ s,

$$Q = \begin{bmatrix} 10^{10} & 0 & 0 & 0 \\ 0 & 1 & 0 & 0 \\ 0 & 0 & 0.00001 & 0 \\ 0 & 0 & 0 & 0 \end{bmatrix}, R = 0.01.$$

Figure 5 shows that the when the inverted pendulum released from a disturbed condition close to upright condition, pendulum tries to converge to upright position. Once it reaches the desired state, the controller maintains the position throughout the remaining duration (Fig. 6).

5 Conclusion

This paper had briefly discussed the design of a reaction sphere pendulum. The key objectives of this project were to simulate the systems stabilization behaviour in MATLAB. The objective was achieved by deriving the modified equation of motion

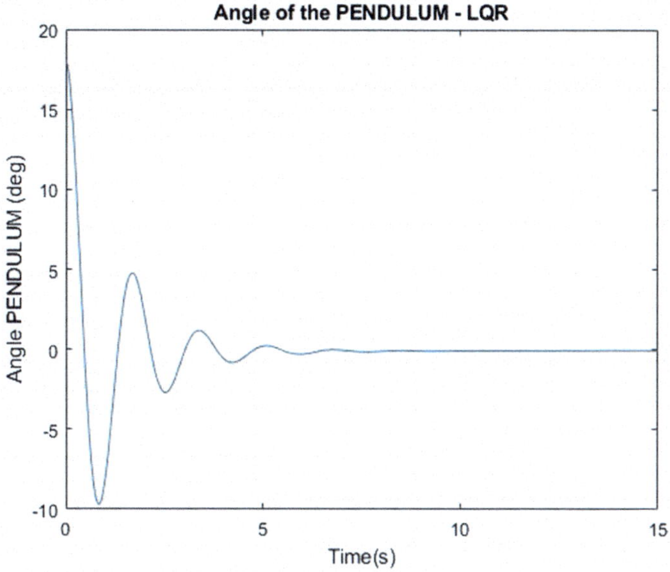

Fig. 5 Simulation output with LQR

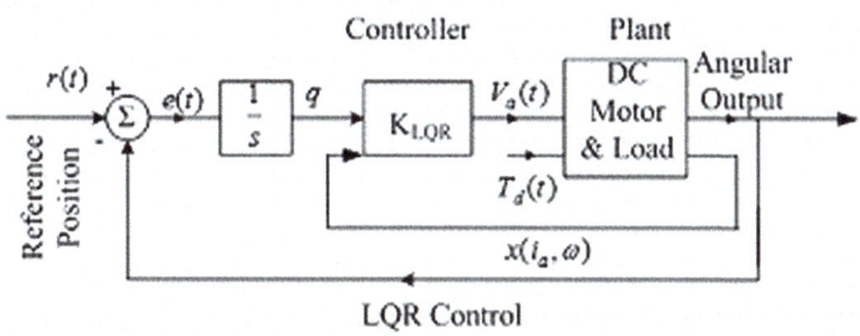

Fig. 6 Block diagram of LQR controller

and then developing a MATLAB code to simulate the system response. Primarily, system controllability was verified using MATLAB, and then, system stability was observed by monitoring the angle of pendulum. LQR feedback controller was used to stabilize the system.

Simulation results showed that reaction sphere pendulum was able to stabilize in its upright position, and it continued to converge to that desired position after disturbances.

Bibliography

1. Houck AC, Katzschmann RK, Souza Ramos JA (2013) Futura pendulum. Technical report, Department of Mechanical Engineering, Massachusetts Institute of Technology
2. Block DJ, Astrom KJ, Mark W (2007) The reaction wheel pendulum. Morgan and Playpool Publishers
3. Lupu ES (2015) Cubic structure capable of balancing. Technical report, University Polithenica of Bucharest
4. Gajamohan M, Muehlebach M, Widmer T, D'Andrea R (2013) The CUBLI: a reaction wheel based inverted pendulum. Technical report, Institute of Dynamics System and Control, Zurich
5. Fantoni I, Lozano R (2002) Non-linear control for under actuated mechanical system, Springer
6. Romanishin JW, Gilpin K, Rus D (2013) M-blocks: momentum driven, magnetic modular robots. IEEE/RSJ
7. Bartholomew K (2017) Self-stabilized monopod. Technical report, Cal Poly Pomona
8. Muehlebach M, Gajamohan M, D'Andrea R (2015) Nonlinear analysis and control of a reaction wheel-based 3D inverted pendulum
9. Brevik P (2017) Two axis reaction wheel inverted pendulum. Technical report, Norwegian University of Science and Technology

Topology Optimization and Modal Analysis of Nanosatellite Structure

K. J. Vinay Kumar, K. Sesha Sai CharanTej, Nabin Kumar Jana, Diva Sharma, and Balasubramanian Esakki

Abstract Design of a nanosatellite structure to accommodate various subsystems components in the preferred position and orientation and also provide ease of an access is of paramount interest. The provision for payloads, attaining desired field of view of camera, experiencing minimal vibration due to static, dynamic and impact loads, modular design for internal placement of components as per the requirement, efficient electrical connectivity and load paths, incorporating the concept of bus structure with lowest mass and expandability of structure are the critical requirements for designing satellite structure. Considering these aspects three a structure with vertical partitioning is considered to be an effective choice for easier accessibility of subsystem components, assembling and disassembling of individual components. The mass properties and centre of gravity of the designed structure are verified with theoretical as well as solid modelling. Finite element analysis (FEA) is performed to examine the strength characteristics of the designed structure. Topology optimization studies are conducted to reduce the weight and optimized model is undergone static analysis to withstand the loading conditions. Natural frequency of the structure is determined through modal analysis for the optimized model.

Keywords Nanosatellite structure · Structural analysis · Modal · Harmonic · Topology optimization · Natural frequency

---

K. J. Vinay Kumar · K. Sesha Sai CharanTej · N. K. Jana · D. Sharma · B. Esakki (✉)
Department of Mechanical Engineering, Vel Tech Rangarajan Dr. Sagunthala R&D Institute of Science and Technology, Avadi, Chennai 600062, India
e-mail: esak.bala@gmail.com

K. J. Vinay Kumar
e-mail: kjvinaykumar3@gmail.com

K. Sesha Sai CharanTej
e-mail: saicharantej31@gmail.com

N. K. Jana
e-mail: knabin025@gmail.com

D. Sharma
e-mail: divasharmadp@gmail.com

L.-J. Yang et al. (eds.), *Proceedings of ICDMC 2019*, Lecture Notes in Mechanical Engineering,
https://doi.org/10.1007/978-981-15-3631-1_3

1 Introduction

Space technology has been ever increasing progressively to investigate and monitor natural resources such as water resources, agriculture, forest fires, landslides, continuously. In general, the satellite is characterized by small (500–1000 kg), mini (100–500 kg), micro (15–100 kg), nano (1–15 kg), pico (100 g–1 kg) and femto (10–100 g) satellites. The simulation studies for the small satellite structure are important in deciding the structure size, shape and material selection. A satellite structure has to withstand various loads during launch and deployment. For this a structure has to be stiff enough and light in weight as required for a satellite. This can be achieved by using topology optimization which retains strength characteristics by removing a portion of material. Hu et al. [1] designed and analysed the structure subsystem of a satellite that conforms with the demands of verification conditions and launch process. Stevens et al. [2] demonstrate design of the spacecraft configuration, and the analysis and experimentation that define the satellite structural and mass properties. Rayburn et al. [3] demonstrate the feasibility of fabricating and launching three 1 kg satellites that are capable of collecting and returning quality scientific and engineering data for several months. Sachdeva and Gupta [4] describe a nanosatellite system comprising a planetary re-entry vehicle, termed as CanSat. Borse et al. [5] demonstrate the stress analysis for predicating the behaviour of flat thin inflatable membrane structure in MATLAB. Chiplunkar et al. [6] examined the structural subsystem to ensure the robustness of satellite structure so that it survives launch loads. Pappas and Spencer [7] deploy a CubeSat from an integrated PPOD and track it using a visible and infrared camera. Bai et al. [8] performed modal analysis on small structures using the finite element method. This analysis is resulted in selection of structure, determining natural frequency and stiffness of each component in the satellite. Rathinam [9] has studied finite element modelling in UWE-4 to ensure the optimum design of satellite in launch and orbital environments. Static analysis is explored in two different loading directions and the structure is experienced maximum displacement of 0.262 mm at loading directions in vertical arrangements. Finite Element (FE) studies are performed [10] to examine the integrity, stiffness and mass distribution of the structure. However, design of new nano satellite structures [11, 12] and FE analysis [13] are considered to be paraamount of interest to the satellite community. Cubesat [14, 15] configurations and its design validation using FE analysis provided greater platform to develop an efficient structure. Hence, the present study focuses on designing a novel configuration of nano satellite structure, performing topology optimization studies to arrive at optimal weight and identifying the natural frequencies of the designed satellite structure.

2 Topology Optimization

The satellite structure within the launch vehicle must be designed to withstand inertial loads, vibration, viscoelastic loads, thermal and mechanical loads which may occur in outer space. The finite element analysis is carried out for ensuring the capability and strength of the satellite structures. The vertical partition structure is considered as shown in Fig. 1 and meshed model is shown in Fig. 2.

2.1 Material Consideration

The material of the satellite structure is considered to be aluminium alloy from previous satellites. But there are a number of alloys in aluminium which have different properties. The selection of alloy is based on the comparison Table 1. Al 7075 has high yield strength and it is considered for the present study.

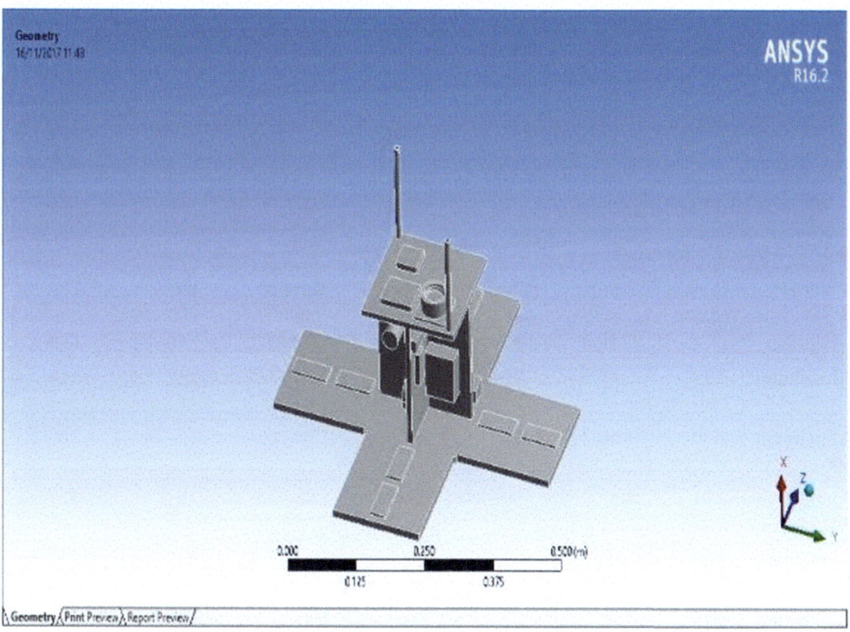

Fig. 1 Conceptual model of nanosatellite structure

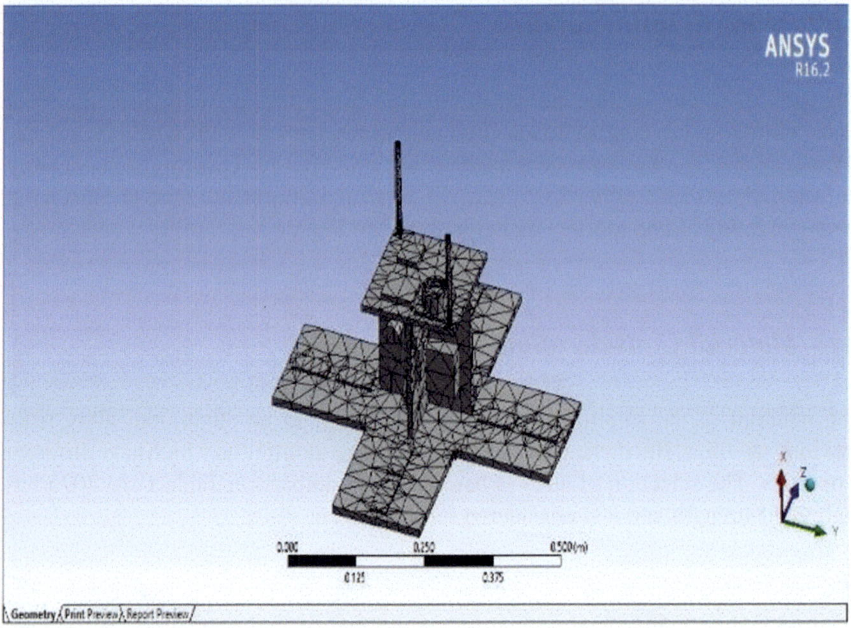

Fig. 2 Finite element model

Table 1 Material properties

Material	Density (g/cm^3)	Tensile strength (MPa)	Yield strength (MPa)	Shear modulus (GPa)	Thermal conductivity (w/mk)	Elongation at break (%)	Fatigue strength (MPa)
Aluminium 6010	2.71	290	170	26	202	24	115
Aluminium 6061	2.70	124–249	276	26	151–202	12–13	96.5
Aluminium 7075	2.81	480	430–480	26.9	130	79	159
Aluminium 5052	2.68	195–290	193	25.9	138	12	117
Aluminium 2040	2.78	469	97	28	190	10–25	–
Aluminium 6063	2.69	145–186	16	25.8	201	10–33	68.9
Aluminium 1100	2.71	110	105	–	218	12	34.5
Aluminium 3003	2.73	95–135	125	25.0	162	10	55

2.2 Moment of Inertia

The moment of inertia calculation of the satellite structure is determined from the CAD model which is given in Table 2. It will be useful for the design of control algorithms for the inertial properties of the satellite structure to achieve stability.

The structure of the satellite is analysed for static loads experienced during launch and deployment conditions. The launch interface has been fixed and acceleration load of 11 g in the longitudinal and 6 g in the lateral direction is applied.

Considering the maximum deformation (Fig. 3) and stress occurred (Fig. 4) as constraint and limit the mass of the structure 800 g as objective function, topology optimization is performed in ANSYS platform.

The static structural analysis is carried out to determine the stress distribution and total deformation experienced by the satellite structure when the static load is applied. In this analysis, the FE ring which is at the bottom of structure is fixed in all directions. In static analysis, the launch interface has been used as a fixed support and an acceleration load was given. This resulted in the stresses as shown in Fig. 2. The launch interface plate has been assigned as a fixed support. The loads given are 11 g in the longitudinal direction and 6 g in the lateral direction. The deformation is only

Table 2 Moment of inertia comparison

Moment of inertia	Software calculation (kg m^2)
Ixx	0.00432
Iyy	0.00633
Izz	0.006332

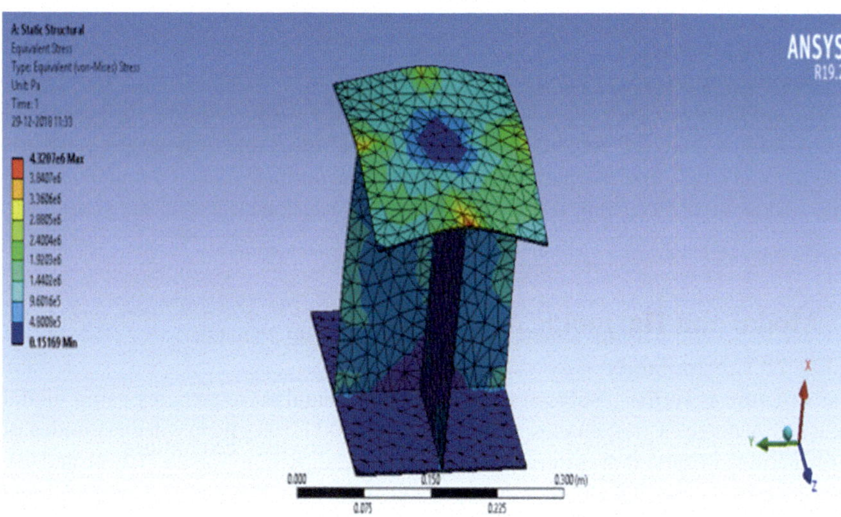

Fig. 3 Static analysis—total deformation

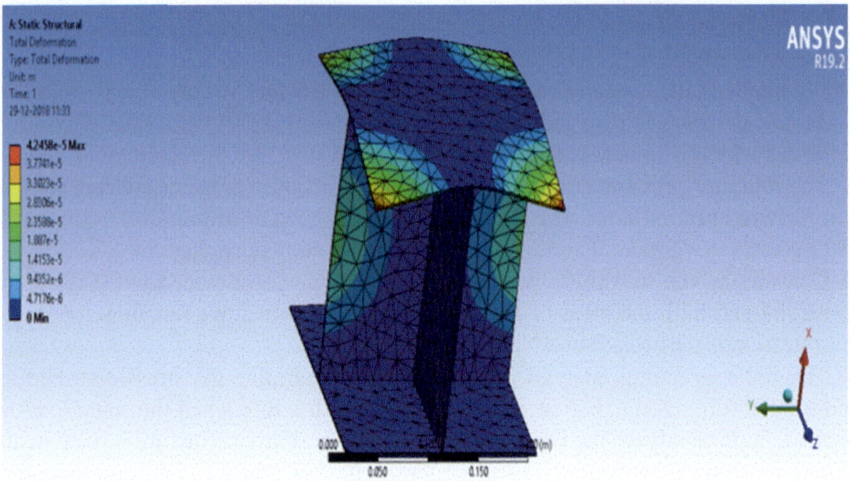

Fig. 4 Static analysis—equivalent stress

on one of the two partitions and the top plate alone. This structure is optimized using topology optimization to distribute the material without disturbing the structural integrity, also reducing the mass. The results obtained were shown below.

The material was removed on the top plate and vertical partitions as depicted in Fig. 5. However, there are components need to be placed on the top plate hence the material in this section is not removed in our design. Only the partition plates were modified and material was removed keeping in view the connections between each compartment and wiring system.

The modified structure after topology optimization is shown in Fig. 6. This structure is further analysed for the static loading conditions and structural strength is evaluated. The optimized structure experienced maximum displacement of 0.045 mm (Fig. 7) and stress occurred of 5.3 MPa (Fig. 8) which is well within the yield limit.

3 Modal and Harmonic Analysis

The structure is further analysed to determine the natural frequencies using modal analysis. A range of frequencies between 1 and 1000 Hz is given. Various modes of deformations observed at different frequencies are shown in Figs. 9, 10, 11 and 12. The antenna started deformation at 198 Hz in 9th mode. The top plate is deformed to the maximum at a frequency of 576 Hz. The frequency of each mode has been given in Table 3.

The structure is analysed for harmonic response using the same loading conditions as static analysis. The results obtained are shown in Figs. 13 and 14.

4 Conclusion

Topology optimization resulted in a weight reduction of 12% thereby achieving the required weight of the structure to be 8% of the overall satellite weight as 800 g. The maximum stress formed in the structure is 4 MPa before optimization and 5 MPa after optimization, which is much lower than the yield strength (276 MPa) of the material. The modal and harmonic response analyses results show that the antenna deformation takes place only from the 9th natural frequency and the solar panels 18th and 20th mode. These results show that the designed structure will experience less deformation and stress distribution. It is ensured that the structure can withstand static and dynamic loading conditions. The Al 7075 can withstand necessary loading occur at the launch and transportation of the structure.

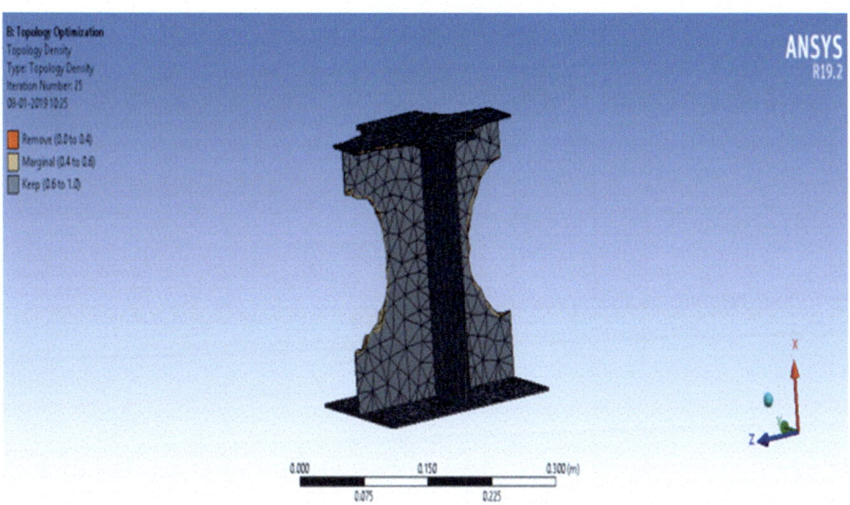

Fig. 5 Topology optimization

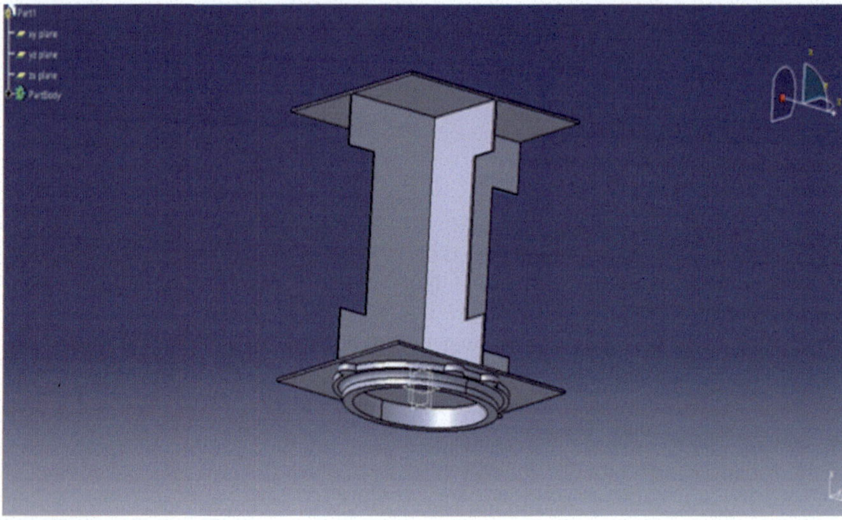

Fig. 6 Optimized structure

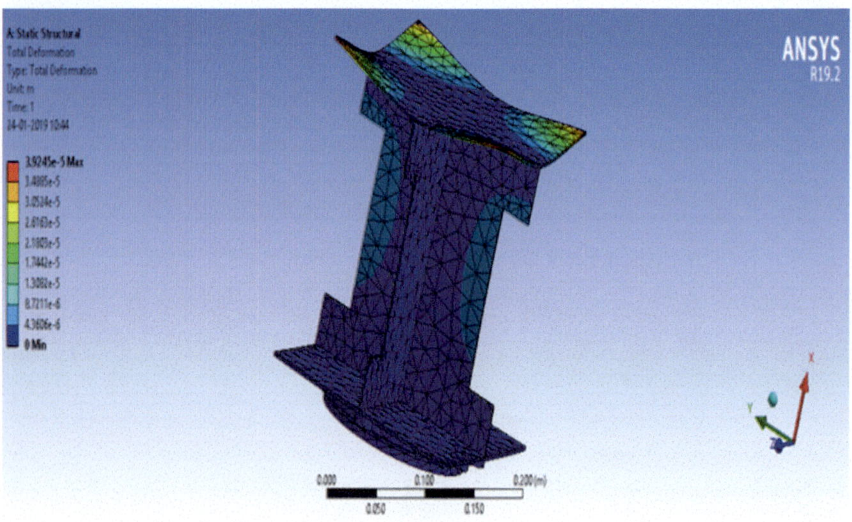

Fig. 7 Total deformation after optimization

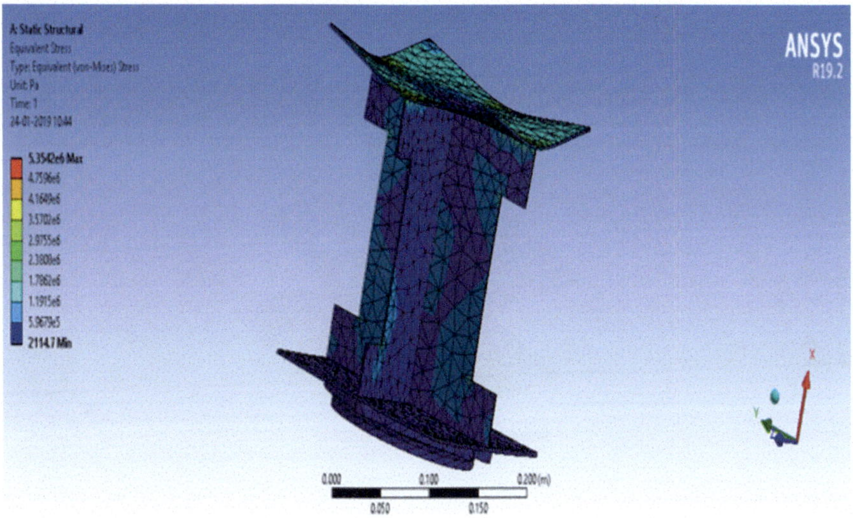

Fig. 8 Equivalent stress after optimization

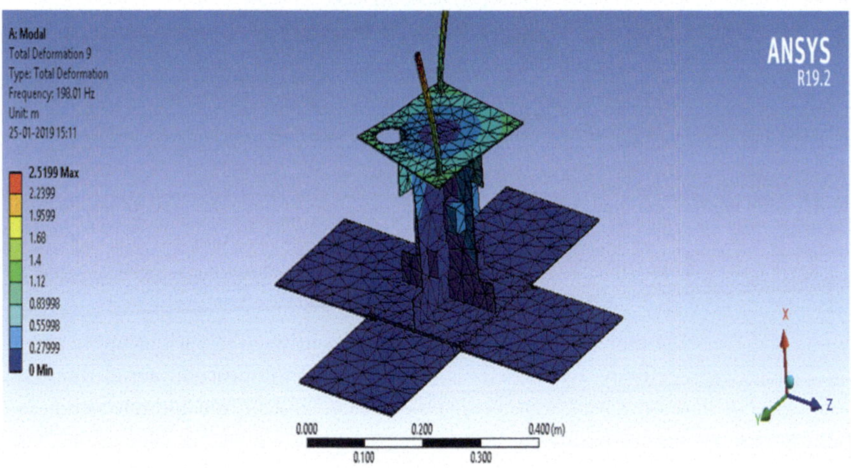

Fig. 9 9th mode

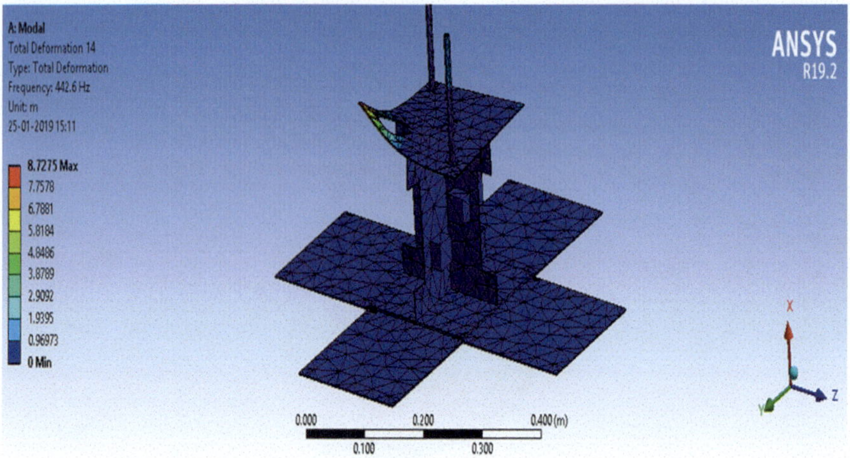

Fig. 10 15th mode

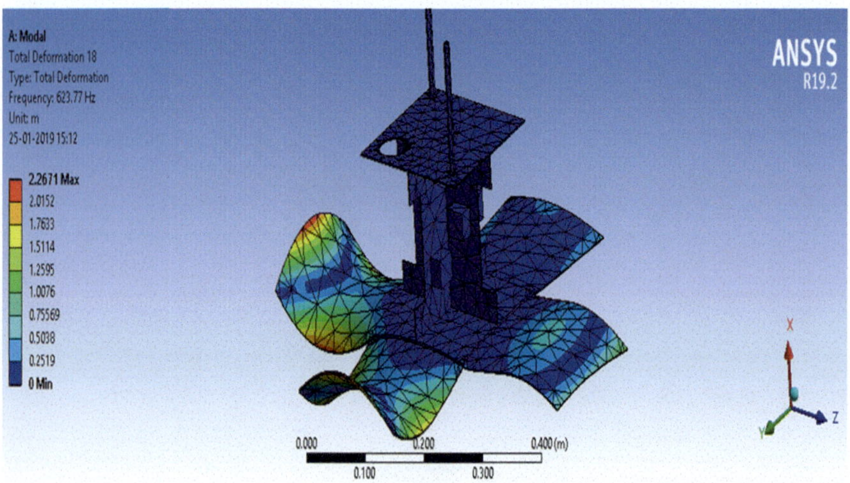

Fig. 11 18th mode

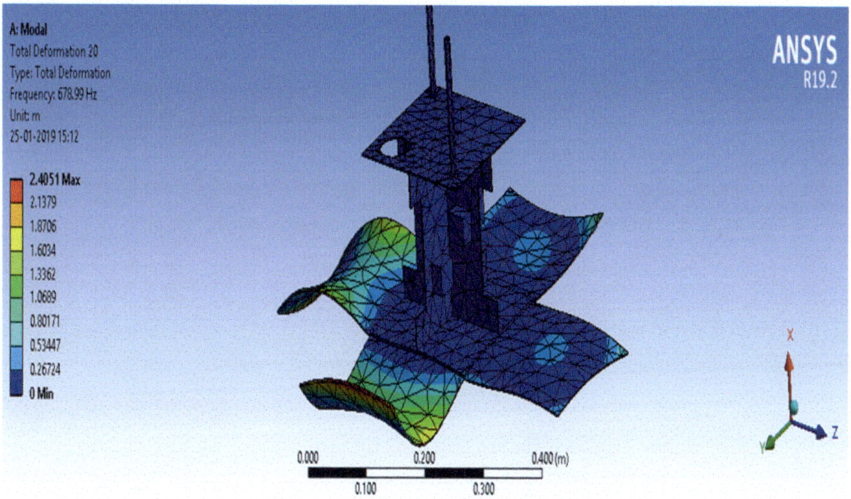

Fig. 12 20th mode

Table 3 Natural frequencies

Mode	Frequency (Hz)
1	58.159
2	59.487
3	60.049
4	61.787
5	158.12
6	166.24
7	170.64
8	191.88
9	198.01
10	208.78
11	214.55
12	261.52
13	310.48
14	442.60
15	556.21
16	557.91
17	576.29
18	623.77
19	660.50
20	678.99

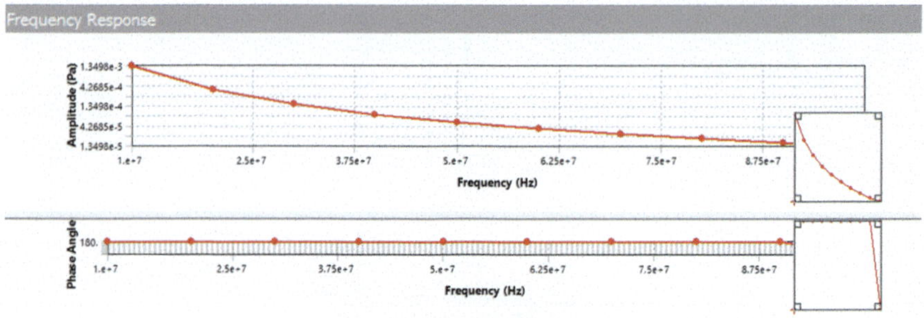

Fig. 13 Frequency response 1

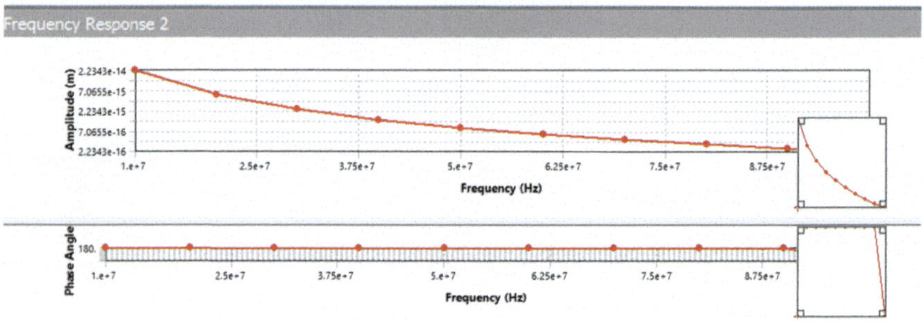

Fig. 14 Frequency response 2

References

1. Hu C-F, Lu S-T, Jenq ST, Juang J-C, Miau J-J (2008) Analysis and design of LEAP structure system. In: The fourth Asian space conference, Taipei, Taiwan, Oct 2008, pp 1–3
2. Stevens CL, Schwartz JL, Hall CD Design and system identification of a nanosatellite structure. University of Washington, Utah State University, and Virginia Tech
3. Rayburn CD, Spence HE, Petschek HE, Bellino M, Vickers J, Murphy M Constellation pathfinder: a university nanosatellite. Boston University Center for Space Physics
4. Sachdeva C, Gupta M (2017) A nano-satellite system for atmospheric monitoring and ground imaging. In: 55th AIAA aerospace sciences meeting, 9–13 Jan 2017
5. Borse DR, Upadhyay SH, Singh KS (2014) Finite element analysis of deployable space structures. IOSR J Mech Civ Eng (IOSR-JMCE), 11(6) Ver. III
6. Chiplunkar A, Pai R, Subramanyam A, Analysis and validation of the structure of pratham, Indian Institute of Technology, Bombay, Mumbai 400076
7. Pappas TD, Spencer DA Design, analysis, and testing of the prox-1 satellite structure. Georgia Institute of Technology, Center for Space Systems
8. Bai ZF et al (2008) Modal analysis for small satellite system with finite element method. In: 2nd international symposium on systems and control in aerospace and astronautics, 2008. ISSCAA 2008. IEEE
9. Rathinam A (2015) Design and development of UWE-4: integration of electric propulsion units, structural analysis and orbital heating analysis
10. Abdelal GF, Abuelfoutouh N, Gad AH Finite element analysis for satellite structures
11. Mulay SS et al (2012) Attitude determination and control of pratham, indian institute of technology bombay's first student satellite. Adv Astronaut Sci 145:1509–1528

12. Jilla CD, Miller D (1997) Satellite design: past, present and future. Int J Small Satell Eng 1
13. Sedighi M, Mohammadi B On the static and dynamic analysis of small satellite (Mesbah). Mechanical Engineering Department, IROST, Theran
14. Heidt H, Puig-Suari J, Moore AS, Nakasuka S, Twiggs RJ (2001) CubeSat: a new generation of picosatellite. In: Proceeding the 14th annual AIAA/USU conference on small satellites. Logan
15. Quiroz-Garfias, Silva-Navarro, Rodriguez-Cortes (2007) Finite element analysis and design of a cubesat class picosatellite structure. J Int Conf Electr Electron Eng (ICEEE 2007) 07:294–297

Experimental Investigation of Twin Elliptic Orifice Using High Speed Jet Facility

S. Parameshwari, Pradeep Kumar, S. Thanigaiarasu, and E. Rathakrishnan

Abstract Essential knowledge of elliptic jets creates an importance in the field of aerospace industry, for example, low jet noise in aircraft/rocket engines, reduced nature of plume infrared (IR) signature in fighter airplanes, and high combustion at combustion chamber by increased rate of fuel–air mixing. The jets which come out from a twin elliptic orifice are non-circular shape that initiate the rapid growth of mixing. Therefore, this current study focuses the aerodynamic mixing of elliptic jet flows for twin elliptic orifice. Two types of orifice jets, namely elliptic orifice facing minor axis and elliptic orifice facing major axis, are investigated at nozzle pressure ratios of 2, 3, and 4. Distances between the orifices kept as 1, 2, and 3 mm. Increasing the proximity between the orifices shows jet mixing rate getting slower. The jet facing major axis plane was efficient in mixing compared jet facing minor axis plane at measured nozzle pressure ratios.

Keywords Twin elliptic jet · Aspect ratio · Jet entrainment

1 Introduction

Recent years, jet study gained an importance in the aerospace field that primarily helps into the aircraft and rocket system/design and development. In this century, studies on the jet flows increased its potential for the future applications of rocket. Because of its complex phenomena, many researchers attracted toward the jet flows ranging from subsonic to supersonic speed. Brown and Roshko [1] reported an extensive work of planar mixing layers of axisymmetric jets. This resulted in a two-dimensional structure which causes entrainment and mixing process of subsonic and supersonic

S. Parameshwari (✉) · P. Kumar
Karunya Institute of Technology and Sciences, Coimbatore, India
e-mail: paramaero29@gmail.com

S. Thanigaiarasu
Madras Institute of Technology, Chennai, India

E. Rathakrishnan
Indian Institute of Technology Kanpur, Kanpur, India

L.-J. Yang et al. (eds.), *Proceedings of ICDMC 2019*,
Lecture Notes in Mechanical Engineering,
https://doi.org/10.1007/978-981-15-3631-1_4

shear layers that depend on compressibility effects across its shear layer. Formation of vortex rings and its merging creates growth of shear layer and increased entrainment rate in the circular jet. Along the jet direction, formation of streamwise vortices increases the aerodynamic mixing of fluid particle which entrains atmosphere fluid particle into jet environment by Liepmann and Kharib [2].

Constant research effort has taken on the axisymmetric jets by Bradbury and Khadem [3] to control the jet decay through insertion of small tabs into the axisymmetric nozzle exit. This resulted in decreased core length of jet potential core from its original characteristic decay. Subsequent researchers studied the circular jets mixing by introducing control methods like tabs, wires, shifted tabs, etc. by the researchers Lovaraju et al. [4], Maruthupandiyan and Rathakrishnan [5], and Menon and Skews [6]. However, employing tabs or wires at nozzle exit has associated the penalty of thrust reduction that is not good feature in the jet engine application view. Recovery of nozzle thrust might be achieved through mixing enhancement of non-circular geometries. Munday et al. [7] did a supportive work of underexpanded square and rectangular jets that result in shock interactions, which modify the Mach disk shape and Mach reflections of jet. Apart from that, popular techniques like Shadowgraph, stereoscopic particle image velocimetry (SPIV), and particle image velocimetry (PIV) techniques supported to visualize the shock patterns and behavior of jet field by these researchers—New et al. [8], New and Tsolovos [9], and Verma et al. [10]. The shock patterns and vortex dynamics of elliptic jet critically depend on the presence of boundary layer through nozzle exit. Again, this is associated with the thrust loss, and introduction of tabs leads into drag increment that affects the propulsion systems efficiency. Present study focused to overcome these penalties by introduction of elliptic orifice (absence of boundary layer) split into two with nominal proximity.

2 Experimental Methods

2.1 Experimental Facility and Experimental Model

The present experiments correspond to twin elliptic jets obtained at Open Jet Facility of Madras Institute of Technology (Fig. 1). Compressed air is supplied to settling chamber through a passage of pipes and gate (control) valve. However, small length of pipe is used in between chamber and control valve, and pressure-regulating valve helps to reduce flow perturbation that is coming out from the control valve. In addition, wire mesh screen placed inside the settling chamber supports in the reduction of flow turbulence just before reaching into the orifice entrance. The nozzle pressure ratio (ratio of total pressure to atmospheric pressure) is maintained with the help of control valve which controls the settling chamber pressure by Chauhan et al. [11].

A twin orifice is used, with a AR2. The orifices were kept at 1, 2 and 3 mm distance from each other. The definition of equivalent diameter (Deq) defines the

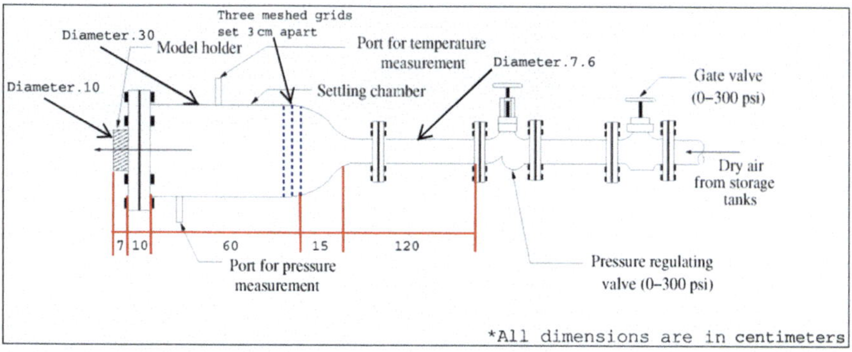

Fig. 1 Layout of open jet facility

ratio of diameter of circular orifice to elliptic orifice total exit area was 10 mm in all cases. The nozzle pressure ratio of 2, 3, and 4 was considered for the elliptic jet study. This work aims at investigating the aerodynamic behavior of jet mixing and entrainment of twin orifice sonic jets at different proximity level with a nominal pressure gradient. Layout diagram shows the geometrical details and its dimension of the twin elliptic orifice along minor axis and then along major axis used as shown in Fig. 2. A photographic view of the experimental setup with elliptic model mounted on the settling chamber is shown in Fig. 3. The orifice fitted with the settling chamber was displayed in the photographic view of open jet facility setup.

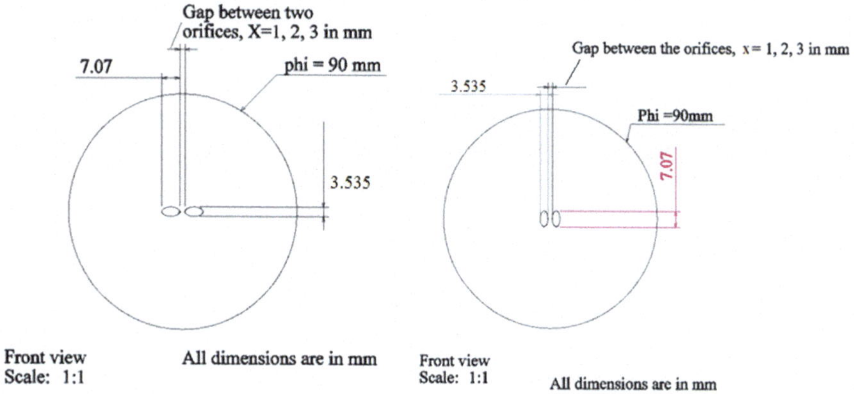

Fig. 2 Aspect ratio two twin elliptic orifice facing minor axes (left) and major axes (right)

Fig. 3 Photographic view of open jet facility

2.2 *Instrumentation*

The axis centerline pitot probe pressure distribution along the major and minor axes of elliptical jet was recorded through a pressure (16-channel Pressure Systems, Inc., 9010) transducer of 0–2.1 MPa range. The user-friendly software provided that the user-friendly software interfaces the transducer and computer. The software helps to acquire data of pressure readings from 16 channels simultaneously, which displayed on computer monitor. This software has provision to select the pressure from its menu. The pressure transducer helps to select an option of average number of samples from the settings. The transducer accuracy (after re-zero calibration) is maintained to be $\pm 0.15\%$ of full scale by Rathakrishnan [12].

The pitot probe pressure readings were measured with the help of pitot tube (inner diameter of 0.4 mm and outer diameter of 0.6 mm), attached on the fixed 3D traverse. This traverse has a resolution of 0.1 mm linear translation in all directions. The ratio of elliptic orifice exit to pitot probe was $(10/0.6)2 = 136.11$, and it is above the nominal limit probe blockage. Rathakrishnan [12] has explained about the blockage limit of mentioned probe, which is limit of 64. The sensing probe is kept perpendicular to the direction of jet axis for all pressure readings by Bajpai and Rathakrishnan [13]. Therefore, the pressure measurements have taken from pitot probe not leading to any error readings due to dominated viscous effect, which present in the pitot tube. However, the jet field was unsteady in the presence of pressure waves present in the jet spread path. So, the initial measured value is the mean value of pitot pressure. Hence, an option was available in the transducer to read the pitot pressure with an averaging of 250 samples per second. That is, each pressure recorded is the average reading of 250 samples per second. The accuracy of measured readings was maintained to repeat within the range of $\pm 3\%$.

3 Results and Discussion

3.1 Centerline Pressure Decay

The jet mixing characteristics are measured with the help of centerline pressure decay plots, and the jet mixing is an indication of entrainment of ambient fluid from the surrounding into the jet field. The centerline pressure decay is the region of jet field which extends up to which orifice velocity is unaffected from the orifice axis in the case of subsonic jets. For supersonic jets, the supersonic flow prevails from the axial extent of jet core in Verma et al. [10]. Unlike the circular jets, the elliptic ones were stand alone in this definition because geometry of orifice is non-uniform shape which has both major axis and minor axis which lead to faster jet spread. Thus, jet spread directly implies to jet mixing. Therefore, to analyze the characteristics of orifice jet, total pressure readings have been extended with an 20Deq from the orifice exit along the jet spread axis. These pressure readings (P0t) are dimensionless with the chamber pressure readings (P0s) and plotted as function of dimensionless axial distance (*x*/Deq). Three different proximity levels of 1, 2, and 3 mm were kept in between the orifices.

The axis centerline pitot pressure decay of jets was measured from the axis centerpoint of the twin elliptic orifice at different NPRs for proximity of 1–3 mm; the increments of 1 are compared in Figs. 4, 5 and 6. In the present investigation, the twin orifice has been chosen with nominal thickness. The jet issuing out from the orifice is sonic and slowly becomes free stream condition after multiple expansions. From the graphs of axis centerline pressure decay (Figs. 4, 5 and 6), it results that the axis remains unaffected up to 0.5Deq at all proximity level, and this indicates clearly that the jet from the orifices was not started to mixing. This region may call as an zone of non-initiated mixing. The mixing begins at 0.5Deq and ends at 10Deq at all measured NPRs for all proximity level. The above results also reveal that jet

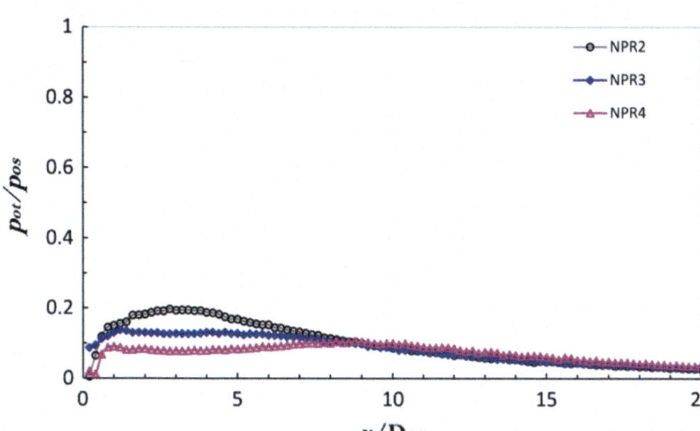

Fig. 4 Axis centerline pressure decay of AR2 elliptic orifice facing minor axis ith 1 mm proximity

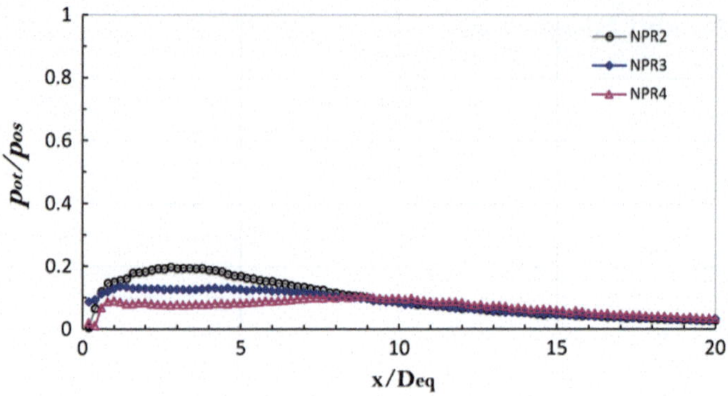

Fig. 5 Axis centerline pressure decay of AR2 elliptic orifice with 2 mm proximity

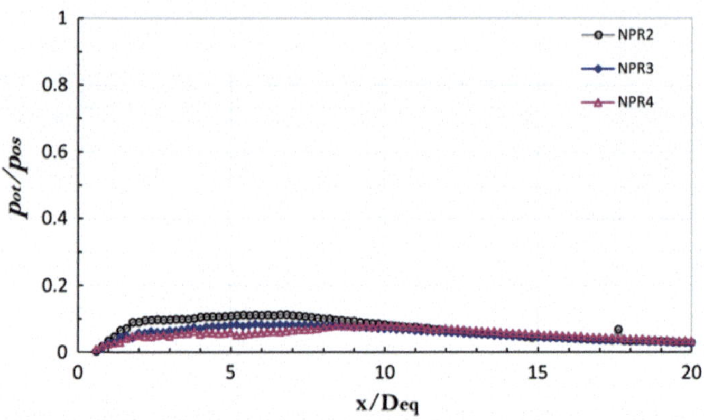

Fig. 6 Axis centerline pressure decay of AR2 elliptic orifice facing minor axis with 3 mm proximity

influences by weak waves along its direction. Since the orifice is facing minor axis in which the jet issuing out from the orifice consist of weak waves along minor axis compared to major axis. This leads to faster entrainment of fluid from the outer atmosphere into jet stream. Also, from the present study of NPRs, there is no nominal difference identified due to absence of boundary layer of orifice. At NPR 2 (Figs. 4, 5, and 6), the elliptic orifice jet mixing is faster than NPR3 and NPR4 in core region. Additionally, the jet mixing is faster at the lowest NPR of 2 within the small distance when distance increases slower down the mixing in the high NPRs.

In the case of the elliptic orifice facing major axis (Fig. 7), the mixing begins at x/Deq = 0 and ends at 6Deq for the first proximity level of $x = 0.1$Deq which corresponds to the nozzle pressure ratio of 2–4. This also conforms that elliptic jet has highest mixing characteristics than the axisymmetric jet in Bajpai and Rathakrishnan [13]. The maximum decay of elliptic jet was possible after the 6Deq and then normalized to near zero. But at NPR4, highest amplitude of pitot pressure oscillations is present compared to other NPR2 and NPR3. Corresponding to these three proximity

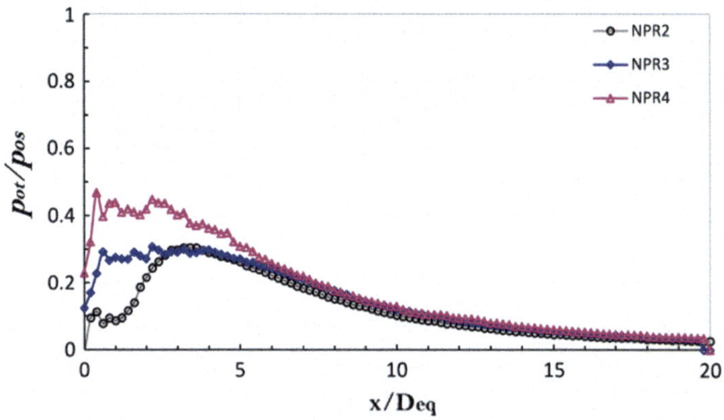

Fig. 7 Axis centerline pressure decay of AR2 elliptic orifice facing major axis with 1 mm proximity

levels at NPR2 from the Fig. 7, the elliptic orifice jet has highest mixing compared to other nozzle pressure ratios. In addition to this, the jet mixing rate slows down at high NPR level than NPR2 from the analyzed results.

Figure 8 reveals that the aerodynamic of jet mixing shows quite similar trend for the measured NPRs. Highest amplitude pitot pressure oscillations present at NPR4 compared to the other NPRs. This indicates that the proximity between the orifices plays significant role in jet mixing of twin orifice jets. Also elliptic jet mixing closes at 6Deq, 6.2Deq, and 7Deq for NPR of 2, 3, and 4, correspondingly at medium proximity range of $x = 0.2$Deq. This also evidenced the increased mixing rate in the twin orifice facing major axis than the twin orifice facing minor axis of aspect ratio two. From Fig. 9, it is seen that the mixing ends (curve drops point) are in between 3 and 5 of equivalent diameter. This clearly draws a conclusion of possible decreased potential core length compared to first and second proximity gaps.

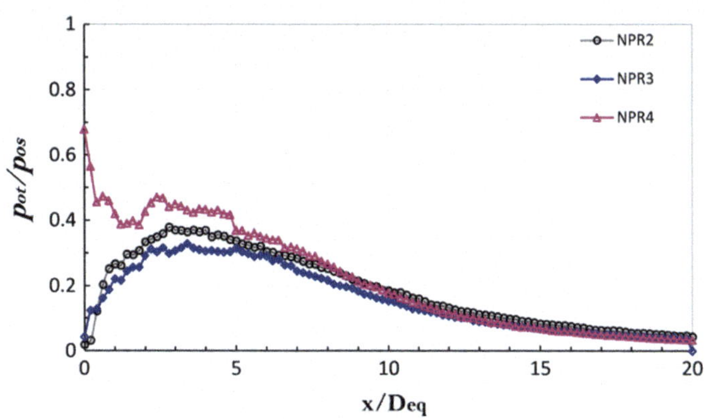

Fig. 8 Axis centerline pressure decay of AR2 elliptic orifice facing major axis with 2 mm proximity

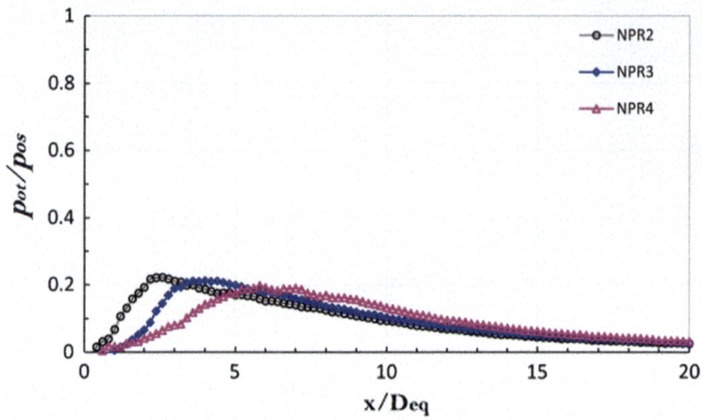

Fig. 9 Axis centerline pressure decay of AR2 elliptic orifice facing major axis with 3 mm proximity

From the results and discussion, the faster mixing is occur at the low NPR and small proximity between the orifices as compared to other proximity range. This is evidenced that rate of mixing slows down as distance between the orifice increases. Apart from that, elliptic orifice facing major axis enhances higher mixing (high amplitude of pressure oscillations) than the elliptic orifice facing minor axis. Further studies such as Shadowgraph flow visualization techniques may require to conform the flow structure in the elliptic jet flows.

4 Conclusions

The results of these experiments demonstrate that the elliptic jet mixing slows down in the maximum measured NPR of four. The amplitude of pitot pressure oscillations presence leads to decreased nature of mixing at highest measured nozzle pressure ratio. In addition, twin orifice jet shows maximum effectiveness of mixing when the two orifices are close to each other. This study suggests that the proximity between the orifices has a significant effect in the aerodynamic jet mixing of twin elliptic orifice jets.

References

1. Brown GL, Roshko A (1974) On density effects and large structure in turbulent mixing layers. J Fluid Mech 60(4):775–816. https://doi.org/10.1017/s002211207400190X
2. Liepmann D, Gharib M (1992) The role of streamwise vorticity in the nearfield entrainment of round jets. J Fluid Mech 245:643–668. https://doi.org/10.1017/s0022112092000612
3. Bradbury LJS, Khadem AH (1975) The distortion of a jet by tabs. J Fluid Mech 70(4):801–813. https://doi.org/10.1017/S0022112075002352

4. Lovaraju P, Paparao KPV, Rathakrishnan E (2004) Shifted cross-wire for supersonic jet control. In: 40th AIAA/ASME/SAE/ASEEJOINT propulsion conference & exhibit 2004, paper no. AIAA, Fort Lauderdale, Florida, 2004–4080. https://doi.org/10.2514/6.2004-4080
5. Maruthupandiyan K, Rathakrishnan E (2016) Supersonic jet control with shifted tabs. In: Proceedings of institutions in mechanical engineering part G: journal of aerospace engineering 232(3):1–17 (2016). https://doi.org/10.1177/0954410016679197
6. Menon N, Skews BW (2010) Shockwave configurations and flow structures in non-axisymmetric underexpanded sonic jets. Shock Waves 20:175–190. https://doi.org/10.1007/s00193-010-0257-z
7. Munday D, Mihaescu M, Gutmark E (2011) Experimental and numerical study of jets from elliptic nozzles with conic plug. AIAA J 49(3):554–564. https://doi.org/10.2514/1.J050587
8. New TH, Lim TT, Luo SC (2004) A flow field study of an elliptic jet in cross flow using DPIV technique. Exp Fluids 36:604–618. https://doi.org/10.1007/s00348-003-0733-7
9. New TH, Tsovolos DA (2009) Digital particle image velocimetry study on jets issuing from hybrid inclined nozzles. Flow Turbul Combust 83:485–509. https://doi.org/10.1007/s10494-009-920-4
10. Verma SB, Venkatakrishnan L, Ramesh G (2007) 2-D PIV study of near-field development from a 2:1 elliptic jet with tabs. In: 37th AIAA fluid dynamics conference and exhibit. Miami, Florida, p 449
11. Chauhan V, Kumar SMA, Rathakrishnan E (2015) Mixing characteristics of underexpanded elliptic sonic jets from orifice and nozzle. J Propuls Power 31(2):496–504. https://doi.org/10.2514/1.b35451
12. Rathakrishnan E (2016) Instrumentation, measurements and experiments in fluids, 2nd edn. CRC Press, Taylor & Francis Group, Singapore. ISBN: 978-1-315-39486-2
13. Bajpai A, Rathakrishnan E (2017) Control of a supersonic elliptical jet. Aeronaut J 114:1–17. https://doi.org/10.1017/aer.2017.114

Study of No_x Treatment with Selective Catalytic Reduction and Diesel Exhaust Fluid with Emphasis on Importance of Mixer in Flow

Ibraheem Raza Khan, Y. Lethwala, Aayush Chawla, and S. Jaichandar

Abstract This paper deals with chemical reactions of nitrogen oxide formation at higher temperatures, when the stable nitrogen disintegrates and forms unstable compounds. It discusses the effect of the unstable nitrogen compounds on human health and environment. To meet Bharat Stage VI norms, which will be implemented in the year 2020, NOx reduction techniques such as alteration in engine operation, changes in engine design, alteration in fuel, after treatment of exhaust have been discussed. Focus on after treatment of exhaust gas, by use of selective catalytic reduction, with problems of urea decomposition, and the effect of further higher temperature on urea was given. Properties of diesel exhaust fluid and factors affecting NO_X reduction such as droplet size, droplet behaviour, urea decomposition, urea deposits, injector positions, vaporization rate, and mixer designs are also discussed. Test method to obtain the best dynamic or static mixture increasing the urea exhaust mixing for better ammonia formation and NOx reduction was also carried out.

Keywords NOx · Emission · SCR · Urea · NH_3 · BS VI

1 Introduction

Engines in which combustion is initialized by compression of diesel fuel are diesel engines, and in these engines, fuel–air mixture is used which on compression absorbs necessary amount of energy so that the fuel can ignite. When surplus air is abounding via a turbocharged intercooled intake method, while the temperature in the combustion chamber is high, nitrogen in the intake air reacts according to the

I. R. Khan (✉) · Y. Lethwala · A. Chawla · S. Jaichandar
Department of Automobile Engineering, Vel-Tech Rangarajan Dr. Sagunthala R&D Institute of Science and Technology, Avadi, Chennai, India
e-mail: ibraheemk39@gmail.com

L.-J. Yang et al. (eds.), *Proceedings of ICDMC 2019*,
Lecture Notes in Mechanical Engineering,
https://doi.org/10.1007/978-981-15-3631-1_5

Table 1 Nitrogen oxides (NO_X)

Formula	Name	Nitrogen valence	Properties
N_2O	Nitrous oxide	1	Colourless gas, water soluble
NO, N_2O_2	Nitric oxide, dinitrogen dioxide	2	Colourless gas, slightly water soluble
N_2O_3	Dinitrogen trioxide	3	Black solid water soluble, decomposes in water
NO_2	Nitrogen dioxide, dinitrogen tetroxide	4	Red-brown gas very water soluble, decomposes in water
N_2O_5	Dinitrogen pentaoxide	5	White solid very water soluble, decomposes in water

extensive Zeldovich's mechanism [1], and the reactions are as follows.

$$\begin{aligned} O + N_2 &= NO + N \\ N + O_2 &= NO + O \\ N + OH &= NO + H \end{aligned} \tag{1}$$

Formation of NO_X: Continuation of nitrogen in the atmosphere is in the structure of N_2 (di-atomic modulus of nitrogen), a extremely stable state. For the duration of combustion, temperature reaches 800–1000 °C within the cylinder. At this elevated temperature, the di-atomic particle nitrogen N_2 into mono-atomic nitrogen which is extremely reactive. This mono-atomic nitrogen (N) reacts with the oxygen which is previously present in the cylinder and forms oxides of nitrogen. Oxides of nitrogen usually occur in the structure of NO and NO_2. These are normally created at high temperature. Elevated temperature and accessibility of free oxygen are the main two reasons for the development of NO and NO_2. Many other nitrogen oxides like N_2O_4, N_2O, N_2O_3, and N_2O_5 also created in low concentration, but they decompose impulsively at ambient circumstances of NO_2 [2] (Table 1).

When the combustion temperature is high nitrogen compound which is N_2 gas that disintegrates into free radical of nitrogen which is highly reactive, the present technology offers us to inject inferior quality of air in which the content of oxygen will be less. In simple terms, it is the exhaust gas recirculated into the intake manifold now this exhaust gas is first food so that the heat energy within the exhaust gas does not add up the combustion temperature. Generally, the set-up is called EGR which stands for exhaust gas recirculation, a simple working can be understood why as a part of the exhaust is taken back via exhaust gas recirculation set-up which calculates the amount of NO_X generated or the temperature in the combustion chamber. Its calculations of amount of EGR of the exhaust which has to be injected are free calculated while engine designs are fed into electronic control unit. Amount of exhaust is injected into the intake manifold, and as the fresh charge is blended with exhaust gases which are not rich in oxygen, the combustion temperature reduces and the possibility of NO_X formation also reduced. On the other hand, because of this interior combustion the amount generated will be high so basically is a trade-off NO_X formation and hydrocarbon.

1.1 Effect on Health

Unstable compounds of nitrogen are very toxic to human. Because of the high temperature of the combustion chamber and uncontrolled combustion please free radicals even after the implementation of BS VI there is a huge lot of improvement required to reduce the NO_X emission. Compounds of nitrogen are an NO_2 decrease in human immunity. Compounds of nitrogen can cause acid rain acidification of soil and other undesired chemical reactions even formation of HNO_3 [3].

Acidification of water will reduce availability of water for drinking, and also, it will affect the marine life the possibility of acid rain as these are unstable compounds and will react with oxygen and hydrogen at a higher extent. Because the uncontrolled NO_X formation certain results have been seen in the past some of them have been seen in Delhi see and around cities where there is formation of smog and there have been reports of acid rain in Agra [4]. There are also reports of landlocked lakes being converted into acid ponds because of acid of rain. With the introduction of BS VI, there are norms which state for the reduction in rock formation per kilometre, complete combustion is highly desired for a better fuel economy higher temperatures are needed, but on the other hand, they will be NO_X formation if the compression temperature goes high for this there is a introduction of after treatment setup which is SCR + D e f, which stands for selective catalytic reactor + diesel exhaust fluid this has to be implemented by all the Companies selling vehicles in India by April 2020 further details are discussed below.

1.2 Emission Regulation for Bharat Stage VI

BS VI emission standard adds for MNN class vehicle which has gross vehicle weight GVW you not more than 3.5 turn for 1 April 2020, as per GSR 889 dated 16 September 2016 [5] (Tables 2, 3 and 4).

There are many techniques which are tried to organize NO_x emission from diesel engine. The subsequent methods may be employed whichever as a single technique or as a grouping.

1. Alteration of engine operations
2. Changes in the engine design
3. Alteration of fuels
4. After treatment of exhaust gases.

Table 2 Bharat stage VI emission standards

		Reference mass (kg)	Mass of carbon monoxide (CO)	Mass of total hydrocarbons (THC)	Mass of non-methane hydrocarbons (NMHC)	Mass of oxides of nitrogen (NO_x)	Combined mass of hydrocarbons an oxides of nitrogen (THC = NO_x)
			mg/km	mg/km	mg/km	mg/km	mg/km
Category	Class		CI	CI	CI	CI	CI
M_1 and M_2	–	All	500	–	–	80	170
N_1	II	1305 < RM ≤ 1760	630	–	–	105	195
	III	1760 < RM	740	–	–	125	215
N_2	–	All	740	–	–	125	215

CI Compression ignition

Table 3 OBD 1 BS6 [9]

		Reference mass (kg)	Mass of carbon monoxide (CO)	Mass of non-methane hydrocarbons (NMHC)	Mass of oxides of nitrogen (NO*x*)
			mg/km	mg/km	mg/km
Category	Class		CI	CI	CI
M_1 and M_2	–	All	1750	290	180
N_1	II	1305 < RM ≤ 1760	2200	320	220
	III	1760 < RM	2500	350	280
N_2	–	All	2500	350	280

CI Compression ignition

Table 4 OBD 2 BS6

		Reference mass (Kg)	Mass of carbon monoxide (CO)	Mass of non-methane hydrocarbons (NMHC)	Mass of oxides of nitrogen (NO*x*)
			mg/km	mg/km	mg/km
Category	Class		CI	CI	CI
M_1 and M_2	–	All	1750	290	140
N_1	II	1305 < RM ≤ 1760	2200	320	180
	III	1760 < RM	2500	350	220
N_2	–	All	2500	350	220

CI Compression ignition

2 After Treatment of Exhaust Gases

A. **Selective catalytic reduction SCR**:

It is the after treatment attachment which is generally placed doc. In this set-up, urea is injected which because of high-temperature exhaust converts into ammonia gas; this ammonia gas mixes with exhaust gases which have NO_X in it, and then, these mixtures of ammonia and exhaust move to SCR where uh generally of vanadium catalyst with the largest surface area is placed the nitrogen within the ammonia gas that reacts with unstable NO_X compounds and forms nitrogen. The quantity exhaust fluid varies with the formation of NO_X. It is well understood that the ammonia which is already present in it on the SCR which reacts with NO_X, and not the current ammonia which is injected. The injector injects the aqueous urea which is water + urea at a higher pressure in small droplet form with the flow of the exhaust. The temperature of the exhaust helps the urea to get convert into ammonia gas [6].

B. **Urea decomposition**:

DEF is a solution of water and urea generally with 32.5% of urea and 67.5 water buy weight. The formula of urea is the formula of urea is CO (NH_2). This is used because of its ability to disintegrate into ammonia. This solution is injected via injector with the flow of the exhaust in the exhaust pipe before the SCR. First because of the heat the urea and water solution evaporates and ammonia and isocyanic acid are formed further, and because of vapours of water, this isocyanic acid converts into ammonia and carbon dioxide.

$$CO(NH_2)_2 + \text{Heat} \rightarrow NH_3 + HNCO \tag{2}$$

$$HNCO + H_2O \rightarrow NH_3 + CO_2 \tag{3}$$

C. **NO_X Reaction with NH_3**:

Now, we will see NO_X reaction with ammonia, and these reactions happen on the surface of catalyst which is in the SCR. Ammonia reacts with nitrogen oxide and nitrogen dioxide to form nitrogen gas and water for the ammonia reacts with nitrogen oxide + oxygen which will give again nitrogen and water, this reaction keeps on happening till there is no more nitrogen oxide available, and still, there is ammonia are also available in which the reactions are stated below; these reactions are all endothermic which means they need heat energy to react because a wedge, a catalyst, is required [7].

$$\begin{aligned} &2NH_3 + NO + NO_2 \rightarrow 2N_2 + 3H_2O \\ &4NH_3 + 4NO + O_2 \rightarrow 4N_2 + 6H_2O \\ &8NH_3 + 6NO_2 + O_2 \rightarrow 7N_2 + 12H_2O \\ &4NH_3 + 6NO \rightarrow 5N_2 + 6H_2O \end{aligned} \tag{4}$$

At higher temperatures, these unwanted reactions occur.

$$\begin{aligned} &4NH_3 + 4NO + 3O_2 = 4N_2O + 6H_2O \\ &4NH_3 + 5O_2 = 4NO + 6H_2O \end{aligned} \tag{5}$$

3 DEF/Ad Blue Specification

See Table 5.

- **Factors Affecting NO*x* Reduction**:

Table 5 Diesel exhaust fluid data [10]

Urea % by weight	31.8–33.2
Alkalinity as NH_3% by weight maximum	0.2
Biuret % by weight maximum	0.3
Insoluble, ppm maximum	20
Aldehyde, ppm maximum	5
Phosphate PO_4, ppm maximum	0.5
Aluminium, ppm maximum	0.5
Calcium, ppm maximum	0.5
Iron, ppm maximum	0.5
Copper, ppm maximum	0.2
Zinc, ppm maximum	0.2
Chromium, ppm maximum	0.2
Nickel, ppm maximum	0.2
Magnesium, ppm maximum	0.5
Sodium, ppm maximum	0.5
Potassium, ppm maximum	0.5
Density at 68 °F (20 °C), lbs/gal	9.07–9.12
Refractive index at 68 °F (20 °C)	1.3814–1.3843
Salt-out temperature, °F (°C)	12(−11)
Recommended storage temperature, °F(°C)	40–80 (4.5–26.6)

1. Droplet behaviour—In gas phase disintegrate of a dewdrop is by shear stress and vaporization. On the surface, break-up is by impingement, liquid film vaporization, sometimes deposits are formed.

- Urea deposit, decomposition of DEF, a common problem, has been seen in urea injection, which is the position of urea as white matter inside the exhaust pipe [8]. This position of urea clogs the pipe an increase back pressure this happens when there is improper injection hydrolysis evaporation of urea, sometimes it also happens because of very high temperatures in which isocyanic acid does not form any further ammonia but forms amides torrent melamine and other deposits.

$$\begin{aligned}
&CO(NH_2)_2 + HNCO \rightarrow NH(CONH_2)_2 \\
&NH(CONH_2)_2 + HNCO \rightarrow \text{Triuret} \\
&3HNCO \rightarrow C_3N_3(OH)_3 \\
&NH(CONH_2)_2 + C_3N_3(OH)_3 \rightarrow \text{Ammeline} + 2H_2O \\
&NH(CONH_2)_2 + C_3N_3(OH)_3 \rightarrow \text{Ammelide} + H_2O
\end{aligned} \tag{6}$$

2 Enhance mixing of spray and exhaust gases and evenly distributing the mixture on the SCR catalyst.

- The conventional approach results in having one fixed mixer orientation angle for the entire operating range of engine and can be valid for smaller engines operating with limited exhaust mass flow and temperatures. However, for bigger engines with wider range of mass flow and temperature, there is a compromise for the mixer orientation angle. This can result in higher or lower spray load on the mixer over the entire operating range, thereby leading to issues like deposition, incomplete utilization of urea solution, etc. Proposed idea is to use the exhaust brake flap as a mixing enhancer of NH_3 + exhaust gas, the system is to be tested with different projection on the plate via CFD analysis.

3 Orientation of injector—the urea injector can be oriented with the flow or against the flow we can also use a static mixer or a dynamic mixer, usually we are using a static mixture, and in this case, we can also use a dynamic mixture. A dynamic mixture can be explained in a better way with the help of the last diagram. Here in the diagram, we can see that exhaust brake flap is welded with fins over it when the exhaust brake is switched off, the flap is in open condition and there is a very less back pressure created in this situation, and the friends over the flap help in better mixture of urea and exhaust gas.

Now, we will see what will happen if the injector is placed against the flow when the injector is placed against the flow, the urea is injected against the flow of the exhaust gas. Now in this case, there is a possibility of reducing the length of the mixture which is required to form ammonia exhaust mixture. The working can be understood why the following steps are. Step 1—a constant flow exhaust will be flowing. Step 2—urea will be injected against the floor after detecting the NO_X formation, and in this case, the urea will evaporate and will convert into ammonia gas which will then move to SCR. Problem in this case is that DEF contains urea and water.

- **Analysis Result** (Fig. 1):

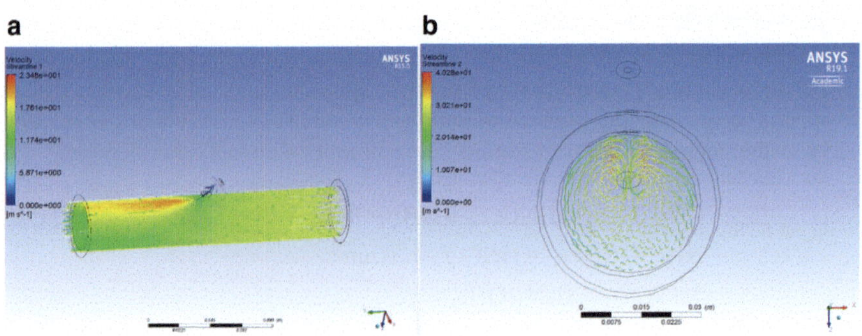

Fig. 1 **a** Injection of urea and which flows without mixing with exhaust. **b** Cross-sectional view of exhaust pipe after injection of urea into exhaust flow, and figure shows the minimal turbulence created. Low turbulence reduces the uniformity index (which defines the quality of mixture)

4 Conclusion

From the group study of the paper, a conclusion can be drawn that NO_X is a harmful by-product of exhaust, as we need complete combustion of a fuel, the temperature increases because of which stable nitrogen reacts. This unstable state of nitrogen compound reacts in the form of acid rain and also affects the human immunity system. Emission regulations in Bharat Stage IV were stringent, but with the introduction of Bharat Stage VI norms, the production of NO_X per kilometre should further be reduced. This will come in action from 1 April 2020. There are many after treatment methods such as exhaust gas recirculation, but it is a trade-off between carbon and nitrogen unstable compounds. With the introduction of selective catalytic reduction + diesel exhaust fluid which is also known as Ad blue NO_X can be reduced. Further, it is discussed the orientation of the injector, and the use of static and dynamic mixer is also discussed. Closing the conclusion can be made in a way that dynamic mixture which is the exhaust brake flap mounted fins, this future ahead as the orientation of the flap can be changed with the changing velocity of the exhaust and a better exhaust urea mixture can be obtained which will completely eliminate reduces the NOx coming out of the exhaust.

References

1. Wallin L (2014) Investigation of urea deposit in vehicles. Division of Combustion Chalmers University of Technology, Göteborg, Sweden
2. Clean Air Technology Center (1999) Nitrogen oxides (NOx), How they are controlled. U.S. Environmental Protection Agency, Research Triangle Park, North Carolina
3. WHO (2000) Nitrogen dioxide. WHO Regional Office for Europe, Copenhagen, Denmark
4. Gautam PSP (2010) Vehicular pollution. Central Pollution Control Board (Ministry of Environment & Forests, Govt. of India), Delhi
5. Ministry of Road Transport and Highways (2016) The gazette of India. Gov of India, New Delhi, 22 Feb 2016
6. The Automotive Research Association of India (ARAI), Indian emission regulation. ARAI, PUNE, 16 Jan 2017
7. Cummins (2009) Diesel Exhaust Fluid (DEF). Cummins Filtration, USA
8. Chithambaramasari K (2011) Reduction of NO_x in D.I. diesel engine emission urea injection. Int J Adv Therm Sci Eng 2(2):97–106
9. Dyno Nobel Inc. (2015) DEF. Dyno Nobel Inc., Utah USA, 20 Aug 2015
10. Thondavadi A (2016) Effects of diesel exhaust fluid injection configurations, deposit formation in the SCR system of a diesel engine. SAE Int, 2016-28-0109:1–4

Advancements in Automotive Applications of Fuel Cells—A Comprehensive Review

Newel Francis Thomas, Rishabh Jain, Nishanth Sharma, and S. Jaichandar

Abstract A fuel cell is an alternate source of power supply, and it has nearly zero emissions and has higher efficiency when compared to the conventional sources of power. This type of a power supply unit has a vast base of application; further, it has many automotive applications too. Since the growing risk of environmental hazards and depletion of the conventional fuels being used now, and the changing emission norms, a better fuel-efficient and environment safe power supply is required. The goal of this paper is to understand the various types of fuel cells that are used in this field and the various applications that it has and to understand the different experiments which were conducted previously, how these types of power units are better than the conventional power units, and various ways in which these types of power supply can be used and applied in the current day scenario; also, various future aspects of the same have been discussed.

Keywords Fuel cells · Hydrogen · FCV · PEMFC · Emissions

1 Introduction

Fuel cells are in the field of automobiles for quite some time, the first fuel cell was invented almost 150 years ago [1], and the idea however was the same that is to replace the conventional fuel system with something that is more reliable and more eco-friendly. Replacing is a bit difficult but when used in a hybrid system, the efficiency is higher and also better work output is received. In the past century, the development in the field of fuel cell has rocketed to sky high that which was initially used for space programs soon found its way to the commercial market into vehicles of daily usage [1]. In the early 1990s when the vehicles with fuel cell-fitted power supply came out, despite the efficiency it gave production of raw materials such as hydrogen and storing it was difficult and thereby very costly. It was soon thought that fuel cell

N. F. Thomas (✉) · R. Jain · N. Sharma · S. Jaichandar
Vel Tech Rangarajan Dr Sagunthala R&D Institute of Science and Technology, Avadi, Chennai, India
e-mail: newel.thomas7@gmail.com

L.-J. Yang et al. (eds.), *Proceedings of ICDMC 2019*,
Lecture Notes in Mechanical Engineering,
https://doi.org/10.1007/978-981-15-3631-1_6

may not be an economic answer to the energy crisis. But in recent decades, there has been massive development in fuel cells, research and development has shown that the cost price of production of hydrogen and usage has come down drastically by 75% [2], and further down the line it will even get lesser due to the technological advancements taking place every day. This paper discusses the ways in which such a power system can be used to increase the efficiency in the vehicle and ways in which such a knowledge can be used to improve the power generation aspects in vehicles and gives a free way to the ultimate goal of maximum efficiency without any hindrance.

2 Working of Fuel Cells

A fuel cell is an electrochemical device, which means that the chemical energy in the fuel is converted to electrical energy. This energy which is formed is a low voltage and also direct current. It works on the principle of electrochemical reactions. The system contains the reactants (fuel and the oxidizer), electrodes (positive and negative), and the electrolyte. The reactants are fed through the porous electrodes, and the reactants react with the electrolyte and produce a voltage. Now when an external load is connected, the ions travel from one electrode to the other through the electrolyte and also electrons are conducted through the load. The operation of a fuel cell is continuous as long as the fuel is supplied, the oxidizer is supplied, the current flows through the load, and the structural integrity of the chemical is maintained. The emf of fuel cells is at the order of 1 v. Many cells are therefore connected in a series, and it is called a module [1].

The Gibbs free energy change is given as

$$\Delta G = \Delta H - T\,\Delta s$$

where ΔG is Gibbs free energy, T is the absolute temperature of oxidation, ΔH is the change in enthalpy, and ΔS is the change in entropy.

ΔG determines the maximum possible electromotive force (emf) of the cell in terms of

$$E = -\Delta G/nF$$

where E is the emf, F is the Faraday value, and n is the number of Faradays transferred.

To function, the ion exchange membrane must be kept moist. Therefore, some of the water produced by the fuel cell is used as a humidifier for the hydrogen and the oxygen coming inside the fuel cell. The water which is left is exhausted from the fuel cell. Some heat is also emitted from the fuel cell. The heat is either released to the outside or captured and used to heat the fuel cell or can be used for other purposes, such as heating the passenger compartment if needed in cold temperatures.

2.1 Diagram—From [3]

See Figs. 1 and 2.

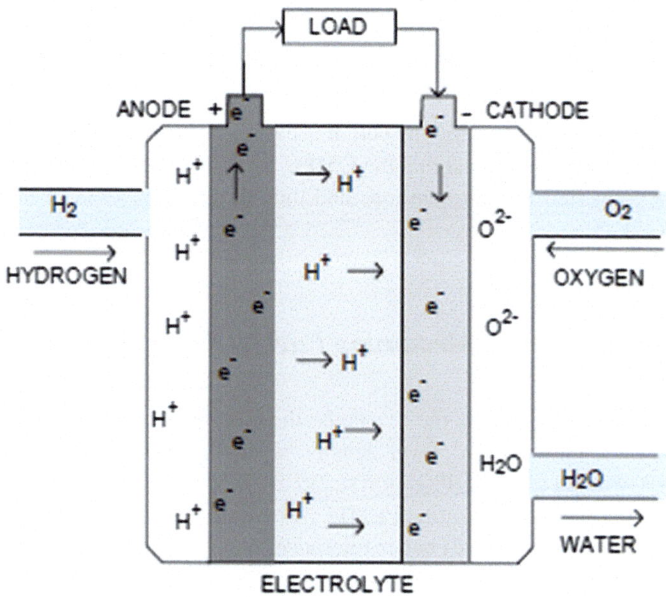

Fig. 1 Fuel cell working

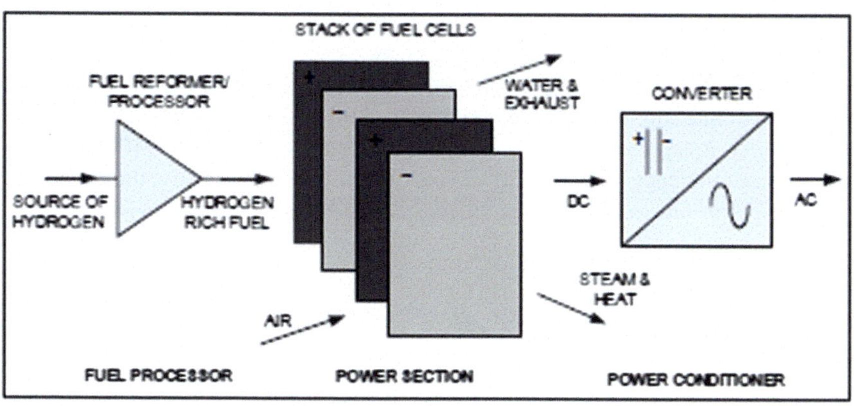

Fig. 2 Assembly of a fuel cell

3 Types of Fuel Cells

The types of fuel cells based on the electrolyte used are as follows.

3.1 Direct Methanol Fuel Cell

The DMFC is a low-temperature fuel cell, it is based on the PEMFC, and it is similar in all aspects, except for the fact that the DMFC uses methanol as the fuel. Thus, the methanol is converted to hydrogen ions and then the usual process of the reaction takes place [4].

3.2 Proton Exchange Membrane Fuel Cell

This type of fuel cell contains the reactants, the electrolyte, and a membrane, and this type of fuel cell has high power density, is smaller in size, and works at low temperatures of about to 90 °C. Platinum stacks are used as the catalyst usually. There is also a high-temperature operating PEMFC, which operates at around 200 °C. This high-temperature PEMFC gives better tolerance to carbon monoxide and also does not need an external cooling system such as water flow to keep it humidified [4].

3.3 Alkaline Fuel Cell

This type of fuel cell was one of the first types of fuel cell to use hydrogen as the fuel. It contains potassium hydroxide as electrolyte. And it uses platinum as the catalyst, due to which it can easily operate in low temperatures of about 50–100 °C. They also have an efficiency of around 50%. This type of fuel cell is costly; thus, it is only used in high-profile sectors such as the military and space exploration [4].

3.4 Phosphoric Acid Fuel Cell

This type of fuel cell again operates at low temperatures of about 160–220 °C. Its efficiency is around 37% but when combined with another power generation source (hybrid system) it can give an efficiency of about 87%. Compared to other fuel cells, this gives less output efficiency therefore not suitable for vehicles, but can be used in stationary places such as hospitals and factories [4].

3.5 *Molten Carbonate Fuel Cell*

This fuel cell operates at high temperature around 650 °C. They also provide high efficiency that is around 55% singularly, and when in a hybrid system it can reach about 87%. Due to high temperatures, no carbon poisoning happens. Nickel is basically used as the catalyst [4].

3.6 *Solid Oxide Fuel Cell*

The electrolyte used in this kind of fuel cell is a solid, and it is non-porous metal oxide. The electrolyte used is yttria-stabilized with zirconia. The cost of production is low, and also it can withstand high temperatures from the range of 550–1000 °C. In this, the fuel can be reformed internally, and the excess heat can also be used for preheating the incoming air [4].

3.7 *Microbial Fuel Cell*

In this, biochemical energy is converted to electrical energy. The catalyst used is microorganisms, and it can be either bacteria or yeast. In this, the chemical transformation is into protons, H^+, and electrons, and H^+ crosses the membrane made of Nafion arriving at cathode. Protons there react with O_2^{2-}, reach external circuit, and produce water [4] (Table 1).

4 Output

When compared to an IC engine, the values of a fuel cell are as follows (Fig. 3).

5 Application of Fuel Cells

Fuel cells are used abundantly in automotive industry from primary power source to secondary to backup, and given above were the different types of fuel cells. So now further we are going to discuss the various areas in which different types of fuel cells are used, and its applications.

Table 1 Classification of fuel cell [4]

Fuel cell types	Polymer electrolyte membrane (PEM)	Alkaline fuel cell (AFC)	Phosphoric acid fuel cell (PAFC)	Molten carbonate fuel cell (MCFC)	Solid oxide fuel cell (SOFC)	Microbial fuel cell
Common electrolyte	Perfluoro sulfonic acid	Aqueous solution of potassium hydroxide soaked in a matrix	Phosphoric acid soaked in a matrix	Solution of lithium, sodium, and/or potassium carbonates	Yttria-stabilized zirconia	Microbes
Operating temperature	50–100 °C 122–212° typically 80 °C	90–100 °C 194–212 °F	150–200 °C 302–392 °F	600–700 °C 1112–1292 °F	700–1000 °C 1202–1832 °F	Ambient temperature
Anode reactions	$H_2 \rightarrow 2H^- + 2e^\bullet$	$H_2 \rightarrow 2OH^\bullet \rightarrow 2H_2o + 2e^\bullet$	$H_2 \rightarrow 2H- + 2e^\bullet$	$H_2 + CO_3^{2\bullet} \rightarrow H_2O + CO_2 + 2e^\bullet$	$H_2 + O^2 \rightarrow H_2O + 2e^\bullet$	
Cathode reactions	‘$1/2O_2 + 2H^- + 2e^\bullet \rightarrow H_2$	$1/2O_2 + H_2O + 2e^\bullet \rightarrow 2OH^\bullet$	$1/2O_2 + 2H^- + 2e^\bullet \rightarrow H_2O$	‘$O_2 + CO_2 + 2e^\bullet \rightarrow CO_3^{2\bullet}$	$1/2O_2 + 2e^\bullet \rightarrow O^{2\bullet}$	
Efficiency	60% transportation 35% stationary	60%	40%	45–50%	60%	50%
Applications	• Backup power • Portable	• Military • Space	• Distributed generation	• Electric utility • Distributed	• Auxiliary power • Electric utility	Bioenergy process

Fig. 3 ICE versus fuel cell [10]

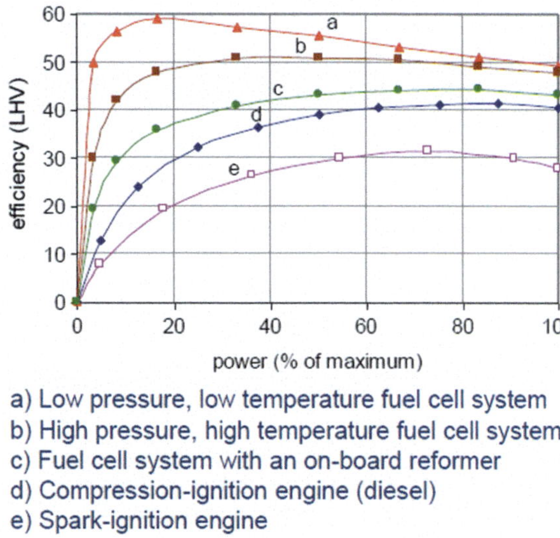

5.1 Transportation

Since the beginning, vehicles have been powered by chemicals (fuels) for power generation of steam, and the hunt for a low emission fuel has always been a topic of research in the automotive field.

Scientists have been searching for alternative that can replace the conventional fuel for transportation. Fuel cell vehicles came into the transportation arena in the late 1990s. One of the most notable types of fuel cells into production was the PEMFC due to its characteristics of high power density, low temperatures of operation, high efficiency, thus giving faster start-up time.

USCAR launched a fuel cell vehicle (FCV) in 2002 to spread awareness and encourage the FCVs.

However, challenges like durability, high cost estimates, and hydrogen storage led to waning research and questions were raised toward the viability of fuel cell of transportation but still recent studies show that the FCV offers potential for larger size and lower cost for long-range vehicles and also faster rate of refueling.

Study by Jih-Sheng Lai et al. [2] Department of Energy's (DOE) Hydrogen and Fuel Cell Multi-Year Research, Development, and Demonstration Plan (MYRDDP) has established cost targets for the FCVs; they have been tracking it since 2002. The cost as of 2017 based on 500,000 vehicles being produced is said to be $53/Kw with a goal to reach around $30/Kw. This shows a drastic cost reduction of about 75% in 2002 when it was $200/Kw. For example if taken into consideration a fuel cell system of 80 Kw with a 5 kg, 700 bar hydrogen tank, which should give a range of 300 miles, the cost target of this system would be approximately $4900.

5.2 Cars

Zehra Ural Bayrak et al. [5] Daimler-Benz built a series of PEMFC cars; in 1997, they released a methanol-powered car which had a range of around 640 km. Toyota built another hydrogen-powered car which used metal hydride storage fuel cell, and it was a hybrid with battery as alternate source. In 1997 again on the same RAV4 platform, Toyota again created a methanol fuel vehicle. The FEVER prototype is being carried on by Renault and PSA Peugeot. All other major manufacturers are also in the quest of fuel cell vehicles. Ballard Power, Plug Power, and International Fuel Cells, these companies are working on 50–100 Kw fuel systems for cars.

The NECAR 5 is a prototype vehicle of DaimlerChrysler. It is fueled with methanol. It is a DMFC system, the methanol is converted to hydrogen onboard, and thus the reactions take place and the power supplied. The vehicle has no pollutants, the efficiency is higher than conventional fueled vehicles, and the output of carbon dioxide is considerably very low [5].

Jürgen Garche et al. [6] Hyundai started innovative worldwide leasing program for the ix35 car, the Toyota Mirai, the latest in the segment has a 100 kW PEFC stack with a power density of 2 kW/kg and has two hydrogen tanks of 700 bar which thus provides a range of 650 km, and it is still a hybrid with a Ni–MH battery. Recently, Toyota has also reported that the platinum demand has gone done to less than 30 g for a vehicle propulsion system.

5.3 Buses

Transit buses were the early application of fuel cells, since it is a mass transport and it operates in cities and places where pollution is already a major problem; thus, changing to such vehicles would be a huge help. It produces zero emissions and is also economic, but still at many places fuel cells are used as a hybrid system than an independent system. In 1997, Ballard provides a 205 kW PEMC units for a fleet of buses [5].

Clean Urban Transport for Europe (CUTE) was carried out for Europe from 2003 to 2006, where 27 Mercedes Citaro buses were equipped with a 250 kW Ballard PEFC and 40 kg of compressed hydrogen at 350 bar which gave a range of 200 km was used for transport. Cost ranging for buses in the year 2018–2022 is expected to be $400–$600 K [6].

5.4 Electric Battery Hybrid

Cairns et al. [7] As we already know that fuel cells can produce low powers for a long amount of time, it also operates at a low temperature, but there comes a problem when specific power of higher range (100 W/lb.) is required. At such times, we need a hybrid system which can provide the needful power. Thus due to this, we combine the fuel cell with high specific secondary cell which can provide us the required values. Thus, this makes the system hybrid. Thus, the hybrid design provides a combination of high specific power capability and also high specific energy of fuel cells. The hybrid design of the power system consists of two types in it—one is the series-type hybrid system, and the other is parallel-type hybrid system. In a series-type hybrid system, the auxiliary power unit is connected in series with the electric motor and the battery which is also connected to the drive system. In a parallel system, the battery is connected parallel to the main auxiliary power and the drive, and only when needed the secondary power will be activated (Fig. 4).

Ahluwalia et al. [16] Experiment was carried out for a mid-sized sedan vehicle, gross weight 1695 kg, the drag coefficient of 0.32, 2.2 m^2 frontal area, coefficient of rolling friction of 0.009. At the rated power point, the PEFC stack operates at 2.5 atm and 80 °C to yield an overall efficiency of 50%. Thus in the layout, a two-way DC motor is used which gives 95% average efficiency used to step up the voltage to match the PEFC stack voltage or step down during regenerative braking. The mechanical energy at the motor shaft is transmitted to the wheel with one-step gear reduction and then final drive. The study was conducted to check the power needed to reach the top speed that is 100 mph and to accelerate from 0 to 60 mph in 10 s.

Shinji Aso et al. [8] Toyota created a similar FCHV which consisted of a parallel hybrid system; in this, the main power supply was connected parallel with a high-voltage battery. The fuel cell stack is directly connected to an inverter which is then connected to the electric motor and then the drive. The parallel power source (battery)

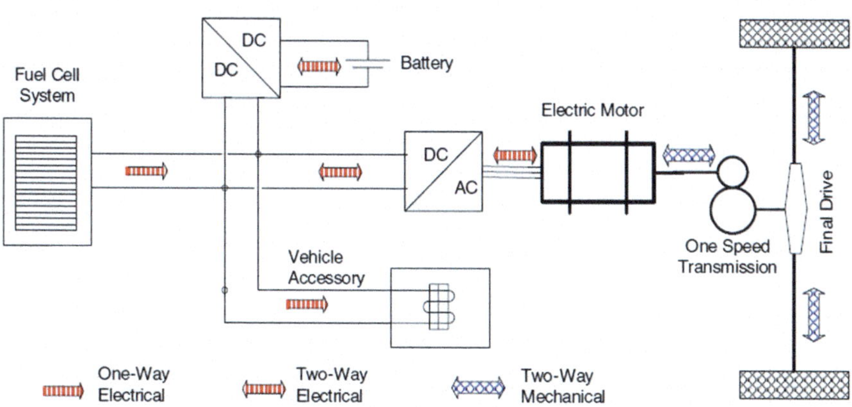

Fig. 4 Electric hybrid drive train [16]

is connected to a converter, which means in times of high power requirements the battery is used and the rest of the time the fuel cell is used. The power regeneration and the efficiency of the system are increased by the addition of the battery, thus making it a hybrid.

Peled et al. [9] Honda created a similar system in which a lithium-ion battery is used, the power output is 100 kW, it has a 288v lithium-ion battery, 4.1 kg hydrogen at 500 psi, and the range 280 km. The Kia new Borrego FCHV again was a hybrid fuel cell vehicle, and it had an 115 kW fuel cell with a lithium-ion battery and had a range of 315 miles. GM HydroGen 4 FCHV was a FCHV too, and it went for testing for 400,000 miles and has a range of 200 miles and 0–60 mphin 10 s. Mercedes blue zero hydrogen concept was again a FCHV, and it had a range of about 400 km for full tank of hydrogen and 100 km for 17.5 kWh on the lithium-ion battery.

5.5 IC Engine-Combined Fuel Cell

Edwards et al. [10] Here, an experiment is conducted to reduce the energy that is destroyed from the engine and which is left out as exhaust; nearly 40% is due to heat transfer and 20% exhaust, and 30% is wasted in combustion; thus, the main motive of the experiment is to reduce the level of energy wasted and thereby increase the efficiency. Thus to solve this, experiments add two power sources for power production, the conventional IC gasoline engine, and the FCV. Thus in such cases, an approximate of 70% efficiency is expected form the whole system. So in this, the process is still in the development stage. The proposed system combines both the systems in such a way that the waste is very low. Thus for that, a normal gasoline engine is taken, then a few of the cylinders are fed with rich fuel equivalence ration, and then the result from that can be used as a hydrogen reformer for hydrocarbons from the fuel; it is later passed through a shift reactor to reduce the carbon monoxide contents, and the rest is fed into the fuel cell for the reaction to occur to produce electricity. Then, any unreacted fuel leaving the fuel cell can again enter into the other few cylinders of the gasoline engine, combust, and give full combustion without any loss (Fig. 5).

5.6 Military Vehicles

Andrukaitis et al. [11] Onboard regenerative design, this is an idea of regeneration of the power while on a mission or operating condition. So by using other electronic components, regeneration can be accessed. The fuel cell works on hydrogen and oxygen, and gives out water. So, the idea is that if you make a closed system, there might be advantages in fuel regeneration. In this closed system, an electrolyzer is

Fig. 5 IC-combined fuel cell [10]

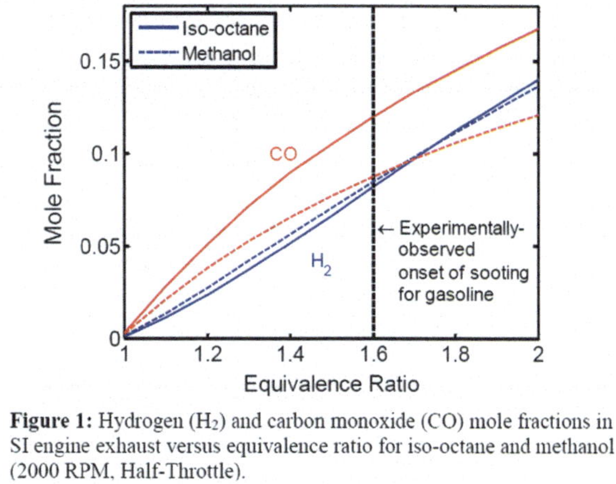

used and the overall reaction occured is reversed. Thus distributing hydrogen and oxygen separately and again feeding it back to the fuel cell, this may increase the range of the vehicle and result in less fuel depletion, thus giving out less vapors and fuel signatures (Fig. 6).

Das et al. [12] Auxiliary power units are used in a vast applications including military, therefore it is necessary that the system be able to withstand certain conditions which relate to the battle fields such as rough terrains, harsh weathers and climate, humidity, and shocks and vibrations, in short has to be sturdy and rigid in various conditions. They should give out very less thermal signatures or any other emission of smoke, light, sound. It should be highly reliable, and also human intervention should be minimum. Such conditions are required in military for silent watch protocol. NMRL has developed 10 kW generator car that uses methanol reformer for hydrogen generation.

Campbell et al. [13] Some of the fuel cells which are used in military applications are H3 350 methanol power system by SER energy, and it has a power output of 350 W.

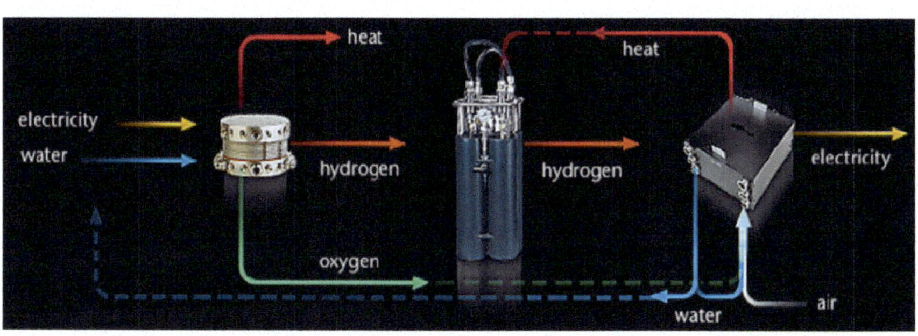

Fig. 6 Military vehicle regenerative design [11]

Power pac is the product made by the company Power cell and it has two variants 2kW and 4kW, which is specially used in military for its low sound applications which maybe used in military operations such as silent watch. Emily 2200 by SFC energy, it is a 90 W and 12/24 VDC system. The Defender series by Ultra Electronics AMI consists of the D300 and D245XR fuel cells. The D300 is a 10.9 kg, 300 W system that can output 12, 14, 24 or 28 V. Nordic power is used as an APU and can make 1 kW. Power core by Topsoe fuel cell can make 2 kW which is again used as an APU. Delphi solid oxide fuel cell auxiliary power unit, it produces 5 kW. Python by Merlin Power Systems produces 1.2 kW at 28 V.

5.7 *Material Handling Systems*

Garche et al. [6] One of the areas where power systems are used and where fuel cells fit exactly is the material handling systems. In this, we require low power with long duration, which is the exact thing which we get usually from a fuel cell, and the major market for these systems is the forklifts. Forklifts need very less power but need for a long duration of time. The operating costs are low too here. Apart from just forklifts, other applications are tow tractors, lift trucks, pallet trucks, and other small-scale vehicles. Plug Power Inc. made a typical power system for these applications called GenDrive which worked at 3 and 14 kW according to the use with fuel capacity of 0.7–3.4 kg of hydrogen at 350 bar pressure.

Bobbett et al. [14] Another application of similar low-powered, low-temperature operating system is used in golf karts. Golf karts usually require small power just to navigate around the golf course and carry passengers and a few loads. Thus in the study, they have used a system which houses a 2 kW fuel cell air cooled, which is developed by Energy Research Corporation.

5.8 *Autonomous Vehicles*

Chraim et al. [15] Autonomous vehicles are vehicles which can operate on its own. It can travel choosing a particular path and reach a destination without any human interference. Thus for such systems, long-life auxiliary power supplies are used so that the range is huge and longer distances can be covered, without being hindered in performance. The shortest path and the least usage of fuel are what give the higher efficiency that is expected from a vehicle of such caliber. Thus, wireless network sensors play a major role in this type of vehicle, the applications of autonomous driving vehicles are vast, and research is going on in major fields to exploit the technology and get better results in the system.

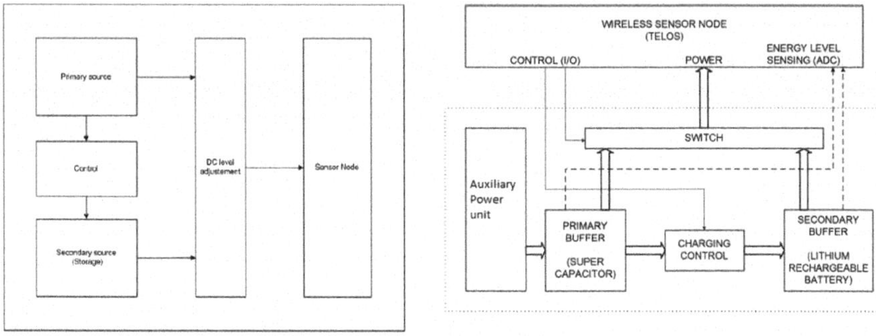

Fig. 7 Autonomous vehicle layout [15]

Wireless network systems, this plays as the most important thing in an autonomous vehicle, as the power and other factors can be provided to the vehicle, but for the vehicle to take decisions on its own and act accordingly is more important. In this, the auxiliary power unit is directly connected to the CPU, the CPU is connected to the various peripherals that are connected with the system, and also the CPU is connected to the communication system. Now when the vehicle moves, all the peripherals are monitored by the CPU and approximate conditions are checked and understood for the best optimal output (Fig. 7).

6 Future Aspects

- Substitution of platinum in a PEMFC would be a significant reduction in cost and would make FCVs more affordable. Research is ongoing in using carbon nanofibers which might be cheaper, lighter, and even more efficient when compared to the current platinum.
- Further onboard hydrogen formation can be achieved by recycling the exhaust vapors through a unit which again separate the hydrogen from the water and thus can give an added power output which can be stored for further use. However, the challenge for this concept is that the additional hydrogen which is generated should be cost-wise feasible when considering the additional costs that would go into the setup of an electrolyzer and a reformer. Also, the time taken for the reformation should be low and effective for the system to be brought into use.
- Further thermoelectric generators coupled with fuel cells can provide power at high operating temperatures for certain types of fuel cells.

7 Conclusion

So from the above paper, the applications of the different types of fuel cells in the different automotive fields are given briefly, the ways it has been used before and also the future applications in which they might be used. Fuel cells are the need of the hour, and shifting to a greener and more economic and even more efficient form of power supply should be the first priority, due to the climate changes and depletion of natural resources. Fuel cells in hybrid with the conventional system are also a viable option. Hydrogen as we all know is the most abundant element on the earth, we must use it as an advantage, and thus for the years to come hydrogen fuel cells are the safest and the best mode of power supplying.

References

1. Giri NK (2012) Alternate energy (sources, applications and technologies), first edition, ISBN: 978-81-7409-304-2
2. Lai J-S, Michael W. Ellis (2017) Fuel cell power systems and applications. In: Proceedings of IEEE 2017
3. Balakrishnan J (2007) Fuel cell technology. In: 2007 Third international conference on Information and automation for sustainability
4. Guaitolini SVM, Yahyaoui I, Fardin JF, Encarnação LF, Tadeo F (2018) A review of fuel cell and energy cogeneration technology. In: The 9th International Renewable Energy Congress (IREC 2018)
5. Bayrak ZU, Gencoglu MT (2013) Application areas of fuel cells. In: International Conference on Renewable Energy Research and Applications Madrid, Spain, 20–23 October 2013
6. Garche J, Jörissen L (2015) Applications of fuel cell technology: status and perspectives. The Electrochemical Society Interface Summer 2015. https://www.electrochem.org/
7. Cairns EJ, Shimorake H (1969) Recent advances in fuel cells and their application to new hybrid system. Adv Chem 90, American Chemical Society
8. Aso S, Kizaki M, Nonobe Y (2007) Development of fuel cell hybrid vehicles in Toyota. 1-4244-0844-X/07/$20.00 ©2007 IEEE
9. Peled E (2009) Fuel-cell-hybrid vehicle (FCHV), IFCBC February 2009
10. Edwards CF, Donohue MA, Fyffe JR, Regalbuto C. Exploration of a fuel cell/Internal combustion engine combined cycle for high efficiency power generation (A GCEP Exploratory Project), gcep.stanford.edu
11. Andrukaitis EE (2010) Fuel cells as tactical auxiliary power in military vehicles. ECS Transactions 26(1):445–454. 10.1149/1.3429017 © The Electrochemical Society
12. Das JN (2017) Fuel cell technologies for defence applications. In: Raghavan KV , Ghosh P (eds) Energy Engineering, Springer Nature Singapore Pte Ltd. https://doi.org/10.1007/978-981-10-3102-1_2
13. Campbell B, Crase S, Sims B (2014) Review of fuel cell technologies for military land vehicles. Commonwealth of Australia AR-016-098 September2014
14. Bobbett E, McCormick JB, Lynn DK, Kerwln WJ, Derouin CR, Ii. Salazar P (1980) Fuel cell powered golf kart. In: third international electric vehicle exposition and conference, St. Louis, MO, May 1980
15. Chraim F, Karaki S (2010) Fuel cell applications in wireless sensor networks. 978-1-4244-2833-5/10/$25.00 ©2010 IEEE
16. Ahluwalia R, Wang X, Rousseau A (2004) Fuel economy of hybrid fuel cell vehicles. J Power Sources

Energy Generation from Piezoelectric Material in Automobile

Aman Akotkar, Anand Kumar Sinsh, and S. Jaichandar

Abstract The world around us is converting itself to an advanced place with a great acceleration as compared to the last few centuries. With all the advancements, automotive sector is also on a great level but the continuous research topic is always energy conservation and generation. Energy can be harvested through many ways but keeping the negative impacts into consideration harvesting green and clean energy resource is the main focus. Piezoelectric material is one such material which can be used. In this article, we are describing to use a piezoelectric material to generate electricity for increment of the battery efficiency. To harness energy from piezoelectric sensor, vibration or pressure is required. We can lodge the piezoelectric material at a particular depth inside the tyres around the circumference, and the part which is in contact with the ground deforms the material and provides the required pressure which generates voltage. The movement of the vehicle will make the piezoelectric material work, and as the part or line of contact changes the material will be back to its neutral state; with the rotation of the tyres, this repeats.

Keywords Piezoelectric material · Harness energy

1 Introduction

For an automobile to move, energy is needed which we get from fuel or battery, but when we talk about the design and lighting system of the vehicle then it drains some extra energy. So, there is an urgent need of some reliable source of energy which can:

Generate the energy from the vehicle itself and then use it for different purposes.

The table below shows the amount of energy needed in a vehicle for various outputs.

A. Akotkar (✉) · A. K. Sinsh · S. Jaichandar
Department of Automobile, Vel Tech Rangarajan Dr Sagunthala R&D Institute of Science and Technology, Chennai 600062, India
e-mail: amanakotkar@gmail.com

L.-J. Yang et al. (eds.), *Proceedings of ICDMC 2019*,
Lecture Notes in Mechanical Engineering,
https://doi.org/10.1007/978-981-15-3631-1_7

S. no.	Purpose	Energy
1	Headlight	
2	Air conditioner	
3	Music system	
4	Inner lights	

Some of this can be fulfilled by the piezoelectric generator which will increase the efficiency of the battery without applying any extra force on the vehicle. This will increase the life of the battery and decrease the battery drainage.

If we go for future scopes, it might replace the dynamo, which will again increase the efficiency of the fuel or battery as it is the biggest energy drainer.

With the invention of piezoelectricity, it became useful for many applications; let it be a small domestic lighter, a microphone or for medical purposes like ultrasound. And this paper is to use it on a new level.

2 Piezoelectric Materials

The atomic arrangement of some materials is in such a specific way that if we squeeze some crystals (such as quartz), a flow of electric current takes place and the opposite also occurs; i.e., on the flow of electric current through them it vibrates. This phenomenon is known as piezoelectricity. The generation of an electrical potential (or voltage) and flow of charges on applying mechanical stress to a material is known as piezoelectricity.

When we observe it practically, the crystal behaves like a battery with opposite polarities on both the sides; a current starts flowing if we connect both the ends in an electric circuit. And in reverse piezoelectric effect on applying voltage, a mechanical stress is experienced.

3 Working Principle

In a normal condition the charges are balanced in a piezoelectric crystal and they act as a neutral object as both the positive and negative charges are present they cancel each other. If we squeeze the crystal, this balance is disturbed. Now, net positive and net negative charges start appearing on the opposite faces. This leads to potential difference across its opposite faces.

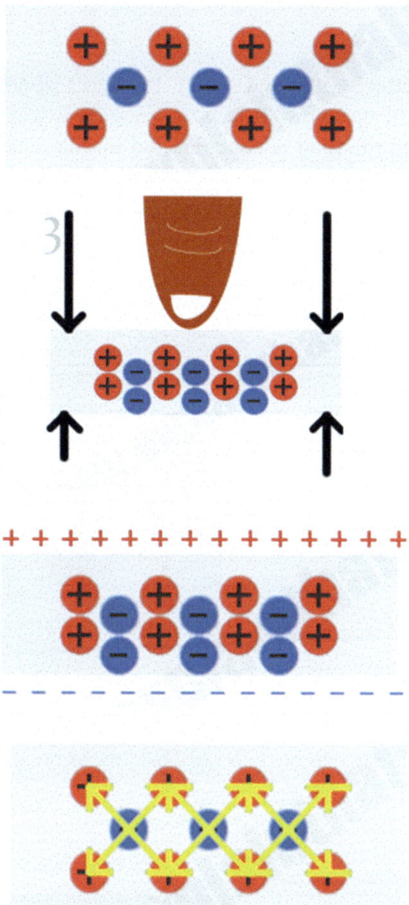

3.1 *Effect of Temperature on the Material*

The piezoelectric materials change its properties with increase in temperature. But it is experimentally observed that if the working temperature does not exceed half of the Curie temperature of the particular material the change in property is least. With further increase in temperature, the material loses its electrical properties and tends to become neutral as it becomes more symmetric. As our work is on tyres, it is obvious that it cannot even exceed 80 °C. So, it will not have much impact on the material. The graph is plotted between *d*33 and the Currie temperature of different piezoelectric materials.

3.2 Insertion

The piezoelectric material is inserted inside the tyres approximately 1.5 cm or depending on the size of the tyres. It is fitted in such a way that the pressure on the point of contact with the road is concentrated on the piezoelectric material for better outcome by the help of a trapezoid shape casting inside the tyres without altering the properties of the tyres (Figs. 1 and 2).

Suppose if we assume that the weight of the vehicle is 2000 N, then

$$F = m * g = 2000\,\mathrm{N}$$

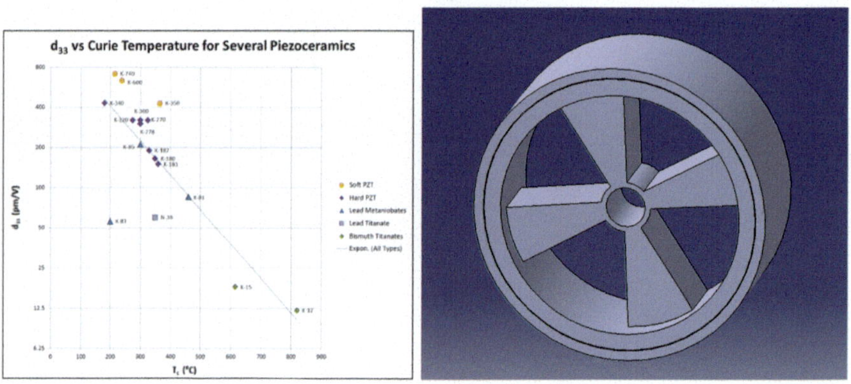

Fig. 1 Complete model

Fig. 2 Piezoelectric plate arrangement along the circumference

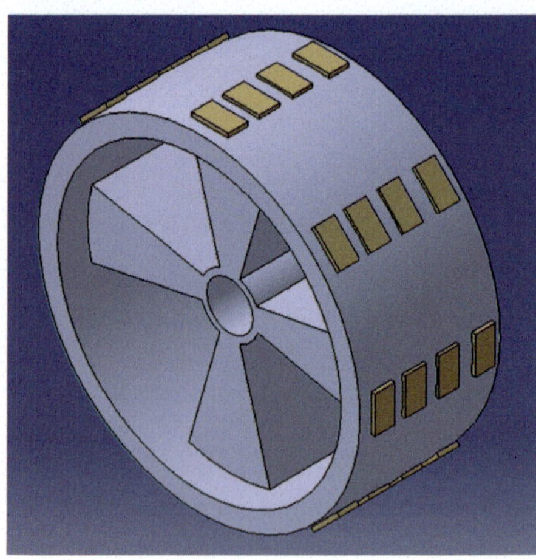

which is distributed to all the four wheels in equal proportions if the weight of the vehicle is equally distributed. But taking the general condition of unequal distribution, let the total weight experienced by the front wheels be 1400 N and on the back wheels be 600 N.

Now, we know that

$$\text{Pressure} = F/A$$

where A is the area of the part of contact with the road.

If we consider one of the front wheels, then

$$\text{Force} = 1400/2 = 700\,\text{N}.$$

Assuming the area of contact as 5 square centimeters and adjusting 4 piezoelectric materials in this area, the force on each material will be approximately

$$F(\text{each}) = 700\,\text{N}/4 = 175\,\text{N}$$

This force will exert pressure on one material as

$$\text{Pressure} = 175/5 = 35\,\text{pa}$$

This pressure is absorbed by the material to generate voltage which is controlled by an electronic circuit and charge the battery.

The voltage which is produced by the material is represented by the equation

$$V = \left(-g33 * h * \frac{f}{A}\right) = -\left(g33 * h * \left(\frac{f}{d^2 * \frac{\pi}{4}}\right)\right) \quad (1)$$

where

g33 is the piezoelectric constant.
V is the expected peak strain due to deformation of the material.
H is the length or thickness of the piezoelectric ceramic.
F is force that is printed on the piezoelectric ceramic (*N*).
d is effective diameter of the piezoelectric ceramic.
A is area of the piezoelectric ceramic.

Assigning the values to the above

$$g33 = 25 \times 10^{-3}\,\text{Vm/N}$$
$$h = 1.25\,\text{mm}$$
$$f = 175\,\text{N (from above)}$$
$$d = 10\,\text{mm}$$

Substituting these values in (1), we get

$$V = 69\,\text{V}$$

And this is just by one so if we consider all the four and also the ones along the periphery we will get the more voltage which is continuously produced until the wheel is rotating. Considering some of the losses, let us assume that 50 V is the practical value. And the resistance and some other factors will depend on the type of piezoelectric material used. With the help of a capacitor, this voltage will be used to charge the battery.

4 Conclusion

This method can reach many heights with more advancements as we know that piezoelectric material is also used in many other ways for energy generation an extra movable battery can also be charged with this method which can be used for other purposes apart from automotives. And we can even replace the dynamo which applies more load on the engine to charge the battery. This method will increase the efficiency of the engine. This method is not even costly, so it can easily be brought to action.

Evaluation of Microalgae Biodiesel Blend Along with DTBP as an Ignition Enhancer on Diesel Engine Attributes

A. Gurusamy, V. Gnanamoorthi, P. Purushothaman, P. Mebin Samuel, and A. A. Muhammad Irfan

Abstract Biodiesel in an overall view has high oxygen content due to which it attains an attractive position to be a high-quality alternate fuel for a direct injection diesel engine. The combustion process along with the particulate matter will find a betterment when it is blended. In this work, commercially used pure diesel fuel (PD) is blended with 20% of an algae-based biodiesel, and also, di-tertiary-butyl peroxide is added in three different proportions (1, 3 and 5%). The performance and emission outcomes of all these blends under examination are studied using a diesel engine. All six parameters related to emission and performance are studied. From the results, it was seen that with the addition of DTBP in B20 blend increases the thermal efficiency and also the emission parameter values were found to be reduced, which provides a promising insight on the usage of DTBP additive in a diesel engine.

Keywords DTBP · Biodiesel · Microalgae · Performance · Emission

1 Introduction

The dwindling of the petroleum products day by day due to the increased demands and also the emission norms provide a greater threat for its successful usage. To

---

A. Gurusamy (✉)
Department of Automobile Engineering, Pace Institute of Technology and Sciences, Ongole, India
e-mail: agsmeice@gmail.com

V. Gnanamoorthi
Department of Mechanical Engineering, University College of Engineering, Villupuram, India

P. Purushothaman
Department of Mechanical Engineering, Agni College of Technology, Chennai, India

P. Mebin Samuel
Department of Automobile Engineering, Madras Institute of Technology, Chennai, India

A. A. Muhammad Irfan
Department of Mechanical Engineering, Mohamed Sathak A J College of Engineering, Chennai, India

L.-J. Yang et al. (eds.), *Proceedings of ICDMC 2019*,
Lecture Notes in Mechanical Engineering,
https://doi.org/10.1007/978-981-15-3631-1_8

fight this demand increase, bioenergy is implemented as a potential alternative. As a general fact, biodiesel has increased level of fatty acid content which is also indicated by many researchers (since vegetable oil or animal fats are used to produce biodiesel). It has higher oxygen, also renewable and biodegradable; due to this, the emission and greenhouse gases are reduced.

As a third-generation biofuel, algae biofuel has received significant interest and also importance, as this can be grown in any liquid ranging from wastewater to freshwater. This is mainly due to the improved yield rate and also the photosynthesis. The yield rate is higher since 70% of lipid is available in algae [1, 2]. The potential of microalgae in biodiesel production study done by Chen [3] between algae and other fuel shows that carbon supply, water supply and harvesting are the main parameters to be considered for calculating the overall cost of production of oil. In addition to the minimal land cost needed for algae, mixing and infrastructure cost are higher though it is shadowed by the higher yield of algae [4]. Photosynthesis process using sunlight is one of the major needs for the algae growth. Thus, being a renewable energy, algae fuel gets immense attention to meet the world-level energy demand. Algae oil can be considered as a major tool for reducing emissions. In addition to all the benefits, algae biomass after the extraction of oil does not pollute the environment [5].

2 Materials with Methods

2.1 *Biodiesel Production*

The purification of algae was done through a serial dilution process followed by plating for all the samples collected from many varieties (various water conditions). The dry microalgae (Botryococcus braunii) biomass is produced by filtration and drying after sample collection. Microalgae oil is separated from algal biomass by using solvent-expeller extraction method. Oil extraction was done using two solvents, namely hexane and ethers. Due to the introduction of these solvents, the algae membrane wall was ruptured and the oil was released. This oil is then placed inside an incubator and maintained at 80 °C to remove the excess of hexane solvents. In this process, 400 ml of oil was initially obtained, and finally, after the separation of hexane, only 380 ml of oil was left behind. This 380 ml of oil is the pure algae oil.

So as to remove the excess fatty acids, transesterification process is carried out using methanol and NaOH as catalyst. At a reaction temperature of about 70 °C, the mixture was constantly mixed at a speed of 800 rpm using a magnetic stirrer for a period of 1 h. Then, the mixture is allowed to cool to room temperature, and after a period of 8 h, algae methyl ester is separated from glycerin formed due to transesterification process using an apparatus called separatory funnel. Further, the algae methyl ester was processed for washing using hot water at a ratio of 1:5. This washing is repeated for three times to ensure the removal of dirt or any unwanted particles. Finally, the crude algae methyl ester was subjected to drying process. Here,

Table 1 Properties of test fuels

S. No	Properties	Diesel	B20	Measurement standards
1	Density @15 °C kg/m^3	876	816	ASTM D1298
2	Kinematic viscosity @40 °C in CST	2.9	3.120	ASTM D445
3	Calorific value (MJ/Kg)	44.43	40.42	ASTM D240
4	Cetane Index	48	47	ASTM D976
5	Flash point °C	60–80	60.49	ASTM D92
6	Sulfur content (%)	0.05	0.38	ASTM D5453
7	Ash content (%)	0.010	0.020	ASTM D482

the oil was subjected to an elevated temperature of 378 K for one hour after being held in a separatory funnel for 30 min and filtered. This is done to ensure the removal of water molecules in the oil and obtain purified microalgae methyl ester (MAME). Table 1 shows the comparison of diesel fuel properties with that of the measured values of B20 algae methyl ester blend.

Piloto-Rodriguez et al. [6] reviewed the implementation of algae methyl ester blend with diesel ranging from 5% till 100% and put forward that the values of B20 blend are comparable with that of diesel fuel with CO_2 and NO_X being on the higher side. Milano et al. [7] processed microalgae fuel to a substitute of diesel fuel in terms of power generation. The current stage of microalgae cultivation, including methods, production cost, energy ration, harvesting processes, adds their way to increase the cost of microalgae oil. Al-lwayzy et al. [8] performed a test on a diesel engine using microalgae Chlorella protothecoides biodiesel and suggested that the oil can be used purely without the need of blending it with diesel fuel and also in its blended form. Satputaley et al. [9] examined the combustion, emission and performance of both microalgae and the methyesther of the same, and proposed that the methyl ester of microalgae has better performance than the oil variant.

2.2 *Di-Tertiary-Butyl Peroxide*

The rate of initiation to form free radical and limiting the aromatic content of petroleum diesel fuel can be obtained through hydrotreating or adding an ignition enhancer [10]. Additive DTBP similarly to 2-EHN was recognized as an ignition enhancer for spontaneous ignition of diesel [11]. Kumar et al. [12] had investigated the effect of DTBP on the ethanol–cotton seed methyl ester blends in DICI engine. They observed that by adding DTBP fuel consumption, NO_X, CO, and UBHC were reduced significantly, and peak pressure of combustion is increased. Vallinayagam et al. [13] equated the results between both isoamyl nitrate (IAN) and DTBP on pine oil-neat diesel fuel blended in the ratio of 1:2 in a single-cylinder CI engine and proposed that HC and CO reduce and NOx increases. But, while adding DTBP, NO_X

Table 2 Properties of di-tertiary-butyl peroxide

S. No	Properties	DTBP
1	Composition	$C_8H_{18}O_2$
2	Flash point (°C)	6
3	Density (@20 °C) kg/m^3	0.79
4	Autoignition temperature (°C)	165
5	Molecular weight (g/mol)	146

emission is retarded when compared with biodiesel blend and biodiesel with isoamyl nitrate (IAN). The properties of DTBP are given in Table 2.

2.3 *Experimental Setup*

A four-stroke, single-cylinder DI diesel engine was used for the experimental analysis, whose specification is given in Table 3, and Fig. 1 gives a schematic diagram. The emission measurement is done with the help of AVL 444 di-gas analyzer, in which NOx and HC are denoted using ppm and amount of CO was denoted in % values. Smoke meter denoting smoke value in the Hartridge Smoke Unit is used for the measurement of smoke values. The reading was taken for different values of engine load ranging from 20 to 100% with 20% increment in each step for all the blends under test. In order to avoid any uncertainty error, the experiments were repeated for five times, and the average values were considered for all the further calculations. Uncertainty analysis is also done to show the change in values for each parameter under study, which is shown in Table 4. Overall uncertainty is measured to be 2.2%.

Table 3 Specification of the engine

Details	Specification
Type	Four stroke, Kirloskar TV1 model, compression ignition, direct injection and water cooled
Rated power and speed	5.2 kW and 1500 rpm
Number of cylinder	Single cylinder
Compression ratio	17.5:1
Bore& stroke	87.5 and 110 mm
Method of loading	Eddy current dynamometer
Type of injection	Mechanical pump-nozzle injection
Injection timing	23° before TDC
Injection pressure	220 bar
Combustion chamber	Hemispherical open type

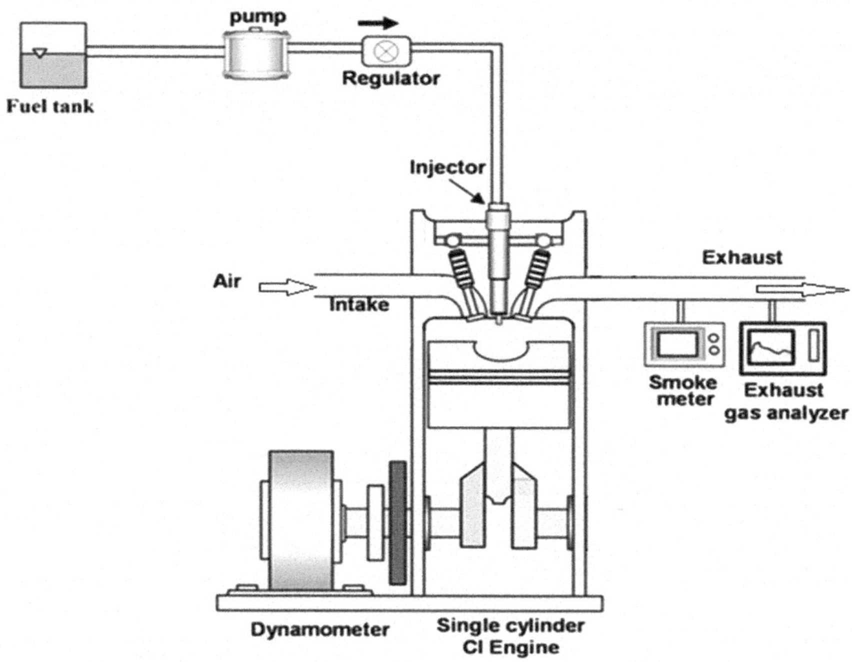

Fig. 1 Schematic diagram of the experimental setup

3 Results and Discussions

3.1 Brake Specific Energy Consumption

It is the amount of input energy needed to develop 1 KW power, and it is defined as the product of BSFC and calorific value of fuel. [14]. BSEC was examined, and the results show that 12.33, 13.12, 12.80, 12.65, 12.60 MJ/kWh for D100, B20 and B20 with 1, 3 and 5% DTBP at 100% load. Figure 2 shows the % differences in BSEC using B20 and B20 with 1, 3 and 5% DTBP compared to PD which is 6.4, 4, 2.9 and 1% at 100% load. The difference arose due to the physical and chemical properties of the fuel like heating and mass flow values. Other parameters such as density, viscosity and fuel injection method are also the main factors affecting the BSEC value of fuel [15, 16]. By increasing the percentage of DTBP in B20 blend, BSEC value is decreased, because of that increase in cetane value of mixture, which results from the engine consuming less fuel to one kilowatt power. BSEC value is equivalent to pure diesel for the blend of B20 with 5% blend [17].

Table 4 List of instruments and their accuracy and percentage uncertainties

Measurement	Type and manufacturer	Accuracy (%)	Uncertainty	Measurement technique
Load	Strain gauge, Sensotronics Sanmar	±10 N	±0.2	Load cell
Speed	Kubler, Germany	±10 rpm	±0.1	Magnetic pickup principle
Fuel flow measurement	Differential pressure Transmitter	±0.1 cc	±1	Volumetric measurement
CO	AVL exhaust gas analyzer, Austria	±0.02%	±0.2	NDIR technique
HC	AVL exhaust gas analyzer, Austria	±0.03%	±0.1	NDIR technique
NO_X	AVL exhaust gas analyzer, Austria	±12 ppm	±0.2	NDIR technique
Smoke	AVL smoke meter	±1 HSU	±1	Opacimeter
EGT indicator	Make Wika	±1 °C	±0.15	Thermocouple
Pressure Pick up	PCB, piezotronics	±0.1 kg	±0.1	Magnetic pickup principle
Crank angle encoder	Kubler, Germany	±1 deg	±0.2	Magnetic pickup principle

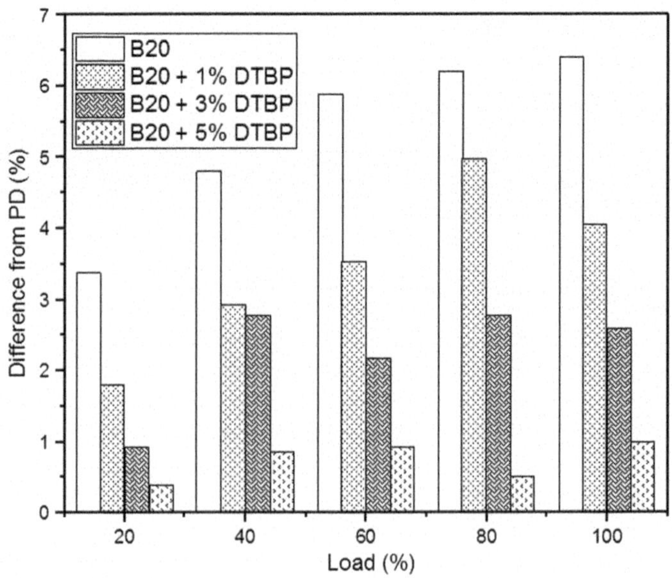

Fig. 2 Brake specific energy consumption percentage differences compared with PD

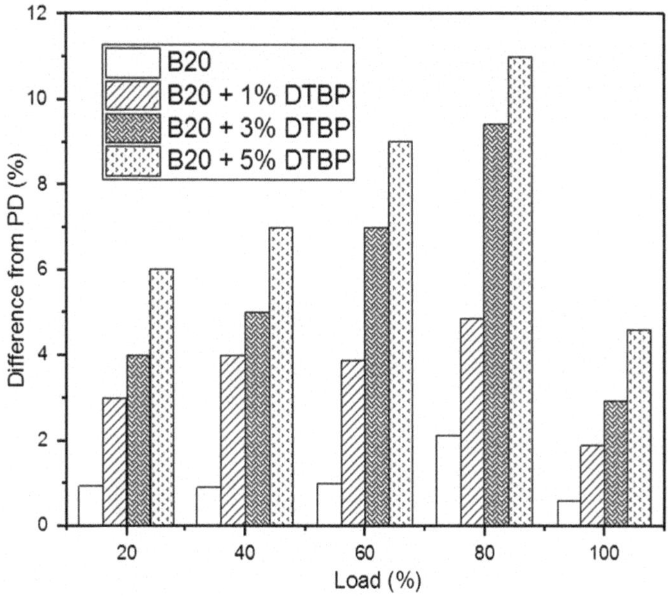

Fig. 3 Brake thermal efficiency percentage differences compared with PD

3.2 *Brake Thermal Efficiency*

It is the amount of work converted from the power from the fuel by the engine [18]. According to the Fig. 3, the percentage of differences in BTE using B20 and B20 with 1, 3 and 5% DTBP compared to PD are 0.6, 1.9, 2.9 and 4.6% at full load and BTE was found to be 29.54, 29.72, 30.1, 30.4 and 30.9% for D100, B20 and B20 with 1, 3 and 5% DTBP at full load condition. BTE is improved for B20 due to the proper use of heat energy, and also, the oxygen content existing in the microalgae methyl ester helps in improving the combustion process due to their improved evaporation factor [19]. The use of ignition enhancer (DTBP) substantially increases brake thermal efficiency, resulted in the reduction of IG delay period and advances in CT of B20 with 1, 3 and 5% [20]. By increasing the percentage of DTPB, brake thermal efficiency value is increased notably.

3.3 *Carbon Monoxide Emission*

Due to oxygen deficiency and lesser time period of CO to CO_2 conversion during combustion, CO emission is formed in IC engine. Figure 4 illustrates the percentage of differences in CO using B20 and B20 with 1, 3 and 5% DTBP compared to PD, which is −14, −28, −39 and −46% at full load and CO emission for D100, B20

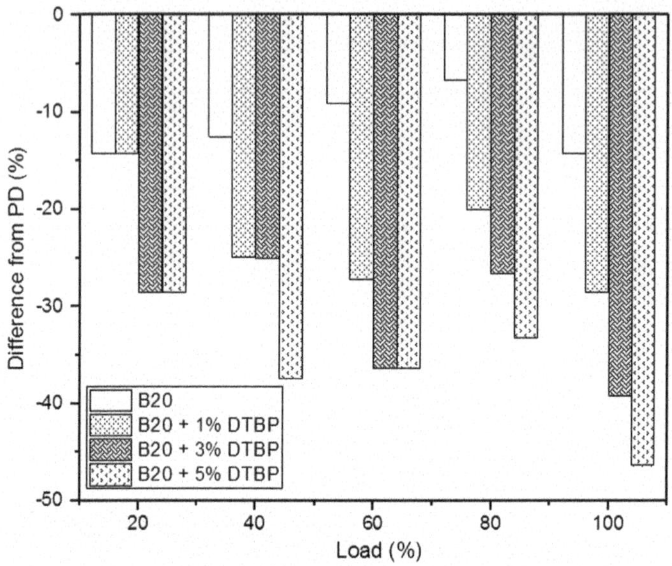

Fig. 4 Carbon monoxide emission percentage differences compared with PD

and B20 with 1, 3 and 5% DTBP fuel blends is 0.28%, 0.24%, 0.2%, 0.17% and 0.15% by volume at full load respectively. A negative sign indicates the reduction in percentage. CO emission is decreased to compare with B20. When compared to that of diesel, CO emission was reduced considerably. Since higher oxygen content present in the blends of biodiesel [21]. The involvement of DTBP improves the combustion process by increasing the combustion even in fuel rich zones; thereby, reduction in emission is achieved [22].

3.4 Unburned Hydrocarbon Emission

UBHC is a result of too lean or too rich fuel–air mixture inside the combustion chamber which could not get combusted or sustain the combustion, thereby oxidizing partially [23]. UBHC emission was found to be 53, 51, 47, 45 and 40 ppm for D100, B20 and B20 with 1, 3 and 5% DTBP fuel blends at 100% load. Figure 5 shows the percentage of differences in UBHC using B20 and B20 with 1, 3 and 5% DTBP compared to PD which is −3, −11, −15 and −25% at full load. Figure 5 acknowledges that HC emission is directly proportional with load for diesel and microalgae methyl ester(MAME); this is due to equivalence ratio improvement [24]. Emission of HC is greatly reduced considering B20 and B20 with 1, 3 and 5% DTBP. This study explains that HC reduction is attained with the help of oxygenated fuel. For B20 with 1, 3 and 5% DTBP, higher viscosity of B20 blend is neutralized with the reduced viscosity of DTBP additive [25].

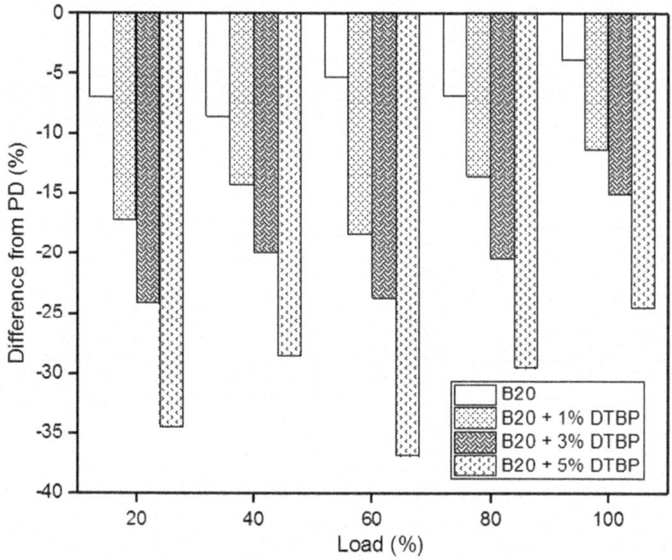

Fig. 5 Unburned hydrocarbon emission percentage differences compared with PD

3.5 Oxides of Nitrogen Emission

NOx is a combination of both NO and NO_2, which depends upon various parameters such as temperature, reaction time and surplus air [26]. NO_X emission for D100, B20 and B20 with 1, 3 and 5% DTBP fuel blends is found to be 925, 1045, 1022, 998 and 912 ppm at 100% load. Figure 6 demonstrates the percentage of differences in NO_X using B20 and B20 with 1, 3 and 5% DTBP compared to PD which is 13, 10, 8 and −1.5% at full load. NO_X is increased to compare with diesel with B20 at all load condition, due to increased O_2 concentration [27]. With the addition of 1, 3 and 5% of DTBP in B20, NO_X emission is decreased gradually, due to advances in combustion timing for B20 blend.

3.6 Smoke Emission

Smoke emission was found to be 53HSU, 42HSU, 39HSU, 35HSU and 24HSU for D100, B20 and B20 with 1, 3 and 5% DTBP fuel blends at 100% load. Figure 7 indicates the percentage of differences in smoke using B20 and B20 with 1, 3 and 5% DTBP compared to PD which is −20, −26, −34 and −45% at full load. Smoke emission is reduced for B20 and B20 with the addition of 1, 3 and 5% DTBP to compare with diesel. This is further supported by the presence of oxygen in fuel [27, 28].

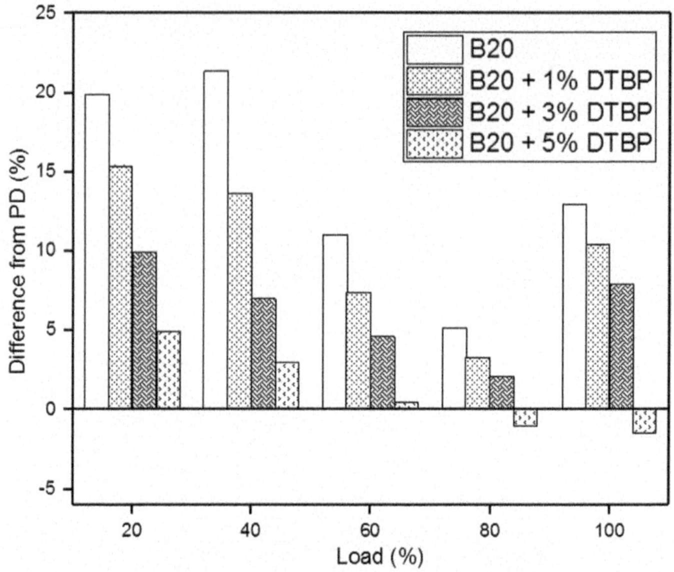

Fig. 6 Oxides of nitrogen emission percentage differences compared with PD

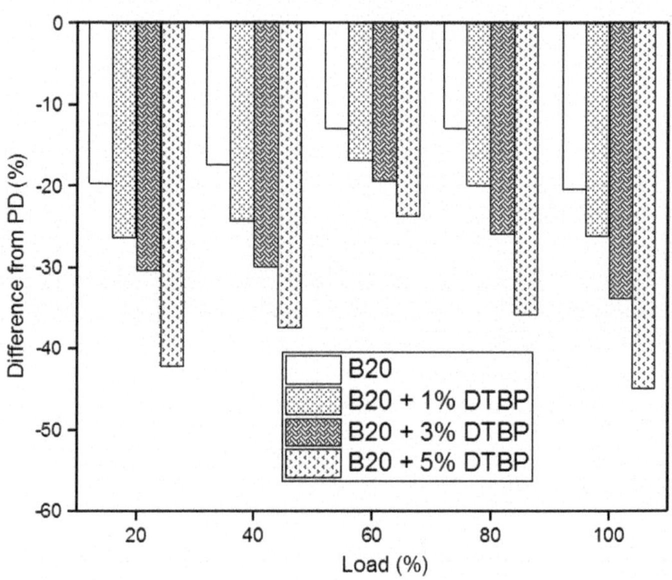

Fig. 7 Smoke emission percentage differences compared with PD

4 Conclusion

The effect of DTBP addition with microalgae methyl ester (MAME) biodiesel blend (B20) in a diesel engine and thereby the effects of DTBP addition (1, 3 and 5%) in MAME (B20) on engine's working behavior characters (performance, emissions) were examined. The results from the study are further given below:

1. BSEC of B20 with 5% DTBP is equivalent to pure diesel (PD) and produced maximum brake thermal efficiency.
2. The emission tests exposed that smoke, UBHC, CO and NOx emissions improved for B20 fuel blend with the addition of DTBP.
3. The DTBP additive can produce satisfactory results when blended with a biodiesel–diesel blend.
4. B20 blends of MAME was able to run properly and also produced satisfactory results when blended with DTBP without any major changes in the engine.

References

1. Raheem A, Prinsen P, Vuppaladadiyam AK, Zhao M, Luque R (2018) A review on sustainable microalgae based biofuel and bioenergy production: Recent developments. J Clean Prod 181:42–59. https://doi.org/10.1016/j.jclepro.2018.01.125
2. Bagul SY, Chakdar H, Pandiyan K, Das K (2018) Conservation and application of microalgae for biofuel production. In: Sharma S, Varma A (eds) Microbial resource conservation. Soil Biology, vol 54. Springer, Cham. https://doi.org/10.1007/978-3-319-96971-8_12
3. Chen J, Li J, Dong W, Zhang X, Tyagi RD, Drogui P, Surampalli RY (2018) The potential of microalgae in biodiesel production. Renew Sustain Energy Rev 90:336–346. https://doi.org/10.1016/j.rser.2018.03.073
4. Shuba ES, Kifle D (2018) Microalgae to biofuels: 'Promising' alternative and renewable energy, 534 review. Renew Sustain Energy Rev 81:743–755. https://doi.org/10.1016/j.rser.2017.08.042
5. Collotta M, Champagne P, Mabee W, Tomasoni G, Alberti M (2019) Life cycle analysis of the production of biodiesel from microalgae. In: Basosi R, Cellura M, Longo S, Parisi M (eds) Life cycle assessment of energy systems and sustainable energy technologies. Green energy and technology. Springer, Cham. https://doi.org/10.1007/978-3-319-93740-3_10
6. Piloto-Rodríguez R, Sánchez-Borroto Y, Melo-Espinosa EA, Verhelst S (2017) Assessment of diesel engine performance when fueled with biodiesel from algae and microalgae: an overview. Renew Sustain Energy Rev 69:833–842. https://doi.org/10.1016/j.rser.2016.11.015
7. Milano J, Ong HC, Masjuki HH, Chong WT, Lam MK, Loh PK, Vellayan V (2016) Microalgae biofuels as an alternative to fossil fuel for power generation. Renew Sustain Energy Rev 58:180–197. https://doi.org/10.1016/j.rser.2015.12.150
8. Al-lwayzy SH, Yusaf T (2017) Diesel engine performance and exhaust gas emissions using microalgae chlorella protothecoides biodiesel. Renew Energy 101:690–701. https://doi.org/10.1016/j.renene.2016.09.035
9. Satputaley SS, Zodpe DB, Deshpande NV (2016) Performance, combustion and emission study on CI engine using microalgae oil and microalgae oil methyl esters. J Energy Inst https://doi.org/10.1016/j.joei.2016.05.011
10. Mofijur M, Rasul MG, Hyde J, Azad AK, Mamat R, Bhuiya MM (2016) Role of biofuel and their binary (diesel–biodiesel) and ternary (ethanol–biodiesel–diesel) blends on internal

combustion engines emission reduction. Renew Sustain Energy Rev 53:265–278. https://doi.org/10.1016/j.rser.2015.08.046
11. Lee SW, Cho YS, Baik DS (2010) Effect of cetane enhancer on spray and combustion characteristics of compressed ignition type LPG fuel. Int J Automot Technol 11(3):381–386. https://doi.org/10.1007/s12239-010-0047-8
12. Kumar KS, Raj RTK (2016) Effect of Di-Tertiary Butyl Peroxide on the performance, combustion and emission characteristics of ethanol blended cotton seed methyl ester fuelled automotive diesel engine. Energy Convers Manag 127:1–10. https://doi.org/10.1016/j.enconman.2016.08.091
13. Vallinayagam R, Vedharaj S, Yang WM, Saravanan CG, Lee PS, Chua KJ, Chou SK (2014) Impact of ignition promoting additives on the characteristics of a diesel engine powered by pine oil–diesel blend. Fuel 117:278–285. https://doi.org/10.1016/j.fuel.2013.09.076
14. Barie NJ, Humke AL (1981) Performance and emission characteristics of a naturally aspirated diesel engine with vegetable oil fuels. Soc Automot Eng. Paper No: 810262. https://doi.org/10.4271/810262
15. Al-Iwayzy SH, Yusaf TF, Al-Juboori RA (2014) Biofuels from the fresh water microalgae Chlorella vulgaris (FWM-CV) for diesel engines. Energies 7:1829–1851. https://doi.org/10.3390/en7031829
16. Scragg AG, Morrison J, Shales SW (2003) The use of a fuel containing Chlorella vulgaris in a diesel engine. Enzyme Microbial Technol 33:884–889. https://doi.org/10.1016/j.enzmictec.2003.01.001
17. Atmanli A (2016) Effects of a cetane improver on fuel properties and engine characteristics of a diesel engine fueled with the blends of diesel, hazelnut oil and higher carbon alcohol. Fuel 172:209–217. https://doi.org/10.1016/j.fuel.2016.01.013
18. Rakopoulos DC, Rakopoulos CD, Giakoumis EG, Dimaratos AM, Kyritsis DC (2010) Effects of butanol—diesel fuel blends on the performance and emissions of a high-speed DI diesel engine. Energy Conversion Manage 51:1989–1997. https://doi.org/10.1016/j.enconman.2010.02.032
19. Chen C, Ma X, Liu K (2011) Thermogravimetric analysis of microalgae combustion under different oxygen supply concentrations. Elsevier, Appl Energy 88:3189–3196. https://doi.org/10.1016/j.apenergy.2011.03.003
20. Mack JH, Dibble RW, Buchholz BA, Floweers DL (2005) The effect of the Di-Tertiary Butyl Peroxide additive on HCCI combustion of fuel blends of ethanol and Diethyl ether. SAE Tech Pap. 2005-01-2135. https://doi.org/10.4271/2005-01-2135
21. Tuccar G, Aydin K (2013) Evaluation of methyl ester of microalgae oil as fuel in a diesel engine. Fuel 112:203–207. https://doi.org/10.1016/j.fuel.2013.05.016
22. Liu S, Zhu Z, Zhang Z, Gao G, Wei Y (2010) Effect of a cetane number (CN) improver on combustion and emission characteristics of a compression-ignition (CI) engine fueled with an ethanol diesel blend. Energy Fuels 24:2449–2454. https://doi.org/10.1021/ef901543m
23. Buyukkaya E (2010) Effects of biodiesel on a DI diesel engine performance, emission and combustion characteristics. Fuel 89:3099–3105. https://doi.org/10.1016/j.fuel.2010.05.034
24. Wahlen BD, Morgan MR, McCurdy AT, Willis RM, Morgan MD, Dye DJ, Bugbee B, Wood BD, Seefeldt LC (2013) Biodiesel from microalgae, yeast, and bacteria: engine performance and exhaust emissions. Energy Fuels 27:220–228. https://doi.org/10.1021/ef3012382
25. Nandi MK (1996) The performance of di-tertiary-butyl peroxide as cetane improver in diesel fuels. Am Chem Soc Div Fuel Chem 41:863–867
26. Hoekman SK, Robbins C (2012) Review of the effects of biodiesel on NOx emissions. Fuel Process Technol 96:237–249. https://doi.org/10.1016/j.fuproc.2011.12.036
27. Mwangi JK, Wen-Jhy L, Liang-Ming W, Tser SW, Wei-Hsin C, Jo-Shu C (2015) Microalgae oil: algae cultivation and harvest, algae residue torrefaction and diesel engine emissions tests. Aerosol Air Qual Res 15:901–914. https://doi.org/10.4209/aaqr.2014.10.0268
28. Wu YU, Gen CHEN (2011) Zuohua HUANG, Influence of cetane number improver on performance and emissions of a common-rail diesel engine fueled with biodiesel-methanol blend. Frontiers Energy 5(4):412–418. https://doi.org/10.1007/s11708-011-0163-9

Experimental Investigation on the Emission Level of a Single Cylinder Petrol Engine with Manifolds of Different Geometry

K. Raja, Amala Justus Selvam, and P. L. Rupesh

Abstract Petrol engines have been considered as the most popular engine to be used in automobiles due to their simplicity and easy operations. The main problem of SI engine is to get the homogeneous mixture of fuel and air. Due to the improper mixing of fuel–air mixture, emissions become a great concern for the researchers. This air movement and ensuing in-barrel stream field structure created are extraordinarily impacted by the design of intake manifolds. The study aims at the usage of intake manifolds with different geometries such as convergent, divergent and venture in a single cylinder SI engine. The emission levels of CO, HC and NO_X have been tested for the manifolds with different geometry, and the results have been compared with the existing manifold. The observed emission level for the convergent manifold is less, and it is comparable with the government norms.

Keywords SI engine · Intake manifold · Venture · Convergent · CO and HC

1 Introduction

Internal combustion engines consist of an integral part in which the fuel and an oxidizer undergo combustion. A direct impulsive force is applied to the flywheel and the crankshaft of the engine due to the expansion of the high temperature and high-pressure gases produced by combustion [1, 2].

A spark (SI engines) or compression (CI engines) of air–fuel mixture in the combustion chamber leads to their ignition, and the combustion starts. An enormous amount of heat is produced due to the combustion of air–fuel mixture, and the

K. Raja (✉) · P. L. Rupesh
Department of Mechanical Engineering, Vel Tech Rangarajan Dr. Sagunthala R&D Institute of Science & Technology, Chennai, India
e-mail: rajak@veltech.edu.in

A. J. Selvam
Department of Automobile Engineering, Vel Tech Rangarajan Dr. Sagunthala R&D Institute of Science & Technology, Chennai, India

L.-J. Yang et al. (eds.), *Proceedings of ICDMC 2019*,
Lecture Notes in Mechanical Engineering,
https://doi.org/10.1007/978-981-15-3631-1_9

pressure produced by the expansion of combustible gases induces the reciprocating motion of the piston. The crankshaft undergoes rotary motion due to the piston power. A transmission system is used to transmit the rotary motion to the wheels and produce propulsion in the vehicle [3–6]. The air–fuel mixture in correct proportion leads to proper and complete combustion which gives better performance and less emission [7]. The shape of the intake manifold is one of the factors which influence the mixing of air–fuel mixture in correct proportion for efficient combustion. The internal flow structure of the manifold influences the flow of air-fuel mixture.

1.1 Government Emission Norms

Emission norms in India as shown in Table 1 have become strict due to the Environment Protection Act (EPA), Motor Vehicles Rules and Air Act. In 1996, these laws were applicable to the vehicles at the manufacturing stage as well as for vehicles in use [7, 8]. In addition to this, the government took initial steps in reducing the emissions through engine optimization.

Ramalingam et al. [5] have performed research on a bio-diesel operated diesel engine and observed the performance improvement and exhaust emissions reduction. He also concluded that increase of CR with B20 improved the performance and reduction in exhaust emissions except NOx emission. Mahesh et al. [9] have conducted an experimental research on a four-stroke SI engine with different geometry shapes, and a computational analysis of these manifolds has been observed, and an optimized manifold has been selected based on outlet velocity. Yerrennagoudarua et al. [10] have modified an exhaust system of a two-wheeler for emission control, and the experimental results show that a usage of a flexible exhaust pipe in a EGR setup reduces various inclination emissions. An experimental research has been conducted by Gowthaman et al. [7] on improved intake manifold design for IC engine emission control, and the emission test on engine whose intake manifold has various inner sections has shown decreased emissions. The present study deals with the

Table 1 Government norms regarding vehicular emissions

S. No.	Vehicle type	CO %	HC (n-hexane equivalent, ppm)
1	Two- and three-wheelers (mfd. Before March 2000)	4.5	9000
2	Two- and three-wheelers (mfd. after march 2000)—two stroke	3.5	6000
3	Two- and three-wheelers (mfd. after March 2000)—four stroke	3.5	4500
4	BS-II compliant four-wheelers	0.5	750
5	Four-wheelers other than BS-II compliant ones	3.0	1500

fabrication of intake manifold of different shapes such as convergent, venture and divergent and to use them in an experimental setup for performance and emission test.

2 Experimental Setup

The experimental setup consists of a **Bajaj CT100** as shown in Fig. 1a mounted on an engine testing platform, coupled with a rope brake dynamometer. This acts as a perfect setup for performing both the performance test as well as the emission test.

The experimental setup is started and throttled to a rate such that the shaft speed goes to as high as 500 rpm. The test is carried under constant rpm and varying loads condition. Figure 1b shows a AVL 5-gas analyzer in which the purge and O_2 test are carried out. An O_2 level of 21% indicates that the analyzer is functioning properly and the test may be conducted. The one end of the probe is attached to the analyzer and the other end into the silencer of the vehicle.

2.1 *Fabrication of Modified Manifolds*

The manifolds designed whose dimensions given in Table 2 were set for fabrication

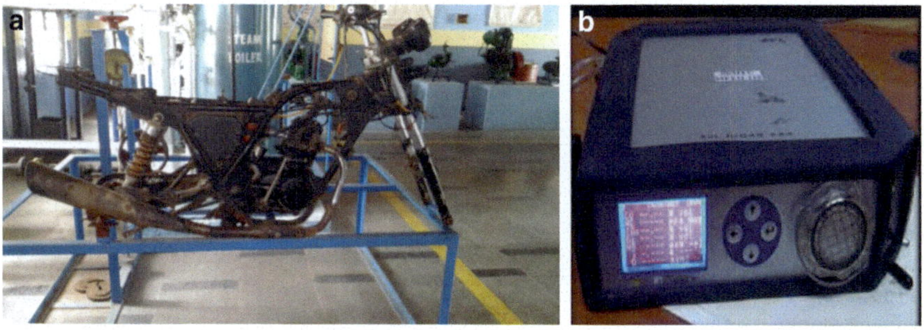

Fig. 1 **a** Experimental test rig. **b** AVL 5-way exhaust gas analyzer

Table 2 Dimensions of different manifolds

Type	Intake diameter (mm)	Throat (mm)	Outlet diameter (mm)	Length (mm)
Existing	21	–	21	75
Venture	21	19	21	75
Convergent	23	–	21	75
Divergent	21	–	23	75

and optimization. The manifolds were classified into three types as based on their internal flow structure as venture, convergent, divergent. The designs were optimized using Computational Fluid Dynamics software [11], and the flow velocity analysis of all the manifolds was performed.

The intake manifold of the above shapes is designed in CATIA according to the dimensions shown in Table 2. The cross section of the manifold is shown in Fig. 2.

These models were fabricated as depicted in Fig. 3a using casting technique. The

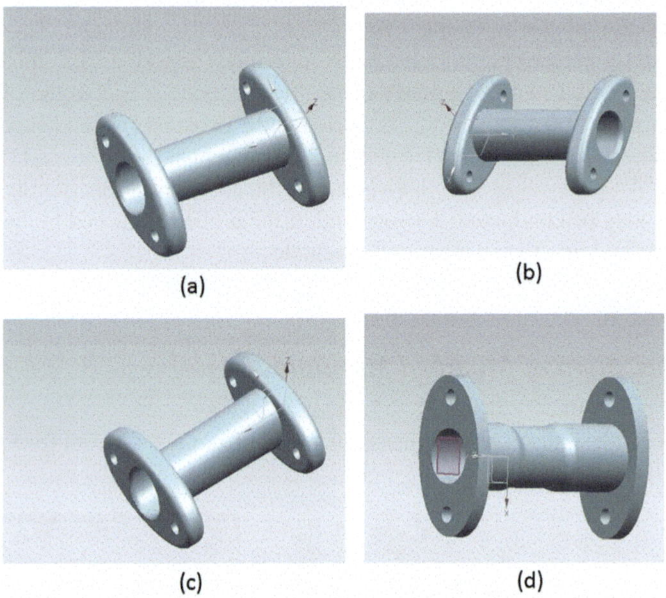

Fig. 2 **a** Existing **b** Convergent **c** Divergent and **d** Venture

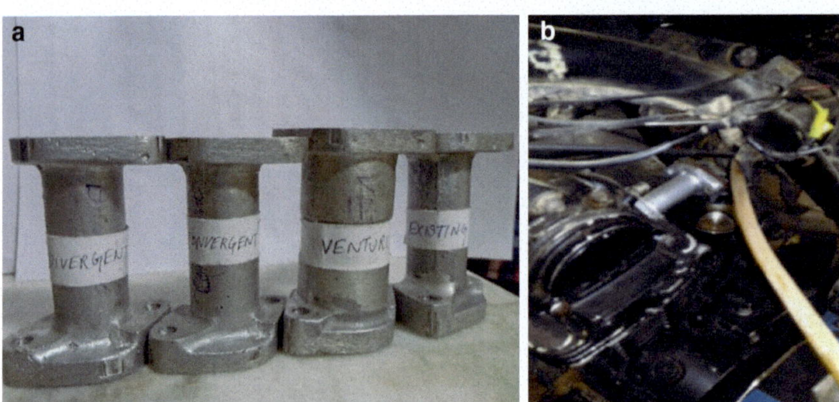

Fig. 3 **a** Fabricated manifolds **b** Manifold installation

components obtained using casting were as per the intended design. Figure 3b shows a manifold installed in the setup for emission test.

3 Results and Discussion

The emission level of the engine at different load conditions such as no load (0%), part load (50%) and full load (100%) has been evaluated at the shaft speed of 500 rpm by mounting a rope brake dynamometer onto the setup. From our study, we have concluded that changing the shape of the inlet manifold, there has been a considerable variation in the levels of pollutants in the emission. Upon compiling, we have come up with the following order of manifolds based on their ability to reduce emissions:

- Convergent.
- Venture.
- Existing.
- Divergent.

3.1 Variation in CO Emissions

Figure 4 indicates the variation in carbon mono oxide emissions at different loads for different types of manifolds usage in the experimental test rig. The emission results show that as the load increases, the emission level of CO also increases. Among the other manifolds, convergent manifold shows a decreased emission level due to the swirl action. The swirl motion in the convergent manifold is formed due to the sudden contraction in the area of the manifold. The decrease in the area leads to increase

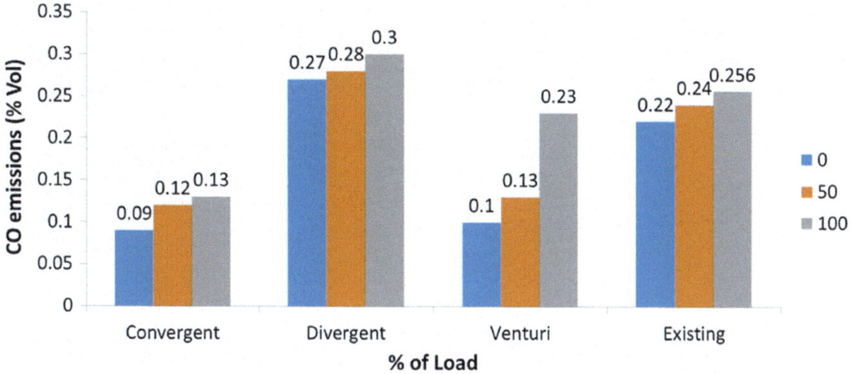

Fig. 4 Comparison of CO levels (% Vol.)

in the flow velocity of the air–fuel mixture which makes the mixture to swirl and provides efficient combustion. The efficient combustion leads to less emissions of CO.

3.2 Variation in HC and NO_X Emissions

The comparison of the emission results of hydrocarbon has been depicted in Fig. 5. Due to the swirl action and turbulent flow, the manifold with convergent geometry emits less amount of HC with comparison to other geometries. The normal and the divergent manifold emits in greater level due to the absence of throat geometry.

The bar graph shown in Fig. 6 compares the NO_X emissions of fabricated manifolds. The NO_X level has been reduced to a considerable amount of 185 ppm by

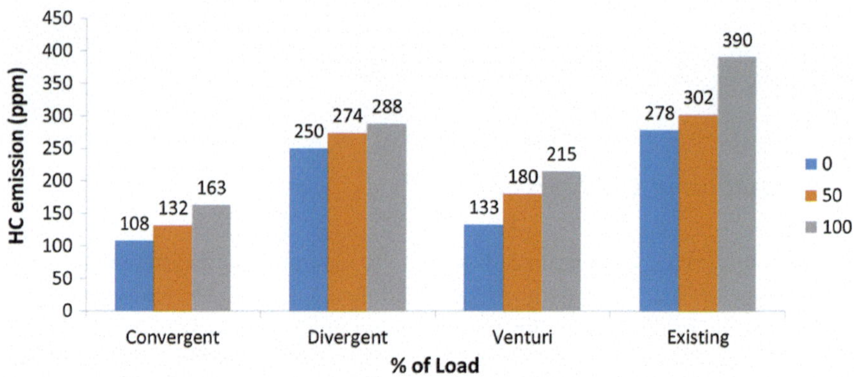

Fig. 5 Comparison of HC levels (ppm.)

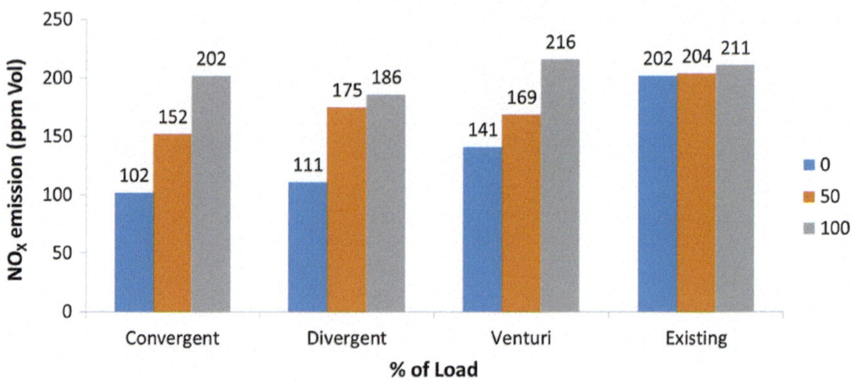

Fig. 6 Comparison of NO_x levels (ppm Vol)

using the convergent manifold at the maximum load of 4 kg. This reduction occurs due to the turbulent motion of air and fuel mixture inside the intake manifold.

4 Conclusion

The experimental evaluation of the emissions of CO, HC and NO_X using manifolds of different geometry in a test rig containing a Bajaj CT 100 engine at different load conditions was performed in the above work. The results indicate that the CO emission (0.13% of Vol.), HC emission (163 ppm HEX) and NO_X emission (185 ppm Vol.) are less by using a convergent manifold in the test setup. The venture manifold also shows a similar result, but due to its complex structure, it was not suggested. The air velocity was increased in the convergent manifold, and the emission is reduced due to the swirl motion. The emission result for the convergent manifold also agrees with the government emission norms.

References

1. Hayder A. Ahmada AKA (2018) Renew Sustain Energy Rev 82:324–342
2. Dhinesh B, Annamalai M (2018) A study on performance, combustion and emission behaviour of diesel engine powered by novel nanonerium oleander biofuel. J Clean Prod. https://doi.org/10.1016/j.jclepro.2018.06.002
3. Ceviz MA, Akin M, (2010) Design of a new SI engine intake manifold with variable length plenum. Energy Conversion Manage 51(11):2239–2244
4. Ismaila TM, Ramzya K, Abelwhabb MN, Elnaghib BE, El-Salamc MA, Ismaild MI (2018) Performance of hybrid compression ignition engine using hydroxy (HHO) from dry cell. Energy Conversion Manage 155:287–300
5. Ramalingam S, Rajendran S, Ganesan P (2017) Performance improvement and exhaust emissions reduction in biodiesel operated diesel engine through the use of operating parameters and catalytic converter. Renew Sustain Energy Rev 3215–3222
6. Fangsuwannaraka K, Wanrikoa P, Fangsuwannarak T (2016) Effect of bio-polymer additive on the fuel properties of palm biodiesel and on engine performance analysis and exhaust emission. Energy Procedia 100:227–236
7. Gowthaman S, Sathiyagnanam AP (2018) Analysis the optimum inlet air temperature for controlling homogeneous charge compression ignition (HCCI) engine. Alexandria Eng J. https://doi.org/10.1016/j.aej.2017.08.011
8. Elfasakhany A (2016) Performance and emissions of spark-ignition engine using ethanol–methanol–gasoline, n-butanol–iso-butanol–gasoline and iso-butanol ethanol–gasoline blends-A comparative study. http://dx.doi.org/10.1016/j.jestch.2016.09.009
9. Mahesh S, Ramadurai G, Nagendra SMS. Real-world emissions of gaseous pollutants from diesel passenger cars using portable emission measurement systems. Sustain Cities Soc. https://doi.org/10.1016/j.scs.2018.05.025
10. Yerrennagoudarua H, Manjunatha KB, Razac A, Kantharaj B Rd, Mujahede A, Irshad KF (2018) Analysis and comparison of performance and emissions of compression ignition engine fuelled with diesel and different bio-fuels blended with Methanol. Mater Proc 5:5175–5185
11. Raja K , Selvam AJ, Rupesh PL, Thamaraikannan M (2018) A computational study of an intake manifold for four stroke S.I. Engine. Int J Mech Prod Eng Res Dev, 2018, 8(4):531–540

Experimental Analysis of Hevea Brasiliensis Methyl Ester Diesel Blend with Antioxidant Additive in a Di-diesel Engine

A. A. Muhammad Irfan, Sivanandi Periyasamy, and A. Gurusamy

Abstract To replacing diesel fuel non-edible feedstock is a probable resource for the alternative fuel creation with taken into account of ecological and food versus fuel demand. Biodiesel is a capable replacement to diesel fuel, due to renewable, non-hazardous, transportable, widely existing, recyclable, ecological, and free from sulfur and aromatic matter. The experimental investigation was carried out, to investigate the response of Hevea brasiliensis methyl ester diesel blend with antioxidant additive in a di-diesel engine. The performance and emission characteristics were determined for the diesel engine powered with Hevea brasiliensis biodiesel blend. With an aid of ASTM standards, tert-butylhydroquinone (TBHQ) antioxidant added in biodiesel blend. The performance and emission distinctiveness were resolute for antioxidant additive added blend. The outcome of antioxidant additive on the performance and emission of diesel engine were analyzed and concluded with base fuel. The addition of antioxidant increased 8.9% average brake thermal efficiency, increased 4.98% average mechanical efficiency, and reduced 8.9% average brake specific fuel consumption. The addition of antioxidant reduced oxides of nitrogen (NO_x) emission, but increased carbon monoxide (CO), carbon dioxide (CO_2), and hydrocarbon (HC) emissions compared to Hevea brasiliensis biodiesel blend.

Keywords Hevea brasiliensis · Transesterified · Antioxidant · Tert-butylhydroquinone · Brake specific fuel consumption · Brake thermal efficiency · Emission characteristics

A. A. Muhammad Irfan (✉)
Department of Mechanical Engineering, Mohamed Sathak A J College of Engineering, Chennai, India
e-mail: mechmuhammad@gmail.com

S. Periyasamy
Department of Mechanical Engineering, Government College of Technology, Coimbatore, India
e-mail: speriyasamy@gct.ac.in

A. Gurusamy
Department of Automobile Engineering, Pace Institute of Technology and Sciences, Ongole, India

L.-J. Yang et al. (eds.), *Proceedings of ICDMC 2019*,
Lecture Notes in Mechanical Engineering,
https://doi.org/10.1007/978-981-15-3631-1_10

Nomenclature

HBME	Hevea brasiliensis methyl ester
TBHQ	Tert-butylhydroquinone
D80 + HBME20	Diesel 80% + Hevea brasiliensis methyl ester 20%
D70 + HBME20 + TBHQ10	Diesel 70% + Hevea brasiliensis methyl ester 20% + tert-butylhydroquinone 10%
B20	Diesel 80% + biodiesel 20%
BSFC	Brake-specific fuel consumption
BTE	Brake thermal efficiency
ME	Mechanical efficiency
CV	Calorific value of fuel
NOx	Oxides of nitrogen
HC	Hydro carbon
CO	Carbon monoxide
CO_2	Carbon dioxide

1 Introduction

IC engine responded good for biodiesel on a place of diesel. Reduction of emissions in diesel engines is a current researcher's agenda by introducing biodiesel. Researchers compared the diesel with various feedstock biodiesels to find out better alternative biodiesel. Hevea brasiliensis is the botanical name rubber seed.

Atabani et al. [1] identified the biggest Hevea brasiliensis producers are Thailand, Indonesia, Malaysia, India, China around the earth with 35%, 23%, 12%, 9%, 7% shares, respectively. Takase et al. [2] recognized the Hevea brasiliensis oil taken in a calendar year is 70–500 kg/ha and oil collected from seeds nearly 50–60% and 40–50% from its kernel. Ramadhas et al. [3] documented Hevea brasiliensis requiring acid esterification and transesterification due to a high acid value of Hevea brasiliensis.

Ramadhas et al. [4] utilized the biodiesel of Hevea brasiliensis in diesel engine and conducted the performance and emissions test. Their conclusion showed thermal efficiency of biodiesel blends are higher while lower concentrations. Still, NO_x emission is also projected to increases with increase in biodiesel blends.

Erol Ileri et al. [5] concluded that decreasing NO_x emissions and improving oxidation stability by using antioxidants additives in biodiesel blends. Erol Ileri et al. [6] concluded that reduction and increased of NO_x and CO, respectively, by using antioxidants in biodiesel blends. Rashid et al. [7] reported by addition the antioxidant with Calophyllum inophyllum methyl ester (B20), BTE and BP improved while decreasing BSFC. At the same time, antioxidants reduced NO_x, growth of HC and CO for all biodiesel blends. For this reason, aromatic amine antioxidants can be used with B20 blends, in compression, ignition engine exclusive of any engine alterations.

Jain et al. [8] compared the efficiency of eight artificial antioxidants in dissimilar biodiesels and they concluded that only three antioxidants considerably improved the stability of biodiesel in the increasing order of propyl gallate, pyrogallol, and TBHQ. David et al. [9] predicted the result, in induction time of methylic biodiesel later than add-ons of increasing concentrations of TBHQ. Oxidation stability is directly proportional to the concentrations of TBHQ. Palash et al., 2014 [10] concluded that NO_x emissions reduced notably by using antioxidant additive with a small degradation in engine power and BSFC as well as CO and HC emissions. Rashedul et al. [11] evaluated that the performance and emission characteristics as well as properties of biodiesel were improved notably by using additives.

2 Materials and Methods

2.1 *Experimental Setup*

The investigational setup of single-cylinder four-stroke engine as shown in Fig. 1 is loading by eddy current dynamometer with direct injection. Assessment of performance and emission characteristics of required samples are taken by using this engine with water cooling concept. The engine specification of this engine was mentioned in the below Table 1.

Fig. 1 Photograph of experimental setup

Table 1 Engine specification

Manufacturer	Kirloskar oil Engine Ltd., India
Method of injection	Direct injection diesel engine
Number of cylinders	Single cylinder
Number of strokes	Four stroke
Cooling type	Water cooling
Engine speed	1500 rpm
Brake power	3.5 kW @ 1500 rpm
Bore diameter	80 mm
Stroke length	110 mm
Type of loading	Eddy current dynamometer
Starting method	Manual cranking
Compression ratio	16.5:1
Orifice diameter	0.02 m
Dynamometer Arm length	0.185

2.2 *Preparation of Biodiesel*

Hevea brasiliensis methyl ester was collected only after two esterification process. Acid esterification is the first stage, carried out at 650 rpm, 55–57 °C for 90 min with 0.7% of weight of H_2SO_4, and oil to methyl alcohol about 2:1 volume ratio in the magnetic stirrer. The second stage is performed at 650 rpm, 60 °C for 60 min with 0.5 wt% NaOH pellets with volume ratio of 10:1 acid esterified oil to methanol in the magnetic stirrer. The Hevea brasiliensis methyl ester was derived from the above process. But it contains impurities as well as glycerin. Outcome of pure methyl esters only happens after several washing processes by distilled water.

2.3 *Properties of Hevea Brasiliensis Methyl Ester and TBHQ*

The major fuel properties were determined as per the ASTM standards and evaluated values are given in Table 2.

Table 2 Properties of sample fuel

Property parameters	Unit	Diesel	HBME	TBHQ
Density	g/cc	0.833	0.886	0.905
Flash point	°C	68	167	171
Fire point	°C	63	181	189
Kinematic viscosity	cSt	3.22	5.78	5.98

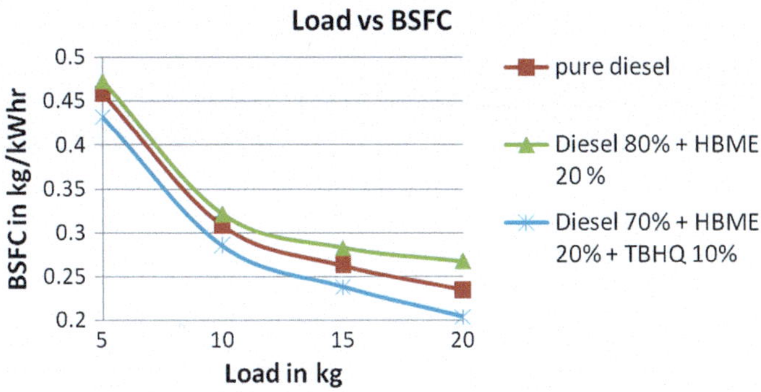

Fig. 2 Brake-specific fuel consumption

3 Results and Discussions

3.1 Performance Characteristics

Performance of sample biodiesel blends was taken with and without TBHQ. The engine was maintained at stable speed for all loads of 1500 rpm. While taking the performance test, 10 cc fuel utilization time taken was followed throughout the various blends.

3.1.1 Brake Specific Fuel Consumption

Figure 2 shows that load versus BSFC curve for different biodiesel blends. The BSFC for D70 + HBME20 + TBHQ10 blend was found to be lower. The BSFC for D80 + CHBME20 is higher than all additive added blend and pure diesel. BSFC for HBME20 blend was slightly increased due to less heating content of Hevea brasiliensis biodiesel. Due to the higher power output, BSFC was decreased on the addition of TBHQ.

3.1.2 Brake Thermal Efficiency

The base fuel has lesser efficiency than blended fuels. The effect of various blends on BTE was shown in Fig. 3. It was clearly indicated that the addition of TBHQ shows highest BTE at all working conditions. Also at higher loads, the increase in BTE was observed for all blends. Reason of this happens because of, the fuel viscosity decreases for the blends, made finer fuel particles which in turn makes the atomization process easier and efficient. This aids to the huge improvement of combustion by mixing of fuel with oxidizer.

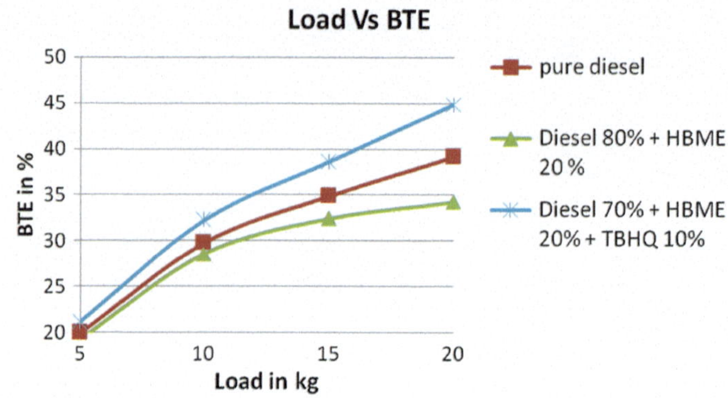

Fig. 3 Brake thermal efficiency

3.1.3 Mechanical Efficiency

From experiments, it was observed that increases in mechanical efficiency of the various while load increases as shown in Fig. 4. It was due to that the fuel viscosity decreases, made finer fuel particles which in turn makes the atomization process easier and efficient. This helps to the improvement of combustion by mixing of fuel with oxidizer.

3.2 Emission Characteristics

AVL gas analyzer was used to analyzing the emission characteristics for various biodiesel blends at different loading condition. HC, NO_x, CO, and CO_2, emissions are measured by using this analyzer.

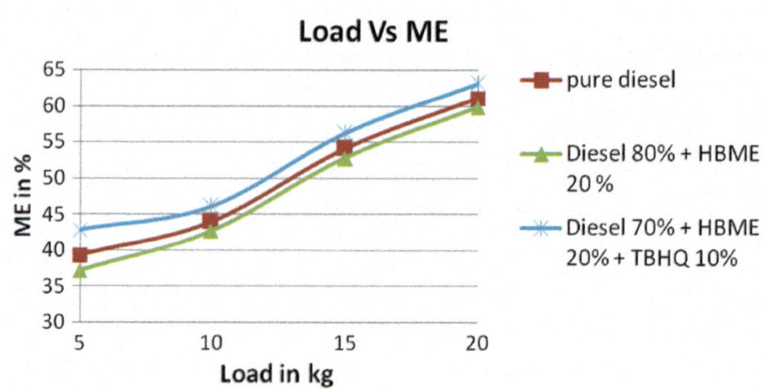

Fig. 4 Mechanical efficiency

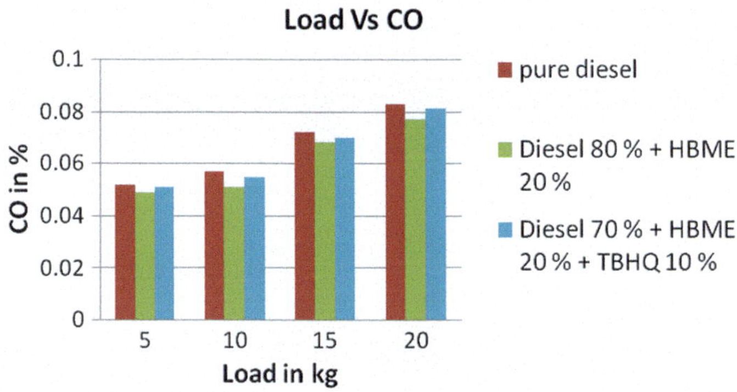

Fig. 5 CO emissions

3.2.1 CO Emission

As shown in Fig. 5, biodiesel addition will reduce the CO formation. Due to the lower participation of fuel at lesser loads, less CO emission was noted. Here, CO directly proportional to the engine loads. Results obtained from various biodiesel blends indicate that reduced CO emission by reason of containing high oxygen and highest cetane number of the Hevea brasiliensis blends. Enhancement in fuel borne oxygen gives complete combustion range of CO, thus lower the CO emission considerably. TBHQ mixed with HBME20 gives highest unburnt CO emission compared to HBME20 blend. However, the level was within emission standards.

3.2.2 CO_2 Emission

As shown in Fig. 6, the biodiesel concentration level will minimize the CO_2 emission. Here, CO_2 directly proportional to the engine loads. TBHQ blend shows reduced CO_2 emission, but it slightly higher than HBME20. This is because of the lower operating temperature due to its high latent heat of vaporization.

3.2.3 HC Emission

HC emission was slightly decreased with blend addition which was shown in Fig. 7. Due to the lower participation of fuel at lesser loads, less HC emission was noted. It will slightly increase with engine load increases. Results obtained from various biodiesel blends indicate that reduced HC emission by reason of containing high oxygen and highest cetane number of the Hevea brasiliensis blends. Enhancement in fuel borne oxygen gives complete combustion range of HC, thus lower the HC

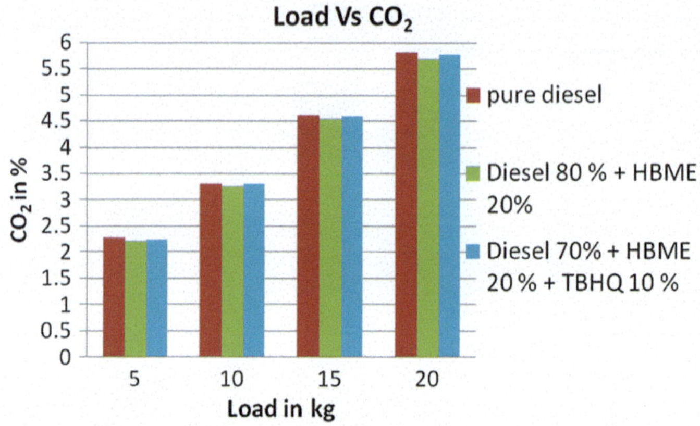

Fig. 6 CO_2 emission

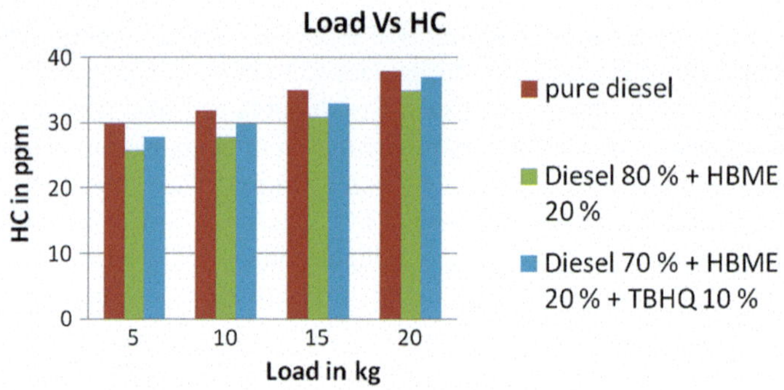

Fig. 7 Hydrocarbon emission

emission considerably. Unburnt HC emission was increased because of Accumulation of TBHQ to HBME20 compared to HBME20 blend. However, the level was within emission standards.

3.2.4 NO_x Emission

Load affects the NO_x emission with linear manner as shown in Fig. 8. Shorter ignition delay and oxygenated fuel produce the convenient combustion. It affects NO_x emission slightly above the diesel. The positive result shows on the addition of TBHQ to HBME20 to controlling the NO_x. Reduction of NOx emission occurs by phenolic hydroxyl groups present in the TBHQ.

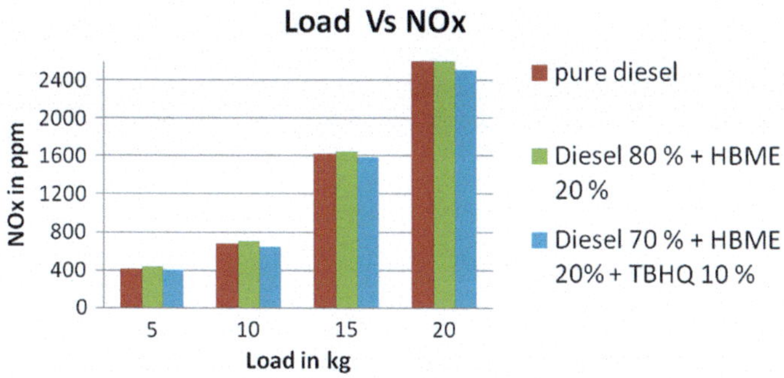

Fig. 8 NOx emission

4 Conclusion

The experimental analysis clearly indicates the following conclusions; the biodiesel prepared from Hevea brasiliensis oil shows the feasible to be a feedstock for biodiesel production. The investigation in terms of performance and emission was taken at D80 + HBME20 blend with and without the presence of tert-butylhydroquinone (TBHQ) antioxidant additive in direct injection four-stroke single-cylinder diesel engine. TBHQ blend gives highest BTE at all loading conditions. The increase in average BTE was 8.9% instead of diesel. Average BSFC of D80 + HBME20 increased by 6.67% instead of diesel. There was 8.9% reduction in average BSFC for D70 + HBME20 + TBHQ10 blend compared to diesel. BSFC increases for D80 + HBME20 blend was due to lower heating content of Hevea brasiliensis biodiesel. Average HC and CO emissions of D80 + HBME20 decreased by 11.29% and 7.27%, respectively, compared to diesel. Reduction in emission of HC and CO was due to the collective end product of its high oxygen content and superior cetane number. Average NO_x emission of D80 + HBME20 increased by 2.67%, while average NO_x emission of D70 + HBME20 + TBHQ10 reduced by 4.07% in comparison with diesel. ASTM D6751 biodiesel standard was satisfied with HBME's fuel properties. Hence, HBME can be recommended instead of diesel fuel. It deals with the universal concerns of energy disaster and ecological deprivation problems.

References

1. Atabani AE, Silitonga AS, Ong HC, Mahlia TMI, Masjuki HH, Badruddin IA, Fayaz H (2013) Non-edible vegetable oils: a critical evaluation of oil extraction, fatty acid compositions, biodiesel production, characteristics, engine performance and emissions production. Renew Sustain Energy Rev 18:211–245. https://doi.org/10.1016/j.rser.2012.10.013
2. Takase M, Zhao T, Zhang M, Chen Y, Liu H, Yang L, Wu X (2014) An expatriate review of neem, Jatropha, rubber and Karanja as multipurpose non-edible biodiesel resources and comparison of their fuel, engine and emission properties. Renew Sustain Energy Rev 43:495–520. https://doi.org/10.1016/j.rser.2014.11.049

3. Ramadhas AS, Jayaraj S, Muraleedharan C (2005) Biodiesel production from high FFA rubber seed oil. Fuel 84:335–340. https://doi.org/10.1016/j.fuel.2004.09.016
4. Ramadhas AS, Muraleedharan C, Jayaraj S (2005) Performance and emission evaluation of a diesel engine fueled with methyl esters of rubber seed oil. Renew. Energy 30(12):1789–1800. https://doi.org/10.1016/j.renene.2005.01.009
5. Ileri Erol, Kocar Gunnur (2014) Experimental investigation of the effect of antioxidant additives on NO_x emissions of a diesel engine using biodiesel. Fuel 125:44–49. https://doi.org/10.1016/j.fuel.2014.02.007
6. Ileri Erol, Kocar Gunnur (2013) Effects of antioxidant additives on engine performance and exhaust emissions of a diesel engine fueled with canola oil methyl ester–diesel blend. Energy Convers Manag 76:145–154. https://doi.org/10.1016/j.enconman.2013.07.037
7. Rashed MM, Kalam MA, Masjuki HH, Habibullah M, Imdadul HK, Shahin MM, Rahman MM (2016) Improving oxidation stability and NOx reduction of biodiesel blends using aromatic and synthetic antioxidant in a light duty diesel engine. Industr Prod 89:273–284. https://doi.org/10.1016/j.indcrop.2016.05.008
8. Jain S, Sharma MP (2010) Stability of biodiesel and its blends: A review. Renew Sustain Energy Rev 14:667–678. https://doi.org/10.1016/j.rser.2009.10.011
9. Fernandes David M, Serqueira Dalyelli S, Portela Flaysner M, Assunção Rosana MN, Munoz Rodrigo AA, Terrones Manuel GH (2012) Preparation and characterization of methylic and ethylic biodiesel from cottonseed oil and effect of tert-butylhydroquinone on its oxidative stability. Fuel 97:658–661. https://doi.org/10.1016/j.fuel.2012.01.067
10. Palash SM, Kalam MA, Masjuki HH, Arbab MI, Masum BM, Sanjid A (2014) Impacts of NOx reducing antioxidant additive on performance and emissions of a multi-cylinder diesel engine fueled with Jatropha biodiesel blends. Energy Convers Manag 77:577–585. https://doi.org/10.1016/j.enconman.2013.10.016
11. Rashedul HK, Masjuki HH, Kalam MA, Ashraful AM, Ashrafur Rahman SM, Shahir SA (2014) The effect of additives on properties, performance and emission of biodiesel fuelled compression ignition engine. Energy Convers Manag 88:348–364. https://doi.org/10.1016/j.enconman.2014.08.034

Recent Application and Preservation of Bamboo as Sustainable Material

Amol Ashok Kamble and Suppiah Subramaniam

Abstract Nowadays construction industry is the main consumer of energy and materials. For sustainable development, we have to use proper resources. Bamboo is a fast-growing fiber structure which has high tensile strength so we can replace steel with bamboo in Bamboo Cement Concrete (BCC). We have more information about vegetable fibers like bamboo can be used either alone or as reinforcement in different types of material, such as soil and cement composites. These studies led to the establishment broad idea about bamboos types and its various recent applications. It also contains in detail of different preservative technique which can increase the considerable life of bamboo.

Keywords Sustainable material · Application of bamboo · Types of bamboo · Preservative technique · Properties of bamboo

1 Introduction

In India, since long time ago bamboo is used as building material. As bamboo is natural material can be agricultural by-products in various crops like rice, sugarcane, and it can be cultivated along a bank of a river. As bamboo is renewable material, we can achieve sustainability by using it. Bamboo is an easily available economical material; therefore, it is largely used in the rural area. Full growth of bamboo is completed in just a few months and its full strength is gained in just a few years. Among different type of wood, bamboo has high tensile and compressive strength. Bamboo is the fast-growing plant in the world. Even some types of bamboo can grow 890 mm per day.

A. A. Kamble (✉)
Smt. Kashibai Navale College of Engineering, Pune, Maharastra, India
e-mail: amol13kamble@gmail.com

S. Subramaniam
Vel Tech, Rangarajan Dr. Sagunthala R&D Institute of Science and Technology, Chennai, India

L.-J. Yang et al. (eds.), *Proceedings of ICDMC 2019*,
Lecture Notes in Mechanical Engineering,
https://doi.org/10.1007/978-981-15-3631-1_11

Bamboo has good tensile strength so it can be used as tensile material in BRC. Bamboo is lightweight so it can be used by not trained worker without heavy machinery. It has light therefore easy to transport. It is a fast-growing cultivated plant so it is pollution free eco-friendly material. It has traded on the surface so it has good bond strength.

2 Application of Bamboo

Due to such properties bamboo has a large scope of uses. Some of the uses of various bamboos are as follow. In Orissa, India, bamboo is used as reinforcement in road construction. In China, bamboo is used for bridge construction. In China, the root of black bamboo is used to treat kidney diseases and cancer. By UNESCO, 70 ha of bamboo can be used to produce 1000 bamboo house. So recently lots of schools are constructed using bamboo very economically. For various scaffolding and tents, bamboos are used as it is economical, eco-friendly, and high tensile strength. Due to bending properties and attractiveness, bamboos are used in furniture. Bamboos are used in making clothes which have more antibacterial properties. Bamboo has good strength so it used as formwork in contraction industries (Fig. 1).

3 Type of Bamboo

Bamboo's species are occurring in the grass family. Bamboos are divided into lots of categories. One hundred thirty-six species of bamboo are observed in India. The following table contains some species which are commonly observed in various parts of old with their dimension and their description (Table 1).

Bamboos which are used for construction

Bamboo used for construction must be mature and 3–4-year old. Such bamboo shrinks less and has less sugar content which leads to less chance of insect infection. Thicker and more mature Bambusa bamboos are used for structural column and beam.

4 Preservation of Bamboo

The main problem of bamboo to use in various products is the durability of bamboo. The life of bamboo is shorter than another type of wood. The problem of durability is most significant due to the hollowness of Culm. If the fungi or insect attacks on outer layer more damage occurs in the bamboo. The hollow portion provides safe space for the insect to hide in Culm so they are most effective in bamboo.

Fig. 1 Recent applications of Bamboo

Rough guideline for life of bamboo

If bamboos are open to air and in contact with soil, the life of bamboos can be in between 1 and 3 year. If bamboos are undercover and free from contact with soil, the life of bamboos can be in between 4 and 6 year. If bamboos are under very good storage and use condition, the life of bamboos can be in between 10 and 15 year.

For increasing the life of bamboo precaution can be done like placing a bamboo in the dry place, avoid soil contact, select the proper type of bamboo to use for construction, and use proper place for storage.

By such precaution, some life of bamboo can increase but for more lifetimes we have to preserve bamboos with proper treatment. Lots of bamboos used in Indian rural area for construction are untreated; therefore, it has durability around 1–2 years. Since bamboo is considered poor people material.

Therefore, proper preservation is required to protect the degradation of bamboo. The selection of proper treatment is depending on various factors as, for example,

Table 1 Some commonly observed species which are in various parts of world with their dimension and description

Species	Max. height (m)	Max. diameter (mm)	Brief species description	Availability
Acidosasa	4.6–12	19–51	Yellow-brown	China
Ampelocalamus scandens	9.1	7.6	Hangs loosely, or is scrambling. Nodes are slightly swollen. Branches are equal and smaller than culms	Southern China
Apoclada simplex			A one-member genus	South-eastern Brazil
Arthrostylidium			Climbing bamboos found in the New World	Central America
Arundinaria	1.5–9.1	5.1–28	Can grow at temperatures as low as − 30 °C in soggy soil	USA
Bambusa	7.6–30	100–180	Fast growing, with thick walls and strong therefore often used in construction. It is also used to make paper in India	India, Australia
Bashania	3–6	23–51	Spreads quickly and vigorously	China
Borinda	3.7–6.1	18–25	Grows in mountainous regions	Bhutan, Tibet, Yunnan, and Sichuan
Cathariostachys	9.1	51	Tall	India, China, Madagascar, Malaysia

(continued)

Table 1 (continued)

Species	Max. height (m)	Max. diameter (mm)	Brief species description	Availability
Chimonobambusa	1.8–7.6	13–38	New shoots start in fall or winter. Sometimes with thorny nodes, and quadrangular culms	China, Japan, Vietnam, Myanmar, and the Himalayas
Chusquea	3.7–7.6	20–32	Grows at a very high altitude	USA
Dendrocalamus	10–30	120–300	Shoots are large and often eaten. A fast-growing black culmed bamboo. Culms are sometimes solid. It flowers frequently. Often used for fishing rods	USA, common in India, China
Fargesia	2.4–7.3	13–33	Grows at elevations above 6000 feet (1800 m) above sea level	China
Himalayacalamus	6.1–9.1	32–51	Found growing at lower altitudes of the Himalayan region	India, Nepal
Sasa	1.5–4.6	3.8–13	Low-growing	Japan
Schizostachyum	15	8	Used to make the Saluang	Indonesia
Semiarundinaria	4.6–12	8.4–25	Young culms are hairy. The leaves are about 7 inches (180 mm) long and 1 inch (25 mm) wide	China and Japan

(continued)

Table 1 (continued)

Species	Max. height (m)	Max. diameter (mm)	Brief species description	Availability
Shibataea	2.1 m	6.4	It is a unique shorter bamboo with dark green leaves	China
Thyrsostachys	13	8	It is a genus of Chinese and Indochinese bamboo in the grass family	Yunnan, Myanmar, Thailand; Assam(India)
Yushania	3.7–20	19–130	Previously known as Sinarundinaria alpine	African, India, China

bamboos are green or dry, bamboos are round or split, where is it used? (in the ground, exposed to air or undercover) It is used for structural or non-structural uses.

Considering above point one can use proper treatment of preservation of bamboo for increasing life of bamboo up to 50 years. There are some defects in bamboos and treatment for that which affects the durability of bamboo.

I. ***Bamboo insect infection***

As discussed earlier, the most probability is that untreated bamboo can get attacked by the insect. This can minimize the considerable life of bamboo. Insects can obtain their food from bamboo and degrade bamboo as it contains a large amount of starch and another carbohydrate. Therefore, bamboo must treat chemically to avoid infection.

Beetles and termites are the most common insect which attacks bamboo. They do not require any specific condition to attack. Beetles can be prevented by a boron-based wood preservative powder. Termites are attacks on bamboo where ground contact is available. For termite's treatment, one should adopt the construction technique in which ground contact is avoided. To avoid insect attack one can soak bamboo in a boric acid or borax solution.

II. ***Removal of Bamboo mold***

When bamboo is not 100% dry and it shipped in the ocean in a humid condition, the mostly mold forms on a surface of bamboo. It can occur only once or twice up all moisture contains is evaporated from bamboo. This can be treated with flame torch, i.e., heat treatment or thicker bamboo can be treated with a boron solution. The most effective method for mold removing is to clean up mold on bamboo with brush and clean clothes and use lemon oil or solution of vinegar and water to clean mold.

After removal of mold use three layers of polyurethane for better performance of bamboo.

III. ***Leaching of bamboo***

Storing bamboo in water is nothing but leaching of bamboo. It is the traditional method of bamboo preservation in many rural areas in Asia. This method is suitably used when bamboo is transporting from one place to other over long duration up to 3–4 weeks. As bamboo is in leaching processes its starch contain get decreased and due to that durability is increased. It is highly advisable to use chemical treatment at a time of leaching as leaching alone is not guaranteed to increase in durability of bamboo. One can be used boric acid in leaching water. In leaching, process water can be changed after 2–3 days to avoid growth of a microorganism in water which can attack the bamboo. Excessive time in water can cause the stain in the epidermis of bamboo which can change the mechanical as well as physical properties of bamboo. The treated bamboo with this method further required additional drying treatment at a time of use.

IV. ***Chemical bamboo preservation***

Chemical bamboo preservation is a very useful method as it ensures the durability of bamboo for a long time. Chemical used for treatment can be long term or short term which depends on the method of treatment. Lots of chemicals used for preservation are toxic so it must use with great care to get good performance, environment requirements, and safety.

There are fixing and non-fixing types of chemical preservatives. The non-fixing preservative will leach out when exposed in the rain so cant used for outdoor uses.

i. *Non-fixing type preservatives*:

It mainly contains boron salts, which are effective on borers, termites, and fungi. Boron salt is dissolved in water. This water is evaporated from the surface of bamboo after treatment. The boron salt gets deposited on the surface on bamboo. The boron salt is non-toxic so it can be used for container or bags which are in contact with food material.

Boric acid Borax
It is the most commonly used method of the non-fixing type of preservative as it is non-toxic and eco-friendly methods. It can be used as a combination of boric acid and borax in the ratio 1:1.5. It is alkaline salt which is known as disodium octaborate tetrahydrate ($Na_2B8O13 \times 4H_2O$). Boric acid and borax can be used with sodium dichromate with ratio 2:2:0.5. It may be submerged or sprayed on bamboo with 4 to 5% concentration with water.

ii. *Fixing type preservatives*:

This type of preservative different type of salts is combined together which interact with each other in the presence of bamboo and become chemically fixed. The degree of fixation and efficiency is depending on concentration and types of salts used for fixation. The following are some example.

Copper Chrome Arsenic (CCA):
It is a heavy duty broad spectrum which used for bamboo preservative patented as AsCu. It has found that it provides protection for 50 years or more. It is a mixture of arsenic pentoxide + copper sulfate + sodium dichromate in the ratio 1:3:4. The concentration for indoor (not exposed to weather or ground contact) uses is 6% and for outdoor (structures exposed to weather and in ground contact) uses is 10% with water.

Copper Chrome Boron (CCB):
It is the best alternative for CCA but it is less effective and has less degree of fixation due to boron component. It is a mixture of boric acid + copper sulfate + sodium dichromate in the ratio 1.5:3:4. The concentration for indoor (not exposed to weather or ground contact) uses is 6–8% and for outdoor (structures exposed to weather and in ground contact) uses is 8–10% with water.

Zinc Chrome:
Zinc chloride is adversely helping paints and other finishes as it is highly hygroscopic and in rainy season treated bamboo will give a wet look. It is a mixture of zinc chloride + sodium dichromate in the ratio 1:1. The concentration for outdoor uses (structures exposed to weather and in ground contact) is 10% with water.

Copper Chrome Acetic:
It is a mixture of zinc copper sulfate + sodium dichromate + acetic acid in the ratio 5.6:5.6:0.25. The concentration for outdoor uses (structures exposed to weather and in ground contact) is 8% with water.

Fire Retardant Preservative:
In this treatment, materials are safe against fire as well as decay and insect attack. It is a mixture of zinc boric acid + copper sulfate + zinc chloride + sodium dichromate in the ratio 3:1:5:6. The concentration for indoor and outdoor uses with water is 25%.

V. ***Drying Bamboo Poles***

Drying bamboo process requires more time as it contains 50–60% of water in it. The time required is depending on the surrounding condition and type of species of bamboo. When bamboo dries, it contracts and shrinks. It reduces its diameter up to 10–15% and its thickness gets reduced up to 15–17%. Green bamboo cannot be used for construction as it contracts and a joint between them can loosen due to shrinkage.

The most economical method of bamboo drying is air drying. Once the bamboo is harvested and chemically treated it should be stacked and undercover. The bamboo should be away from ground contact to avoid insect attack. There should not be direct sunlight to avoid rapid shrinkage which results in cracks in bamboo. There must be good ventilation. The stacking can be done by horizontal and vertical ways. Vertical stacking gives faster drying and less chance of fungal attack. Horizontal stacking is the slower process but it is suitable for large bamboo as it does not cause bending in the pole. In this process, bamboos are horizontally placed on some height.

Bamboo must be rotated in a longitudinal direction to avoid concentrated shrinkage and drying. This method of preservation contains 6–12 weeks and its time

Table 2 Mechanical properties of bamboo versus mild steel (all units are in N/MM^2)

Sr No.	Properties	Bamboo	Mild Steel
1	Density	700–800	7700
2	Young modulus (E)	9000–7000	206,000
3	Bending strength	100	155
4	Shear strength	8	115
5	Compressive strength	40–80	155
6	Tensile strength	160	190

depends on initial moisture content, sunlight, bamboo nature, environmental condition, and speed of surrounding air. Bamboo can be dry by another method like oven drying methods.

VI. ***Smoking of bamboo***

In this method, bamboo is placed over the fire for a short duration as it can get the coating of ash on the surface of bamboo. This coating is toxic in nature which helps to avoid insects attack and can improve the durability of bamboo. In this method, the surface crack occurs which can increase the attacks' rate.

5 Mechanical Properties of Bamboo

See Table 2.

6 Comparison

Mainly, the use of bamboo as replacement of steel is due sustainability of bamboo as it is the fast-growing plant which will not contribute to environmental pollution. The energy required to produce steel is 50 times more than required for bamboo. The cost of bamboo reinforced concrete is 35–40% lesser than RCC. The tensile strength of bamboo is greater in terms of strength to specific weight. This ratio is 6 times more in bamboo than steel. Bamboo has a natural coating on the surface of it which provided good resistance to corrosion due to water. This is a serious problem in steel. The compressive strength of bamboo is 40–50% of mild steel. The tensile strength of bamboo is quite equal to steel. Even in some species tensile strength is greater than steel also.

7 Conclusion

There are various types of bamboos which can be used for various uses. Hard, thick, and three- to four-year-old bamboos can be used for construction. The green bamboo must avoid construction uses as it undergoes large shrinkage.

Bamboo is a fast-growing cultivated plant. This can be agricultural by-products in various crops like sugarcane. Therefore, it is used as the best eco-friendly material.

Untreated bamboo has less durability so for a long time uses bamboo must be properly treated. The proper treatment for bamboo can improve the life of bamboo up to 50 years.

The bamboo has 60% tensile strength as compared to mild steel but as compare to costing bamboo is cheaper than mild steel. So we can use bamboo economically for various structural as well as non-structural uses.

Bibliography

1. Sharath Shekar HS (2018) Green composites: a review. Mater Today Proc 5:2518–2526
2. Dey Abhijeet (2018) Experimental study of bamboo reinforced concrete beams having various frictional properties. Mater Today Proc 5:436–444
3. Thakur Ashish (2018) Characterization and evaluation of mechanical behavior of Epoxy- CNT-Bamboo matrix hybrid composites. Mater Today Proc 5:3971–3980
4. Lau K (2018) Properties of natural fiber composites for structural engineering applications. Compos Part B 136:222–233
5. Li W (2017) Axial load behavior of structural bamboo filled with concrete and cement mortar. Constr Build Mater 148:273–287
6. Paraskeva TS (2017) Design and experimental verification of easily constructible bamboo footbridges for rural areas. Eng Struct 143:540–548
7. Wei Y (2017) Flexural performance of bamboo scrimber beams strengthened with fiber-reinforced polymer Constr Build Mater 142:66–82
8. Zhong Y (2017) Bending properties evaluation of newly designed reinforced bamboo scrimber composite beam. Constr Build Mater 143:61–70
9. Chen G (2017) Water effects on the deformation and fracture behaviors of the multi-scaled cellular fibrous bamboo. S1742-7061, 30621-9
10. Jeeva Chithambaram S (2017) Flexural behavior of bamboo based Ferro cement slab panels with flyash. Constr Build Mater 134:641–648
11. Costa MME (2017) Influence of physical and chemical treatments on the mechanical properties of bamboo fibers. Procedia Eng 200:457–464
12. Dixon PG (2017) 3D printed structures for modeling the Young's modulus of bamboo parenchyma. Acta Biomaterialia
13. Karthik S (2017) Strength properties of bamboo and steel reinforced concrete containing manufactured sand and mineral admixtures. Eng Sci 29:400–406
14. Rahman Nabihah (2017) Enhanced bamboo composite with protective coating for structural concrete application. Energy Procedia 143:167–172
15. Karthik S, Mohan Rao P, Awoyera PO (2017) Strength properties of bamboo and steel reinforced concrete containing manufactured sand and mineral admixtures. Eng Sci 29
16. Kumar BP (2017), Microstructure and mechanical properties of aluminum metal matrix composites with addition of bamboo leaf ash by stir casting method. Trans Nonferrous Met Soc China 27:2555 – 2572

17. Puri V (2016) Bamboo reinforced prefabricated wall panels for low cost housing. S2352-7102. 30312-6
18. Khan Z (2016) Fracture behavior of bamboo fiber reinforced epoxy composites. S1359-8368. 31528-1
19. Subramaniam S (2015) Bamboo reinforced concrete slabs for fence walls. J Sustainable Dev 8(9)
20. Akmaluddin (2015) Flexural behavior of steel reinforced lightweight concrete slab with bamboo permanent formworks. Procedia Eng 125
21. Ikponmwosa E (2015) Structural behaviour of bamboo-reinforced foamed concrete slab containing polyvinyl wastes (PW) as partial replacement of fine aggregate. Eng Sci
22. Terai M (2011) Fracture behavior and mechanical properties of bamboo reinforced concrete members Procedia Eng 10
23. Indian Standard IS: 10262 (2009) Recommended guidelines for concrete mix design Indian standards institution. Manak Bhavan, 9, Bahadur Shah Zafar Marg, New Delhi-110 002. India
24. Ghavami K (2005) Bamboo as reinforcement in structural concrete elements. Cem Concr Compos 27
25. Indian Standard IS: 456(2000).Code of practice for plain and reinforced concrete (4th Revision) Indian Standards Institution, Manak Bhavan, 9, Bahadur Shah Zafar Marg, New Delhi-110 002. India

Development of Eco-friendly Alkaline Activated Concrete

Dhavamani Doss Sakthidoss and Thirugnanasambandam Senniappan

Abstract Sustainability is an emerging term in the field of innovative materials for productions, especially in construction arena. There are numerous complications such as global warming, material exhaustion, and high energy loss. These are the freezing issues ascend in the world due to the urbanization enlargement. In this research, conventional concrete and alkaline activated concrete are cast to find the suitability of new sustainable material for making concrete. Cement is replaced by silica- and aluminum-rich materials such as fly ash and Ground-Granulated Blast-furnace Slag (GGBS). Further, river sand is replaced with M Sand. By replacing these materials, the reduction of CO_2 emission and exhaustion of river sand can be minimized. Three different combinations of materials are tried for M40 grade with the mix ratio of 1:1.63:2.64. The experimental investigations exhibit that the alkaline activated concrete with river sand and M Sand is better in compressive strength than conventional concrete. Alkaline solution is used in concrete matrix for activating the binding property of alkaline activated concrete.

Keywords Alkaline solution · Binding property · CO_2 emission · Suitability of materials

1 Introduction

The materials are the important phenomenon which develops the lifestyle of the human beings. The lifestyle of the people before a century and today is not similar. This dissimilarity in their life is because of the kind of materials available at that particular era. So the nature of the human life mainly depends upon the kind of materials. At the moment, the availability of natural raw materials is a million dollar question, especially in the field of construction. An important accomplishment in the construction industry is the development of concrete. It is generally made up of cement, river sand and crushed stone. At present, there are a lot of setbacks

D. D. Sakthidoss (✉) · T. Senniappan
Annamalai University, Annamalainagar, Chidambaram 608002, India
e-mail: dhavamanidossphd@gmail.com

L.-J. Yang et al. (eds.), *Proceedings of ICDMC 2019*,
Lecture Notes in Mechanical Engineering,
https://doi.org/10.1007/978-981-15-3631-1_12

in producing concrete. The production of the concrete is mainly influenced by the availability of raw materials. The increase in the use of these raw materials will make an imbalance in ecosystem. At present, the availability of these raw materials is becoming scarce. Cement is enriched with calcium and silica. This makes it an inevitable material as a binder in concrete. The increased utilization of cement in the construction field creates vulnerable effects to the environment [1, 2]. Production of cement increases the rate of emission of carbon dioxide in the atmosphere. The rate of emission of carbon dioxide is 100% from the cement industry. Due to this, the pollution rate is increased, resulting in global warming [2].

The production of cement also involves high utilization of natural limestone and energy. The energy used for calcination process is very high next to energy used in the steel plant. To resolve these problems, evolution of new eco-friendly material must occur. On the other side, the continuous usage of sand from river for construction lead to serious issues in the ground water table. It takes hundreds of years to form river sand by weathering of rock. The continuous usage of river sand endangers its availability. It is mandatory to reduce the use of natural river sand to maintain the balance in environment. Now the focus of the construction industries turned on finding an eco-friendly material for producing concrete. Since the need for concrete is vastly increasing due to the development of infrastructure, the main purpose of this research is to enhance the concrete production without affecting the environment by developing sustainable concrete called alkaline activated concrete. It is produced by using silica- and aluminum-rich materials such as fly ash and GGBS. Cement is replaced by fly ash and GGBS. Simultaneously, the river sand is replaced by M Sand. In this concrete, an alkaline reagent solution is required to activate the binding property of the binder material (fly ash and GGBS); hence, the concrete is known as alkaline activated concrete. Development of this concrete will be appropriate to find solutions to resolve the difficulties in the traditional concrete production.

2 Alkaline Activated Concrete

Alkaline activated concrete is a mixture of alkaline activated cement (fly ash + GGBS), fine aggregate, coarse aggregate, and alkaline reagent solution. Figures 1, 2, 3, 4, and 5 show the materials used for making alkaline activated concrete. The concrete is known as alkaline activated concrete since the binder used in this concrete is activated by alkaline reagents. The alkaline reagent is a mixture of potassium or sodium hydroxide and potassium or sodium silicate solution with required amount of water. The alkaline reagent is used to trigger the binding ability of alkaline activated cement or silica–aluminum-rich cement with other filler materials. The alkaline activated concrete is also known as geopolymer concrete. The name geopolymer was characterized by Professor Joseph Davidovits in 1978 [3, 4]. The alkaline activated concrete has a lot of advantages over normal cement concrete such as quick hardening, eco-friendly, fire resistance, and ambient curing [1, 3, 5].

Fig. 1 Fly ash

Fig. 2 GGBS

Fig. 3 River sand

Fig. 4 M-Sand

Fig. 5 Alkaline reagent solution

2.1 *Alkaline Activated Cement*

Alkaline activated cement is generally classified into four types, such as the slag-based geopolymer cement, the rock-based geopolymer cement, the fly-ash-based geopolymer cement, and the ferro-silicate-based geopolymer cement. In this, fly-ash-based geopolymer cement is further classified as alkali activated fly ash geopolymer cement and slag/fly-ash-based geopolymer cement. Slag/fly-ash-based geopolymer cement is chosen to achieve the hardening of concrete in ambient conditions [3]. The hardening of geopolymer cement is entirely different from the Ordinary Portland Cement. The hardening of alkaline activated concrete is not attained by the hydration process. In alkaline activated concrete, hardening is attained by polycondensation process where sodium oligo (sialate–siloxo) into sodium poly (sialate–siloxo), a cross-linked network [2, 4, 6]. The reaction of polycondensation of sodium-based geopolymer concrete is shown in Eqs. 1 and 2 [5]. Combination of fly ash and GGBS is added in a ratio of 50:50. Fly ash and GGBS are silica–aluminum-rich material which will readily react with alkaline reagents to form a binder to the alkaline activated concrete.

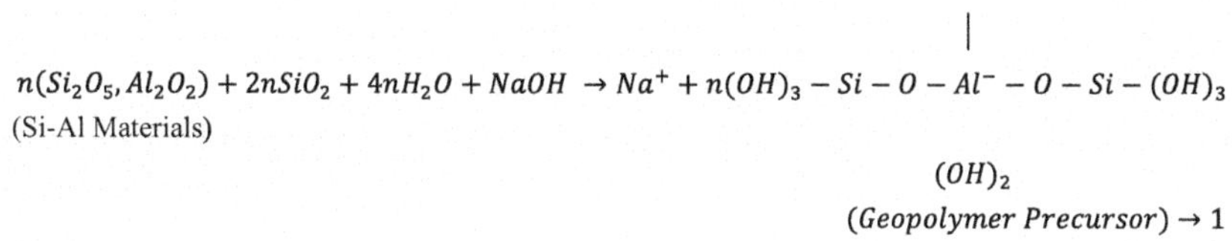

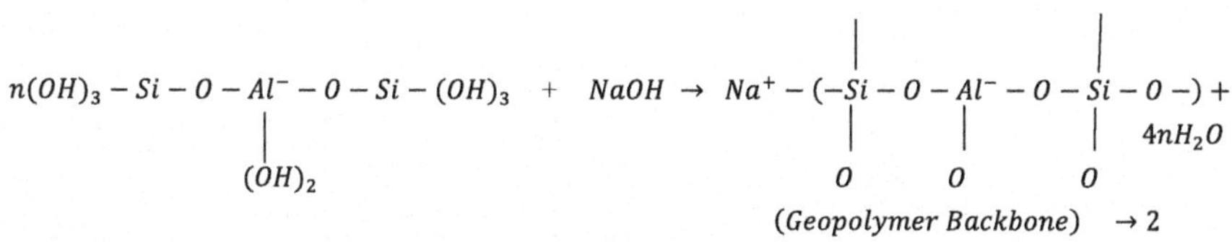

Fly Ash Fly ash is generally a byproduct, but it is considered as hazardous waste material from the thermal power plants. The occurrence of fly ash is abundant in India. The safety disposal of the fly ash is quite complicated, and it is dumped in the thermal power plants. The fly ash is a smaller than cement particles. It leads to a danger of mixing of mixing of fly ash in the atmosphere and creates harmful effects to the surroundings. To rectify these, fly ash is used in concrete production as a replacement for cement. Fly ash is classified into class *C*-type fly ash and class *F*-type fly ash. Class *C*-type fly ash consists of high calcium content more than 10%, and it is known as high-calcium fly ash. Class *F*-type fly ash consists of calcium content lesser than 10%, and it is known as low-calcium fly ash. Among these two types of fly ash, low-calcium fly ash is used for making alkaline activated concrete [7]. Low-calcium fly ash is collected from the Mettur thermal power plant. The specific gravity of fly ash is 2.1. The particle size of fly ash ranges from 4 to 16 μm. The bulk density of fly ash is 995 kg/m^3, and the specific surface area of fly ash is 410 m^2/kg.

Ground-Granulated Blast-Furnace Slag GGBS is a byproduct from the steel industries. GGBS is an efficient material for the replacement of Ordinary Portland Cement. GGBS consists of ample amount of calcium silicates and other constituent of a cementitious material. GGBS is developed simultaneously with iron in the blast furnace in molten condition [8]. GGBS was added with the geopolymer cement to enable hardening of cement in ambient temperature, and it also improves the mechanical property of the cement paste [9]. The GGBS is brought from steel manufacturing industries (JSW brand). The specific gravity of GGBS is 2.65. The particle sizes of GGBS range from 0.4 to 40 μm. The bulk density of GGBS is 1220 kg/m^3, and the specific surface area of GGBS is 422 m^2/kg.

2.2 *Ordinary Portland Cement (OPC)*

OPC is used to cast conventional concrete specimens for the comparison of results with alkaline activated concrete. OPC grade 53 Zuari cement is used for casting specimens. The specific gravity of cement used is 3.15. The particle size of OPC is ranges from 6.6 to 31 μm. The specific area of OPC is 337 m^2/kg. OPC used for casting concrete is confirming IS 12269:1987. The production of concrete using OPC is one of the main contributors of carbon dioxide emission and high energy loss. The hardening process of OPC concrete is attained by hydration process where calcium silicate is converted into calcium di-silicate hydrate (CSH) gel and lime ($Ca(OH)_2$) [3].

2.3 Alkaline Reagent Solution (ARS)

Alkaline reagent solution is mandatory for the use of alkaline activated concrete [10]. Mostly, alkaline reagent is differentiated into two categories such as hostile product and friendly product. User-friendly alkaline reagent is chosen for casting concrete specimen. The usage of hostile reagent requires usage of glasses, gloves, and mask which are must for safety considerations [3]; hence, it is only suitable for laboratory purposes. Till now for casting alkaline activated concrete, user-friendly reagent product is advisable. The alkaline reagent solution used is a fusion of sodium hydroxide solution and sodium silicate solution with a molar concentration of 10 M [4]. The ratio of alkaline reagent solution to fly ash is chosen as 2.5. 10 molarity of ARS is obtained by dissolving 400 grams of sodium hydroxide pellets in 1000 ml of water. The concentration of the solution will influence the compressive strength of the alkaline activated concrete specimen. Higher addition of water to the alkaline reagent solution will decrease the strength of the concrete. For cost-effective and superior performances, sodium hydroxide and sodium silicate solution are utilized [11]. The ARS will kindle the binding capability of fly ash. The ARS concentration relies upon the amount of sodium hydroxide liquefied in the solution. The properties of sodium hydroxide and sodium silicate solution are given in Tables 1 and 2.

Table 1 Specification of Sodium Silicate Solution

Specification	
Chemical formula	$Na_2O \times SiO_2$ (colorless)
pH	Neutral
Assay of Na_2O	7.5–8.5%
Assay SiO_2	25–28%
Wt per ml 20 °C	1.35 g/ml
Appearance	Liquid (Gel)

Table 2 Specification of Sodium Hydroxide

Specification	
Purity	97% min
Sulfate (SO_4)	0.05% max
Potassium (K)	0.1% max
Zinc (Zn)	0.02% max
Leda (Pb)	0.001% max
Chloride (Cl)	0.1% max
Carbonate (Na_2CO_3)	2% max
Silicates (SiO_3)	0.05% max
10% Aqueous solution	Clear and colorless
Iron (Fe)	0.001% max

2.4 Fine Aggregate

Fine aggregates used for casting concrete specimen are river sand and manufacturing sand (M-Sand). Generally, river sand is used in the production of concrete as fine aggregate. The continuous usage of sand for the past decades creates a vacuum in its occurrence. The continuous degradation of natural river sand affects the environment severely. To minimize the usage of river sand, a new material called M-Sand is recommended as the replacement of river sand. In this research, both the river sand and the M-Sand are used for casting alkaline activated concrete.

River Sand Locally available river sand is used for casting concrete specimens. The specific gravity of river sand is 2.68 and falls under the zone II. River sand is used and confirming the IS 383:1970 [12]. The fineness modulus of sand is 3.

Manufacturing Sand (M-Sand) The usage of M-Sand has been started for practice in construction industries to overcome the drawbacks [13] in using river sands such as extinction of river sand and rough packing of particles in concrete matrix. It is manufactured in industries, which will ensure proper particle shape and size of the material. The impurities present in the M-Sand can be removed. The cubical shape of the M-Sand will ensure the packing of the matrix and will increase the mechanical properties of the alkaline activated concrete. The specific gravity of M-Sand is 2.70. The particle sizes of M-Sand range from 0.4 to 40 μm. The bulk density of M-Sand is 1820 kg/m^3, the fineness of modulus of M-Sand is 2.85, and M-Sand chosen is confirming the IS: 383-1970 [12] and falls under the category zone II.

2.5 Crushed Stone Aggregate (CSA)

The natural CSA is used. Clean hard and dense crushed granite stone of size passing 20 mm and retained 12.5-mm sieves were taken for the study. The specific gravity of CSA used was 2.70. The fineness modulus of CSA used was 6.75 and the bulk density of 1535 kg/m^3. The CSA chosen for test confirms the IS: 2386-1963 parts IV and V.

2.6 *Superplasticizer (SP)*

The Conplast 430 with a specific gravity of 1.2 was used for casting of concrete. The SP used was a naphthalene-based agent. The percentage of SP used was 0.50 by weight of cementitious material.

3 Investigational Report

3.1 *Mix Proportion*

Three mixes with three different materials are successfully cast. The grade of concrete was proportioned for M40 grade. OPC, river sand, and crushed stones are used for conventional concrete (CC). Alkaline activated cement, river sand, and crushed stone are used for casting alkaline activated concrete with river sand (GPCRS). Alkaline activated cement, M-Sand, and crushed stones are used for casting alkaline activated concrete with M-Sand (GPCMS). The ratio was not changed for all the mix proportions as 1: 1.63: 2.64. The alkaline reagent solution was chosen as 2.5. The molarity of the solution is 10 M. The alkaline reagent solution to fly ash ratio is taken as 0.38. The alkaline reagent solution was prepared before 24 hours prior to mixing of alkaline activated concrete. The sodium hydroxide solution was prepared by mixing sodium hydroxide pellets with sufficient water. Then, sodium silicate solution has been mixed with the sodium hydroxide after an ample period. The mixed solution is kept in room temperature for 24 hours.

3.2 *Casting and Mixing of Control Concrete (CC) and GPCRS (Alkaline Activated Concrete with River Sand) and GPCMS (Alkaline Activated Concrete with M-Sand)*

CC was cast by using materials such as OPC and river sand mixed for 2 min under dry condition. 75% of the water is mixed with concrete materials initially, and the remaining 25% of water is mixed with SP. Then, the water with SP was mixed well, and it is applied to the concrete mix and mixed thoroughly. Then, concrete was poured into steel molds of size 150 mm × 150 mm × 150 mm. Slump value of 65 is

Fig. 6 Casting of Concrete Cube Specimens

obtained through slump cone test in the laboratory. For GPCRS, river sand is mixed with the concrete mix and M-Sand is used for GPCMS. In alkaline activated concrete mix SP is added with 25% of ARS. Other than this, the mix procedure was same for both conventional and alkaline activated concrete. True slump was obtained. Figure 6 shows the mixing and casting of cube specimens.

3.3 Curing and Testing of Specimens

The compressive strength is determined by testing cube specimens under uniaxial compression. The conventional cube specimens are exposed to water curing for 28 days. The alkaline activated concrete is exposed to ambient curing for 24 hours [10]. The conventional cubes take 28 days for complete (90%) hardening, while alkaline activated concrete takes 24 hours for hardening. The compressive strength of conventional concrete specimens at 7, 14, and 28 days is given in Tables 3, 4, and 5. Figures 7 and 8 show the ambient curing of the concrete specimen. The compressive strength of alkaline activated concrete with river sand and M-Sand is given in Tables 6 and 7.

The compressive strength test is successfully conducted on the concrete specimens, and the results were shown in the Tables 3, 4, 5, 6, and 7. Among them, alkaline activated concrete with M-Sand shows better strength.

Table 3 (7-day Compressive Strength of Conventional Concrete)

SI. No.	Cube specimen size (mm)	Weight of specimen (kg)	Ultimate load (kN)	Compressive strength (N/mm^2)	Average compressive strength (N/mm^2)
1	150 × 150 × 150	8.593	797	35.42	35.73
2	150 × 150 × 150	8.693	814	36.18	
3	150 × 150 × 150	8.883	801	35.60	

Table 4 (14-day Compressive Strength of Conventional Concrete)

SI. No.	Cube specimen size (mm)	Weight of specimen (kg)	Ultimate load (kN)	Compressive strength (N/mm^2)	Average compressive strength (N/mm^2)
1	150 × 150 × 150	8.641	960	42.67	43.19
2	150 × 150 × 150	8.751	977	43.42	
3	150 × 150 × 50	8.692	978	43.47	

Table 5 (28-day Compressive Strength of Conventional Concrete)

SI. No.	Cube specimen size (mm)	Weight of specimen (kg)	Ultimate load (kN)	Compressive strength (N/mm^2)	Average compressive strength (N/mm^2)
1	150 × 150 × 150	8.884	1056	46.93	48.41
2	150 × 150 × 150	8.684	1095	48.66	
3	150 × 150 × 150	8.784	1117	49.64	

Fig. 7 Ambient Curing

Fig. 8 Testing of Concrete Specimens

Table 6 3-day Compressive Strength of Alkaline Activated Concrete with River Sand

SI. No.	Cube specimen size (mm)	Weight of specimen (kg)	Ultimate load (kN)	Compressive strength (N/mm^2)	Average compressive strength (N/mm^2)
1	100 × 100 × 100	2.534	542	54.20	56.03
2	100 × 100 × 100	2.682	565	56.50	
3	100 × 100 × 100	2.931	574	57.40	

Table 7 3-day Compressive Strength of Alkaline Activated Concrete with M Sand

SI. No.	Cube specimen size (mm)	Weight of specimen (kg)	Ultimate load (kN)	Compressive strength (N/mm^2)	Average compressive strength (N/mm^2)
1	100 × 100 × 100	2.923	605.00	60.50	61.82
2	100 × 100 × 100	3.012	612.00	61.20	
3	100 × 100 × 100	2.956	637.60	63.76	

4 Results and Discussions

The experimental result shows that the alkaline activated concrete performs well. The alkaline activated concrete attains higher value than the conventional concrete. Alkaline activated concrete achieved its maximum strength at 3 days. The conventional

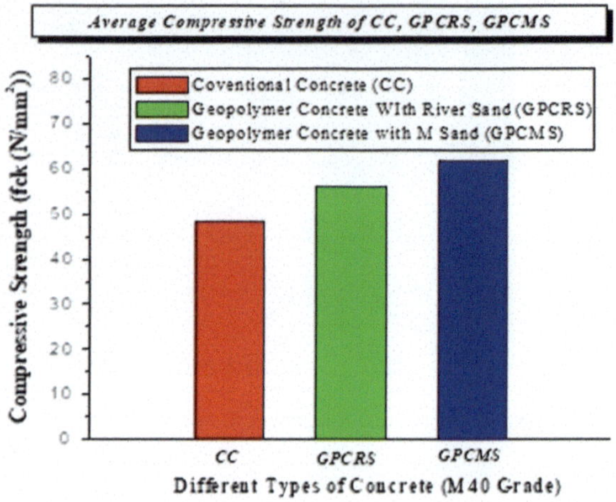

Fig. 9 Comparison of M40 Concrete (CC vs. GPCR) versus GPC(MS)

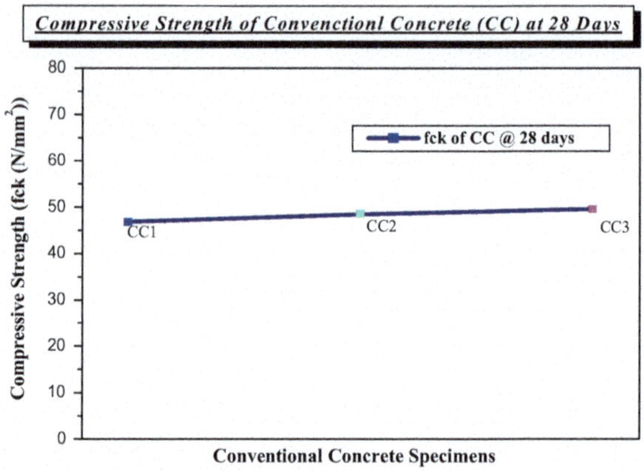

Fig. 10 Compressive Strength of 28-Day Conventional Concrete

concrete achieved its maximum strength at 28 days, where traditional concrete attains in 28 days. Figures 9, 10, 11, 12, 13, and 14 show that the representation of results is obtained by the compressive strength on concrete specimens of conventional concrete and alkaline activated concrete with river sand and M-Sand. Alkaline activated concrete with M-Sand shows higher strength than other two types of concrete shown in Fig. 9.

5 Conclusions

Alkaline activated concrete with M-Sand and river sand is successfully cast and compared with the compressive strength of conventional concrete. From the results, it is found that the alkaline activated concrete with M-Sand achieved higher compressive strength compared to conventional and alkaline activated concrete with river

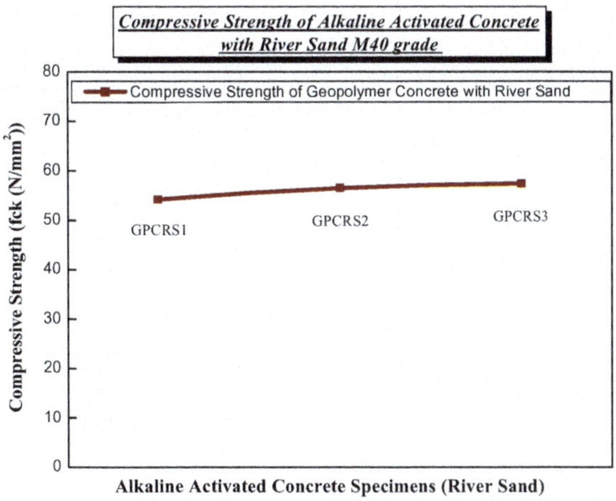

Fig. 11 Compressive Strength of M40 Grade Alkaline Activated Concrete with River Sand

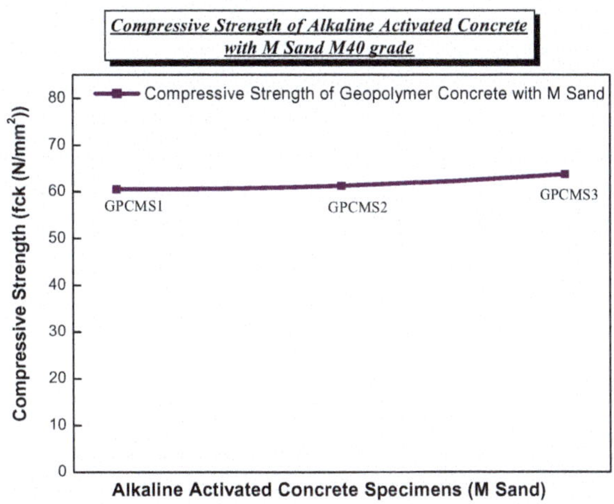

Fig. 12 Compressive Strength of M40 Grade Alkaline Activated concrete with M-Sand

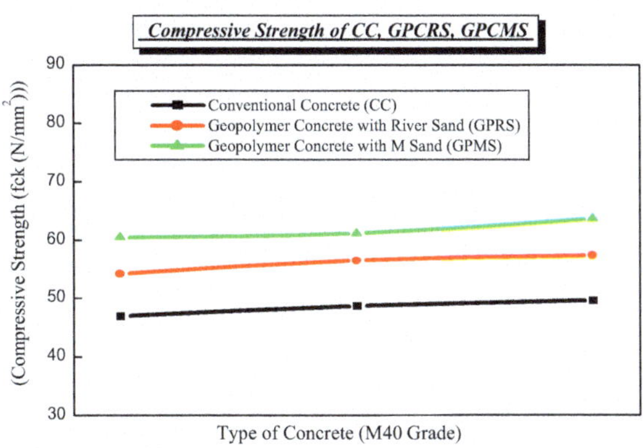

Fig. 13 Comparison of fck of M40 Grade Concrete (CC vs. GPC(S) vs. GPC(M-S))

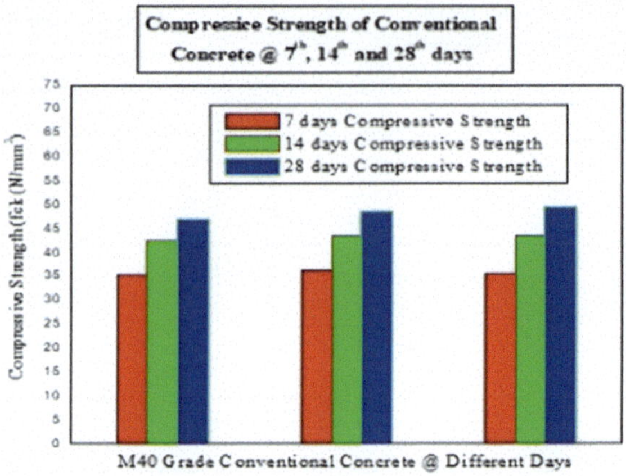

Fig. 14 Comparison of fck values of M40 Grade Conventional Concrete @ 7, 14, and 28 Days

sand. At same time, the alkaline activated concrete with river sand achieved higher strength than the conventional concrete but lesser strength compared to alkaline activated concrete with M-Sand. The increase in strength in alkaline activated concrete is attributed to the presence of silica–aluminum-rich cementitious materials. The smaller particle size of fly ash, GGBS, and the addition of alkaline reagent solution improves the binding property of the alkaline activated concrete. The proper particle size also influences the strength of the concrete. The water curing is neglected for alkaline aggregated concrete.

The experimental results convey that the alkaline activated concrete shows better compressive strength, and the following conclusions are made:

1. The alkaline activated concrete hardens and achieved strength under ambient curing after 24 hours.
2. The compressive strength of cement concrete is 35.73 N/mm^2, 43.19 N/mm^2, 48.41 N/mm^2 on 7, 14, and 28 days respectively.
3. The compressive strength of alkaline activated concrete with river sand is 56.03 N/mm^2. It is 15.74% higher than cement concrete.
4. The compressive strength of alkaline activated concrete with M-Sand is 61.82 N/mm^2. It is 27.7% higher than cement concrete and 10.33% higher than alkaline activated concrete with river sand.

References

1. Thirugnanasambandam S, Antony Jeyasehar C (2019) Ambient cured geopolymer concrete products. Lect Notes Civil eng 25:811–828
2. McCaffrey R (2002) Climate change and the cement industry. Global cement and lime magazine (Environmental Special Issue), 15–19. Available at www.propus.com/gcl2002
3. Davidovits J (2013) Geopolymer cement, a review. Institut Geopolymere

4. Davidovits J. (1999) Chemistry of geopolymeric systems, terminology. In: Geopolymer '99 international conference. France
5. Vijiya Rangan B (2010) Fly ash-based geopolymer concrete. IN: Proceedings of the international workshop on geopolymer cement and concrete. Allied Publishers Private Limted, Mumbai, India, pp 68–106
6. Antony Jeyasehar C, Saravanan G, Salahuddin M, Thirugnanasambandam S (2012) Development of fly ash based geopolymer precast concrete elements. Asian J Civil Eng 14(4):605–616
7. Malhotra VM, Ramezanianpour AA (1994) Fly ash in concrete. CANMET, Ontario, Canada, Ottawa
8. Javeed MA (2015) Studies on mix design of sustainable geopolymer concrete. Int J Innovative Res Eng Manag 2(4). ISSN: 2350-0557
9. Kiran T et al (2015) Impact test on geopolymer concrete slabs. Int J Res Eng Technol 4:110–116
10. Annamalai S, Thirugnanasambandam S, Muthumani K (2017) Flexural behaviour of geopolymer concrete beams under ambient temperature. Asian J Civil Eng 18(4):621–631
11. Xu H, van Deventer JSJ (2000) The Geopolymerisation of alumino-silicate minerals. Int J Miner Process 59(3):247–266
12. IS 2386 (1963) Methods of test for aggreagtes for concrete, Part IV Mechanical Properties. Bureau of Indian Standards, New Delhi, India
13. Kumars S Kotian RS (2018) M-SAND, an alternative to the river sand in construction technology. 9(4):98–102

Design Optimization of Suspension and Steering Systems for Commercial Vehicles

V. Vijaykumar and P. Anand

Abstract Steering system is used to provide directional stability for the vehicle, and suspension system is to provide ride comfort and transfer acceleration, braking and cornering loads from road wheels to frame and body. Thus, both steering and suspension systems play a very important role in vehicle ride and handling characteristics. Vehicle drift or the steering pull is the deviation of vehicle path from a straight line while braking and is related to various parameters of suspension, steering, tires, and wheel alignment. Analyses of steering and suspension characteristics such as brake steer and bump steer which affect vehicle drift play an important role. This research deals with multi-body simulation-based optimization of the parameters of suspension and steering systems by focusing on their dynamic interaction between the two systems and thereby helps to achieve the targets (bump steer-4°/m and brake steer-0.01°/kN). Commercial vehicle industry is currently aiming toward the development of vehicles with improved passenger safety and comfort at lower cost. In general, the steering systems of longer front overhang (FOH) vehicles have multiple steering linkages instead of conventional draglink arrangement because of design and packaging constraints. The paper handles the design of high-strength draglink for longer FOH vehicle by eliminating multiple linkages and reducing the cost by Rs. 2000.

Keywords Bump steer · Brake steer · Draglink

V. Vijaykumar (✉)
Vel Tech Rangarajan Dr. Sagunthala R&D Institute of Science and Technology, Avadi, Chennai 600062, India
e-mail: vissvijay@hotmail.com

P. Anand
Department of Mechanical Engineering, Vel Tech Rangarajan Dr. Sagunthala R&D Institute of Science and Technology, Avadi, Chennai 600062, India

L.-J. Yang et al. (eds.), *Proceedings of ICDMC 2019*,
Lecture Notes in Mechanical Engineering,
https://doi.org/10.1007/978-981-15-3631-1_13

1 Introduction

The vehicle should be able to travel in a straight path with no external driver inputs which is important for safe driving. The straight-line path of vehicle should be maintained during the events of bump and braking. Deviation of vehicle from straight path while braking is known as brake steer and during bump is known as bump steer [1]. During such an event, the driver has to apply constant steering correction to maintain a straight-line path. It gives fatigue to driver, and vehicle will drift. This is related to steering and suspension characteristics such as bump steer and brake steer. The vehicle needs to be analyzed for this behavior by the design optimization of steering and suspension system hard points. There is a need to increase the overall length of the vehicle for higher payload and increased passenger capacity. This is achieved by increasing the front overhang, wheelbase, and rear overhang of the vehicle. Designing a steering system for longer wheelbase vehicle is critical as it needs to meet to turning circle diameter (TCD) for ease of maneuverability.

1.1 Objective and Methodology

Suspension hard point will be finalized considering the vehicle axle loads, ride height, and ride requirements. The suspension and steering systems will be modeled by using multi-body kinematic software ADAMS with flex axle. This will be analyzed to meet bump and brake steer targets with different spring configuration and loading conditions of the vehicle. Once the steering drop arm point is finalized from ADAMS analysis, steering linkage design and evaluation will be done to reduce cost without compromising performance and durability.

2 Commercial Vehicle Suspension and Steering Systems

Typical steering and suspension system hard points are shown in Fig. 1. The hard point definitions are given below.

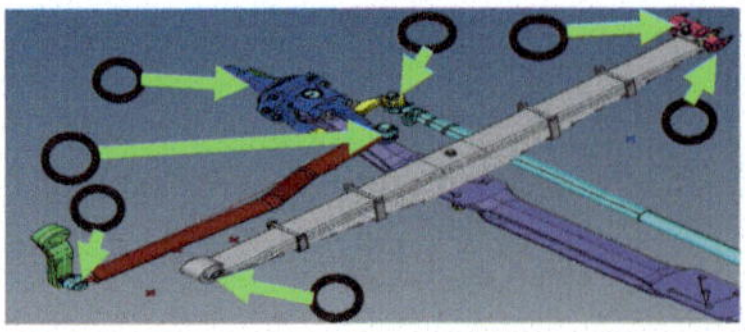

1. Wheel center; 2. Track rod to lever; 3. Steering arm to draglink; 4. Draglink to drop arm; 5.Leaf spring front eye to frame; 6. Leaf spring rear eye to shackle; 7. Shackle to frame

Fig. 1 Suspension-steering system hard points

2.1 Multileaf Spring for Commercial Vehicles

Multileaf spring can be considered as cantilever beam where the one end is fixed at axle and another end is fixed at frame [2]. Load is applied at frame end.

$$\sigma^{\circ} = 6\mathrm{Fx}/bt^2 \tag{1}$$

σ—bending stress, Fx—bending moment, b—leaf width, t—leaf thickness.

If the section modulus varies from fixed end (at axle where the bending moment is more) to frame end where the load is applied, then the spring will be constant stress spring (parabolic spring PA). If the section modulus is constant from fixed end (at axle where the bending moment is more) to frame end where the load is applied, then the spring will be constant strength spring (semi-elliptic spring SE). While finalizing leaf spring design, critical parameters like ride frequency, spring span, weight, and height are considered. Front suspension is designed for 6T and 8T front axle weight (FAW). Ride frequency is the indicator for ride of the vehicle. In general, ride frequency is expressed in Hz or cycles per seconds and it is considered as undamped [3]. Higher spring stiffness leads to higher ride frequency, and higher frequency leads to inferior ride. Lower the frequency, better the ride.

$$F = 1/(2\pi)\sqrt{(K/M)} \tag{2}$$

F—natural frequency (Hz), K—spring rate (N/m), M—mass (Kg).

Lower frequency increases stack height for a given span, and longer span will give lesser stack height as shown in Figs. 2 and 3 and Table 1. Also, lower frequency increases weight for a given span and longer span will have more weight. By considering the weight, cost, and ride height requirement (250 mm), 1650 mm span is selected.

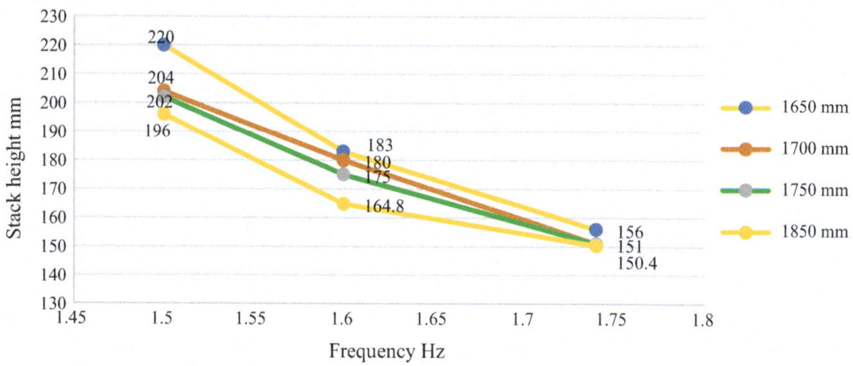

Fig. 2 Frequency versus Stack height

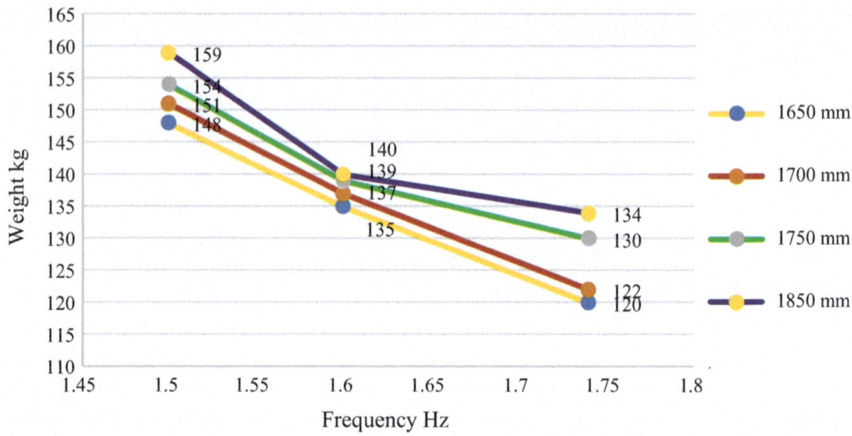

Fig. 3 Frequency versus Weight

Table 1 Leaf spring critical parameters

Span (mm)	Frequency (Hz)	Stack height (mm)	Ride height (mm)	Weight (kg)
1650	1.5	220	309	148
	1.74	154	243	120
1750	1.5	202	291	154
	1.69	160	249	133
1850	1.5	196	285	159
	1.64	160	249	138

2.2 *Multi-body Dynamic Analysis*

Suspension and steering systems are modeled in ADAMS to evaluate bump steer and brake steer. Bump steer occurs as the steering Pitman arm ball joint does not coincide with ideal center. Ideal center is the center of ideal arc of leaf spring as it deflects during bump and rebound. Brake steer happens during vehicle braking. Brake steer happens as the steering arm does not coincide with center of rotation as the springs are subjected to wind up due to brake torque. Targets are set for bump steer and brake steer as ±4°/m and ±0.01°/kN, respectively. Steering arm point is adjusted to meet brake steer, and drop arm point is adjusted to meet bump steer. Steer arm and drop arm points are plotted considering the packaging requirements and target as shown in Fig. 4. Ideal bump steer line is plotted as shown in Fig. 5, and the drop arm point along the ideal bump steer line helps to meet the target. Toe-angle change during wheel travel and bump steer for the points *B*, *C*, *D* in the ideal bump steer line is shown in Fig. 6. and Table 2, respectively. Bump steer comparison is shown in Fig. 7 for different leaf spring combinations. Brake steer comparison for different leaf spring combinations is shown in Table 3. Point C is selected for considering

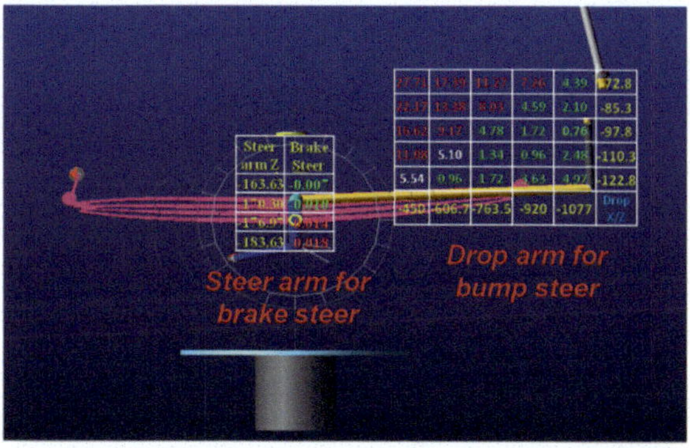

Fig. 4 Bump steer/brake steer-DOE

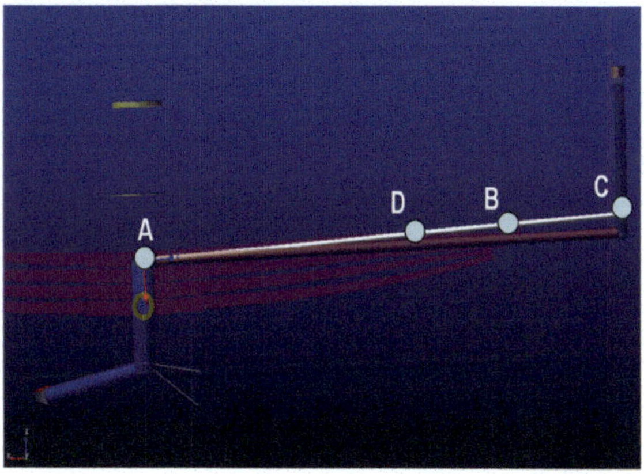

Fig. 5 Ideal Bump steer line

packaging requirements.

3 Steering Linkage Design

High-strength single draglink is proposed for 12 m intercity bus in place of bevel box and relay arm arrangement which has multiple linkages. Existing and proposed arrangements are shown in Figs. 8, 9, and 10.

The objective is to minimize the forces in draglink by designing the linkages optimally without compromising the wheel turn angles to meet the required turning circle diameter of 22 m. The draglink is packaged with 40 mm diameter, considering the packaging requirement and a length of 1750 mm with bend offset of 45 mm. Rankine's formula is used to calculate buckling strength. Maximum force in the drag

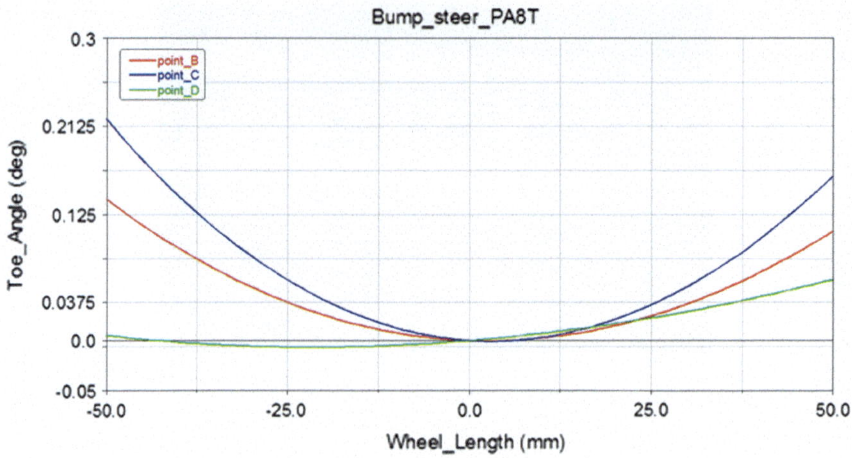

Fig. 6 Toe-angle versus Wheel travel length

Table 2 Bump steer comparison for the points *B*, *C*, *D* in the ideal bump steer line

Bump Steer							
Wheel Travel	Units	Point B		Point C		Point D	
		Rebound	Bump	Rebound	Bump	Rebound	Bump
±20 mm	°/m	−1.2	0.7	−2	1	0.3	0.8
±50 mm	°/m	−2.8	2.2	−3.4	3.3	−0.7	1.2

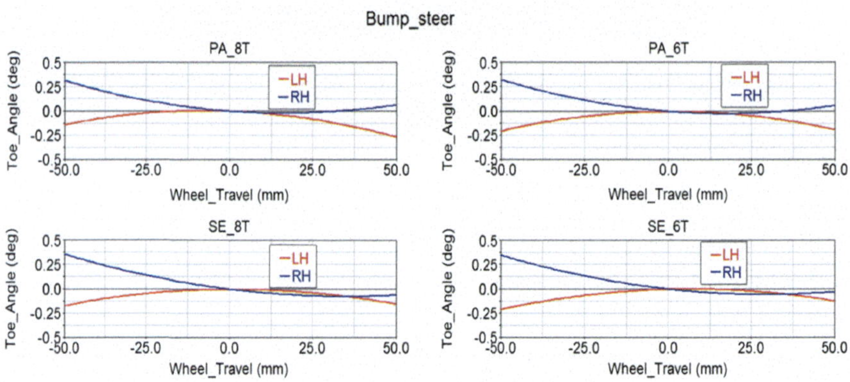

Fig. 7 Bump steer comparison with parabolic and semi-elliptic spring at 6T & 8T axle loads

link is obtained by dividing the steering gear maximum output torque by drop arm pitch. Drop arm pitch was 285 mm, and 4730 Nm is the steering gear maximum torque. Existing draglink material was ST-52-3 NBK DIN-2393 which has 360 MPa

Table 3 Brake steer comparison

Brake steer (°/kN)—Target 0.0°/kN		
Variants	Flex axle	Rigid axle
Parabolic spring (PA)-6T	−0.013	−0.002
Semi-elliptic (SE) spring-6T	−0.011	0.00
Parabolic spring (PA)-8T	−0.012	−0.004
Semi-elliptic spring (SE)-8T	−0.011	−0.0008

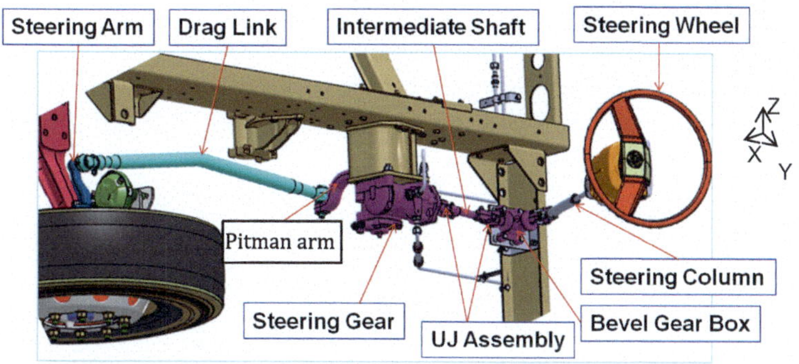

Fig. 8 Existing Bevel box arrangement

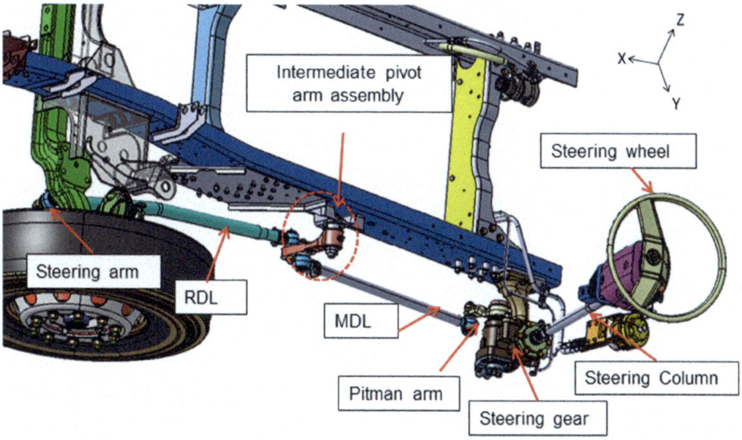

Fig. 9 Existing Relay arm arrangement

yield strength. The proposed draglink material is AISI 1019 M with 700 MPa minimum yield strength. As per the industry standard, the buckling strength should be twice the rated force of the draglink. Maximum force in the link is 16.6 kN. Hence, the buckling strength of the link should be 33.2 kN. Slenderness ratio of the link was verified to calculate buckling strength.

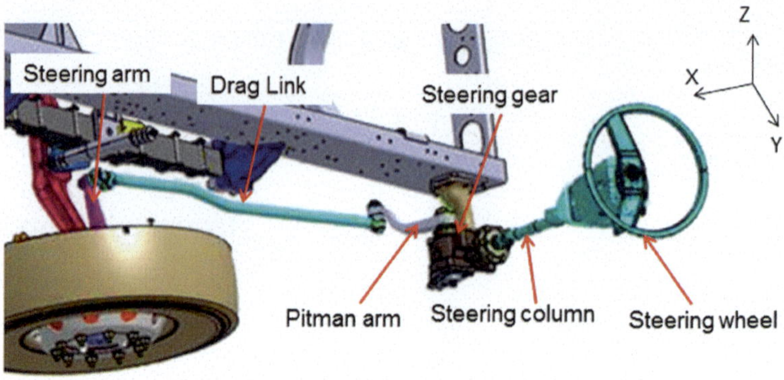

Fig. 10 Proposed High strength draglink arrangement

Slenderness ratio of the tube is given by equation,

$$\mathrm{SL} = L/R,\ R = \sqrt{\frac{I}{A}} \tag{3}$$

Critical slenderness ratio CSL is given by the equation,

$$CSL = \sqrt{\frac{\pi^2 E}{\sigma_y}} \tag{4}$$

Buckling strength is calculated by Rankine's formula,

$$F_b = \frac{\sigma_y \times A}{\left(1 + \left(\frac{\sigma_y}{\pi^2 E}\right) \times \left(\frac{L}{R}\right)^2 + \frac{x \times r}{R^2}\right)} \tag{5}$$

F_b—buckling force(N), x—tube bend offset from neutral axis (m), r—radius of the tube (m), SL—slenderness ratio, L—length of draglink(m), R—radius of gyration(m), I—moment of inertia (m^4), A—cross-sectional area (m^2), E—Young's modulus of elasticity (N/m^2), σ_y—yield strength of material (N/m^2).

(Rankine's formula is preferred when SL is less than CSL.) Buckling strength of the link is calculated with existing and proposed material. Existing material is found to have 26.7 kN with solid bar, and the proposed material with 6 mm thickness is having 34.9 kN which satisfies the requirement.

3.1 Structural Analysis of Draglink

Finite element analysis was done using ANSYS 14.5 as shown in Fig. 11. One end of the tube is constrained with all three translation degrees of freedom and free to rotate in any direction. Compressive load is applied in *X*-axis at other end and translation movements in *Y* and *Z* are constrained. Rotation is allowed about *Y* and *Z* and constrained to rotate about *X*-axis. From Figs. 12 and 13, the link looses stability

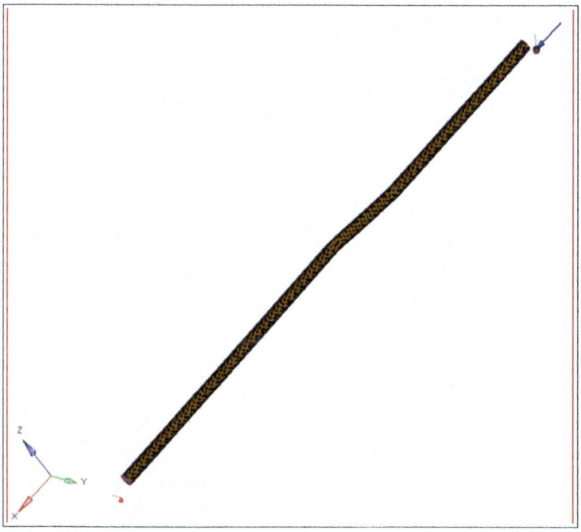

Fig. 11 FE model of draglink

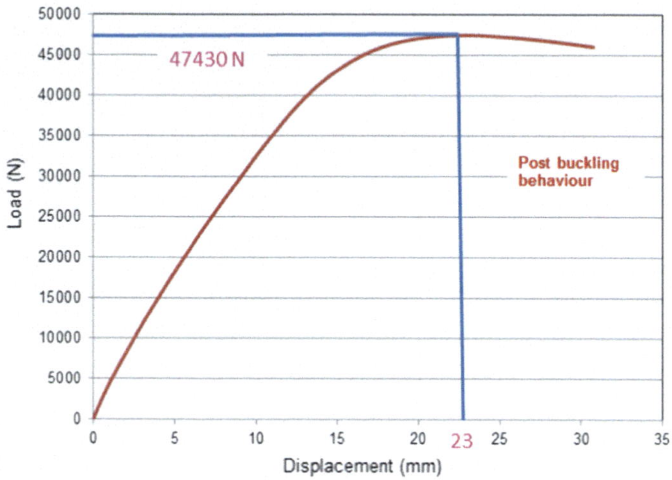

Fig. 12 Force Displacement part of draglink

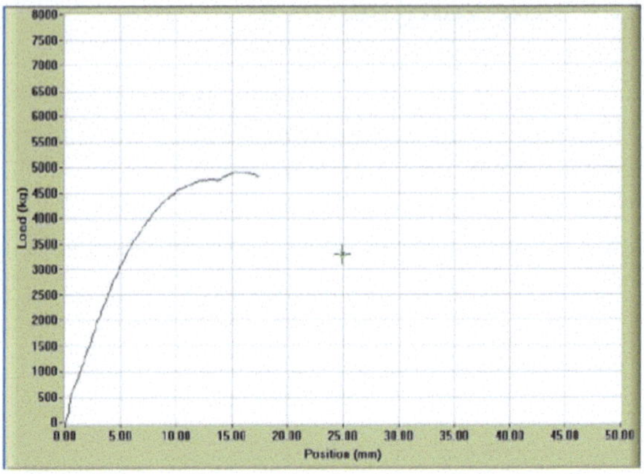

Fig. 13 FE model of draglink

after 47.4 kN where the buckling strength requirement is 33.2 kN minimum. Three numbers of samples were tested as shown in Figs. 14 and 15 for buckling strength, and the results are given in Table 4.

Fig. 14 Test setup

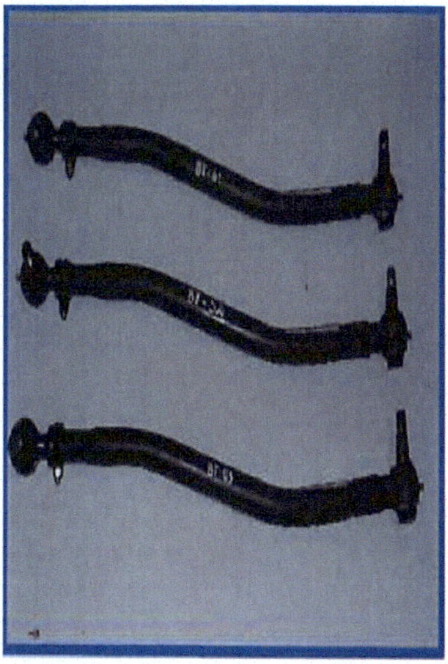

Fig. 15 Tested samples

Table 4 Buckling strength test results

Sample	Required buckling load (kN)	Numerical buckling load (kN)	Actual buckling load (kN)	Remarks
S1	33.2	47.4	49.78	All the samples met the requirement
S2	33.2	47.4	52.4	
S3	33.2	47.4	49.1	

4 Vehicle Steering Performance Evaluation

Vehicle was tested with existing and proposed systems. The proposed high-strength draglink fitment in vehicle is shown in Figs. 16 and 17. TCD of existing and new system was same. The results are given in Table 5.

Steering effort at dynamic condition was measured at two conditions: (1) steering road wheel at maximum condition with hydraulic power support and (2) steering road wheel to meet TCD (20 m radius) without hydraulic power support. Self-centering efficiency is the ability of the road wheels to return from maximum turned position to straight position in 3 s. Since the high-strength link has replaced multiple linkages in the system, steering effort with and without hydraulic power support has come down due to less friction.

Fig. 16 Test vehicle

Fig. 17 Draglink installation

Table 5 Performance evaluation results

Parameter	Existing system	Proposed system
Turning circle diameter (TCD)	22.2 m	22.2 m
Steering effort with hydraulic assistance	5.4 kgf	4.73 kgf
Steering effort without hydraulic assistance	41.2 kgf	37.8 kgf
Self-centering efficiency	53.20%	73.40%

Table 6 Existing system cost

Existing system			
Sl. No.	Description	Qty	Total cost-Rs
1	Gear box assembly	1	9875
2	Bevel box assembly	1	2900
3	UJ assembly	2	775
4	Intermediate shaft	1	585
5	Bevel box bracket	1	193.6
6	Draglink	1	2004
7	Fasteners (bolt, nut, washer, spacer)	4	95.52
System Cost			**16428.12**

Table 7 Proposed system cost

Proposed system			
Sl. No.	Description	Qty	Total Cost-Rs
1	Gear box assembly	1	9875
2	High-strength draglink	1	4000
3	Fasteners (bolt, nut, washer, spacer)	6	385.5
System Cost			**14260.5**

5 Cost and Weight Comparison

Costing is done for the existing and proposed systems as shown in Tables 6 and 7. The proposed high-strength link has Rs. 2100 savings per vehicle. Existing and proposed systems weigh 19.9 and 15.2 kg, respectively. Weight saving per vehicle would be 4.7 kg.

6 Summary and Conclusion

The paper detailed the methodology for finalizing the suspension and steering hard points. Bump steer and brake steer were analyzed in ADAMS for optimizing the hard points. Suspension and steering hard points are arrived based on the targets of bump steer (4°/m) and brake steer (0.01°/kN). High-strength draglink was proposed in place of bevel box and relay arm arrangement which has multiple linkages. Since the proposed system has less number of components, it reduces cost by Rs. 2100 and weight by 4.7 kg. It also saves assembly time during production and service.

References

1. Oz Y, Ozan B, Uyanik E (2012) Steering system optimization of a ford heavy-commercial vehicle using kinematic & compliance analysis. SAE Technical Paper 2012-01-1937. https://doi.org/10.4271/2012-01-1937
2. Rajvardhan RP, Shankapal SR, Vijaykumar SM (2010) Effect of wheel geometry parameters on vehicle steering. SAS TECH J 9.2
3. Durstine JW (1973) The truck steering system from hand wheel to road wheel. No. 730039. SAE Technical Paper

Experimental Study on Vibration Control of Transportation Trailers Used for Spacecraft

Nandikolla Diwakar and S. Balaguru

Abstract This paper presents the experimental study to control the vibrations of trailer used for the spacecraft transportation of launch vehicles (PSLV/GSLV) by regulating the vertical stiffness of air springs, vertical tire stiffness, speed and ballast weight. The spacecraft, after final preparations, are transported to launch pad for stacking using different trailers. The spacecraft experience vibrations during transportation and it is very critical that the vibration levels have to be controlled else if damages the electronic packages and other vital subsystems, which would jeopardize the whole mission. An experimental study was carried out to control the vibrations of trailers by changing the configurations, such as varying the vertical stiffness of air springs for air suspension trailer, vertical stiffness of tires, transportation speed, addition of ballast weight and sudden acceleration/braking. For this study, a 15 t capacity pneumatic suspension full trailer is selected and for the measurement of vibration levels in three directions, accelerometers are fixed to the trailer. The results were plotted and discussed in detail for each configuration. Nevertheless, it allowed for estimation of the trailer dynamic loads under real road conditions.

Keywords P_0 inner air pressure · V_0 air volume at static equilibrium position · Df small amount of axial deformation · K variable factor to consider different vibration conditions · A effective area of cross section · D is the effective diameter of the air spring

1 Introduction

In launch systems spacecraft transportation, the vibration 'g' levels on the vehicle play a vital role as it affects the system components. The air springs in the air suspension trailer act as dampers to control the vibrations and to provide ride comfort for vehicle. The effect of rough road surface on vehicle vibrations, with minimum vibration 'g' levels, is a challenging task as part of satellite launch mission into

N. Diwakar (✉) · S. Balaguru
Department of Mechanical Engineering, Vel Tech Rangarajan Dr. Sagunthala R&D Institute of Science and Technology, Avadi, Chennai 600062, India
e-mail: diwa13@gmail.com

L.-J. Yang et al. (eds.), *Proceedings of ICDMC 2019*,
Lecture Notes in Mechanical Engineering,
https://doi.org/10.1007/978-981-15-3631-1_14

space which is of national importance. Hence, this experimental study was taken up to study how the vibration 'g' levels can be minimized.

The main objective of this experimental study is:

1.1 To study the vibration 'g' levels with respect to air inlet pressure of air springs
1.2 To study the vibration levels for vertical tire stiffness with respect to ballast weight
1.3 To study the speed with respect to vibration 'g' levels and
1.4 To study the ballast weight with respect to 'g' levels

The vertical static stiffness of air spring C_0 is given by

$$C_0 = (p_0 - 1)\frac{dA}{df} + Kp_0\frac{A^2}{V_0} \quad (1)$$

Effective area of cross section,

$$A = \frac{\pi}{4}D^2 \quad (2)$$

It is noticed that from Eq. (1) that p_0,V_0, A and f all vary with the piston movement. We also considered that the rate of the effective area of cross section is relatively easy to control in the structure design.

2 Experimental Methods

A 15 t single axle drawbar pneumatic suspension full trailer as shown in Fig. 1 was selected for the assessment of vibration levels on sprung mass and the trailer was mated with BEML make tow hook mounted hauler. Steel bush was used at the interconnection of tow bar and pin to avoid jerking during braking/acceleration. The trailer was fitted with 4 air springs at each suspension group left and right pneumatic circuits are connected separately. The inlet pressure line of the air springs was fitted

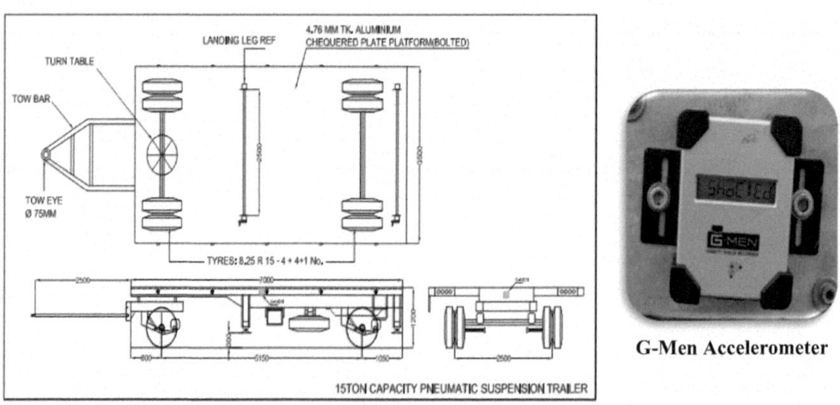

Fig. 1 Schematic of the satellite transportation trailer with G-men accelerometers

Fig. 2 Air pressure regulators in the air spring circuit

with Spava make air pressure regulators as shown in Fig. 2, having a pressure range of 0–12 kg/cm^2 so that the air pressure can be adjusted in the air springs.

2.1 Load with Respect to Air Pressure Calculation

Diameter of the air spring or bellow, D	29	cm
Area of the air spring or bellow, A	660.6	cm^2
Un laden weight of chassis	3500	kg
Weight of the axles	2500	kg
Load on front axle in unladen condition	2915	kg
Load on rear axle in unladen condition	3085	kg
Total weight of the trailer (without payload)	6000	kg
Load on front axle air spring or bellow (of one air spring or bellow)	1457.5	kg
Load on rear axle air spring or bellow (of one air spring or bellow)	1542.5	kg

Front axle		
Payload	5000	kg
Front axle one side	1250	kg
Load on front axle	2707.5	kg
Reqd. air pressure	**4.0**	kg/cm^2

Rear axle		
Payload	5000	kg
Rear axle one side	1250	kg
Load on rear axle	2792.5	kg
Reqd. air pressure	**4.22**	kg/cm^2

From the above calculation, it can be concluded that the minimum pressure required is 4.0 kg/cm^2 to transport the 5000 kg spacecraft with container.

SRIC make, G-Men DR20 model accelerometers—2 nos were used for the measurement of acceleration in three axes (X, Y and Z), having a frequency response of 0–100 Hz, minimum sampling frequency of 1, 2, 5, 10, 20 ms (selectable), measurement resolution 0.2 g and range 0.1–20 g.

Accelerometers were firmly fitted at the left longitudinal center and lateral center of the trailer (Fig. 1) and the vibration levels were measured in three axes. Trials were conducted over the cattle traps and for a distance of 3 km for each configuration as per Table 1.

The measured data was recorded and analyzed for evaluating the effect on sprung mass for various speeds by controlling vertical stiffness of air springs and tires (Fig. 3).

Table 1 Trial configurations

S. No.	Pay load in kg	Air spring or bellow pressure in bar	Tire pressure in bar
1.	0 (No load)	6	6.8
2.		4	6.8
3.	1500	4	6.8
4.		4	4.0
5.	5000	4	6.8
6.		4	4.0
7.	10,000	6	8.0
8.		6	5.5

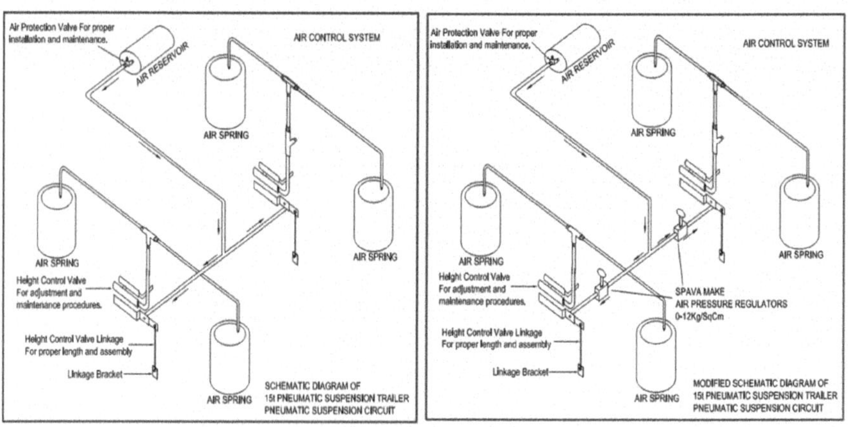

Fig. 3 Schematic and modified schematic diagram of 15 t pneumatic suspension trailer pneumatic suspension circuit

Table 2 '*g*' values in three axes for each trial on different road conditions and at cattle traps

Trial	Load in tons	Air spring or bellow pressure in bar	Tire air pressure in bar	*X* vertical	*Y* longitudinal	*Z* lateral
				g (m/s^2)		
On different road conditions with maximum 30 kmph speed						
1	0	6	6.8	0.8	0.6	0.4
2	0	4	6.8	0.6	0.5	0.2
3	1.5	4	6.8	0.75	0.3	0.2
4	1.5	4	4.0	0.6	0.25	0.1
5	5	4	6.8	0.4	0.2	0.1
6	5	4	4.0	0.3	0.2	0.1
7	10	6	8.0	0.4	0.2	0.1
8	10	6	5.5	0.2	0.1	0.1
At cattle traps						
1	0	6	6.8	2.0	0.5	0.2
2	0	4	6.8	1.7	0.5	0.2
3	1.5	4	6.8	0.85	0.35	0.2
4	1.5	4	4.0	0.7	0.25	0.1
5	5	4	6.8	1.8	0.3	0.2
6	5	4	4.0	0.6	0.2	0.1
7	10	6	8.0	0.4	0.2	0.1
8	10	6	5.5	0.3	0.1	0.2

2.2 *Test Results*

The trials were conducted as per the configurations of Table 1 with two accelerometers on the trailer and results were plotted in Fig. 10, acceleration '*g*' with respect to time (Table 2).

3 Conclusions

Experimental trials were conducted with unladen weight of trailer and with payloads of 1.5 t, 5 t and 10 t in different configurations as per Table 1. By decreasing the air pressure of suspension system air spring from 6 to 4 bar and by decreasing the tire pressure from 8.0 bar to 4 bar, vibration levels were reduced from '1.8 to 0.3 g' with the said payloads and configurations. From Figs. 4, 5, 6, 7, 8, 9 and 10, it can be concluded that the vibration levels '*g*' values are decreasing with increasing payload. At cattle traps near entrance gates, vertical acceleration alone is more than '2 g.'

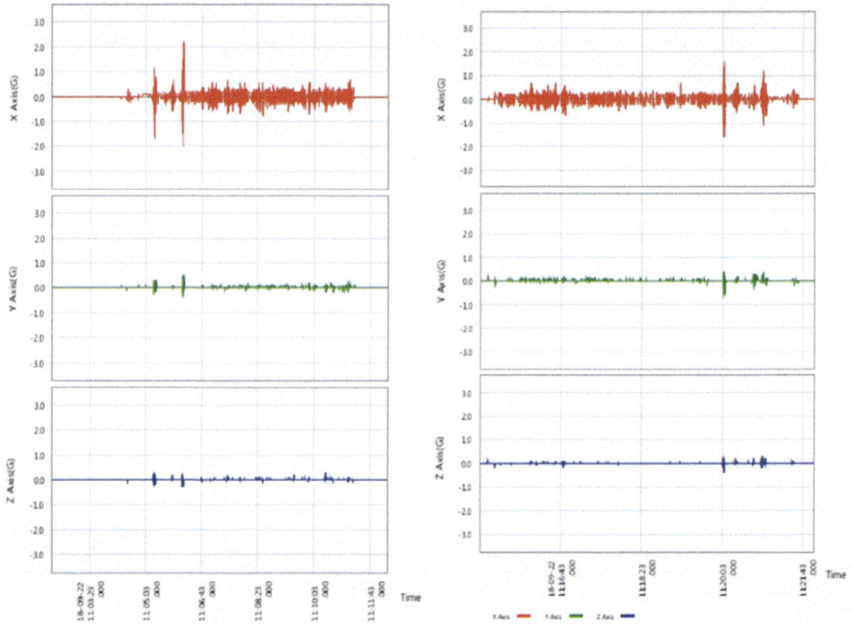

Trial – 1
Unladen trailer with 6 bar air spring or bellow pressure with 6.8 bar tire Pressure with max 30 kmph

Trial – 2
Unladen trailer with 4 bar air spring or bellow pressure with 6.8 bar tire Pressure with max 30 kmph

Fig. 4 Trial 1 and 2 time versus acceleration '*g*' graphs in three axes directions

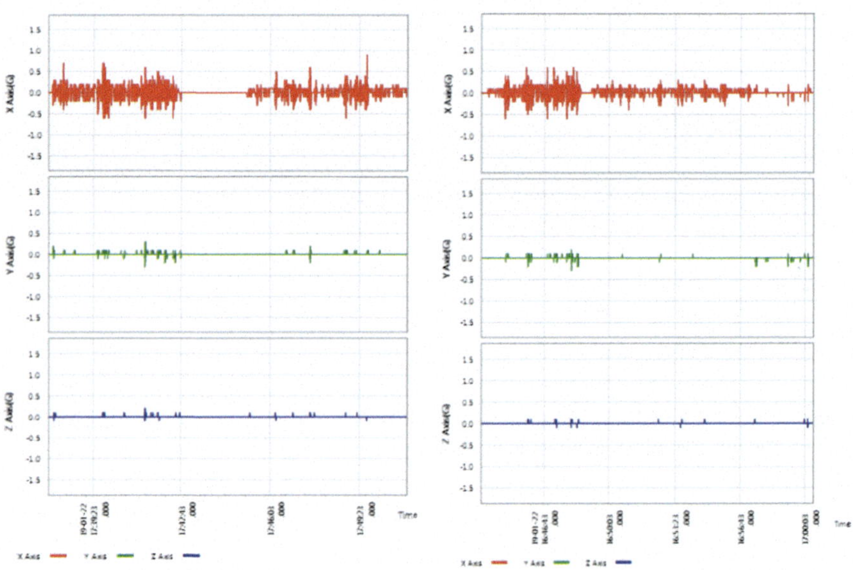

Trial – 3
1.5t Payload on trailer with 4 bar air spring or bellow pressure with 6.8 bar tire Pressure with max 30 kmph

Trial – 4
1.5t Payload on trailer with 4 bar air spring or bellow pressure with 4 bar tire Pressure with max30 kmph

Fig. 5 Trial 3 and 4 time versus acceleration '*g*' graphs in three axes directions

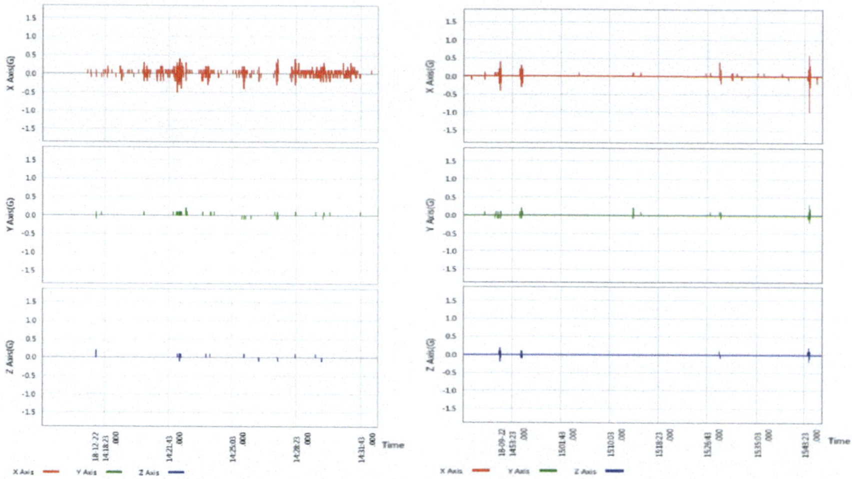

Trial – 5

5t Payload on trailer with 4 bar air spring or bellow pressure with 6.8 bar tire Pressure with max 30 kmph

Trial – 6

5t Payload on trailer with 4 bar air spring or bellow pressure with 4 bar tire Pressure with max 30 kmph

Fig. 6 Trial 5 and 6 time versus acceleration '*g*' graphs in three axes directions

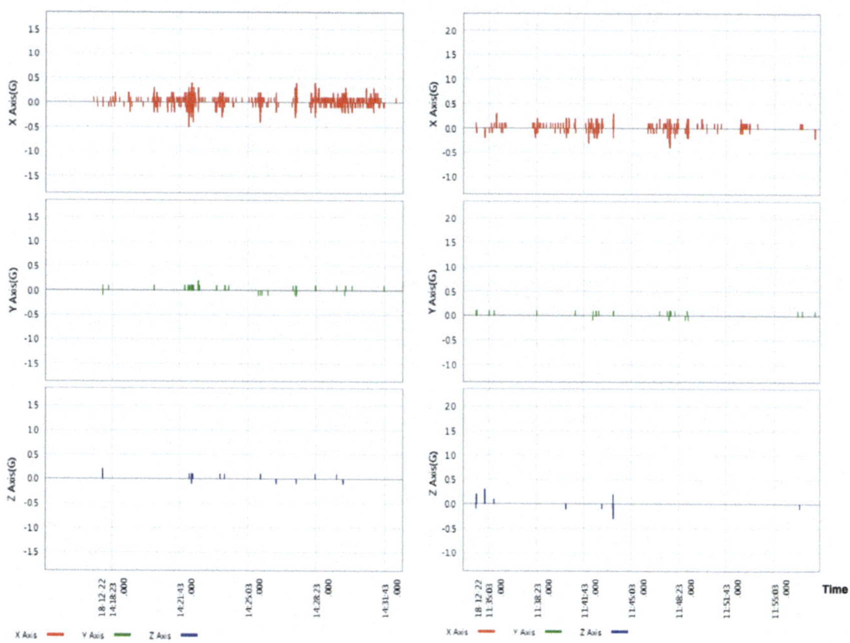

Trial – 7

10t Payload on trailer with 6 bar air spring or bellow pressure with 8 bar tire pressure with max 30 kmph

Trial – 8

10t Payload on trailer with 6 bar air spring or bellow pressure with 5.5 bar tire pressure with max 30 kmph

Fig. 7 Trial 7 and 8 time versus acceleration '*g*' graphs in three axes directions

(a) Unladen Trailer

(b) Trailer with 5t pay load

(c) Trailer with 10t pay load

(d) Trailer with Spacecraft Container to Launch pad

Fig. 8 15 t Pneumatic suspension trailer with different pay loads and spacecraft container

(a) Cattle trap

(b) Culvert above the road

(c) Rough road with ditches and bumps

(d) Road with patch works

Fig. 9 Trials conducted in different road conditions

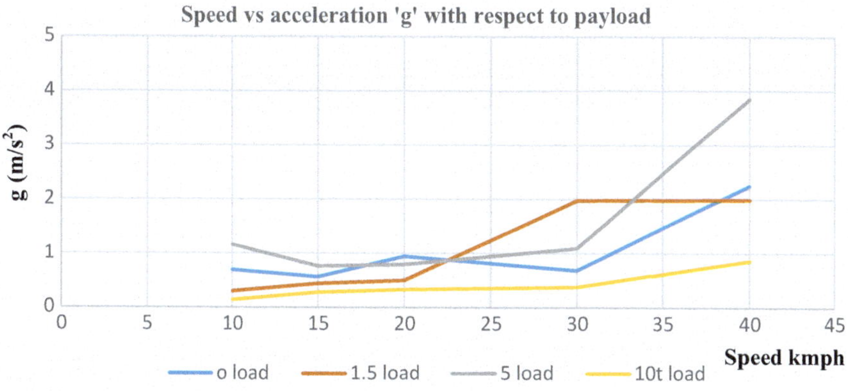

Fig. 10 Summary of test results (*X*-axis: speed & *Y*-axis: '*g*' levels in m/s^2)

Bibliography

1. Turkay S, Akcay H (2005) A study of random vibration characteristics of the quarter-car model. J Sound Vibr 282:111–124
2. Verros G, Natsiavas S, Papadimitriou C (2005) Design optimization of quarter-car models with passive and semi-active suspensions under random road excitation. J Sound Vibr 11:581–606
3. Andrzejewski R, Awrejcewicz J (2005) Nonlinear dynamics of a wheeled vehicle. Springer, New York
4. Verros G, Natsiavias S, Stepan G (2000) Control and dynamics of quarter-car models with dual-rate damping. J Sound Vibr 6:1045–1063
5. Lee J, Thompson DJ, Yoo HH, Lee JM (2000) Vibration analysis of a vehicle body and suspension system using a substructure synthesis method. Int J Vehicle Des 24:360–371
6. Litak G, Borowiec M, Hunicz J, Koszaka G, Niewczas A (2009) Vertical vibrations of a delivery car excited by railway track crossing. Chaos, Solitons Fractals 42:270–276
7. Gavriloski V, Jovanova J (2010) Dynamic behaviour of an air spring element. J Mach Technol Mater 4–5:24–27
8. Balaguru S, Deenadayalan K, Murali Vela, Chellapandi P (2014) Influence of welding speed over dilution for circular grid plate Hardfaced with Colmonoy-5. Appl Mech Mater 565:53–58
9. Abdelghaffar A, Hendy A, Desouky O, Badr Y, Abdulla S, Tafreshi R (2014) Effects of different tire pressures on vibrational transmissibility in vehicles. In: International conference on mechanical engineering and mechatronics Prague, Czech Republic, August 14–15 2014
10. Balaguru S, Murali V, Chellapandi P (2015) Effects of welding speeds on macro and microstructures in Hardfacing of Colmonoy on un—grooved and grooved 316 L(N) SS Base metal. Int J Appl Eng Res 10:25627–25631
11. Dalvi G, Chivkula P, Trivedi J, Abi M, Thomas G (2017) Experimental and simulation study to control stiffness of an air spring or bellow. Int J Mech Prod Eng 5(7). ISSN: 2320-2092
12. Sun J (2017) Calculation of vertical stiffness of air spring or bellow with FEM., 4th ANSA &μETA International Conference
13. Balaguru S (2019) Dynamics of machines. Cengage learning. 6th edn

Thermo Mechanical Analysis of Multipass Butt-Welded Joints by Finite Element Method

T. Raja, P. Anand, M. Sundarraj, M. Karthick, and A. Kannappan

Abstract Welding is an unswerving and competent metal joining process broadly used in industries. Residual stresses and distortion are most important challenges to the safety of system and constructions of butt-welded joints. The difficulties of the archaic measurement of residual stress are time intense and destructive. A three-dimensional fleeting thermomechanical analysis of butt-welded joint has been performed using FEM method in two steps to forecast the residual stresses and distortion. The first step is fleeting thermal analysis that yields dynamic temperature distribution throughout weld and the plates. The next step is the mechanical analysis which yields the residual stress, strain and displacement. It comprises of moving heat source, metal plasticity and elasticity has done during the material deposition, temperature dependent on material properties, element true and death technique was used for filler metal deposition in ANSYS software.

Keywords Dynamic temperature distribution · Residual stresses · Distortion · Butt-welded joints · ANSYS software

1 Introduction

Arc welding method is used to join the plates in butt joints, [1]. The results of tensile stress with the depositions of weld beads and the relation between maximum temperature and the residual stress in the welded metal pass were discussed. X-ray diffraction technique was used to measure stresses in the welded plate, [2]. The nodal displacements were measured in both welding simulations. The elastic-plastic analysis was used to measure structural displacement of the material [3]. In another work, three-dimensional model is used to predict the residual stress and distortion induced in laser beam welding of circumferential-welded pressure vessel simulated in aluminium butt-welded joint, [4]. To measure the residual stress in pressure vessel

T. Raja (✉) · P. Anand · M. Sundarraj · M. Karthick · A. Kannappan
Department of Mechanical Engineering, Vel Tech Rangarajan Dr. Sagunthala R&D Institute of Science and Technology, Chennai 600062, India
e-mail: rajasd28@gmail.com

L.-J. Yang et al. (eds.), *Proceedings of ICDMC 2019*,
Lecture Notes in Mechanical Engineering,
https://doi.org/10.1007/978-981-15-3631-1_15

joint used keyhole drilling method, numerical results are agreed with experimental results and analysed the residual stresses on a welded tee joint plate using contour method, [5]. Plain carbon steel plates and gas metal arc welding process are used. It is observed cutting a welded joint, deformations occurred at the cut surfaces as a result, relaxation of residual stresses are measured, [6]. The TIG welding distortions in austenitic stainless steel welded plate are compared with numerical results. In this analysis, coordinate measuring machine was used to measure the distortion in thin welded plates, [7]. A residual stress and distortion are generated when welding a flat bar-stiffened steel plate. In this analysis, Gaussian surface heat flux was used as a moving heat source. Combined convection and radiation boundary conditions are used for heat losses during and after welding, [8]. From the above motivation, to determine the residual stress, structural and thermal analysis of butt welded joints for 304 duplex stainless steel material plate by applying element.

2 Experimental Method

In this work, the arc welding was used for joining the two plates. The temperature is more in the area near the welding torch; this area expands more than regions further away. During the heat input, stresses in the area near to welded joint are compressively in plastic stage because the thermal growth in this area is controlled by adjacent metal with less temperature and more yield stress. The process of welding has been finished, and welded plate starts to cool; then, its deformation was in reverse direction. As a result of compressive plastic strains was formed in the area near welded region, the deformation of welded plate during and after the joining process is shown in Fig. 1.

Both the residual stresses and deformations are associated with each other. These residual stresses exist in a body if there are no external loads or body forces. Plastic deformation is one of the sources which occur in many manufacturing processes.

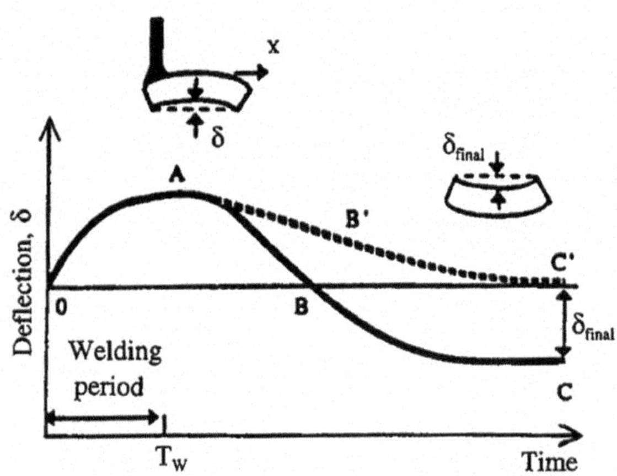

Fig. 1 Deformation during and after weld

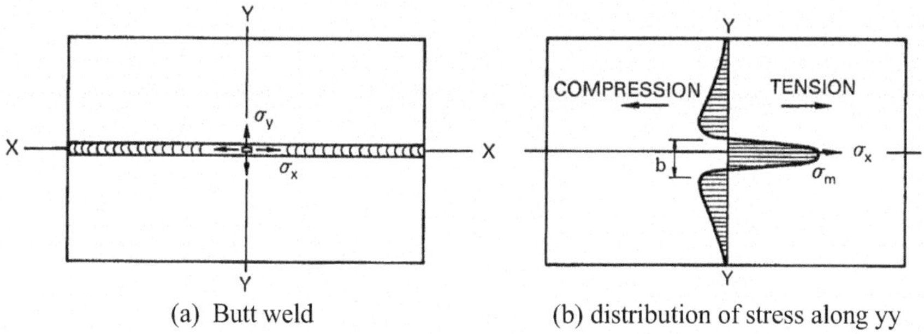

Fig. 2 **a**, **b** Residual stress distribution in butt-welded joint

Non-uniform plastic deformation occurs, if an initially stress-free body is subjected to loading, [9]. The body will elastically unload when these loads are removed. After all loads were removed, the stresses remaining are the residual stresses, which are smaller or equal to than the yield stress existing after the weld has completely cooled. Distortion and residual stresses are two faces of same problem, namely thermally induced plasticity. To control the residual stress and distortion is crucial in manufacture of welded structure. A typical distribution of longitudinal residual stress in butt-welded joint is shown in Fig. 2a, b; these stresses are parallel to the welding direction.

In this experimental study, duplex stainless steels (DSS) are used as welding joint. The matching filler metal is used where a welding was performed, whereas welds made with the filler metal enriched with nickel were used in as-welded condition. The weld metal microstructure from a composition exactly matching that of the parent steel will contain high ferrite content. The increase in nickel is made to improve the as-welded phase balance and increase austenite content. Furthermore, traditional trial and error approach is costly and time consuming. Numerical simulations based on finite element (FE) models provide a very suitable tool for investigating the thermal and mechanical consequences of welding process. FE simulation techniques for welding processes have been a major topic in welding research for several years, [10–12].The availability of 64-bit high-performance computing machines and enhanced finite element computational techniques has made it possible to simulate temperature fields developed from welding process. The computer simulation of welding processes enables the welding engineers to predict transient and residual stress fields and deformation behaviour of welded structures. These can be further used for the evaluation of structural misalignments and premature failures due to in service loads and weld-induced residual stresses.

3 Result and Analysis

3.1 3D Modelling of Welding Joint

In these modelling techniques, solid modelling as well as finite element modelling is created. In solid modelling, the graphical entities are used such as key points, lines, areas and blocks (Fig. 3).

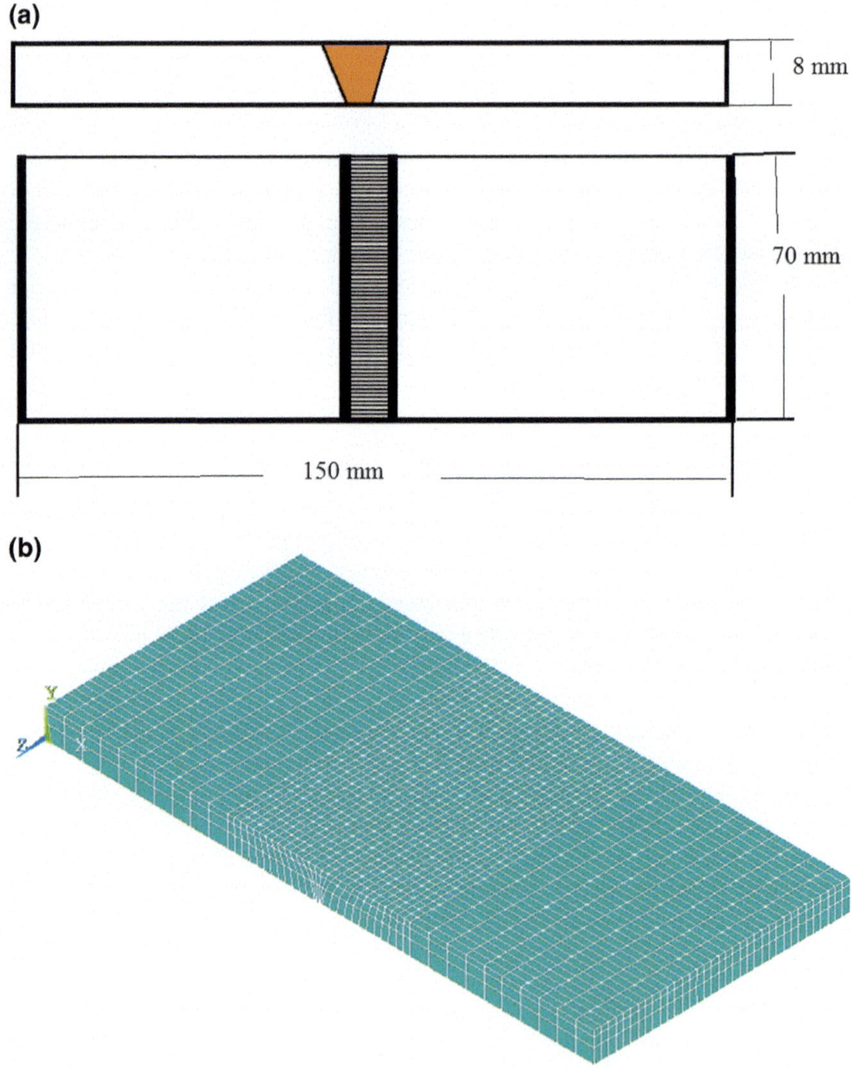

Fig. 3 **a** CAD model of weld plate, **b** finite element model of weld plate

3.2 Heat Source Modelling

A heat input during welding process with particular path on top of the work piece material fusion process was done. Residual stresses, distortion and reduced strength of structures directly from the thermal series are caused by localized strong heat input (Table 1).

General heat conduction for constant thermal conductivity:

$$\frac{\partial^2 T}{\partial x^2} + \frac{\partial^2 T}{\partial y^2} + \frac{\partial^2 T}{\partial z^2} + \frac{q_g}{k} = \frac{\delta c}{k}\frac{\partial T}{\partial \tau}$$

$$\frac{\partial^2 T}{\partial x^2} + \frac{\partial^2 T}{\partial y^2} + \frac{\partial^2 T}{\partial z^2} + \frac{q_g}{k} = 0$$

$$\frac{\partial^2 T}{\partial x^2} + \frac{\partial^2 T}{\partial y^2} + \frac{\partial^2 T}{\partial z^2} = 0$$

The finite differential equation for x-, y- and z-axis is

$$\frac{\partial^2 T}{\partial x^2} = \frac{T(i+1,j,k) - 2T(i,j,k) + T(i-1,j,k)}{\Delta x^2}$$

$$\frac{\partial^2 T}{\partial y^2} = \frac{T(i,j+1,k) - 2T(i,j,k) + T(i,j-1,k)}{\Delta y^2}$$

$$\frac{\partial^2 T}{\partial z^2} = \frac{T(i,j,k+1) - 2T(i,j,k) + T(i,j,k-1)}{\Delta z^2}$$

Figure 4 shows the three-dimensional view of heat sources. From the figures, the heat flux is maximum at centre of the arc. The heat flux value gradually decreased away from the centre arc (Figs. 5 and 6).

Table 1 Experimental results for heat source model

x-coordinate from centre of arc (1×10^{-3} m)	z-coordinate from centre of arc (1×10^{-3} m)	Heat flux (W/m^2)
0	0	198,625,369
1	1	101,977,664
2	2	13,801,176
3	3	492,343
−1	−1	101,977,664
−2	−2	13,801,176
−3	−3	492,343

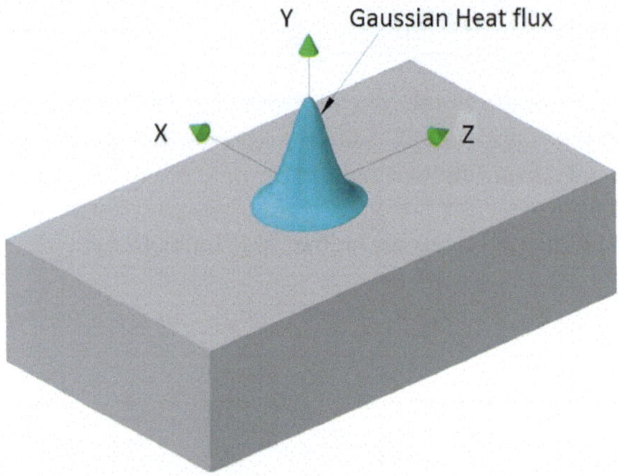

Fig. 4 Distribution of heat source

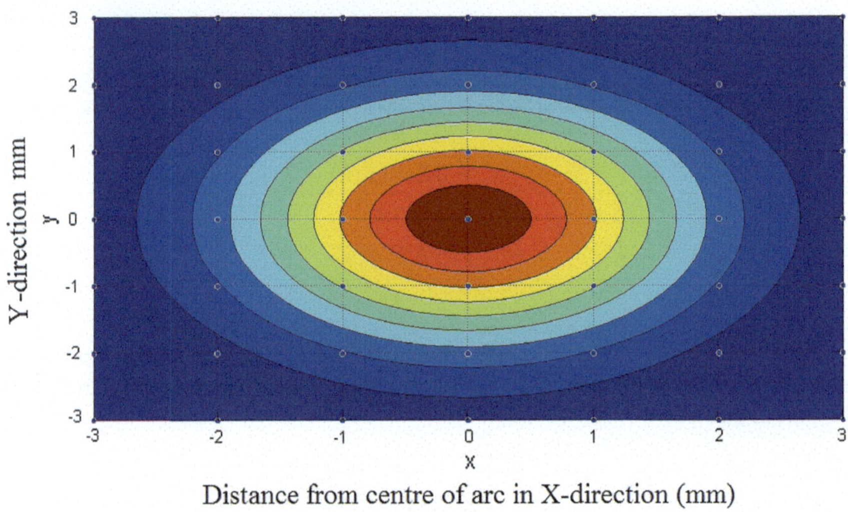

Fig. 5 Two-dimensional view of heat source model

3.3 Factor Birth and Death Process

The element death effect, by the ANSYS software, deactivates the element by multiplying their stiffness by a reduction factor. That the severe factor was set to 1.0E-6 by default, [13, 14]. Similarly, the other elements like specific heat, damping and mass deactivated by the elements are set to zero. The mass and energy of killed elements are not added over the model, and the strain value also is to zero; then, the

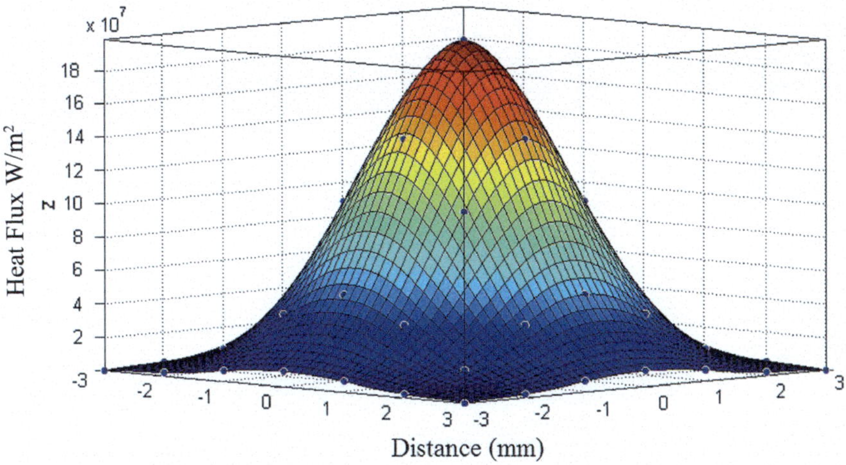

Fig. 6 Three-dimensional view of heat source

process of element deactivated was quick reaction. The element birth process was not actually added to the model; the elements are reactivated like, temperature, mass, stiffness. So that the element birth technique which was the factor of the model is simply reactivated, and the death technique which was the factor is deactivated from the model.

Figure 7 shows the FEM of weld plate with weld pool elements. Figure 8 shows that the weld pool elements are deactivated for filler metal deposition and temperature of plate is 300 K. The temperature distribution analysis during the welding and cooling at different time periods is shown in Figs. 7 and 8.

Welded plate returns to room temperature after cooled for more than one hour. The temperature is suddenly dropped after the welding, but more time is taken to drop temperature from 350 to 300 K.

3.4 Solving the Structural Analysis

The nodal temperatures were obtained from thermal analysis which is applied for the structural analysis to predict the residual stress and distortion in welded joint. In thermal analysis, nodal temperature distribution alone predicted, [15, 16]. In this analysis, full Newton–Rapson method is used to update the stiffness matrix. It is also used in thermal analysis.

The total strain was calculated by assembling the elastic strain, plastic strain and thermal strain. For the elastic-plastic analysis or thermomechanical analysis for calculating the elastic and plastic strains, rate-independent bilinear isentropic

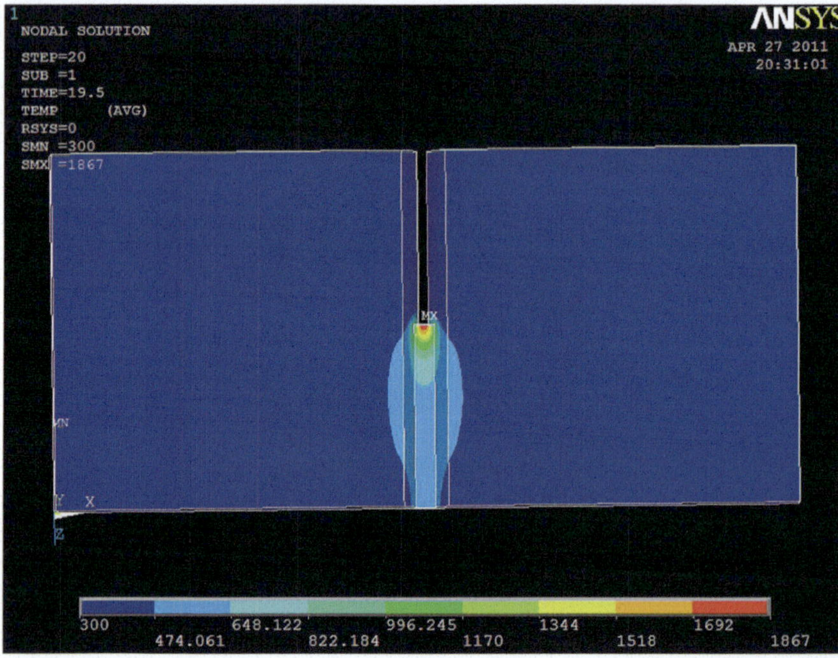

Fig. 7 Temperature distribution of first pass $t = 19.5$ s

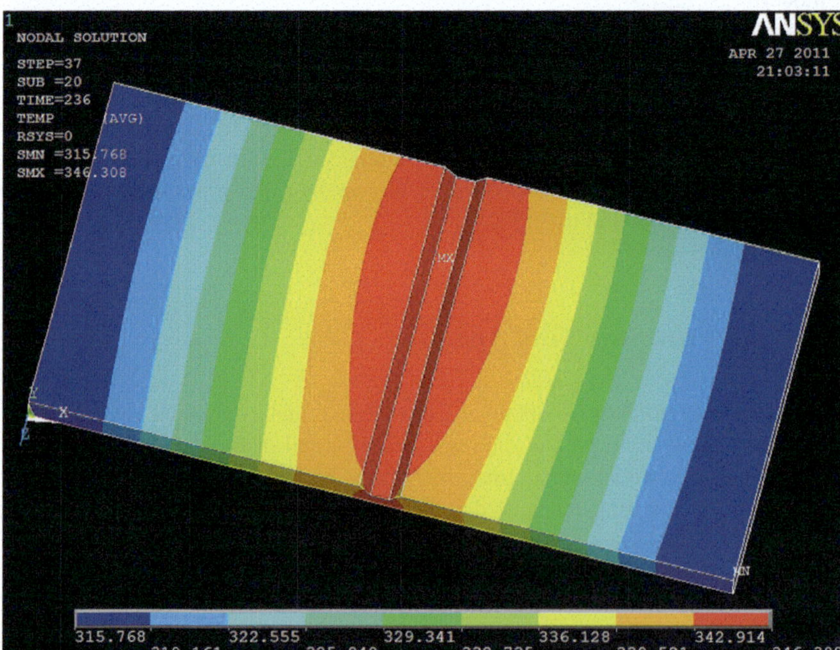

Fig. 8 Temperature distribution during cooling $t = 236$ s

hardening rule was considered. The thermal strain was calculated by its thermal expansion described in equation.

During the heating phase, the metal is heated from room temperature to higher temperature. So, the metal expands during the heating phase due to its coefficient of thermal expansion; distorted weld plate during heating phase is shown in Fig. 8. During the cooling phase, hotter regions in the weld plate produce shrinkage. Due to this shrinkage, the metal is distorted in the opposite direction of heating phase, shown in Fig. 9. This is called as welding distortion. The vertical displacement of welded plate is shown in Fig. 10. In this vertical displacement, both edge plates move upwards because the centre of weld plate is the hotter region during welding. During cooling, the metal in heat-affected zone (HAZ) shrinks, so the edges of plate move upward.

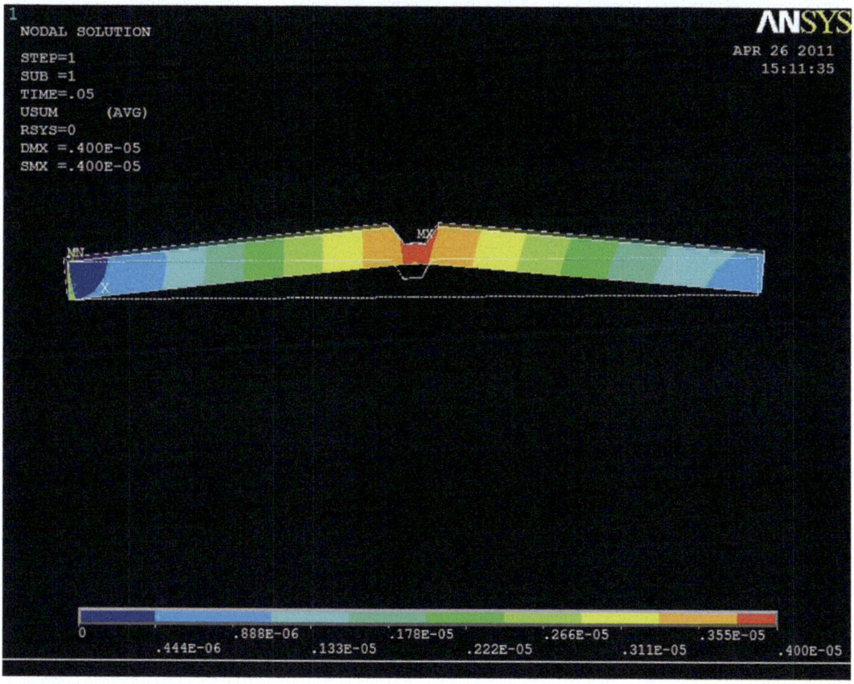

Fig. 9 Distortion during heating

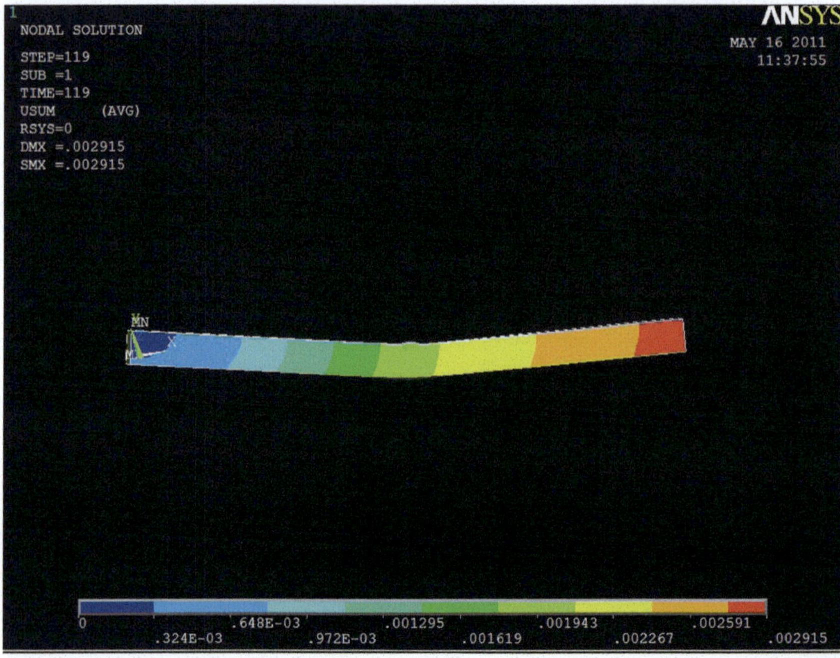

Fig. 10 Distortion after cooling

4 Conclusion

- A three-dimensional transient thermomechanical simulation of welded butt joint has been performed using finite element method. It also includes a moving heat source, material deposit, temperature dependant material properties, metal plasticity and elasticity. Element birth and death technique was used for filler metal deposition.
- During the heating phase, the metal is distorted in opposite direction of final distortion. The vertical displacement of welded plate is shown in Fig. 11. In this vertical displacement, the edges of the plate bend upwards look like *V* shape. The maximum vertical displacements 2.915 mm are shown in Fig. 10.
- The peak value often residual stress by the simulation is 402 MPa and the compressive residual stress 122 MPa. Yield stress of duplex stainless steel at room temperature is 470 MPa. The peak value often residual stress obtained by the simulation is almost equal to the yield stress of the material at normal ambient temperature. From the observation, the compressive residual stress region was a smooth curve than the tensile stress region. Coarser elements were used in weld zone to reduce the CFD weld time.

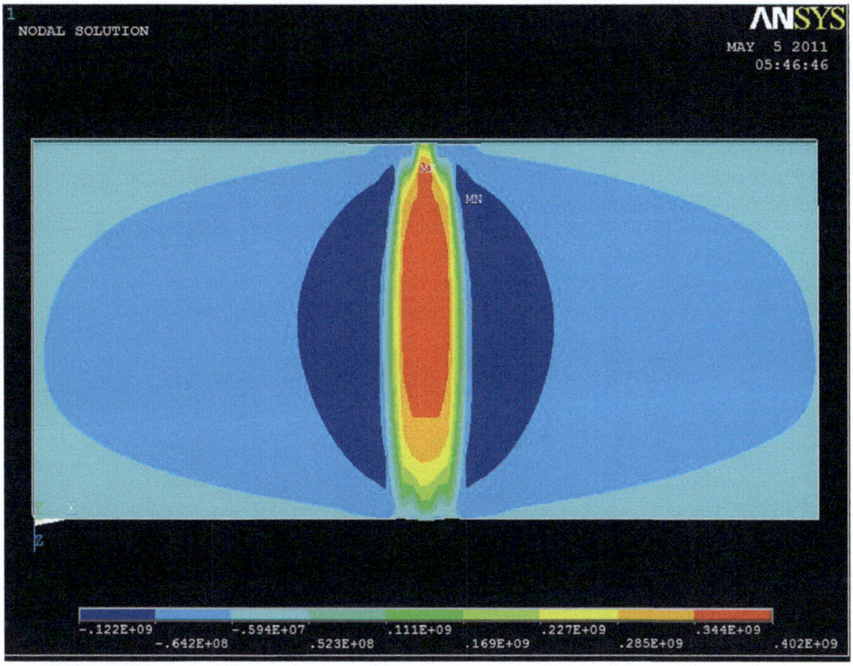

Fig. 11 Residual stress distribution in a welded plate

References

1. Withers PJ (2007) Residual stress and its role in failure. Rep Prog Phys 70(12):2211–2264
2. James MN (2011) Residual stress influences on structural reliability. Eng Fail Anal 18(8):1909–1920
3. Caron C, Heinze C, Schwenk M, Rethmeier SS, Babu J (2010) Lippold, Effect of continuous cooling transformation variations on numerical calculation of welding-induced residual stresses. Weld J 89:151–160
4. Neubert S, Pittner A, Rethmeier M (2017) Influence of non-uniform martensitic transformation on residual stresses and distortion of GMA-welding. J Constr Steel Res 128:193–200
5. Guo Y, Wu D, Ma G, Guo D (2014) Numerical simulation and experimental investigation of residual stresses and distortions in pulsed laser welding of hastelloy C-276 thin sheets. Rare Met Mater Eng 43(11):2663–2668
6. Yaghi AH, Hyde TH, Becker AA, Williams JA, Sun W (2005) Residual stress simulation in welded sections of P91 pipes. J Mat Process Technol 167:480–487
7. Lee CH, Chang KH (2012) Temperature fields and residual stress distributions in dissimilar steel butt welds between carbon and stainless steels. Appl Therm Eng 45–46:33–41
8. Moraitis GA, Labeas GN (2008) Residual stress and distortions calculation of laser beam welding for aluminium lap joints. J Mater Process Technol 198:260–269
9. Moraitis GA, Labeas GN (2009) Prediction of residual stresses and distortions due to laser beam welding of butt joints in pressure vessels. Int J Press Vessel Pip 86:133–142
10. Fallahi A, Jafarpur K, Nami MR (2011) Analysis of welding conditions based on induced thermal irreversibility's in welded structures: Cases of welding sequences and preheating treatment. Sci Iran 18(3):398–406

11. Mrvar P, Medved J, Kastelic S (2011) Welding sequence definition using numerical calculation. Weld J 90(8):148s–151s
12. Gannon L, Liu Y, Pegg N, Smith M (2010) Effect of welding sequence on residual stress and distortion in flat-bar stiffened plates. Mar Struct 23(3):385–404
13. Brown S, Song H (1992) Finite element simulation of welding of large structures. J EngInd 114(44):441–451
14. Deng D (2009) FEM prediction of welding residual stress and distortion in carbon steel considering phase transformation effects. Mater Des 30(2):359–366
15. Yaghi AH, Hyde TH, Becker AA, Sun W (2011) Finite element simulation of welded P91 steel pipe undergoing post-weld heat treatment. Sci Technol Weld Join 16(3):232–238
16. Venkatkumar D, Ravindran D, Selvakumar G (2018) Finite element analysis of heat input effect on temperature, residual stresses and distortion in butt welded plates. Mater Today Proc 5(2):8328–8337

Effect of Aspect Ratio on the Buckling Load of Stiffened Circular Plates

V. Lakshmi Shireen Banu and Veeredhi Vasudeva Rao

Abstract This paper deals with the buckling analysis of stiffened circular plates through numerical simulation techniques using Abaqus 6.5. In the present numerical investigation, the effect of aspect ratio (R/T) of a stiffened circular plate on the buckling strength is determined for different mode shapes. Three different geometric shapes, namely *T*, *I* and HAT, are considered to reinforce the plate by way of stiffeners. For all the cases studied in the present investigation, a circular plate with a radius of 0.5 m is subjected to in-plane loading. The boundary conditions are such that the edge is fixed but released in the radial direction during buckling. The results of the analysis for the radius-to-thickness ratio of the plate, in the range from 200 to 250, with four intermediate steps are presented in the form of buckling loads corresponding to different mode shapes in tabular and graphical form.

Keywords Buckling loads · Circular plates · Ring stiffeners · Aspect ratio · Thin shells

1 Introduction

Thin circular steel plates are widely used in the fabrication of steel structures such as cylindrical pressure vessels, storage containers in oil and gas companies, maritime and aerospace industry. While in service, these circular steel plates are subjected to mechanical loads and temperature variations leading to thermo-mechanical loads. As a result, the plates will undergo deformations due to buckling loads. A simple

V. Lakshmi Shireen Banu (✉)
Department of Civil Engineering, Malla Reddy Engineering College (Autonomous), Hyderabad 500100, India
e-mail: vlsbanu@yahoo.co.in

V. V. Rao
Department of Mechanical and Industrial Engineering, CSET, University of South Africa, Pretoria, Republic of South Africa
e-mail: v_vasudevarao@yahoo.com

L.-J. Yang et al. (eds.), *Proceedings of ICDMC 2019*,
Lecture Notes in Mechanical Engineering,
https://doi.org/10.1007/978-981-15-3631-1_16

approach to increase the buckling strength of the steel plate is to increase its thickness. However, increasing the thickness of the plate proportionately increases the weight and cost of the structure. Hence, to increase the buckling resistance, generally, stiffeners are attached on the face of the plates. For stiffeners, commercially available steel sections with standard geometric shapes like *T*, *I* and HAT are chosen from the open market. Reinforcement of plates with stiffeners increases the buckling strength without adding much to the self-weight for the structure.

Estimation of the buckling strength of a thin plate is possible through analytical solutions. But, it becomes difficult to determine the buckling load when stiffeners are attached to the plate. Therefore, one has to resort to either experiments or numerical simulations. It is known that conducting experiments is not only expensive but also time-consuming process. On the other hand, finite element-based numerical simulation offers a great advantage to incorporate a variety of stiffeners and boundary conditions. These simulations can be repeated several times to investigate the effect of different parameters. In the present investigation, it is proposed to determine the buckling load of a circular steel plate with three different stiffeners attached to the face of the plate through numerical simulations. For this purpose, Abaqus is employed, which is one of the powerful FE-based software tools for numerical simulations.

2 Literature Review

From the published literature, most pertinent studies conducted by the previous investigators are presented here to give a brief background. Rossettos and Yang [1] presented a treatment of asymmetric buckling problem of ring-stiffened circular plates. They indicated a coalescence of the lowest symmetric and asymmetric mode eigenvalues involving ring torsional stiffness. Grondin et al. [2] conducted a parametric study on buckling of stiffened steel plates. They considered the stability of stiffened plates using a finite element model. In their study, *T* -shaped stiffeners are used for buckling analysis and parameters included are (1) initial plate imperfections, (2) direction and magnitude of residual stresses, (3) slenderness ratio of the plate, (4) aspect ratio of the plate and (5) cross-sectional area ratio of plate to the stiffener.

Ma and Wang [3] investigated nonlinear bending and post-buckling of a functionally graded circular plate under direct mechanical and thermal loads. They used classical nonlinear von Karman plate theory, for axisymmetric large deflection and bending of a functionally graded circular plate. The types of loads considered in their investigation are mechanical, thermal and combined thermal–mechanical. Brubak and Hellesland [4] conducted an approximate buckling strength analysis for plates reinforced with stiffeners with arbitrary orientation. Ding and Lee [5] gave an analytical solution for a circular plate that is uniformly loaded and clamped at the edges.

Modarresi and Showkati [6] experimentally investigated the failure of diagonally stiffened circular plates to determine ultimate bending capacity. In their experiments, to strengthen the plate, two types of shallow and deep stiffeners of reverse *L* and rectangular cross sections were attached. They considered simply supported and clamped edge boundary conditions for a circular plate subjected to lateral loads up to their final collapse. They compared experimental results with the nonlinear analysis of stiffened circular plates using FEM. The results are presented through load–deflection and load–strain curves and volumetric charts. Finally, it was concluded that increasing the height of the diagonal stiffener or changing its cross-sectional shape may not be an effective way of increasing the load-bearing capacity of the diagonally stiffened circular plates. It was recommended that, instead of using one diagonal stiffener, it is better to use two perpendicular diagonal stiffeners with the same amount of steel as the first one.

Rao and Rao [7] presented an exact solution for elastic buckling of circular plates resting on internal elastic ring support and elastically restrained against translation at the outer edge. They used classical plate theory to solve the governing differential equations. Further, they introduced a Mathematica code to estimate the buckling load. Tran et al. [8] studied the buckling behaviour of stiffened curved panels under uniform longitudinal compression to address the linear buckling and the ultimate strength of panel which are both influenced by the combined effect of curvature and stiffening. They finally proposed a design methodology for stiffened flat plates to be adopted by European Standards. They concluded that the plate curvature has to be taken into consideration to optimize the design.

Daniel [9] attempted the stability problem of thin circular plates that are stiffened with a thin cylindrical shell on the circumference and also on the plate. In his investigation, an attempt was made to determine the in-plane axisymmetric critical load with a view to show the effect of cylindrical shell stiffener. Daniel [10] summarized the differential equations for the stability of shell-stiffened circular plates and shell-stiffened annular plates assuming the deformations are axisymmetric. Yu et al. [11] focused their attention on the uniaxial ultimate compressive strength of stiffened panels with openings. In their work, experiments were conducted on longitudinal stiffened panels with rectangular openings applying combined loads. Numerical simulations were carried out on stiffened panels considering large elastoplastic deformation behaviour using Abaqus.

From the literature review, it is found that there are a very few papers, but not too many, focusing on the buckling of stiffened circular plates. But, investigations concerning the buckling load as a function of the plate aspect ratio, particularly when the plate is stiffened, are very much limited in the open literature. Therefore, in the present investigation it is proposed to determine the effect of aspect ratio on the buckling load of a stiffened circular plate using the protocols specified in the user's manual of Abaqus 6.5 [12].

3 Analytical Solution

In process industry, thin circular plates are subjected to in-plane radial compressive forces from the supporting structure when there is a change in the working temperature of the medium. As a result, buckling of the circular plates can take place. In the following section, governing differential equations for plates subjected to buckling loads are presented.

Circular plate subjected to uniformly distributed in-plane compressive radial forces:

As shown in Fig. 1, in the present analysis, a circular plate is considered that is subjected to in-plane radial compressive forces (q_r). In the buckling analysis, we include only axisymmetric geometries such as circular plates. Thus, we can use the polar coordinate system represented by r and Ø. Equation (1) that represents buckling of a rectangular plate can be transformed into a circular plate for convenience of analysis.

$$\frac{\partial^4 \mathbf{w}}{\partial \mathbf{x}^4} + 2\frac{\partial^4 \mathbf{w}}{\partial \mathbf{x}^2 \partial \mathbf{y}^2} + \frac{\partial^4 \mathbf{w}}{\partial \mathbf{y}^4} = \frac{1}{\mathbf{D}}\left(\mathbf{N_x}\frac{\partial^4 \mathbf{w}}{\partial \mathbf{x}^2} + 2\mathbf{N_{xy}}\frac{\partial^4 \mathbf{w}}{\partial \mathbf{x}\partial \mathbf{y}} + \mathbf{N_y}\frac{\partial^4 \mathbf{w}}{\partial \mathbf{y}^4}\right) \tag{1}$$

For a specific case, when the structure is in equilibrium, for an axisymmetric loading condition,we have

$$\mathbf{Nx} = \mathbf{Ny} = \mathbf{Nr} = -\mathbf{qr},\ \mathbf{Nxy} = \mathbf{0} \tag{2}$$

Denoting

$$\mu^2 = \frac{q_r}{D} \tag{3}$$

and using the relations between the polar and Cartesian coordinates, we can derive an equation for axisymmetric circular plate subjected to a circumferential compressive force as shown in Fig. 1, as given below.

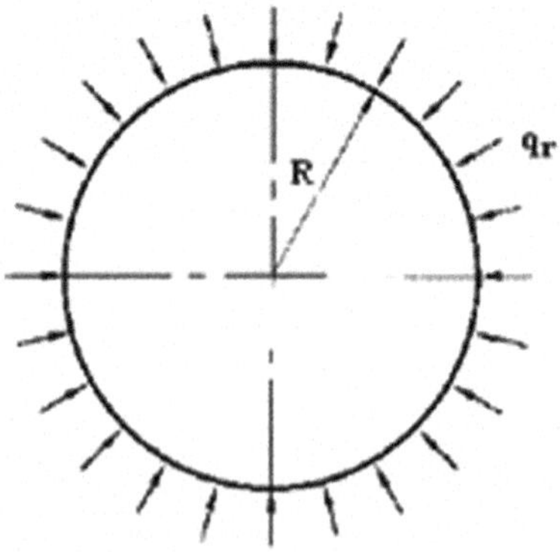

Fig. 1 Circular plate subjected to uniformly distributed in-plane compressive radial forces

$$\frac{d^4w}{dr^4} + \frac{2}{r}\frac{d^3w}{dr^3} - \frac{1}{r^2}\frac{d^2w}{dr^2} + \frac{1}{r^3}\frac{dw}{dr} + \mu^2\left[\frac{d_w^2}{dr^2} + \frac{1}{r}\frac{dw}{dr}\right] = 0 \quad (4)$$

Now, let us introduce the following new variable:

$$\boldsymbol{\rho} = \mu \mathbf{r} \quad (5)$$

that denotes a non-dimensional polar radius. By means of new variable ρ, it is possible to rewrite Eq. (4), into the form of Eq. (6).

$$\frac{d^4w}{d\rho^4} + \frac{2}{\rho}\frac{d^3w}{d\rho^3} + \left(1 - \frac{1}{\rho^2}\right)\frac{d^2w}{d\rho^2} + \frac{1}{\rho}\left(1 + \frac{1}{\rho^2}\right)\frac{dw}{d\rho} = 0 \quad (6)$$

Equation (6) is a linear fourth-order, homogeneous differential equation. The general solution for Eq. (6) is

$$w(\rho) = C_1 + C_2 \ln \rho + C_3 J_0(\rho) + C_4 Y_0(\rho) \quad (7)$$

where $J_0(\rho)$ is Bessel's function of the first kind and $Y_0(\rho)$ is Bessel's function of the second kind both for zeroth order. In Eq. (7), C_1, C_2, C_3 and C_4 are constants of integration. Since $w(\rho)$ must be finitely quantified for all values of ρ, including $\rho = 0$, then the two terms with ρ and $Y_0(\rho)$, having singularities at $\rho = 0$, must be dropped for the solid plate because they approach infinity, when $\rho = 0$. Thus, for the solid circular plate, Eq. (7) must be taken in the form

$$w(\rho) = C_1 + C_3 J_0(\rho) \quad (8)$$

to determine the critical buckling loads q_r. These loads are assumed to be acting at the mid-plane of the circular plate.

4 Finite Element Modelling

FE modelling of the plate with *T*, *I* and HAT stiffeners is carried out using Abaqus software version 6.5. From Abaqus library, a four-node, doubly curved, linear shell element is selected for the geometric modelling of the structure under consideration. To maintain an acceptable aspect ratio for the element, shell elements were preferred over three-dimensional elements. The choice of the shell elements will also ensure a reasonable mesh density. Stiffeners of three different geometric shapes *T*, *I* and HAT are chosen for reinforcement of the circular plate. As shown in Fig. 2, on each plate any one type of ring stiffeners is placed at three different radial locations, the first one at 0.1 m, the second one at 0.25 m and the third one at 0.4 m. Mild steel properties are chosen for plate and stiffeners during the analysis. The in-plane compressive load of 20 kN is considered acting on the circular plate with stiffeners. As shown in

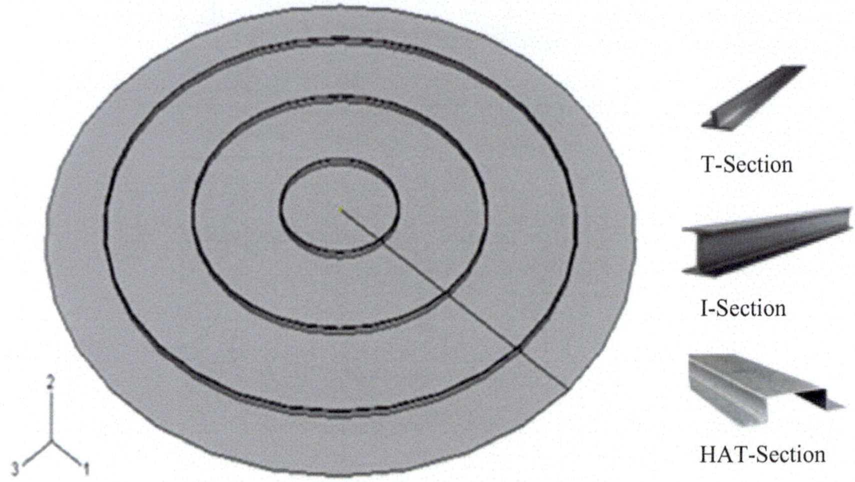

Fig. 2 Circular plate with three commercially available stiffener sections

Fig. 3, loading and the boundary conditions are such that the plate is fixed along the edge while released in the radial direction. Abaqus 6.5 is employed to carry out FE analysis on the structure under consideration.

In the present numerical simulations, radius of the circular plate is fixed at 0.5 m. The thickness of the plate is varied such that the radius-to-thickness (R/T) ratio is in the range of 200–250 with an increment of 10 each. Height and thickness of the stiffeners are held constant at 0.007 m and 0.002 m, respectively.

Material Properties of Mild Steel:

Young's modulus $E = 200$ GPa,
Poisson's ratio = 0.3.

Fig. 3 Loading and the boundary conditions for stiffened circular plate

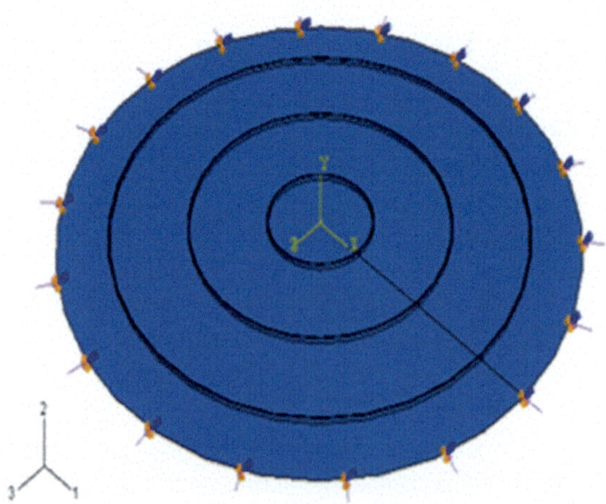

5 Results and Discussion

The results of the simulation studies to determine the buckling load of the stiffened circular plate as a function of aspect ratio in the range of investigation from 200 to 250 in steps of 10 are presented in Tables 1, 2 and 3. It is observed that for a given mode of deformation the buckling load decreases with the increase of aspect ratio. It is also observed that for any given aspect ratio in the range of investigation, the buckling load increases with the progressive mode of deformation from mode-1 to mode-6. Further, the plate with *T*-shaped stiffener has the lowest buckling strength for all aspect ratios and all modes of deformation. The plate with HAT-shaped stiffener showed the largest buckling load for all conditions. The plate with *I*-shaped stiffener has larger buckling loads compared to the plate with *T*-shaped stiffeners but marginally less buckling load when compared to the plate with HAT-shaped stiffeners. However, the buckling loads for *I*-shaped and HAT-shaped stiffeners are closely comparable to each other. One set of results on buckling loads of plate with *T*-, *I*- and HAT-shaped stiffeners are presented in Table 4 for comparison. Typical shapes of deformation of plates with *T*-, *I*- and HAT-shaped stiffeners are shown in Figs. 4, 5 and 6, respectively.

From the results in Figs. 7, 8 and 9, it is observed that the buckling load curves of the stiffened circular plates are seen in pairs. Buckling load curves for mode-1 and

Table 1 Buckling load of the circular plate with *T* stiffener under different modes for various R/T ratios

R/T	Buckling load (kN)					
	Mode-1	Mode-2	Mode-3	Mode-4	Mode-5	Mode-6
200	31.798	40.14	68.706	69.884	85.864	95.466
210	28.214	35.164	60.198	61.538	75.854	83.714
220	25.206	31.01	53.096	54.532	67.464	73.838
230	22.69	27.556	47.19	48.672	60.456	65.57
240	20.596	24.696	42.302	43.794	54.626	58.684
250	18.8638	22.344	38.284	39.756	49.808	52.988

Table 2 Buckling load of the circular plate with *I* stiffener under different modes for various R/T ratios

R/T	Buckling load (kN)					
	Mode-1	Mode-2	Mode-3	Mode-4	Mode-5	Mode-6
200	42.008	46.976	81.142	84.204	108.112	109.956
210	38.042	41.83	72.244	75.116	97.126	97.284
220	34.668	37.504	64.764	67.396	86.568	87.79
230	31.806	33.878	58.496	60.86	77.546	79.874
240	29.384	30.85	53.262	55.352	69.99	73.188
250	27.35	28.336	48.92	50.744	63.706	67.574

Table 3 Buckling load of the circular plate with HAT stiffener under different modes for various R/T ratios

R/T	Buckling load (kN)					
	Mode-1	Mode-2	Mode-3	Mode-4	Mode-5	Mode-6
200	42.1	47.57	82.278	85.08	109.512	111.058
210	38.148	42.442	73.406	76.026	98.426	98.596
220	34.778	38.12	65.928	68.314	87.724	89.3
230	31.916	34.488	59.648	61.77	78.696	81.404
240	29.454	31.452	54.392	56.244	71.122	74.726
250	27.456	28.926	50.024	51.612	64.816	69.114

mode-2 are close to each other and appear like a pair of curves. Likewise, buckling load curves for mode-3 and mode-4, and mode-5 and mode-6 formed another two pairs of curves. This observation is consistent for all three types of stiffeners considered in the present investigation. Figures 10 and 11 show the variation of buckling load as a function of plate aspect ratio for T, I and HAT stiffeners in mode-1 and mode-2, respectively. It is further observed that there is not much of a difference between buckling loads for plates with I and HAT stiffeners.

6 Conclusions

In the present investigation, buckling load for stiffened circular plates is presented for six modes of deformation as a function of plate aspect ratio in both tabular and graphical forms. From the present analysis, it is found that the circular plate with HAT-shaped stiffener offers highest buckling strength. The plate with *T*-shaped stiffener offers lowest buckling strength. However, the plate with I-shaped stiffener offered buckling strength in between the plates with *T*- and HAT-shaped stiffeners. Further, the buckling strength of plates with HAT-shaped and *I*-shaped stiffeners is closely comparable. It is concluded that the geometric shape of the stiffener is very important since it can affect the buckling load and overall buckling behaviour of the plate considerably. Through this investigation, the capabilities of Abaqus software are demonstrated to determine buckling loads of stiffened plates. It is understood that the predictive capabilities of the numerical models and software tools greatly depend on the proper definition of the problem and well-defined boundary conditions specific to a problem.

Table 4 Buckling load of circular plates with stiffeners of different geometric shapes

Stiffener	Buckling load (kN)										
	Mode-1						Mode-2				
	200	210	220	230	240	250	200	210	220	230	240	250
T shape	31.79	28.21	25.20	22.69	20.59	18.86	40.14	35.16	31.01	27.55	24.69	22.34
I shape	42.00	38.04	34.66	31.80	29.38	27.35	46.97	41.83	37.50	33.87	30.85	28.33
HAT shape	42.1	38.14	34.77	31.91	29.45	27.45	47.57	42.44	38.12	34.48	31.45	28.92

Fig. 4 Deformed shape of the circular plate with *T*-shaped stiffener

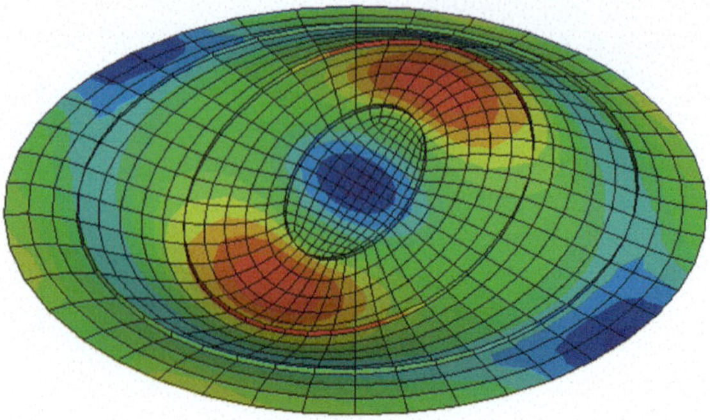

Fig. 5 Deformed shape of the circular plate with *I*-shaped stiffener

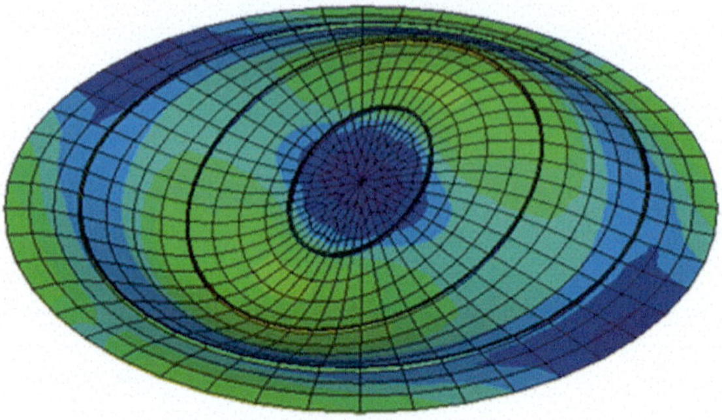

Fig. 6 Deformed shape of the circular plate with HAT-shaped stiffener

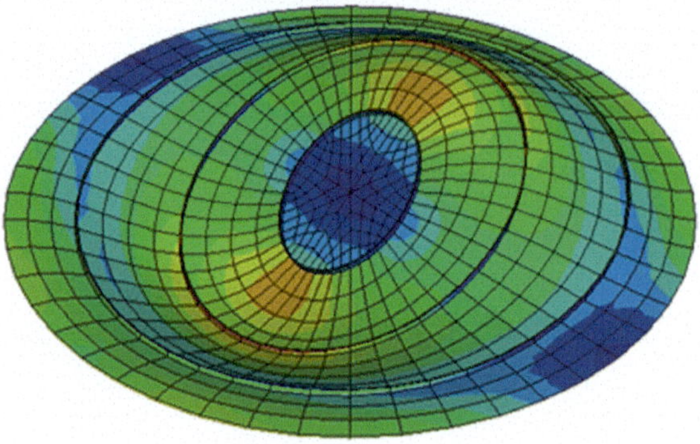

Fig. 7 Buckling load of circular plate with *T* stiffener under different modes for various R/T ratios

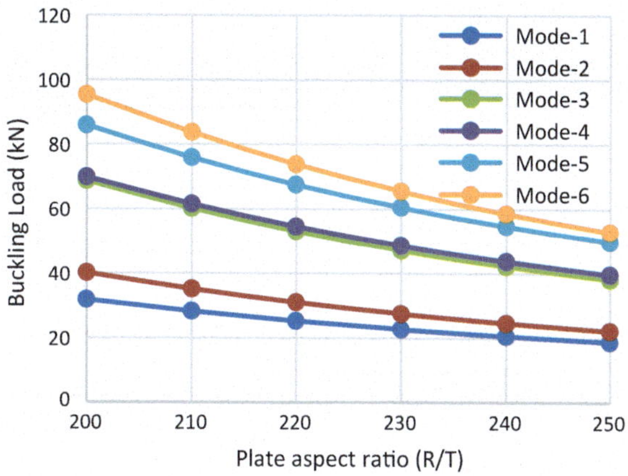

Fig. 8 Buckling load of circular plate with *I* stiffener under different modes for various R/T ratios

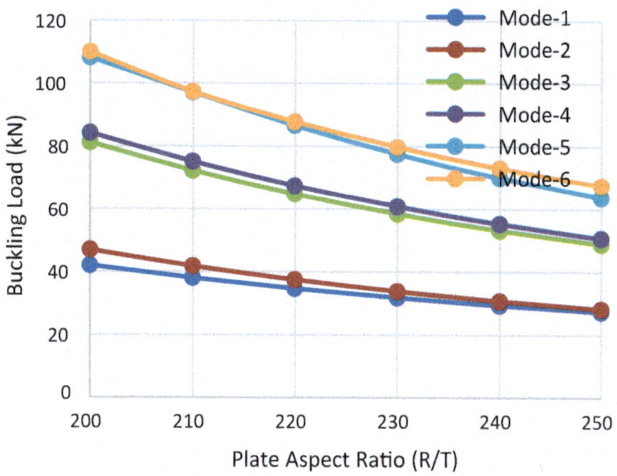

Fig. 9 Buckling load of circular plate with HAT stiffener under different modes for various R/T ratios

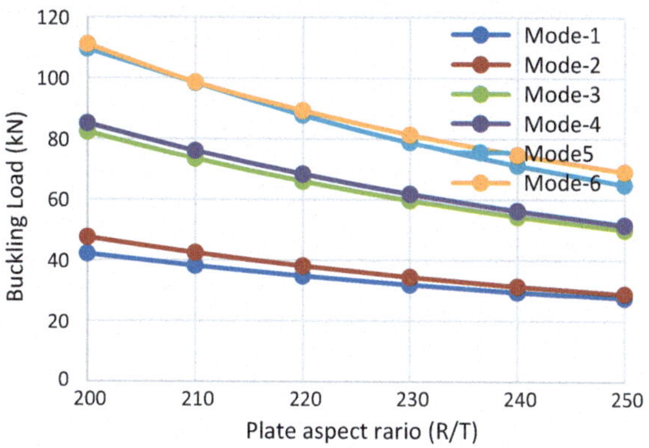

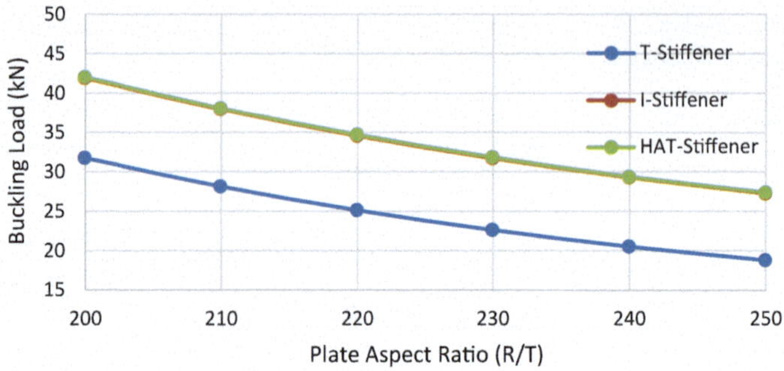

Fig. 10 Buckling load of circular plates in mode-1, with stiffeners of different geometric shapes

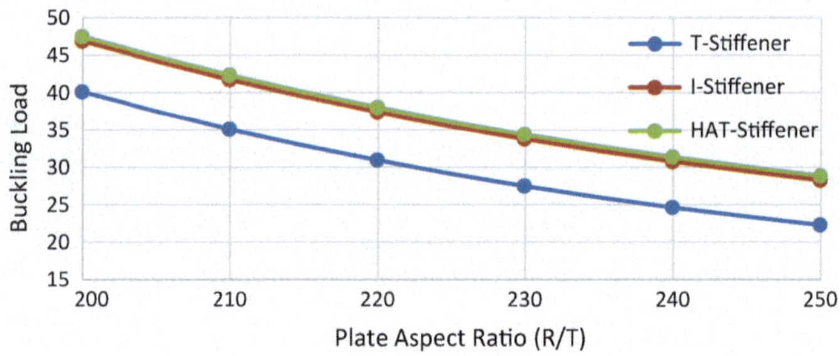

Fig. 11 Buckling load of circular plates in mode-2, with stiffeners of different geometric shapes

Acknowledgements The authors would like to thank their respective organizations MREC, India, and UNISA, Republic of South Africa, for the support extended towards this research work and publishing.

References

1. Rossettos JN Yang G (1984) Asymmetric buckling of ring stiffened circular plates. J Appl Mech 53(2):475–476
2. GrondinGY, Elwi AE, Cheng JJR (1999) Buckling of stiffened steel plates—a parametric Study. J Constr Steel Res 50:151–175
3. Ma LS Wang TJ (2003) Nonlinear bending and post-buckling of a functionally graded circular plate under mechanical and thermal loadings Int J Solids Struct 40:3311–3330
4. Brubak L, Hellesland J (2004) Approximate buckling strength analysis of plates with arbitrarily oriented stiffeners Univ Oslo 72:1163–1168
5. Ding H-j, Lee X-y (2005) Analytic solutions for a uniformly loaded circular plate with clamped edges. J. Zhejiang Univ Sci 72:1163–1168
6. Modarresi H, Showkati H (2011) Experiments on the bending failure of diagonally stiffened circular plates. Mat-Wiss u Werkstofftech 42(5):403–416
7. Rao LB, Rao CK (2012) Buckling of circular plates with an internal elastic ring support and outer edge restrained against translation. J Eng Sci Technol 7(3):393–401

8. Tran KL, Douthe C, Sab K, Dallot J, Davaine L (2014) Buckling of stiffened curved panels under uniform axial compression. J Constr Steel Res 103:140–147
9. Daniel B (2012) Effects of shell-stiffening on the stability of circular plates. Procedia Eng 48:46–55
10. Daniel B (2015) Buckling of circular plates with shell-stiffening on the boundary. J Comput Appl Mech 10(1):3–23
11. Yu C-Y, Feng J-C Chen K (2015) Ultimate uniaxial compressive strength of stiffened panel with opening under lateral pressure. Int J Naval Archit Ocean Eng 7:399–408
12. Hibbit K, Sorenson Inc (1995) ABAQUS User's Manual, Version 6.5

Experimental Study on Tyre Dynamics and Properties of Heavy Load Transporting Vehicle

S. Balaguru and J. Venkataramana

Abstract Tyres of heavy load transporting vehicle are stiffer than the other tyres to withstand heavy loads. Tyre dynamics and properties influence the dynamic qualities of vehicles and have an impact on energy consumption of vehicle (up to 7% due to tyre rolling resistance). Study of dynamics and properties of heavy tyre is nothing but study of performance of the tyre during motion with or without load. The handling performance of vehicle depends on the characteristics of tyre-road contact. The horizontal and vertical forces acting on the vehicle are transferred through tyres as a result of steering, braking and driving. The paper presents the results of the experimental study of tyre properties and dynamic forces and corresponding moments calculated by using various mechanical devices and empirical formulas. Test trails were conducted to explore the changes in dynamics with respect to tyre properties and road conditions by changing payloads on the transporting vehicle.

Keywords Rolling resistance · Energy consumption · Laden and unladen · Dynamics and test trails

1 Introduction

Tyres of the heavy load transporting vehicle are mostly bias (cross ply) tyres, the cords of body plies are laid at an angle of 30°–40° to the tyre axis. Bias tyre of the heavy vehicle had stiff sidewall due to high ply rating, resulting to less deflection of tyre, less road contact area and offers less rolling resistance. Because of these properties, bias tyres are less stable during high speeds compared to radial tyres. Study of the tyre properties and dynamics of the selected heavy load transporting vehicle during movement with road contact is more difficult, compared to the study

S. Balaguru · J. Venkataramana (✉)
Department of Mechanical Engineering, Vel Tech Rangarajan Dr. Sagunthala R&D Institute of Science and Technology, Avadi, Chennai 600054, India
e-mail: ramanajanipella@gmail.com

S. Balaguru
e-mail: iitmguru239@gmail.com

L.-J. Yang et al. (eds.), *Proceedings of ICDMC 2019*,
Lecture Notes in Mechanical Engineering,
https://doi.org/10.1007/978-981-15-3631-1_17

on tyre models. In this experimental study, various properties of the tyre were studied, viz. tyre deflection, hardness of sidewall and tread, change in temperature, dynamic forces on tyre in laden and unladen condition of vehicle.

The objectives of this experimental study are

- To study the tyre dynamics and properties in unladen and laden condition of transporter.
- To study the change in dynamics with respect to properties of tyre and road conditions at tyre-road contact.
- To evaluate the causes for wear out of tyre tread of the heavy load transporting vehicle.
- To evaluate the life period of tyres based on the results of the experimental study.

2 Specification of Heavy Load Transporting Vehicle

The transporting vehicle used for this study is equipped with electronically controlled self lifting and steering systems. These special features like self lifting and steering systems are useful for loading and unloading of heavy loads without the aid of lifting cranes, and also this vehicle is having different steering mode system. It consists of multi-mode driving, lifting and steering functions with special and safety features, viz. pipe break safety valve, tilt angle alarm, display centre of gravity point, etc.

Specification of the transporter:

Transporter make	M/s KAMAG-GERMANY
Self-weight and hauling capacity	45 and 200 Tons
Dimensions	14,000 × 4500 × 1300 mm
No. of axle units	18 (Drive axles: 6, brake axles: 7, idle axles: 5)
No. of tyres	36 (each axle unit having 2 tyres)
Tyre size and capacity	355/65-15 24PR, 7500 kgs @Inflation Pressure 10 bar.

3 Details of Tyre Designation

Designation details of tyre, viz. section width, aspect ratio, rim diameter and ply rating of the selected tyre 355-65/15 24 PR are shown in Fig. 1.

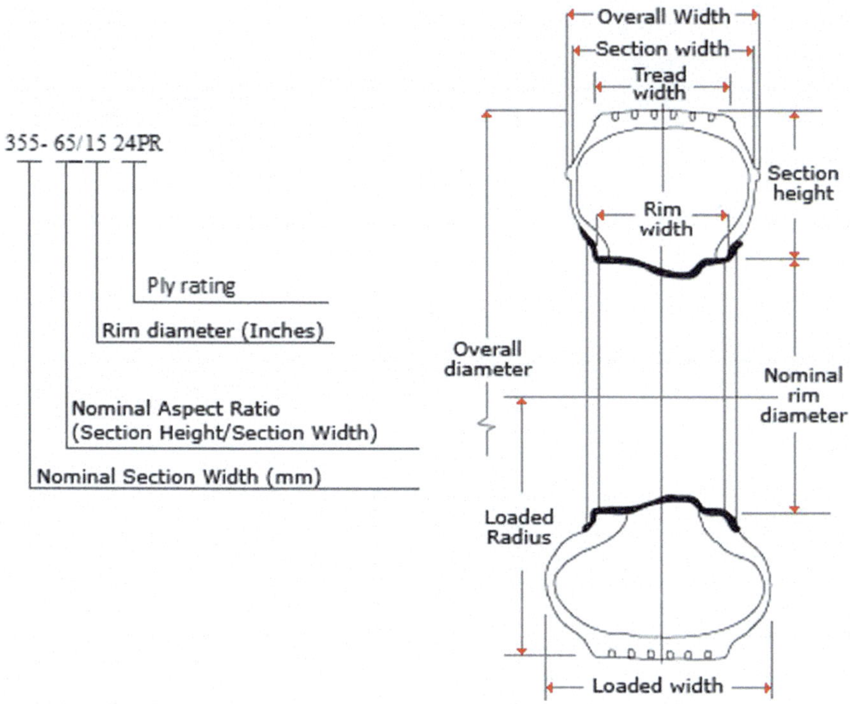

Fig. 1 Designation of tyre

4 Experimental Trails

Replacement philosophy of tyre is based on tyre wear out indication and change in tyre properties, viz. surface cracks, damages, etc. Tyres became brittle due to ageing and weathering effects, and this can be found out by measuring the hardness of tyre, stiffness of tyre and heat generation in the tyres during movement. For carrying out experimental study on this transporter, tyres of drive wheel, brake wheel and idle wheel were replaced with new tyres. During test trails the changes in hardness, deflection, wear out and temperature of tyre during movement with no load and loaded conditions are measured on these tyres. Deflection of tyre indicates the stiffness of the tyre which was measured by arranging fixture on disc by keeping the tyre inflation pressure at 10 bar is given in Table 2. The surface hardness of tyres increases due to ageing factor and wear out. Hardness of the tyre was measured using hardness meter (Shore A) at different loading schemes, and the movement of vehicle is given in Table 3. Wear out of tyre tread and surface temperature during movement of the transporter with no load and loaded conditions were measured using depth gauge and non-contact thermometer are given in Tables 1 and 4.

5 Measurement of Tyre Properties

See Tables 1, 2, 3 and 4.

Table 1 Tyre temperatures measured during the transportation of loads

S. No.	Details of test trail/load on transporter	Tyre temperature (°C) during transportation (@½ an hour interval)				
		@7.30 h	@8.30 h	@9.00 h	@9.30 h	@10.00 h
1	Trail-0 (self-weight only)	25	28	30	33	35
2	Trail-1 (with 25 Ton load)	28	31	35	37	41
3	Trail-2 (with 65 Ton load)	26.5	39	46	45	48
4	Trail-3 (with 165 Ton load)	29	35	42	46	50

Table 2 Deflections values of tyres under various loading schemes

S. No.	Transporter condition (Inflation pressure @10 bar)	Rolling radius ($R1$) (mm)	Loaded radius ($R2$) (mm)	Deflection ($\delta = R1 - R2$) (mm)
1	No load on transporter (deflection due to self-weight 45 Ton)	394	389	5
2	Load of 25 Ton	394	383	11
3	Load of 65 Ton	394	372	22
4	Load of 165 Ton	394	351	43

Table 3 Measurement of hardness values at various location of tyre

S. No.	Tyre location (inflation pressure 10 bar)	Hardness values (Shore A)	
		@Side wall	@Tread area
1	Tyres with disc (not fixed to the vehicle)	50	60
2	Tyres of drive wheel	50	62
3	Tyres of brake wheel	55	70
4	Tyres of idle wheel	50	61
5	Tyres without disc (old tyres)	65	72
6	Tyres of drive with 165 Ton load (Temp. 50 °C)	55	75

Table 4 Wear out of tyre tread after 100 km of vehicle running

S. No.	Details of tyre (inflation pressure 10 bar)	Measurement of depth of tread (mm)	
		Initial (while fixing on vehicle)	After 101 kms running
1	Tyres on drive wheel	10.0	9.2
2	Tyres on brake wheel	10.0	9.5
3	Tyres on idle wheel	10.0	9.5

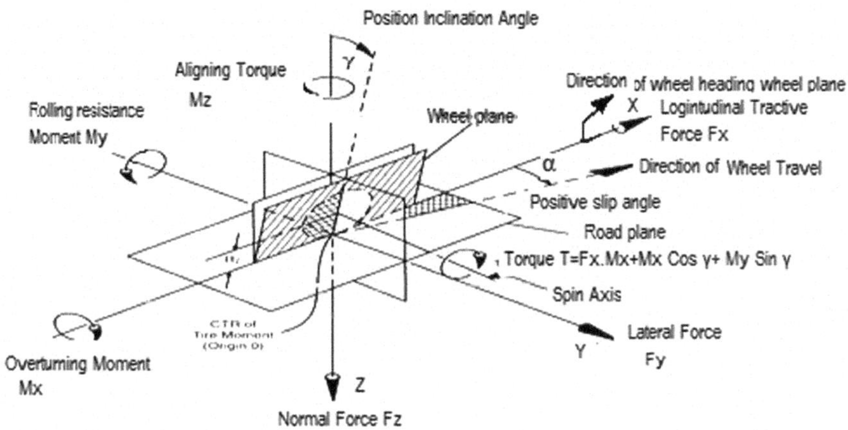

Fig. 2 Forces and moments acting on tyre

6 Dynamics of Tyre

The stability of the heavy load transporting vehicle is based on the tyre dynamics under various loads as shown in Fig. 2. The actual dynamic forces acting on tyre are normal force due to load on the transporter, longitudinal force due to braking/driving and lateral force due to cornering effect during movement of the transporter and slip angle and camber of the tyre, etc. By conducting various test trails with different loads, the dynamic forces of tyre and the corresponding moments were calculated.

7 Calculation of Dynamics of Tyre

7.1 Driving Force/Tractive Effort and Corresponding Moment

Driving or tractive effort of the transporter is calculated as per IS 15543, and driving force is generated at hydraulic motor of drive wheel. Tractive effort at hydraulic motor

Table 5 Dynamic forces and moments of tyre curing movement

S. No.	Details of load on transporter	Dynamic forces of tyre (kgf)				
		Fx	*Fy*	*Fz*	*My*	*Rr*
1	No load	309.25	4665.96	1250	131.25	37.5
2	With 25 Ton	733.63	7336.8	1944.44	306.28	58.33
3	With 65 Ton	1077.94	10265.09	3055.56	481.28	91.66
4	With 165 Ton	1942.12	14931.05	5833.33	700.04	175

of known torque rating and gear ratio is calculated based on the hydraulic pressure generated at drive pump under various loading schemes. Based on the availabilty of test loads trails are conducted. The calculated tractive effort (Fx) values for respective loads are shown in Table 5.

Torque rating of hydraulic motor of the transporter = 0.446 N m/bar .

Gear ratio of drive motor (GR) = 88.2.

$$Th = Pd \times Tr \tag{1}$$

$$Tw = Th \times GR \tag{2}$$

$$Fx = Tw/Re \tag{3}$$

where

Th Torque developed at hydraulic motor
Pd Pressure of drive pump
Tr Torque rating of drive motor
Tw Torque developed at wheel
Re Effective radius of tyre (static roll radius).

Using Eqs. 1, 2 and 3 driving force/tractive effort (Fx) calculation at drive pump pressure (Pd) 30 bar.

$$Th = 30 \times 0.446 = 13.38\,\text{N m}$$

$$Tw = 13.38 \times 88.2 = 1180.12\,\text{N m}$$

$$Fx = 1180.12/0.389 = 3033.73\,\text{N}$$

$$\text{Driving Force, } Fx = 3033.73/9.81 = 309.25\,\text{kgf}$$

Overturning moment, Mx on these tyres is negligible due to more aspect ratio; overall diameter of the tyre is less and also the axle unit was equipped with twin tyres.

7.2 Lateral Force/Cornering Force and Corresponding Moment

Lateral force or the cornering force acts on the transporter, while negotiating curves during movement. The transporting vehicle is having electronically controlled steering system. Lateral force acting on tyres with different loads is calculated by measuring the pressures of steering cylinder of the particular wheel. Steering cylinder pressures are measured as 50 bar, 80 bar, 110 bar and 160 bar for load on transporter as self-weight, 25 Ton, 65 Ton and 165 Ton, respectively. Calculated values of lateral force (Fy) are shown in Table 5.

$$Fy = Ps \times A \tag{4}$$

where

Fy Lateral Force/cornering force
Ps Pressure developed in steering cylinder, bar
A Area of the cylinder piston, mm^2.

For model calculation of lateral force/cornering force while negotiating 25 m turn circle radius curve, steering operated up to 18° and pressure developed in the cylinder (Ps) 50 bar and piston area (A) is 0.0091546 m^2.

$$Fy = 50 \times 100,000 \times 0.0091546$$

$$Fy = 45,773/9.81 = 4665.96\,\text{kgf}$$

Due to load tyre gets deflects leads to increase in contact area of tyre with road hence it decreases the contact pressure of tyre and road.

Rolling moment (My) act at wheel axis is nothing but torque developed on wheel-road contact. The moment values are calculated based on pressure developed at drive pump for different loading schemes, viz. 25 Ton, 65 Ton and 165 Ton and the values are shown in Table 5.

$$My = Pd \times Tr \tag{5}$$

where

My Rolling moment

Pd Drive pump pressure
Tr Torque rating of drive motor.

7.3 Normal Force and Corresponding Moment

Normal force is the load acting on the transporter including the self-weight of the transporter. Normal force of the transporter at various loading schemes is calculated and given in Table 4.

$$Fz = \frac{W}{N} \quad (6)$$

where

Fz Normal force
W Load on transporter (self-weight + payload)
N No. of tyres.

Self-alignment moment (Mz) acts along the vertical axis due to camber in tyre alignment, but the tyres of the transporter were arranged with swivel joint; hence, there is no camber in alignment of tyres.

7.4 Calculation of Rolling Resistance

The rolling resistance depends on the inflation pressure, tread profile, quality of road in tyre contact, etc. Rolling resistance of tyres depends on the inflation pressure and quality of road, the same was tested for different loads 30 kg /ton (self-weight, 25 Ton, 65 Ton and 165 Ton) on the transporter are shown in Table 5.

$$Rw = Rr \times Wt \quad (7)$$

where

Rw Rolling resistance at wheel
Wt Load on each tyre (Total load/No. of tyres).

Rolling resistance of tyres of the transporter with self-weight 45,000 Kg, Rolling resistance (Rr).

$$Rw = 30 \times 1.25 = 37.5\,\text{kg}$$

8 Discussion

The results of the experimental study reveal that the tyre properties deflection, temperature and wear out are varying proportionately with load. Hardness of the tyre-Here hardness of tyre means hardness of the tyre at tread area. The change in dynamic forces with respect to tyre deflection is shown in Fig. 3. For this study rolling resistance of tyre on asphalt is taken as 30 kg/ton, the variation of rolling resistance with deflection is nominal. During the period of experimental study (74 days) inflation pressure of tyre was maintained as 10 bar. Based on the measured data of wear out of selected tyres the expected life period of tyres is calculated, and due to the feature of all steerable wheels of transporter and stand still steering operation, wear out of tyres of all wheels are more and equal to 0.5 mm; for drive wheel tyre it is 0.8 mm and the results are shown in Figs. 4 and 5. Wind force and centripetal forces are negligible due to slow speed of transportation and low speed of transportation while negotiating curves. The angle of steering of wheels of the transporter can directly

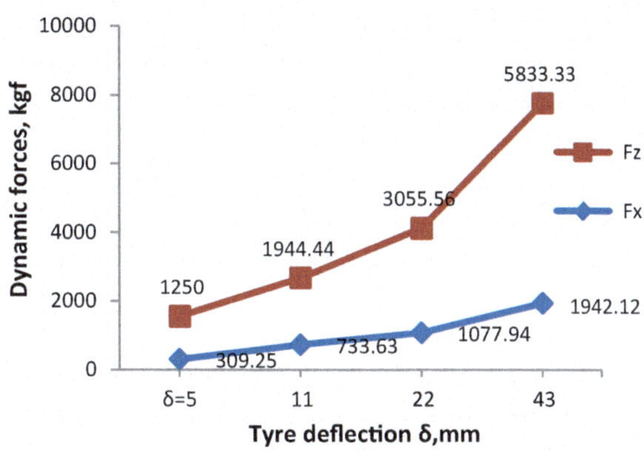

Fig. 3 Dynamic forces versus tyre deflection (δ)

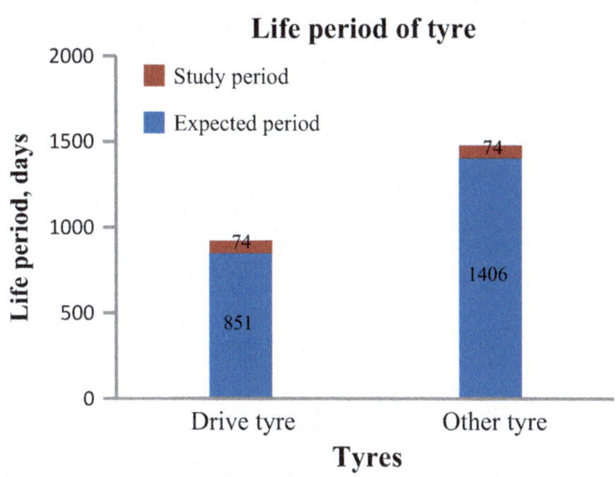

Fig. 4 Graph of tyres versus periodicity

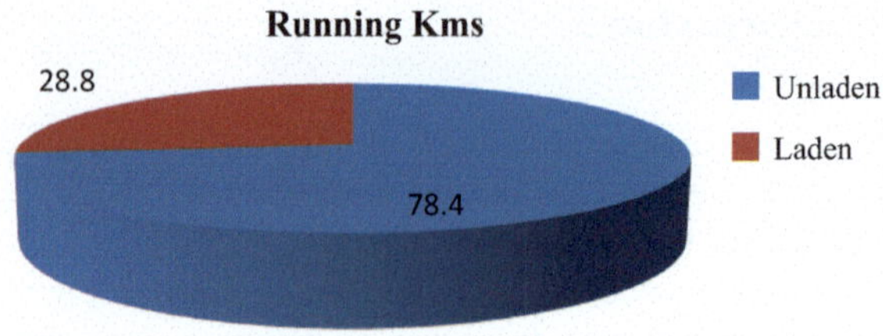

Fig. 5 Details of transporter running kilometres

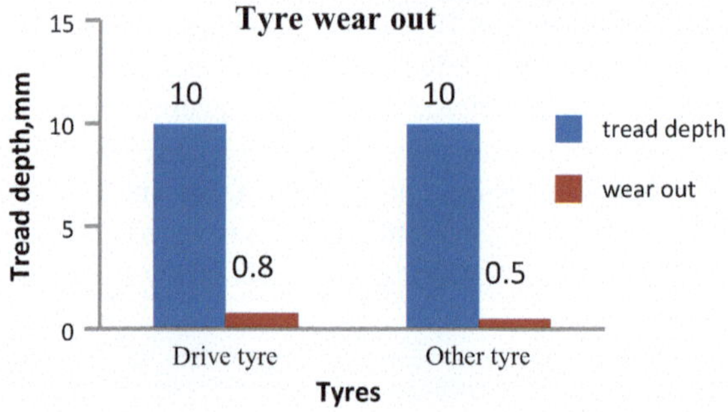

Fig. 6 Graph of wear out of tread versus tyres

read from the dash board of the vehicle. Through this study of tyre properties and dynamics of heavy load transporting vehicle, it is found that the unladen running kilometres are more than laden. During the study, it is observed that the decrease in inflation pressure of tyre (viz. 9.0 bar) caused deflection of tyres 40% more than deflection at suggested pressure of inflation i.e. 10 bar (Figs. 6, 7 and 8).

9 Conclusions

It is concluded that

- Dynamic forces act on tyres which vary with change in the properties of the tyre and quality of the surface in contact with the tyre during movement.
- The experimental trails reveal that the wear out percentage is 5% more on drive wheel tyre than the tyres of other wheels.
- Based on the wear out of tread, life period of tyres on drive wheel and wheels is evaluated, life period of tyre on drive wheel is around 37.5% less than other tyres.
- Steering on standstill and slow speed in loaded condition cause tyres to experience more contact stresses and slip.
- It is found that the unladen running kilometres of transporter are around 40% more than the laden running kilometres.

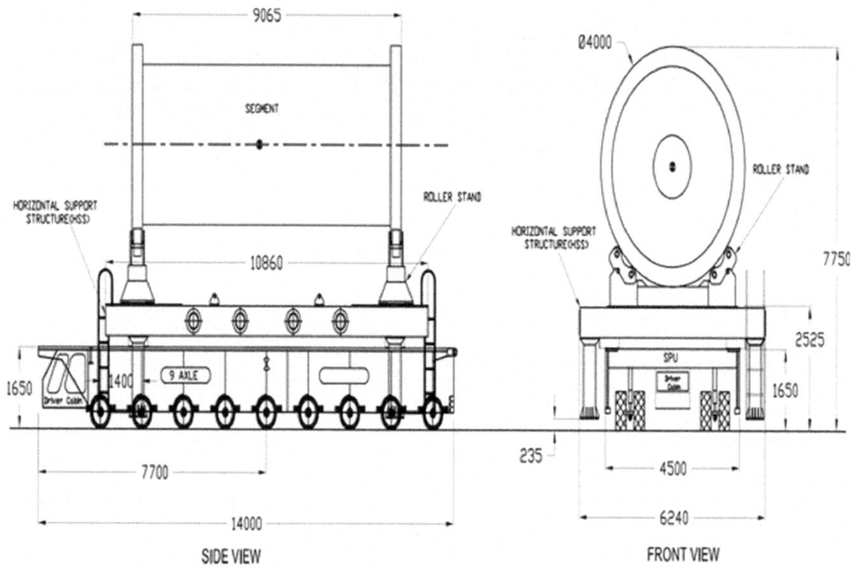

Fig. 7 Schematic sketch of horizontal transportation of 165 Ton load on transporter

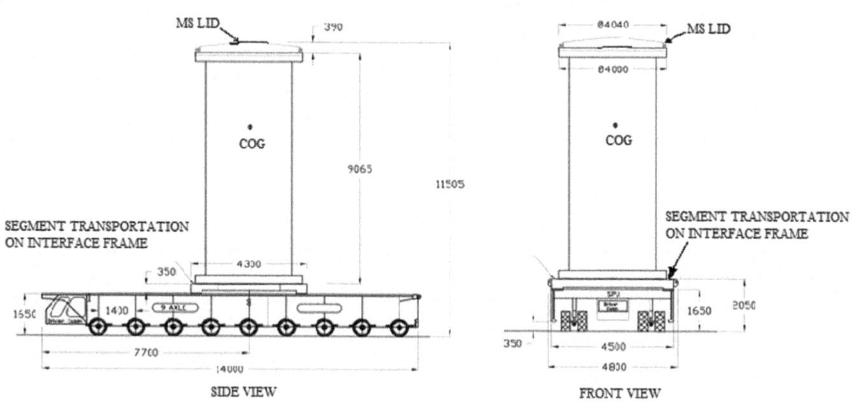

Fig. 8 Schematic sketch of vertical transportation of 125 Ton load on transporter

Bibliography

1. Abd_Elsalam A et al Modal analysis on tyre with respect to different parameters-AASTMT, 24 Sept 2016. Alex
2. Kubba A et al Experimental investigation on the dynamic response of piezoelectric transducer on tyre specimens. TSTCA 2016 and in tyre expo 2017
3. Moisescu A-R et al Investigation of radial modal behavior using finite element analysis for truck tyres without road contact. In: 11th international conference inter disciplinarily in engineering, INTER-ENG 2017, 5–6 Oct 2017. Tirgu-Mures

4. Hackl A (2016) Tyre dynamics: model validation. © Springer
5. Tuononen AJ (2009) On-board estimation of dynamic tyre forces from optically measured tyre carcass deflections. Int J Heavy Veh Syst 16(3):362–378
6. Basics principles of vehicle dynamics-©Springer, 2014, VIII, 275 p. ISBN 978–3-658-03977-6
7. Balaguru S (2014) Variation in residual stresses due to thermal cycling induced on the hardfaced grid plate in PFBR. Appl Mech Mater 591:98–102
8. Balaguru S (2012) Thermo mechanical analysis of SS304 circular grid plate hard faced with colmonoy. Trans Tech publications. ISSN: 229–231
9. Li B, Bei S, Zhao J Research method of tyre contact characteristics based on modal analysis, 17 Jan 2017. School of Automobile and Traffic Engineering, Jiangsu University of Technology, Changzhou 213001
10. Edward R et al (2007) Dynamic mechanical properties of passenger and light truck tire treads
11. Rill G First order tyre dynamics. In: IIIrd European conference on computational mechanics solids, structures and coupled problems in engineering, 5–8 June 2006. C.A. Lisbon
12. Krmela J (2017) Dynamic experiment of parts of car tyres. In: 10th international scientific conference Transbaltica
13. Pauwelussen JP Tyre dynamics, tyre as a vehicle component. In their paper published in HAN University 16 Oct 2007
14. Wallén L (2001) Dynamic tyre models in adaptive slip control. Department of Automatic Control, Lund Institute of Technology
15. Bijarimi M et al (2010) Mechanical properties of industrial tyre rubber compounds, University Malaysia Pahang, Lebuhraya Tun Razak, 26300 Gambang, Kuatan. J Appl Sci 10(13):1345–1348. ISSN 1812-5654
16. New pneumatic radial tires for light vehicles—U.S. Department of Transportation, National Highway Traffic Safety Administration, 6 Mar 2007
17. Buhari'll R et al Dynamic load coefficient of tyre forces from truck axles. Faculty of Civil and Environmental Engineering, University Tun Hussein Onn Malaysia
18. Pacejka HB (2012) Tyre and vehicle dynamics. © Elsevier. ISBN 978-0-08-097016-5
19. Yang S (2013) An overview on vehicle dynamic. © Springer, Berlin Heidelberg
20. Volusia TN et al Modern methodology for tyre design. In: Scientific research institute of the tyre industry (NIIShP). State Unitary Enterprise, Moscow
21. Vivkar DS, Thombare DG (2013) Parametric study and experimental evaluation of vehicle tyre performance. Int J Mech Eng Robot Res
22. Khan V et al (2017) Evaluation of tyre/road noise research in India investigation using statistical pass by method. Int J Parameter Res Technol
23. Hou Y et al Study on the microscopic friction between tire and asphalt pavement based on molecular dynamics simulation. In: National Center for Materials Service Safety, University of Science and Technology Beijing, Beijing 100083, China, Joint USTB-Virginia Tech Lab on Multifunctional Materials, USTB, Beijing, China, Virginia Tech, Blacksburg, VA 24061, United States, 5 Sept 2017

Analytical Modeling and FEM Simulation of the Collapse Voltage of an Angular Ring Metallization-Based MEMS Ultrasonic Transducer

Reshmi Maity, N. P. Maity, Shonkho Suvro, K. Guha, K. Srinivasa Rao, K. Girija Sravani, and S. Baishya

Abstract In this paper, we propose an angular ring metallization pattern for CMUT structure. The metallization is done in such a way that it is on the border of the membrane which forms an annular ring. It is observed that with decreasing area of the electrode, the collapse voltage increases and vice versa. The changes in collapse voltage with electrode area percentage, uncovered portion radius along with the change in gap height and different set of membrane thickness have been evaluated. Variation in collapse voltage with several membrane materials is also been observed. The proposed analytical model is compared with the finite element method (FEM)-based simulation by PZFlex, a commercially available software tool from Weidlinger Associates Inc. Excellent agreements between them are observed.

Keywords CMUT · FEM · Finite element analysis (FEA) · Collapse voltage · PZFlex

1 Introduction

CMUT is a device belonging to the class of micro electro mechanical system (MEMS) [1–4]. As the name suggests, it works as a transducer which converts energy from one form to another. Miniaturization competence of the silicon material-based micromachining procedure prepared the fabrication of CMUT operational with ultrasonic

R. Maity · N. P. Maity (✉) · S. Suvro · S. Baishya
Department of Electronics and Communication Engineering,
Mizoram University (a Central University), Aizawl 796004, India
e-mail: maity_niladri@rediffmail.com

K. Guha · K. Girija Sravani
Department of Electronics and Communication Engineering,
National Institute of Technology, Silchar, Silchar 788010, India

K. Srinivasa Rao
Department of Electronics and Communication Engineering,
K. L. University, Vaddeswaram 522502, India

L.-J. Yang et al. (eds.), *Proceedings of ICDMC 2019*,
Lecture Notes in Mechanical Engineering,
https://doi.org/10.1007/978-981-15-3631-1_18

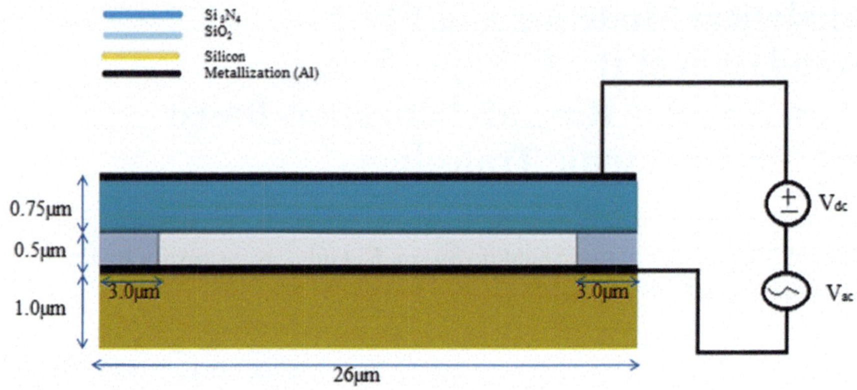

Fig. 1 Schematic diagram of two-dimensional CMUT

frequencies. CMUT devices consist of a thin silicon nitride membrane. It is suspended above a conductive Si substrate using SiO_2 insulating supports. The gap in the middle of the membrane and the substrate material is vacuum-sealed for immersion applications or left unsealed for air as the coupling medium, and it can be as small as 500 Å. Cross-sectional representation diagram of fully metalized CMUT is shown in Fig. 1.

CMUT has been the topic of extensive investigation and improvement in the analytical modeling of the device [5–9]. CMUT devices are consisting of thin membranes. It is fundamentally a parallel plate capacitor [10–12]. There is a gap between the plates. The conductive silicon wafer on which the membrane is fabricated makes up one of the plates of the capacitor. The membrane is usually made of an insulating material, most frequently silicon nitride, as well as it is coated by a metal electrode.

Non-compulsorily, the upper electrode area may be coated using an insulator. Like as low-temperature silicon dioxide (LTO) to deliver electrical isolation from the adjacent medium. CMUT devices are used together for instance of acoustic transmitters and receivers. For the duration of transmission process, electrostatic attraction forces are used to place the membranes into vibration through applying an alternating potential. On the other hand, since electrostatic force is unipolar in characteristics (continuous attraction), the vibration frequency of the membranes is double the applied frequency. If the applied AC voltage frequency is equal to the natural frequency of the membrane, the membrane vibrates with larger amplitude.

Membranes sustained by pillars are used as vibrating components of CMUTs. The residual stress built up throughout the fabrication procedure controls the transducer properties. They are resonance frequency, collapse voltage, and gap distance. Henceforth, it is very significant to estimate and administrate the stress in the very thin film CMUT membranes. Here, in this paper, Sect. 2 describes the proposed electromechanical model for CMUT. Section 3 discusses the analytical model results and validated the FEM simulation results by PZFlex, a commercially available simulation tool. At last, Sect. 4 concludes the finding and investigation of the work.

2 Electromechanical Model

We considered the plates of the capacitor as perfectly straight, ideal metallic contacts, the membrane restoring force as a linear function of displacement and CMUT operating in vacuum. Here, particularly one CMUT component is proposed through the lumped electromechanical model in Fig. 2. Here, it consists of a spring, a mass with a parallel plate capacitor.

- Spring will represent the restoring force of membrane. We are assuming that it can be a linear characteristic of membrane transposition. The restoring force results from the resistance shown by the membrane residual stress.
- Mass accounts for the mass of the membrane.
- Capacitor takes the stand for the electrostatic force.

The spring is actuated by the resultant of capacitive force; that is, force due to capacitor (F_c) is equal to force due to spring (F_s). Let t_g be the gap height; t_m be the membrane thickness; K_m be the spring constant of the membrane; Y_m be the Young's modulus of the membrane; ε_m be the permittivity of the membrane; ρ_m be the density of the membrane; ε_g be the permittivity of the gap; σ_m be the Poisson's ratio of the membrane; and X_m be the displacement of the membrane from the mean position. There are two very strong reasons to opt for partial metallization:

1. Price will decrease as the need for metallization is less. We have used aluminum for the electrode which is cheap but there are also metallizations which are done using gold, silver, etc., which are expensive metals.
2. The displacement of the membrane at collapse is unaffected by the area of metallization.

The metallization is done on the border of the membrane which forms an annular ring as shown in Fig. 3. Let a_{in} be the radius of the membrane which is not covered

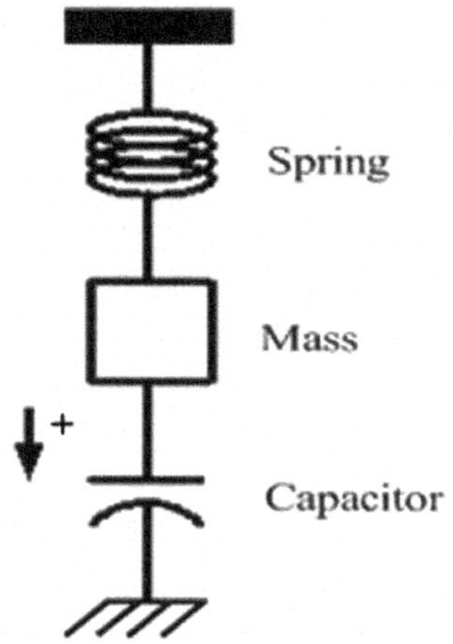

Fig. 2 Lumped electromechanical model for CMUT

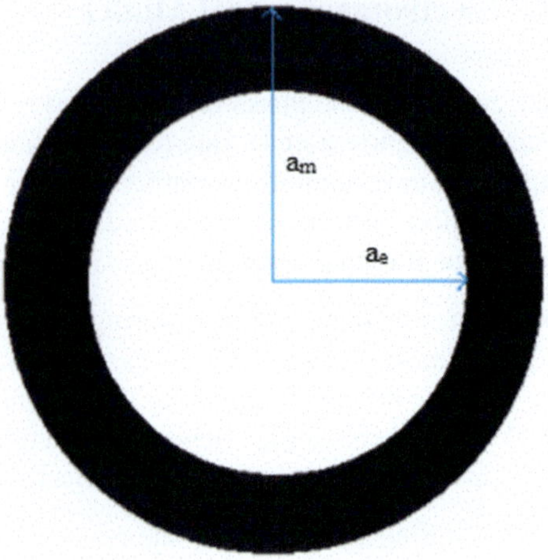

Fig. 3 CMUT with metallization at the border which looks like an annular ring

by the electrode and a_m be the radius of the membrane. Let A_m be the membrane area and A_e be the metal covered portion area.

$$A_e = \pi a_m^2 \left(1 - \left(\frac{a_{in}}{a_m}\right)^2\right) \tag{1}$$

Now, the effective capacitance is given by,

$$C'_{eq} = C_{eq}\left(1 - \left(\frac{a_{in}}{a_m}\right)^2\right) \tag{2}$$

Here, C_{eq} is the equivalent capacitance for the fully metalized membrane which can be evaluated as,

$$C_{eq} = \frac{\varepsilon_g \times \varepsilon_m \times A_m}{\varepsilon_m (t_g - X_m) + (\varepsilon_g \times t_m)} \tag{3}$$

The effective capacitive force is given by,

$$F'_c = \left(\frac{1}{2}\right)\left[\frac{d\left(C'_{eq} V^2\right)}{dX_m}\right] \tag{4}$$

Here, V is the applied DC bias voltage. The force due to spring of the membrane is given by $F_s = -K_m \times X_m$. K_m is the spring constant of the membrane and is given by,

$$K_{\mathrm{m}} = (16) \times \left[\frac{Y_{\mathrm{m}} \times t_{\mathrm{m}}^3}{a_{\mathrm{m}}^2 \left(1 - \sigma_{\mathrm{m}}^2\right)} \right] \tag{5}$$

Under equilibrium condition, $F_{\mathrm{c}}' = F_{\mathrm{s}}$. Hence, the applied bias can be represented as,

$$V = \sqrt{\left(\frac{2 \times K_{\mathrm{m}} X_{\mathrm{m}}}{\varepsilon_{\mathrm{g}} A_{\mathrm{m}}}\right)} \left[\left\{ t_{\mathrm{g}} + \left(\frac{\varepsilon_{\mathrm{g}}}{\varepsilon_{\mathrm{m}}}\right) \times t_{\mathrm{m}} \right\} - X_{\mathrm{m}} \right] \times \left(\frac{a_{\mathrm{m}}}{\sqrt{a_{\mathrm{m}}^2 - a_{\mathrm{in}}^2}} \right) \tag{6}$$

At collapse, the variation of voltage with respect to variation in displacement is zero that is equating, $\mathrm{d}V / \mathrm{d}X_{\mathrm{m}} = 0$, the displacement of the membrane at collapse can be evaluated as,

$$X_{\mathrm{m}} = \left(\frac{1}{3}\right) \left[t_{\mathrm{g}} + \left\{ \left(\frac{\varepsilon_{\mathrm{g}}}{\varepsilon_{\mathrm{m}}}\right) \times t_{\mathrm{m}} \right\} \right] \tag{7}$$

So, the displacement does not change even for partial metallization. Substituting the value in Eq. (6), we get,

$$V_{\mathrm{col}}' = V_{\mathrm{col}} \times \left(\frac{a_{\mathrm{m}}}{\sqrt{a_{\mathrm{m}}^2 - a_{\mathrm{in}}^2}} \right) \tag{8}$$

Here,

$$V_{\mathrm{col}} = \sqrt{\frac{8 K_{\mathrm{m}} \left[t_{\mathrm{g}} + \left\{ \left(\frac{\varepsilon_{\mathrm{g}}}{\varepsilon_{\mathrm{m}}}\right) \times t_{\mathrm{m}} \right\} \right]^3}{27 \varepsilon_{\mathrm{g}} A_{\mathrm{m}}}} \tag{9}$$

is the collapse voltage for a fully metalized membrane.

3 Results and Discussion

The variation of collapse voltage with change in uncovered portion of electrode radius is shown in Fig. 4. When the radius of uncovered portion a_{in} becomes half of the radius of the membrane that means the uncovered portion area becomes one-fourth of the area of the membrane, the percentage change in collapse voltage is 15.47005384%. While the metallization area is decreased by half, the percentage change in collapse voltage is 41.4213% that means the uncovered portion electrode area becomes half of the area of the membrane.

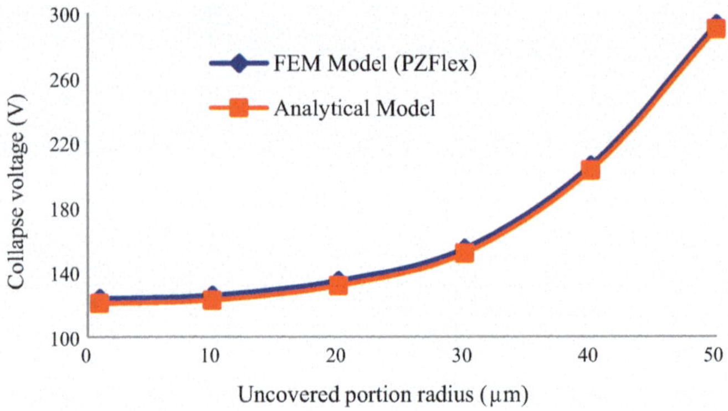

Fig. 4 Collapse voltage with change in uncovered portion radius

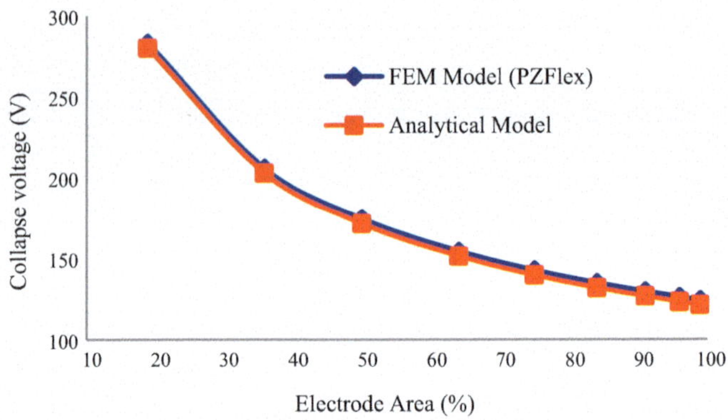

Fig. 5 Collapse voltage with percentage change in electrode area

The change in collapse voltage with electrode area percentage is given in Fig. 5. If we restrict the percentage change in collapse voltage to 20%, then we can decrease the metallization area to nearly 60%.

With the variation in gap height, the change in collapse voltage is shown in Fig. 6. The electrode separation strongly affects the collapse voltage. When the gap height increases, the membrane metallization also affects the collapse voltage significantly.

The change in collapse voltage with the variation in membrane thickness is shown in Fig. 7. The CMUT capacitance is formed by the equivalent capacitance of the membrane and the gap in series. The collapse voltage increases as the capacitive transduction force on the membrane deceases. Hence, as the membrane becomes thicker at the same electrode separation, higher bias is required to displace the membrane from its rest position which results in the rise of the collapse voltage as depicted in the figure. Moreover, membrane metallization is a dominant parameter in determining the voltage at which the membrane collapses.

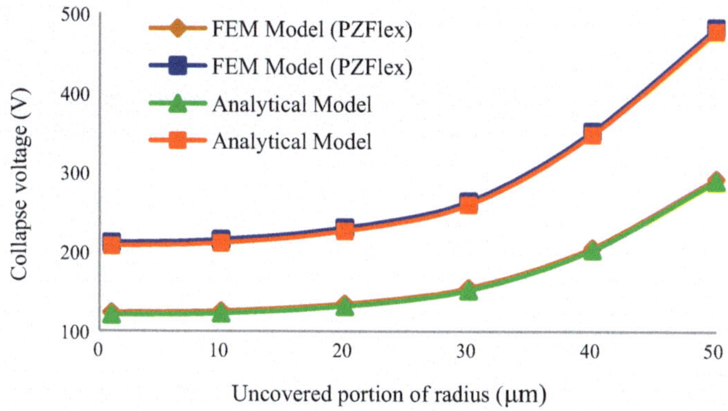

Fig. 6 Collapse voltage with change in uncovered portion radius along with change in gap height

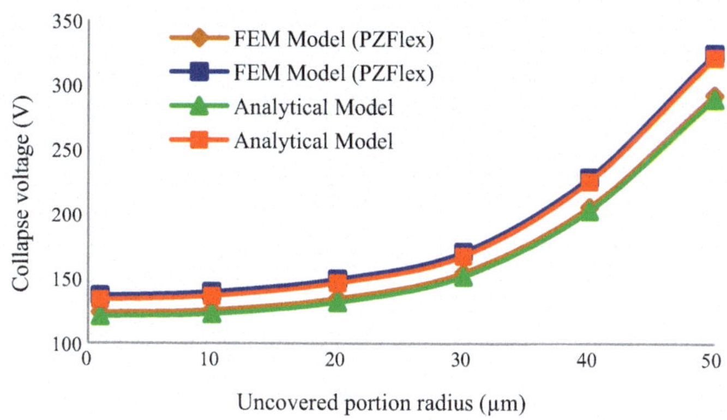

Fig. 7 Collapse voltage with change in uncovered portion radius along with change in membrane thickness

4 Conclusion

This paper analyzes the effect of membrane metallization pattern on the maximum static bias applicable to CMUT. It is observed that decrease of the electrode area increases the collapse voltage and 20% increase of the same can be tolerated for more than 30% decrease of the metallization area. For the same area of metallization, the collapse voltage is affected more by the electrode separation than the membrane thickness. All the parameters can be judiciously modulated to achieve an optimum performance-based ultrasonic device for air and underwater imaging.

Acknowledgements The authors are highly indebted to the University Grants Commission (UGC), Ministry of Human Resource and Development (MHRD), India, for technical help for this work.

References

1. Miao J, Wang H, Li P, Shen W, Xue C, Xiong J (2016) Glass-SOI-based hybrid-bonded capacitive micromachined ultrasonic transducer with hermetic cavities for immersion applications. IEEE J Microelectromech Syst 25(5):976–986
2. Maity R, Maity NP, Srinivasa Rao K, Guha K, Baishya S (2018) A new compact analytical model of nano-electro-mechanical-systems based capacitive micromachined ultrasonic transducers for pulse echo imaging. J Comput Electron 17(3):1334–1342
3. Maity R, Maity NP, Guha K, Baishya S (2018) Analysis of spring softening effect on the collapse voltage of capacitive MEMS ultrasonic transducers. Microsyst Technol. https://doi.org/10.1007/s00542-018-4040-x
4. Maity R, Maity NP, Guha K, Baishya S (2018) Analysis of fringing capacitance effect on the performance of MEMS based micromachined ultrasonic air transducer. IET Micro Nano Lett 13(6):872–877
5. Gurun G, Tekes C, Zahorian J, Xu T, Satir S, Karaman M, Hasler J, Degertekin FL (2014) Single-chip CMUT-on-CMOS front-end system for real-time volumetric IVUS and ICE imaging. IEEE Trans Ultrason Ferroelectr Freq Control 61(2):239–250
6. Chen K, Lee H, Sodini C (2016) A column-row-parallel ASIC architecture for 3-D portable medical ultrasonic imaging. IEEE J Solid-State Circ 51(3):738–752
7. Maity R, Maity NP, Thapa RK, Baishya S (2017) An improved analytical and finite element method model of nanoelectromechanical system based micromachined ultrasonic transducers. Microsyst Technol 23(6):2163–2173
8. Maity R, Maity NP, Baishya S (2017) Circular membrane approximation model with the effect of the finiteness of the electrode's diameter of MEMS capacitive micromachined ultrasonic transducers. Microsyst Technol 23(8):3513–3524
9. Boulmé A, Ngo S, Minonzio J, Legros M, Talmant M, Laugier P, Certon D (2014) A capacitive micromachined ultrasonic transducer probe for assessment of cortical bone. IEEE Trans Ultrason Ferroelectr Freq Control 61(4):710–723
10. Maity R, Maity NP, Baishya S (2020) An efficient model of nanoelectromechanical systems based ultrasonic sensor with fringing field effects. IEEE Sens J 20(4):1746–1753
11. Pal M, Maity NP, Maity R (2019) An improved displacement model for micro-electro-mechanical-system based ultrasonic transducer. Microsyst Technol 25(12):4685–4692
12. Maity R, Maity NP, Thapa RK, Baishya S (2015) Analytical characterization and simulation of a 2-D capacitive micromachined ultrasonic transducer array element. J Comput Theor Nanosci 12(10):3692–3696

Study of 3D Hexagonal Membrane Structure for MEMS-Based Ultrasonic Transducer Using Finite Element Method

Reshmi Maity, N. P. Maity, K. Guha, K. Srinivasa Rao, K. Girija Sravani and S. Baishya

Abstract This paper discussed the modeling approach of capacitive micromachined ultrasonic transducers (CMUTs) of hexagonal membrane structure. The active area of the hexagonal membrane is 1963.49 μm^2. Area of hexagonal has been considered $3\sqrt{3}/2L^2$, where L is the length of every side. Analyses have been brought up by single-cell hexagonal geometries as well as array of the cells through finite element method (FEM) model and done with an industry standard tool, SolidWorks simulation. The comparative analysis of membrane displacement arrays of hexagonal geometries on a single substrate assembled with different distances between each other is also investigated and analyzed the results.

Keywords FEM · Finite element analysis (FEA) · Ultrasonic transducer · CMUT · PZFlex

1 Introduction

Microelectromechanical systems (MEMS) or nano-electromechanical systems (NEMS) are very useful for today's nanoelectronics system which consists of electrical and mechanical components. Dimension varies from micrometer to nanometer. The applications of MEMS/NEMS technology are very extensive due to its major advantages. It is fabricated by microfabrication technology which provides great

R. Maity · N. P. Maity (✉) · S. Baishya
Department of Electronics and Communication Engineering, Mizoram University (A Central University), Aizawl 796004, India
e-mail: maity_niladri@rediffmail.com

K. Guha · K. Girija Sravani
Department of Electronics and Communication Engineering, National Institute of Technology, Silchar, Silchar 788010, India

K. Srinivasa Rao
Department of Electronics and Communication Engineering, K. L. University, Vaddeswaram 522502, India

L.-J. Yang et al. (eds.), *Proceedings of ICDMC 2019*,
Lecture Notes in Mechanical Engineering,
https://doi.org/10.1007/978-981-15-3631-1_19

cost-effective and lesser maintenance cost. It also provides manufacturing of enormous arrays of devices [1]. The key applications are in sensors and actuators. Those are used in industrial applications as well as automobile and medical applications. It provides lesser unit cost. Instead of these applications, MEMS/NEMS technology also creates innovative mechanical and electronics systems. It can regulate the intelligence and trigger mechanical responses.

Sound generated above human hearing range of 20 kHz is known as ultrasound, and ultrasonic transducers are the devices which convert electrical energy into ultrasonic during transmission and ultrasonic to electrical during reception. These may be based on continuous wave reflection type, continuous wave through transmission or pulse echo ultrasonic technique [2]. First, MEMS-based ultrasonic transducer was fabricated during the 1990s at Stanford University which consisted of suspended membranes whose vibrations are used to transmit and receive ultrasonic waves. Ultrasonic transducer may be capacitive micromachined ultrasonic transducer (CMUT), piezoelectric transducer or magnetostrictive type. Ultrasonic technology is a very influential tool. It is applying in medical investigation and industrial development control. It is also useful for environmental mechanism and nondestructive measurement. Despite the fact that the piezoelectric technologies are leading, capacitive micromachined ultrasonic transducers are being discovered as a corresponding technology. CMUT devices are established using electrostatic principle for their functioning. They have a huge amount of benefits such as impedance corresponding with neighboring, superior bandwidth and respectable competence to create great frequency arrays [3, 4]. It has the capability to integrate with electronics. It requires very lesser cost for huge number of production and higher pressure sensitivity. It has the ability to prompt reaction to altering pressure. It can tolerate huge vibrations and lower temperature sensitivity [5]. Micromachining technology permitted this to be a talented substitute for piezoelectric transducers [6]. It consists of fundamentally a parallel plate capacitor. Voltage is applied on one electrode, and last one is grounded [7, 8]. Replacement of the plates produces ultrasonic waves for the duration of transmission, and modification in capacitance value is identified at the time of reception [9]. CMUT is working on the design of front-end electronics, packaging and modular assembly of bigger area. Also, fine-pitch two-dimensional CMUT arrays are also in progress [10–15].

In this paper, the modeling approaches of capacitive micromachined ultrasonic transducers of hexagonal membrane structure are explained in detail. Section 2 describes the structure of a CMUT and the FEM modeling approach of CMUT. Section 3 includes the result analysis and discussion on membrane displacements for single-cell hexagonal geometries including its array of cells. Also, proportional investigation of membrane movements is done for arrays of hexagonal geometries on a single substrate in Sect. 3. Section 4 concludes the important finding and investigation of the work.

2 Modeling of CMUT

The cross-sectional device structure of CMUT is depicted in Fig. 1. In this modeling methodology of the CMUT, the area of the three membrane profiles is reserved constant. Area of hexagonal can be written as $(3\sqrt{3}/2)L^2$, where L is length of every side. The side length is taken as 28.86 μm. The design parameter values for SolidWorks to model the CMUT hexagonal structure are shown in Table 1 (Fig. 2).

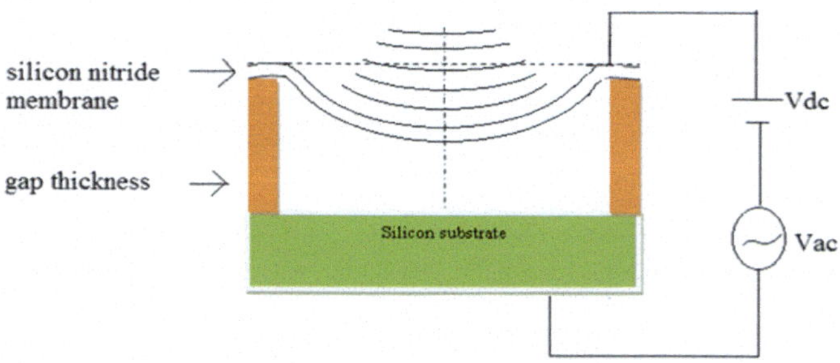

Fig. 1 Cross-sectional device structure of the CMUT

Table 1 Design parameters for SolidWorks

Parameters	Values
Thickness of the membrane	0.75 μm
Sides length of membrane	28.86 μm
Thickness of the gap	0.50 μm
Force	16.894 μN
Pressure	8603.98 N/m^2

Fig. 2 Model name: hexagonal assembly with single cell

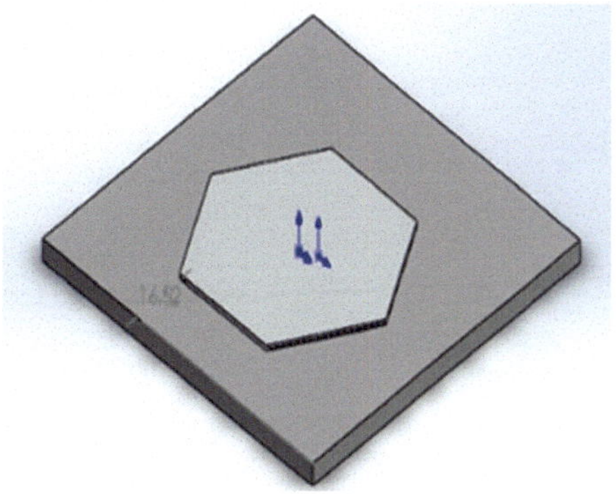

3 Results and Discussion

3.1 Membrane Displacements for Single-Cell Hexagonal Geometries

The model of CMUT is fabricated using thin hexagonal membrane of silicon nitride (0.75 μm). The physical properties of the CMUT are presented in Table 2. Pressure loads and force are applied consistently in the direction of the membrane as shown in Table 3.

Table 2 Three-dimensional model with single-cell physical and material properties

Model reference	Properties	Components
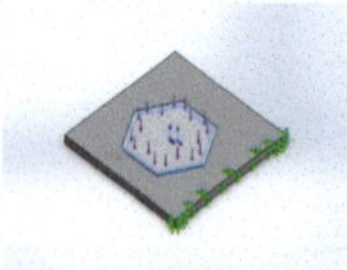	Name: Silicon dioxide Model category: Linear elastic isotropic Yield strength: 1.55×10^8 N/m^2 Tensile strength: 1.1×10^8 N/m^2 Compressive strength: 1.38×10^9 N/m^2 Elastic modulus: 7.3×10^{10} N/m^2 Poisson's ratio: 0.17 Mass density: 2650 kg/m^3 Shear modulus: 3.23×10^{10} N/m^2	Solid Body 1 (Boss-Extrude1) (Base Hexagonal 28-4)
	Name: Silicon nitride Model category: Linear elastic isotropic Tensile strength: 2.8×10^8 N/m^2 Elastic modulus: 3.2×10^{11} N/m^2 Poisson's ratio: 0.263 Mass density: 3270 kg/m^3 Shear modulus: 1.27×10^{11} N/m^2	Solid Body 1 (Boss-Extrude1) (Membrane Hexagonal 28-1)
	Name: Silicon Model category: Linear elastic isotropic Yield strength: 1.2×10^8 N/m^2 Elastic modulus: 1.124×10^{11} N/m^2 Poisson's ratio: 0.28 Mass density: 2330 kg/m^3 Shear modulus: 4.9×10^{10} N/m^2	Solid Body 1 (Boss-Extrude2) (sub05-1)

Table 3 Loads, fixtures and contact

Fixture name	Fixture appearance	Fixture details
Fixed-1	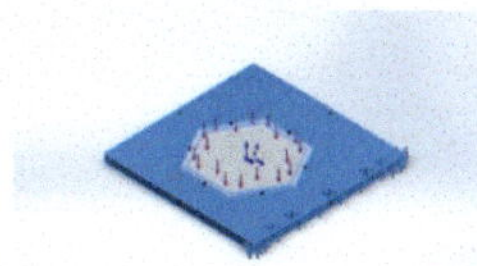	Entities: 17 face(s) Category: Fixed geometry
Load name	Load appearance	Load details
Force-1	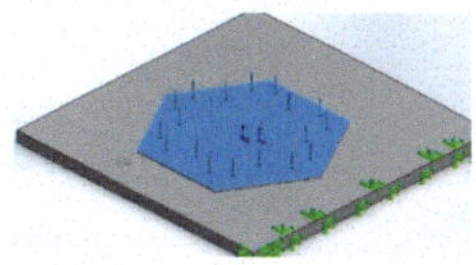	Entities: 1 face(s) Category: Apply normal force Value: 1.6894 $\times$ 10^5 N/m^2 Phase angle: 0 Units: deg
Pressure-1	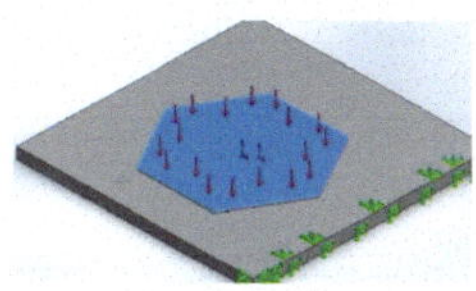	Entities: 1 face(s) Category: Normal to selected face Value: 8603.98 Units: N/m^2 Phase angle: 0 Units: deg
Contact	Contact appearance	Contact properties
Global contact	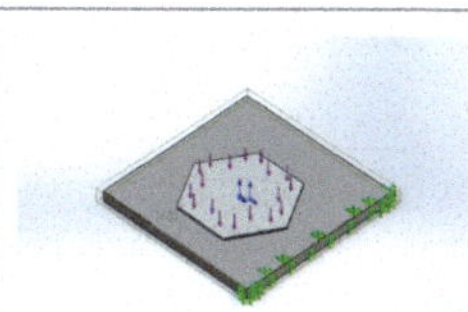	Category: Bonded Components: 1 component(s) Options: Compatible mesh

The pressure load is applied on the membrane as providing a value of 8603.98 N/m^2. The static voltage is taken as 40 V. Signal value is 100 mV. Here, the force is estimated as 16.894 μN. The whole surfaces of the proposed models are static excluding the face 1 (membrane).

The meshed configuration of a hexagonal CMUT with 16,834 nodes and 9653 components is depicted in Fig. 3. Subsequently applying the loads, the membrane twitches distorting by supreme displacement at the middle of membrane. The displacement result of hexagonal membranes with single-cell structure is shown in Table 4.

Fig. 3 Meshed structure of a hexagonal CMUT

Table 4 Displacement result of hexagonal membrane with single cell

Name	Category	Minimum	Maximum
Displacement1	URES: Resultant displacement	0 mm Node: 1	6.56866×10^{-9} m Node: 2547

This enormous movement agrees to a strain of 2.31891×10^{-5}. The rupture strain for silicon nitride material is 3.6×10^{-2}. Consequently, the CMUT is functioned healthy lower than the fracture border and exhaustion would not be an important difficulty.

3.2 Membrane Displacements of Array of Cells

The modeled structure for array of a four-hexagonal membrane cell CMUT is presented in Fig. 4. Physical properties of the similar are presented in Table 5. The pressure loads and force are applied consistently in the direction of the membrane as shown in Table 6.

In this case, also pressure load is applied on the membrane as providing a value of 8603.98 N/m^2. The static voltage is taken as 40 V. Signal value is 100 mV. Here, the force is estimated as 16.894 μN. The whole surfaces of the proposed models are static excluding the face 1 (membrane). A meshed structure of an array of hexagonal CMUT including 18,416 nodes and 10,459 components is depicted in Fig. 5. Subsequently applying the loads, the membrane twitches distorting by supreme displacement at the middle of the membrane. In this case, also the displacement results of hexagonal membranes with array structure are shown in Table 7.

In this case, the enormous movement agrees to a strain of 1.88822×10^{-5}. The rupture strain for silicon nitride material is 3.6×10^{-2}. Consequently in this case, also the CMUT is functioned healthy lower than the fracture border and exhaustion would not be an important difficulty.

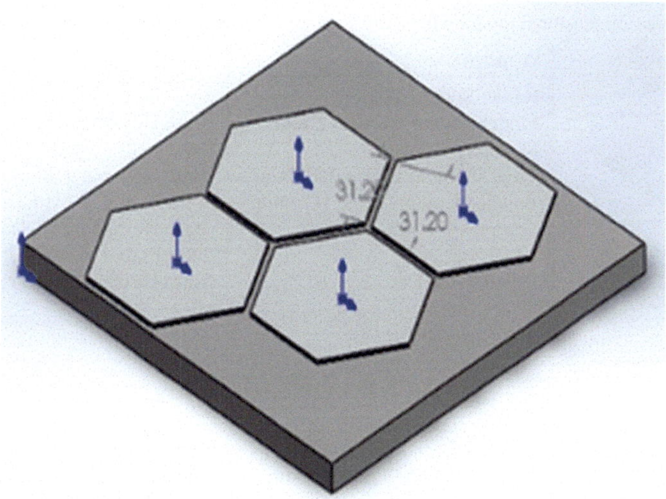

Fig. 4 Model name: hexagonal assembly with array of cells

3.3 Comparative Analysis of Membrane Displacement

The four hexagonal membranes assembled with different distances between each other are shown in Fig. 6.

Pressure loads and force are applied to each membrane face 1. Membrane displacement performance comparison has been shown in Fig. 7. Simulation study of membrane displacement is shown in Table 8. The side length of hexagonal-shaped membrane is taken as 28.86 μm with a thickness of 0.75 μm.

4 Conclusion

Hexagonal membrane has been proposed to suit better for a given application. However, more researches are needed to arrive at a foolproof structure, array and optimum parameters. This work analyzes a comparison between the 3D array structures having different distances between the membranes at each instant. Hexagonal CMUT elements through FEM simulations and attempted to arrive at a conclusion for choosing the best peak displacement. Three-dimensional modeling is carried out by SolidWorks simulation tool.

Table 5 Three-dimensional model with an array of cell physical and material properties

Model reference	Properties	Components
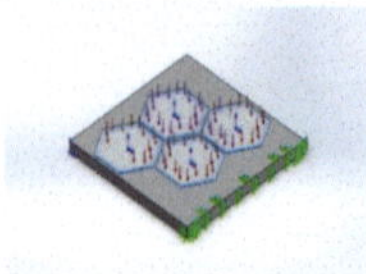	Name: Silicon dioxide Model category: Linear elastic isotropic Yield strength: 1.55×10^8 N/m^2 Tensile strength: 1.1×10^8 N/m^2 Compressive strength: 1.38×10^9 N/m^2 Elastic modulus: 7.3×10^{10} N/m^2 Poisson's ratio: 0.17 Mass density: 2650 kg/m^3 Shear modulus: 3.23×10^{10} N/m^2	Solid Body 1 (Boss-Extrude1) (Base Hexagonal 28-1) Solid Body 1 (Boss Extrude1) (Base Hexagonal 28-2) Solid Body 1 (Boss Extrude1) (Base Hexagonal 28-3) Solid Body 1 (Boss Extrude1) (Base Hexagonal 28-4)
	Name: Silicon nitride Model category: Linear elastic isotropic Tensile strength: 2.8×10^8 N/m^2 Elastic modulus: 3.2×10^{11} N/m^2 Poisson's ratio: 0.263 Mass density: 3270 kg/m^3 Shear modulus: 1.27×10^{11} N/m^2	Solid Body 1 (Boss Extrude1) (Membrane Hexagonal 28-1) Solid Body 1 (Boss Extrude1) (Membrane Hexagonal 28-2) Solid Body 1 (Boss-Extrude1) (Membrane Hexagonal 28-3) Solid Body 1 (Boss Extrude1) (Membrane Hexagonal 28-4)
	Name: Silicon Model category: Linear elastic isotropic Yield strength: 1.2×10^8 N/m^2 Elastic modulus: 1.124×10^{11} N/m^2 Poisson's ratio: 0.28 Mass density: 2330 kg/m^3 Shear modulus: 4.9×10^{10} N/m^2	Solid Body 1 (Boss-Extrude1) (Substrate 5-1)

Table 6 Loads, fixtures and contact

Fixture name	Fixture appearance	Fixture details
Fixed-1		Entities: 49 face(s) Category: Fixed geometry
Load name	Load appearance	Load details
Force-1		Entities: 4 face(s) Category: Apply normal force Value: 1.6894×10^5 N Phase angle: 0

(continued)

Table 6 (continued)

Fixture name	Fixture appearance	Fixture details
Pressure-1	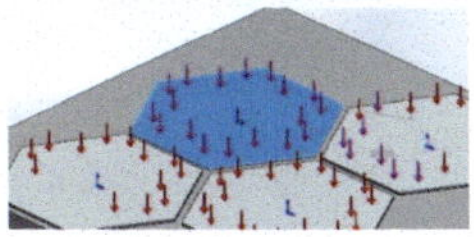	Entities: 1 face(s) Category: Normal to selected face Value: 8603.98 Units: N/m^2 Phase angle: 0
Pressure-2	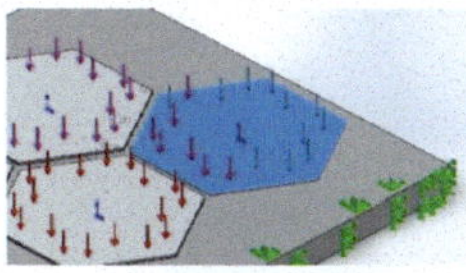	Entities: 1 face(s) Category: Normal to selected face Value: 8603.98 Units: N/m^2 Phase angle: 0
Pressure-3	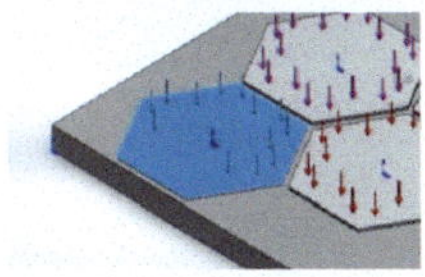	Entities: 1 face(s) Category: Normal to selected face Value: 8603.98 Units: N/m^2 Phase angle: 0
Pressure-4	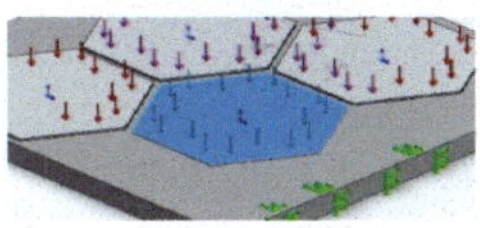	Entities: 1 face(s) Category: Normal to selected face Value: 8603.98 Units: N/m^2 Phase angle: 0
Contact	Contact appearance	Contact properties
Global Contact	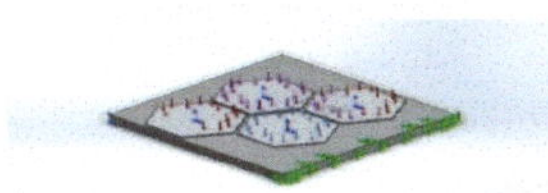	Category: Bonded Components: 1 component(s) Options: Compatible mesh

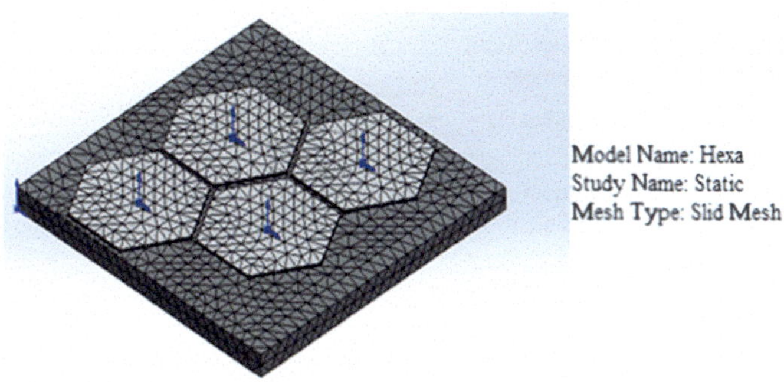

Fig. 5 A meshed configuration of array of hexagonal CMUT

Table 7 Displacement result of hexagonal membrane with an array

Name	Category	Minimum	Maximum
Displacement1	URES: Resultant displacement	0 mm Node: 1	2.19142×10^{-6} mm Node: 2504

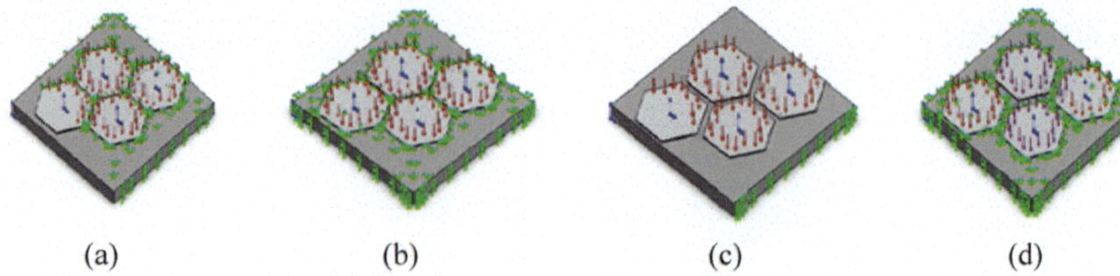

Fig. 6 Distance between the hexagonal membranes: **a** 1 μm, **b** 2 μm, **c** 5 μm, **d** 8 μm

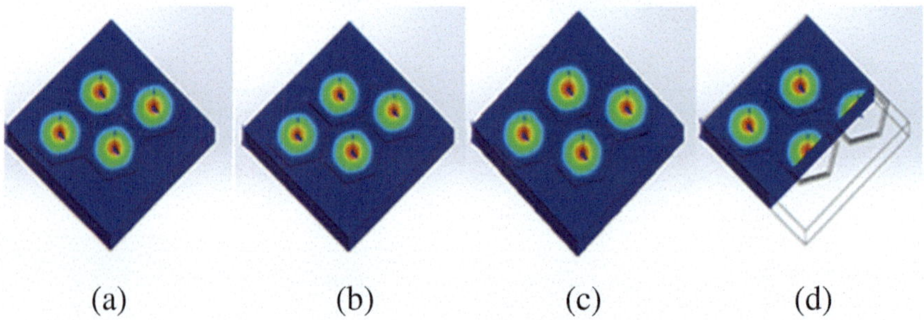

Fig. 7 Displacement of hexagonal membrane array: **a** 1 μm, **b** 2 μm, **c** 5 μm, **d** 8 μm

Table 8 Study of membrane displacement

Name	Category	Minimum displacement at the edges (nm)	Maximum displacement at the middle (nm)
CMUT hexagonal membrane with 1 μm	Subsequent static displacement	0	2.209
CMUT hexagonal membrane with 2 μm	Subsequent static displacement	0	2.279
CMUT hexagonal membrane with 5 μm	Subsequent static displacement	0	2.208
CMUT hexagonal membrane with 8 μm	Subsequent static displacement	0	2.205

Acknowledgements The authors are highly indebted to University Grants Commission (UGC), Ministry of Human Resource and Development (MHRD), India, for technical help for this work.

References

1. Cezar M (2011) Development and microfabrication of capacitive micromachined ultrasound transducers with diamond membranes. MS thesis, Electrical and Electronics Engineering, Middle East Technical University, Ankara, Turkey
2. Chhikara A, Malik S (2014) Parameters affecting performance of capacitive micro machined ultrasonic transducer (CMUT). Int J Appl Sci Eng Res 3(2):333–338
3. Maity R, Maity NP, Srinivasa Rao K, Guha K, Baishya S (2018) A new compact analytical model of nano-electro-mechanical-systems based capacitive micromachined ultrasonic transducers for pulse echo imaging. J Comput Electron 17(3):1334–1342
4. Maity R, Maity NP, Guha K, Baishya S (2018) Analysis of spring softening effect on the collapse voltage of capacitive MEMS ultrasonic transducers. Microsyst Technol. https://doi.org/10.1007/s00542-018-4040-x
5. Malik S, Duhan M, Chhikara A (2015) Designing a capacitive micro-machined ultrasonic transducer cell for high frequency application. Int J Sci Technol Manag 04:563–567
6. Maity R, Maity NP, Guha K, Baishya S (2018) Analysis of fringing capacitance effect on the performance of micr-electromechanical-system-based micromachined ultrasonic air transducer. IET Micro Nano Lett 13(6):872–877
7. Maity, Reshmi, Maity NP, Baishya S (2020) An efficient model of nanoelectromechanical systems based ultrasonic sensor with fringing field effects. IEEE Sens J 20(4):1746–1753
8. Maity R, Maity NP, Thapa RK, Baishya S (2017) An improved analytical and finite element method model of nanoelectromechanical system based micromachined ultrasonic transducers. Microsyst Technol 23(6):2163–2173
9. Zhang W, Zhang H, Jin S, Zeng Z (2016) A two-dimensional CMUT linear array for underwater applications: directivity analysis and design optimization (State Key Laboratory of Precision Measurement Technology and Instrument)
10. Park KK, Kupnik M, Lee HJ, Khuri-Yakub BT, Wygant IO (2010) Modeling and measuring the effects of mutual impedance on multi-cell CMUT configurations. In: IEEE ultrasonics symposium. https://doi.org/10.1109/ULTSYM.2010.5936010
11. Caronti A (2005) Acoustic coupling in capacitive microfabricated ultrasonic transducers: modeling and experiments. IEEE Trans Ultrason Ferroelectr Freq Control 52(12):2220–2234
12. Thomas G (2013) Packaging and modular assembly of large-area and fine-pitch 2-D ultrasonic transducer arrays. IEEE Trans Ultrason Ferroelectr Freq Control 60(7):1356–1375
13. Maity R, Maity NP, Baishya S (2017) Circular membrane approximation model with the effect of the finiteness of the electrode's diameter of MEMS capacitive micromachined ultrasonic transducers. Microsyst Technol 23(8):3513–3524
14. Pal M, Maity NP, Maity R (2019) An improved displacement model for micro-electro-mechanical-system based ultrasonic transducer. Microsyst Technol 25(12):4685–4692
15. Maity R, Gogoi K, Maity NP (2019) Micro-electro-mechanical-system based capacitive ultrasonic transducer as an efficient immersion sensor. Microsyst Technol 25(12):4663–4670

Intelligent Fixture Layout Design for End Milling Process Using Artificial Neural Networks

F. Michael Thomas Rex, D. Ravindran, A. Andrews, and Lenin Nagarajan

Abstract Fixture layout depicts the arrangement of locators and clamps around the workpiece. The position of the locators and clamps is one of the major factors which cause the elastic deformation of the workpiece during machining. The identification of optimal fixture layout design is a feasible solution to ensure the end quality of the workpiece by minimizing the elastic deformation. In this study, the effect of fixture layout on the elastic deformation of the workpiece is analysed based on full factorial experiments using finite element method (FEM). The analytical results are adequate to develop empirical models using response surface methodology (RSM) and artificial neural network (ANN). The result shows that the ANN model has higher modelling capability than RSM. Further, the ANN model is integrated with real-coded genetic algorithm (GA) to find the optimal fixture layout with the consideration of stability constraint. The result of the presented technique is compared with the previously published results, and it shows the effectiveness of the proposed technique as a practical tool to solve fixture layout design problems.

Keywords FEM · RSM · ANN · GA · Fixture layout

1 Introduction

Fixture design plays a vital role in a milling process to ensure the form and dimensional accuracies of components. Proper fixture design is inevitable during the machining of a workpiece of low rigidity. In any batch production environment, the cost of fixtures is about 30% of the total cost of manufacturing [1]. Further, the

F. Michael Thomas Rex (✉) · D. Ravindran · A. Andrews
Department of Mechanical Engineering, National Engineering College, Kovilpatti, Tamil Nadu 628503, India
e-mail: michealrex@hotmail.com

L. Nagarajan
Department of Mechanical Engineering, Vel Tech Rangarajan Dr. Sagunthala R&D Institute of Science and Technology, Chennai, Tamil Nadu 600062, India

L.-J. Yang et al. (eds.), *Proceedings of ICDMC 2019*, Lecture Notes in Mechanical Engineering,
https://doi.org/10.1007/978-981-15-3631-1_20

poor design of fixtures leads to dimensional inaccuracy and rejection of the components due to deformation caused during machining. In recent days, researchers are attracted towards the prediction and analysis of the deformation of components during machining to optimize the fixture layouts. The evolution of numerical methods like FEA is of great help in solving complicated and time-consuming computational analysis in the above area.

Bai et al. [2] developed a memetic algorithm to solve multi-objective fixture layout problem considering location accuracy and stability. Liao [3] presented a GA-based approach to optimize the number of locators, clamps and their positions in a fixture for sheet metal assembly. Kaya and Ozturk [4] proposed frictional contact analysis in the workpiece–fixture layout verification. Chip removal effect was also considered in the analysis using element death technique. Deiab and Elbestawi [5] presented a realistic dynamic simulation model for the face milling process with fixture dynamics. Yu et al. [6] implemented a method to check immobilization in fixture design. Amaral et al. [7] developed an algorithm using ANSYS parametric design language to optimize the fixture support and clamp locations. Kaya [8] developed a tool using GA integrated with ANSYS to optimize the fixture layout. Deng and Melkote [9] presented a particle swarm optimization (PSO)-based technique to determine the minimum clamping force required to ensure dynamic stability. The proposed technique finds the optimum fixture layout design with minimum workpiece deformation. Liu et al. [10] developed a finite element model along with cutting forces for the fixture layout optimization of a thin-walled workpiece. Rex et al. [11] used FEM-based contact analysis to simulate the dynamic behaviour of the workpiece under machining. Siebenaler and Melkote [12] explored the viability of FEM to determine the elastic deformation of the workpiece using ANSYS. Chen et al. [13] presented a multi-objective optimization technique based on GA to find optimal fixture layouts and clamping forces. Fan and Senthil Kumar [14] presented a hybrid technique to optimize fixture layouts using Taguchi method and Monte Carlo statistical method. Wang and Nee [15] applied the ant colony algorithm (ACA) and GA for the fixture layout optimization, and it is observed that ACA is superior to GA. Kumar and Paulraj [16] applied GA along with ANSYS integrated tool to optimize the fixture layout for a drilling application. Sun and Chen [17] proposed an intelligent fixture design by applying case-based reasoning technique. Dou et al. [18] used PSO algorithm, improved particle swarm optimization (IPSO) algorithm, GA and improved genetic algorithm (IGA) to optimize the fixture layout and compared the performances of the above algorithms. Abedini et al. [19] presented a GA-based approach to optimize the locator position to reduce machining error.

Despite the significant efforts directed towards solving fixture layout problems, most of the previous studies considered FEM-GA integrated technique towards the prediction of optimal fixture layout. Since the positions of the locators and clamps are not discrete variables, it is difficult to analyse the effect of all possible positions using FEM. Hence, the application of FEM-GA-based technique is less practicable as it takes more computational effort.

Recently, RSM-based empirical model has been used instead of FEM to predict the elastic deformation of the workpiece while identifying the optimal fixture layout by Sundararaman et al. [20]. In their work, the suitability analysis was not performed while selecting a predictive model, since its performance will not be the same for all the problems. Rex and Ravindran [21] predicted the optimum fixture layout using iterative design of experiments (IDOE). Further, the possibility of using ANN in fixture layout optimization has been discussed.

Recent advances in computer-aided engineering and artificial intelligence facilitate the design and analysis of manufacturing systems in ensuring the end quality of the products. ANN-GA-based artificial intelligence techniques had been successfully applied by many of the researchers in various engineering applications [22, 23].

Hence in this study, the application of using ANN and RSM models is explored to predict the elastic deformation of the workpiece. The novelty of the proposed work is to identify a suitable model by comparing their performances in mapping elastic deformation with the fixture layout. The suitable model is integrated with GA to optimize the fixture layout by considering stability constraint. The proposed methodology for the optimal fixture layout design is shown in Fig. 1.

The proposed optimization procedure is implemented for a slot milling operation considered by Kaya [8]. The axial machining forces developed in the slot milling process are less significant, and it is not considered in the dynamic analysis. Consequently, the analysis is simplified into a two-dimensional plane stress problem. The geometry of the workpiece in two dimensions is shown in Fig. 2.

2 Structural Model of the Machining Process Using FEM

The analysis and prediction of deformation of the workpiece under the influence of machining forces, reactions from locators and clamping forces are performed with the aid of effective modelling using FEM. Finite element software package ANSYS is employed for FEA analysis. The assumptions made during the FEM modelling of the workpiece along with locators, clamps and cutting tool are given below.

- The workpiece is assumed as an elastic body, whereas the locators and clamps are considered as rigid bodies.
- The workpiece is assumed to be subjected to plane stress condition.
- Locators are taken as translational constraints as it restricts the movement of the workpiece against locators.
- Clamping force is assumed to be acting at a point of application, and its location is taken as any one of the nodal points of the workpiece.
- Machining forces are considered as transient forces (time-dependent) and applied over the workpiece.

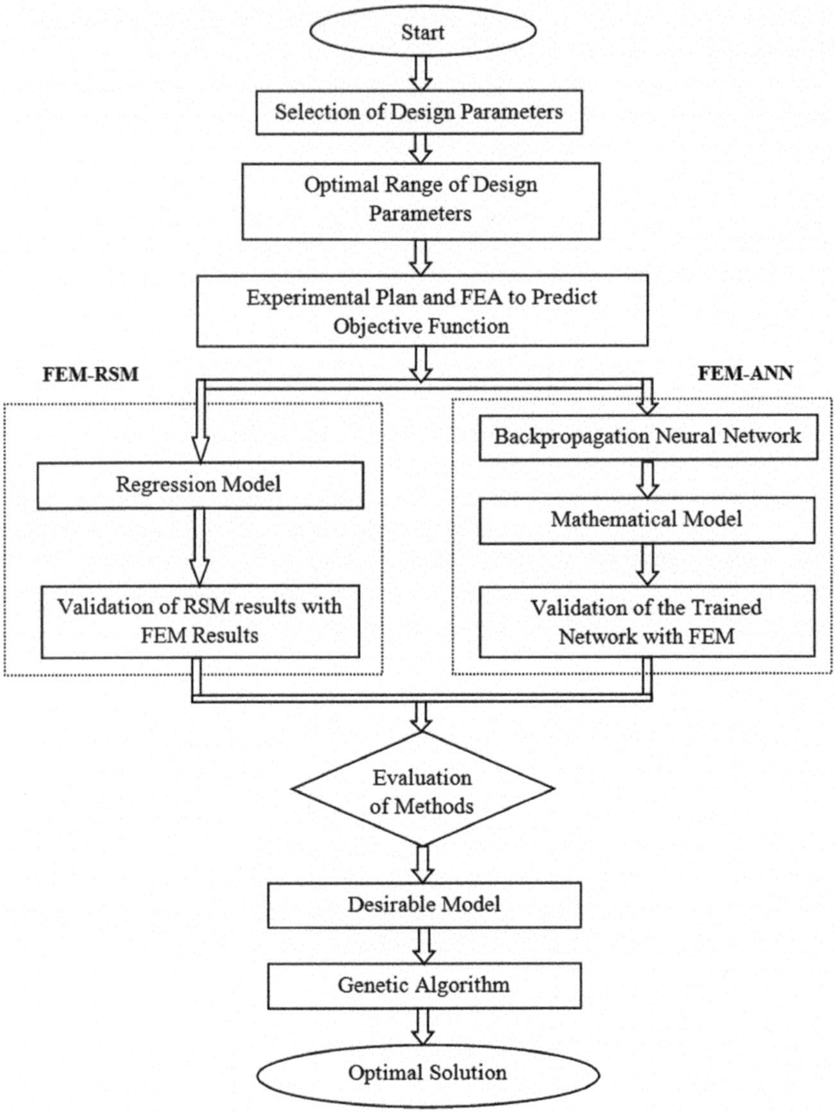

Fig. 1 Fixture layout optimization procedure

2.1 Numerical Description of the Slot Milling Process

In the slot milling process, the workpiece is held in place with the aid of clamps and locators. Hence, the workpiece–fixture system is initially at static condition. The effect of clamping force on the workpiece with the support of locators is considered as static load and is taken into account in the succeeding transient dynamic analysis.

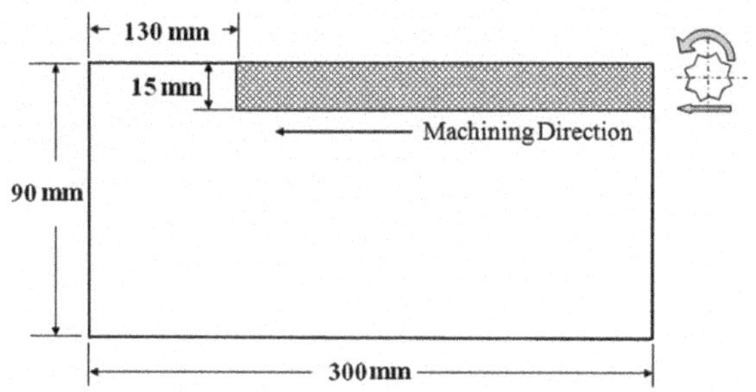

Fig. 2 Workpiece configuration

The second-order linear differential equations for the transient response of a structure at a time 't' and '$(t + \Delta t)$' are given by Eqs. (1) and (2), respectively. The 'Δt' is a small increment of time, and in the present case, it is considered as step size.

$$[M]\{\ddot{u}\} + [C]\{\dot{u}\} + [K]\{u\} = \{F(t)\} \tag{1}$$

$$[M]\{\ddot{u}_{t+\Delta t}\} + [C]\{\dot{u}_{t+\Delta t}\} + [K]\{u_{t+\Delta t}\} = \{F_{t+\Delta t}\} \tag{2}$$

where

$[M]$	Structural mass matrix
$[K]$	Structural stiffness matrix
$\{\ddot{u}\}$	Nodal acceleration vector
$\{u\}$	Nodal displacement vector
$[C]$	Structural damping matrix
$\{F(t)\}$	Applied load vector
$\{F_{t+\Delta t}\}$	Load vector at $(t + \Delta t)$
Δt	Step size.

The workpiece deformation is analysed by applying appropriate machining force with the aid of the above equations. Totally, 14 load steps are considered to simulate the machining process, in which first load step is designed to simulate the static behaviour of the workpiece after the application of clamping force. The final load step is considered to study the system behaviour after the influence of machining force on the workpiece. The rest of the load steps are modelled to depict the workpiece motion during the machining process. Each load step is divided into a suitable number of sub-steps based on Δt. The machining forces are $F_x = 100$ N and $F_y = 286$ N. Clamping forces are $F_1 = 350$ N and $F_2 = 200$ N. Young's modulus and density of the material are considered as $E = 2 \times 10^5$ N/mm^2 and 7800 kg/m^3. The material and machining properties are taken from the work published by Kaya [8]. The workpiece and the slot milling operation are modelled using FEM and are illustrated in Fig. 3.

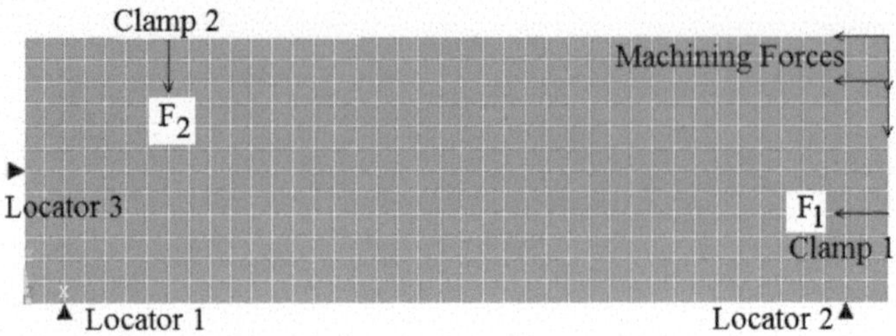

Fig. 3 Finite element model of the workpiece–fixture system with machining force

3 Problem Formulation

The arrangement of locators and clamps in the workpiece–fixture system is shown in Fig. 4. Three locators L_1, L_2 and L_3 are adopted to locate the workpiece concerning the tool. Two clamps C_1 and C_2 are used to restrict the movements of the workpiece against the locators. The position of the three locators and two clamps is measured from the nearer edge and denoted as L_A, L_B, L_C, C_A and C_B. The possible range of the locators and clamps is obtained considering the tool–fixture and fixture–fixture interference constraints and given in Table 1.

Fig. 4 Fixture layout configuration

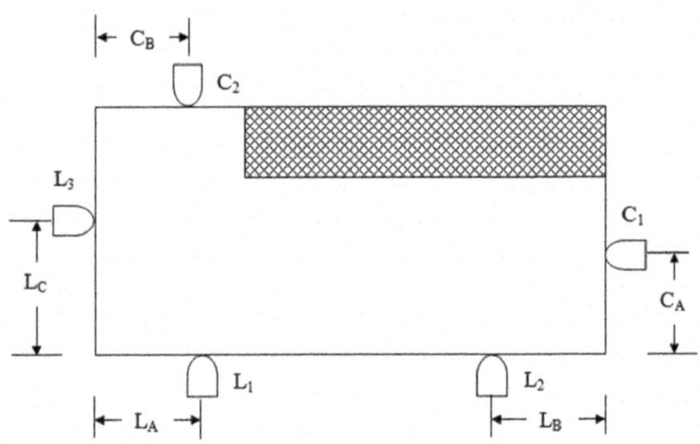

Table 1 Three limits within the position constraint of locators and clamps

Parameters	Unit	Position constraint	Limits		
			1	2	3
L_A	mm	14.444–130	14.444	72.222	130
L_B	mm	14.78–133.04	14.78	73.91	133.04
L_C	mm	15–75	15	45	75
C_A	mm	15–60	15	37.5	60
C_B	mm	14.444–115.56	14.444	65	115.56

3.1 Formulation of Constraint

The range of possible positions of locators and clamps is explicated in the preceding section. The feasible positions of locators and clamps are obtained by applying the following stability constraint given in Eq. (3). It ensures that the feasible fixture layout configurations do not have negative reaction force during machining.

$$\left\{ \begin{matrix} R_{1_k} > 0 \\ R_{2_k} > 0\, 1 \leq k \geq N_L \\ R_{3_k} > 0 \end{matrix} \right\} \tag{3}$$

The terms R_{1_k}, R_{2_k} and R_{3_k} are the reaction forces at locators for the kth load step. The total number of load step is termed as N_L. The static equilibrium principle ($\sum F_x = 0$, $\sum F_y = 0$ and $\sum M_o = 0$) is used to determine the reaction forces at each load step of the machining process.

3.2 Formulation of the Objective Function

The maximum elastic deformation of the workpiece is to be predicted for all feasible fixture layouts to optimize the fixture layout. The layout that shows minimum deformation is selected as an optimal layout. The elastic deformation of the workpiece during machining is obtained in FEM using Eq. 2. The objective function is formulated to find the maximum deformation $U_{\max_j}$ of the workpiece for the specified layout and given in Eq. (4).

$$\left\{ \text{Minimize } U_{\max_j}\ 1 \leq j \geq q \right\} \tag{4}$$

where q = Number of fittest fixture layouts.

$U_{\max_j}$ is the maximum value of deformation obtained during the machining force simulation.

3.3 Experimental Plan and Prediction of Objective Function

The locators and clamps have numerous possible positions within the range of each parameter. It is required to determine the elastic deformation of the workpiece for each fixture layout configuration to determine the optimum fixture layout. In the present work, full factorial (3^5) experiments are planned using three discrete positions within the range of each parameter. Hence, the number of experiments planned is adequate to formulate constructive empirical model. The FEM is used to find the objective

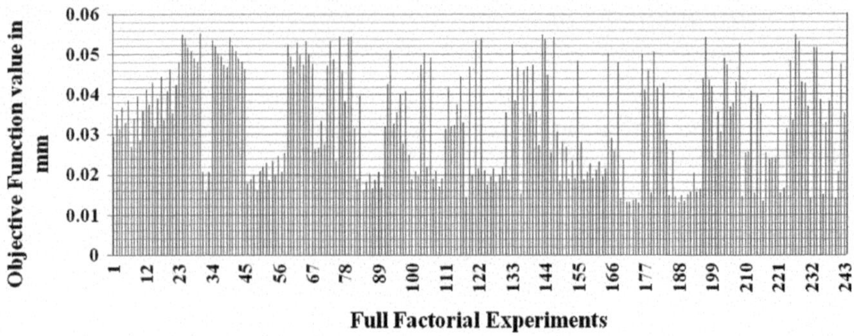

Fig. 5 Objective value prediction for full factorial experiments

value for the full factorial combination of fixture layouts. The objective value of a full factorial design is shown in Fig. 5.

4 Development and Evaluation of Empirical Models

The most efficient empirical model can replace the time-consuming complicated modelling and analysis. However, the selection of suitable empirical model is always crucial, since the prediction accuracy of the model depends on the nature of the problem. In this work, RSM- and ANN-based models are developed using the FEM results, and the models are evaluated against the new set of FEM results to find suitable predictive model. The suitable model is used as an objective function equation in GA.

4.1 Development of RSM-Based Empirical Model

Response surface method employs both mathematical and statistical techniques, which is used to depict the correlation between system responses to the design variables. It is well suited for the problems where the responses do not have a direct relation with design variables. In this work, a mathematical model is derived by mapping the fixture layout and the maximum elastic deformation of the workpiece during machining.

RSM needs a structured initial design and its responses to obtain a mathematical model. It uses a model like linear, quadratic and cubic to approximate the responses. In this work, a statistical software package Design-Expert V7 is used for response surface method analysis. The fittest layouts and its responses are fed to the system as historical data. The quadratic model is fit well for the given data. The derived mathematical model by the statistical software is given below.

Maximum elastic deformation

$$= 0.091029 - L_A 1.44551^{-3} - L_C 3.88211^{-4} + C_A 7.51561^{-5} - C_B 4.35478^{-5} + L_A L_C 1.26386^{-6} - L_A C_A 3.13772^{-6} + L_A C_C 8.45354^{-7} + L_C C_A 1.36527^{-6} + L_C C_B 1.07568^{-6} + C_A C_B 4.50153^{-7} + L_A^2 6.68291^{-6} + L_C^2 2.24811^{-6} + C_A^2 2.05086^{-6} + C_B^2 1.38370^{-8} \quad (5)$$

4.2 *Elastic Deformation Prediction Using ANN*

In recent years, ANN has been successfully applied to numerous manufacturing engineering problems. In this work, it is used to develop an empirical model to predict the elastic deformation of the workpiece for the fixture layout configurations. Feed-forward backpropagation neural network is employed to train the network. It consists of an input layer, an output layer and two hidden layers. The general architecture of the multilayer feed-forward neural network used in this work is shown in Fig. 6. The number of processing units in the input and output layer is equal to that of number of input and output variables. But, there is no general procedure for selecting number of processing units for the hidden layer. Hence, the optimum number is obtained by varying the number of neurons to get minimum mean squared error (MSE). The minimum MSE value is obtained for the two hidden layer networks with ten neurons in each layer, and it is depicted in Fig. 7. The training performance of the network is shown in Fig. 8. The training process is terminated when the maximum epoch is reached. The values of parameters used in the training process are given in Table 2.

The initial weights (IW) and layer weights (LW) are obtained from the trained neural network. The input and layer weights are propagated through activation function to get output from the trained network.

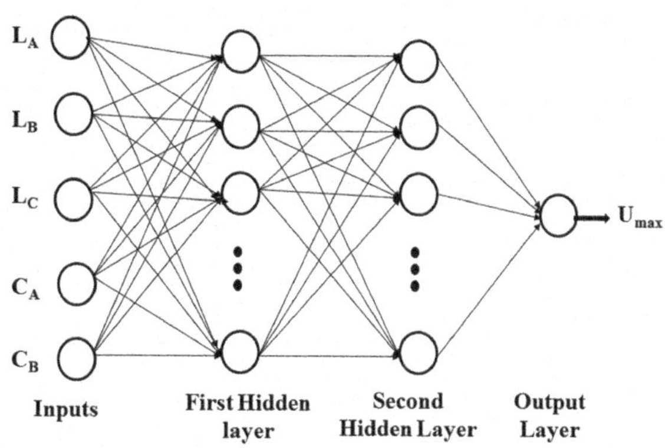

Fig. 6 General architecture of ANN

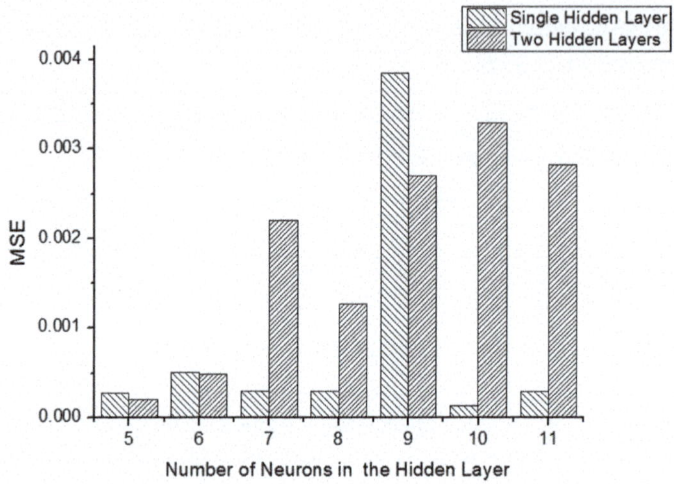

Fig. 7 MSE value distribution for the number of neurons in the hidden layer

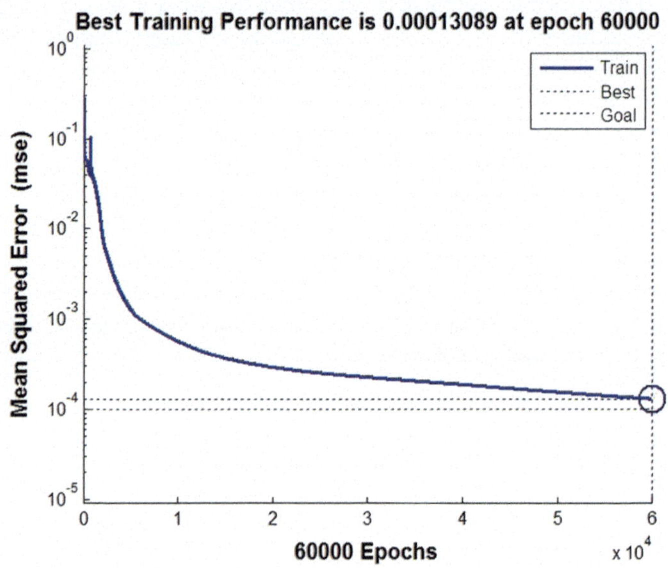

Fig. 8 Training performance of ANN

4.3 Evaluation of Predicted Models

The developed RSM and ANN models are used to find the maximum elastic deformation of the workpiece for new sets of fixture layouts. The new sets of fixture layouts are not used in the training process of RSM and ANN. Table 3 and Fig. 9 show the

Table 2 Parameters used in ANN training

Parameters used in ANN training	Values
Number of hidden layer neuron (*H*)	10
Number of input layer neuron (*I*)	5
Number of output layer neuron (*O*)	1
Number of hidden layers	2
Goal (mean squared error)	0.0001
Number of epochs	60,000
The transfer function for hidden layer and output layer	tansig
The transfer function for input layer	logsig

Table 3 Objective value prediction for the new set of layouts using RSM and ANN

Layout No.	FEM (mm)	ANN (mm)	RSM (mm)
1	0.013115	0.013027	0.012297
2	0.012989	0.012925	0.01245
3	0.01286	0.012839	0.012632
4	0.012571	0.012779	0.012841
5	0.012738	0.012757	0.013078
6	0.01276	0.012784	0.013344
7	0.012843	0.012876	0.013637

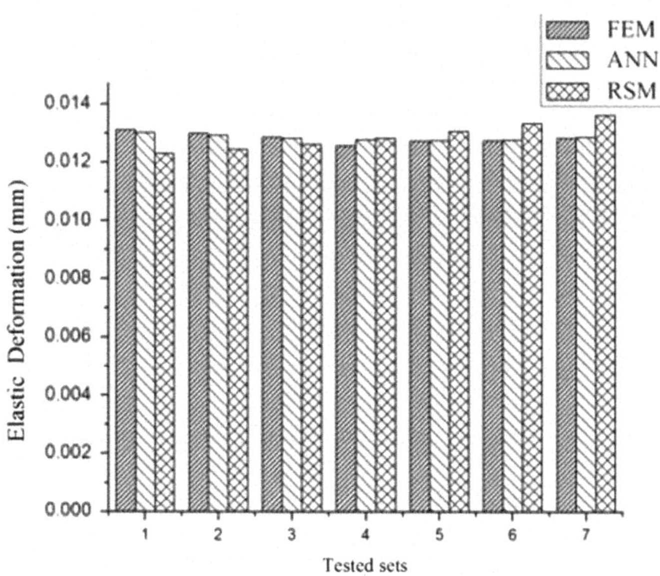

Fig. 9 Validation of predicted model with FEM

Table 4 Prediction capability of RSM and ANN

Parameters	Training data		Tested data	
	RSM (%)	ANN (%)	RSM (%)	ANN (%)
R^2	94.0	99.9	28.3	83.9
R^2_{adj}	93.9	99.9	14.0	80.7

predicted results of the RSM and ANN along with FEM results.

The prediction ability of RSM and ANN is evaluated against the FEM results and shown in Table 4. The developed ANN model has the highest determination coefficients (R^2 and R^2_{adj}) compared to the RSM model for the training and tested data concerning FEM results. The reason attributed that the ANN model is a real substitute for statistical modelling techniques like RSM for the data showing nonlinearities.

5 Optimization Procedure

Genetic algorithm is a robust optimization technique based on the mechanisms of natural evolution and finds application in many engineering optimization problems. It starts with randomly generated initial population and improves from generation to generation by performing biologically inspired operations on it. The tactics behind the GA is survival of the fittest. The fitness value is calculated for each layout in the population using the mathematical model arrived from ANN. The unfit layouts are eliminated from the population and are evolved to the next iteration using GA's biologically inspired operators such as reproduction, crossover and mutation. The population evolution continues until the algorithm reaches the global fitness value for the population. The convergence of GA depends on parameters like population size, the probability of mutation and probability of crossover. The general procedure of GA is given in Fig. 10.

5.1 GA-Based Optimization

A MATLAB-based GA code is developed and implemented to obtain the optimum design parameter. GA randomly generates fixture layouts within the range of the parameters, which constitute the initial population of GA. The layouts are tested with stability constraint and evaluated with the fitness function to find the fitness value. GA uses mutation and crossover operators to produce a new population at every generation. GA generates feasible layouts that satisfy constraints and bounds. The parameters used in GA are illustrated in Table 5.

The objective function model developed using ANN is coupled with GA to predict the optimal setting for the design parameters. The results obtained are presented in

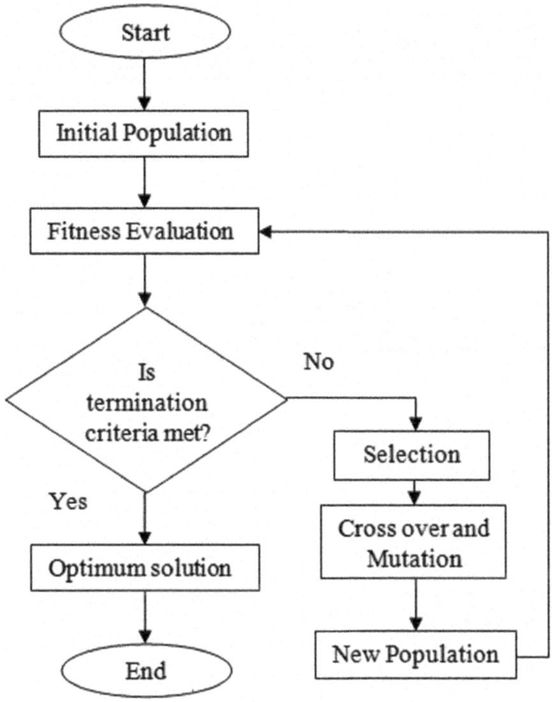

Fig. 10 Flowchart of GA-based optimization technique

Table 5 Parameters in GA

S. No.	Parameter	Value
1	Chromosome encoding	Real value
2	Initial population	50
3	Type of crossover	Scattered
4	Percentage of mutation	1
5	Percentage of crossover	80
6	Termination criteria	100 generation

Table 6. Further, the objective value obtained for the optimal setting of the design parameter is validated with FEM results. The result shows the competence of the FEM-ANN-GA integrated technique.

Table 6 Optimum settings for the parameters

Optimum layout (L_A, L_B, L_C, C_A, C_B) (mm)	ANN-GA predicted response value (mm)	FEM predicted response value (mm)
108.05, 14.78, 45.06, 15.00, 92.97	0.01127	0.01137

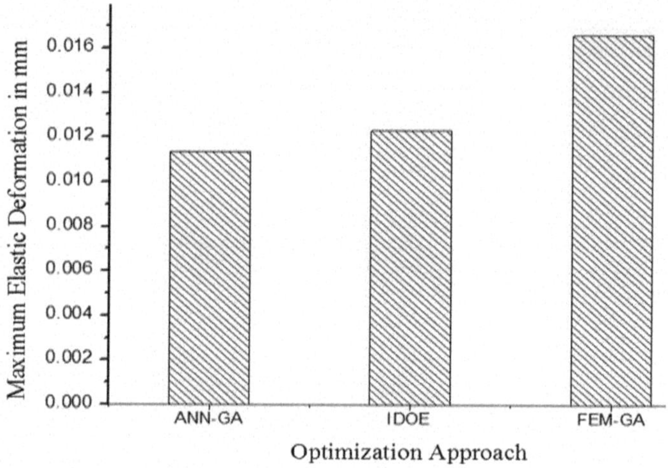

Fig. 11 Comparison of the result of optimization approaches

Further, the optimal objective value found by the ANN-GA approach is compared with the results of IDOE [21] and GA-FEM [8] techniques. The results revealed that ANN-GA approach works well over other approaches, and it is illustrated in Fig. 11. In addition to that, the computation time is considerably reduced due to the use of ANN for FEM calculation.

6 Conclusion

In this study, FEM is successfully used to predict the elastic deformation of the workpiece during machining using transient dynamic analysis. FEM results are used to analyse the influence of fixture layouts on the maximum elastic deformation with the aid of RSM- and ANN-based predictive models. It is found that the ANN model has higher modelling capability than RSM model. It is because the ANN model seems to have better prediction ability in approximating nonlinear correlation between elastic deformation of the workpiece and fixture layout. Hence, the ANN-GA integrated technique is used to identify an optimal fixture layout with a minimum number of trials. The results obtained from the proposed technique are validated with FEM results to ensure its credibility. The computational cost is reduced as the proposed technique determined the optimum fixture layout without the aid of finite element analysis. Hence, it is suggested for designing optimal fixture layout to minimize the deformation of the workpiece.

References

1. Perremans P (1996) Feature-based description of modular fixturing elements: the key to an expert system for the automatic design of the physical fixture. Adv Eng Softw 25:19–27
2. Bai X, Hu F, He G, Ding B (2015) A memetic algorithm for multi-objective fixture layout optimization. Proc IMechE, Part C: J Mech Eng Sci 229:3047–3058
3. Liao YG (2003) A genetic algorithm-based fixture locating positions and clamping schemes optimization. Proc IMechE, Part B: J Eng Manuf 217:1075–1083
4. Kaya N, Ozturk F (2003) The application of chip removal and frictional contact analysis for workpiece–fixture layout verification. Int J Adv Manuf Technol 21:411–419
5. Deiab IM, Elbestawi A (2004) Effect of workpiece/fixture dynamics on the machining process output. Proc IMechE, Part B: J Eng Manuf 218:1541–1553
6. Yu KM, Lam TW, Lee AHC (2003) Immobilization check for fixture design. Proc IMechE, Part B: J Eng Manuf 217:499–512
7. Amaral N, Rencis JJ, Rong Y (2005) Development of a finite element analysis tool for fixture design integrity verification and optimization. Int J Adv Manuf Technol 25:409–419
8. Kaya N (2006) Machining fixture locating and clamping position optimization using genetic algorithms. Comput Ind 57:112–120
9. Deng H, Melkote SN (2006) Determination of minimum clamping forces for dynamically stable fixturing. Int J Mach Tools Manuf 46:847–857
10. Liu S, Zheng L, Zhang ZH, Wen DH (2006) Optimal fixture design in peripheral milling of thin-walled workpiece. Int J Adv Manuf Technol 28:653–658
11. Rex FMT, Ravindran D, Lenin N (2016) Prediction of workpiece elastic deformation using FEM based contact analysis. Appl Mech Mater 852:498–503
12. Siebenaler SP, Melkote SN (2005) Prediction of workpiece deformation in a fixture system using the finite element method. Int J Mach Tools Manuf 46:51–58
13. Chen W, Ni L, Xue J (2007) Deformation control through fixture layout design and clamping force optimization. Int J Adv Manuf Technol 38:860–867
14. Fan L, Kumar AS (2010) Development of robust fixture locating layout for machining workpieces. Proc IMechE, Part B: J Eng Manuf 224:1792–1803
15. Wang BF, Nee AYC (2011) Robust fixture layout with the multi-objective non-dominated ACO/GA approach. CIRP J Manuf Sci Technol 60:183–186
16. Kumar KS, Paulraj G (2012) Geometric error control of workpiece during drilling through optimisation of fixture parameter using a genetic algorithm. Int J Prod Res 50:3450–3469
17. Sun SH, Chen JJ (1996) A fixture design system using case-based reasoning. Eng Appl Artif Intell 9:533–540
18. Dou J, Wang X, Wang L (2012) Machining fixture layout optimisation under dynamic conditions based on evolutionary techniques. Int J Prod Res 50:4294–4315
19. Abedini V, Shakeri M, Siahmargouei MH, Baseri H (2014) Analysis of the influence of machining fixture layout on the workpiece's dimensional accuracy using genetic algorithm. Proc IMechE, Part B: J Eng Manuf 228:1409–1418
20. Sundararaman KA, Padmanaban KP, Sabareeswaran M (2016) Optimization of machining fixture layout using integrated response surface methodology and evolutionary techniques. Proc IMechE, Part C: J Mech Eng Sci 230:2245–2259
21. Rex FMT, Ravindran D (2017) An integrated approach for optimal fixture layout design. Proc IMechE, Part B: J Eng Manuf 231(7):1217–1228
22. Vijay Kumar K, Naveen Sait A (2017) Modelling and optimisation of machining parameters for composite pipes using artificial neural network and genetic algorithm. Int J Interact Des Manuf (IJIDeM) 11:435–443
23. Nagesh DS, Datta GL (2008) Modeling of fillet welded joint of GMAW process: integrated approach using DOE, ANN and GA. Int J Inter Des Manuf (IJIDeM) 2:127–136

Design and Analysis of an Agriculture Solar Panel Support Structure with Tilting Mechanisms

V. Aghilesh, P. Manigandan, Suresh Reddy, K. Raja sekar, and N. Shunmugavelu

Abstract The requirements for solar water pumping system in the agriculture are increased day by day. The performance of the solar electrical power generating system entirely depends on the structural stability of the supporting system. In this paper, an eight-panel solar supporting system is designed and analyzed for its structural efficiency for the high wind loads. The finite-element-based structural analysis is performed using the software package for different tilt positions through tilting mechanism used for everyday application and for the seasonal variation of the sun rays' direction. Quadrilateral and triangular beam elements are used for the mesh generation of the support structure, and the CBUSH and RBE2 are used to model the bolt joints. The wind load is applied as pressure, and the self-weight of the solar panel is applied as a lumped mass and transferred to the main structure through rigid element. The static analysis is performed for the wind loads for the three tilting positions. The results are conforming that the main structural stresses and deformations are within the limits.

Keywords Finite element method · Solar support system · Wind loads · Tilting mechanism

V. Aghilesh
CSIR-National Aerospace Laboratories, Bengaluru, Karnataka 560017, India

P. Manigandan
ESS India LLP, Bengaluru, Karnataka 560016, India

S. Reddy
Seagate Technology, Bengaluru, Karnataka 560087, India

K. Raja sekar (✉) · N. Shunmugavelu
Department of Aeronautical Engineering, Bannari Amman Institute of Technology, Sathyamangalam, Tamil Nadu 638401, India
e-mail: rajasekark@bitsathy.ac.in

L.-J. Yang et al. (eds.), *Proceedings of ICDMC 2019*,
Lecture Notes in Mechanical Engineering,
https://doi.org/10.1007/978-981-15-3631-1_21

1 Introduction

Nowadays, using solar energy for agricultural purpose is continuously increasing to utilize the natural resources available around us. The design of the solar panel support structure is required to move in the desired location to view the sun directly. One such design is made, and the strength of the structure is also confirmed through finite element analysis software [1]. At the same time, the weight of the supporting structures is also important factor for easy transport and handling at the installation site. This way the cost reduction and proper utilization of the material can be achieved. This paper presents the basic design of a solar panel structure with tilting mechanism for seasonal wind. The initial analysis showed the amount of material to be used is on the higher side. To reduce the material cost and the weight, optimization study was made to reduce the weight and to keep the strength of the structure on the safer side. This paper gives the results of the analysis for the structure for different tilt positions. The loads are estimated using wind load design code [2] and validated by computational fluid dynamics (CFD) analysis [3] for the wind passing over the panels. The analysis uses both self-weight of the structure and the wind load. The finite element method is one of the effective tools used in the structural optimization in the mechanical industry. The efficient power production of the solar panel depends on the contact angel between the sun rays and the panel, so the tilting mechanism becomes essential for the solar system. In agricultural application, we cannot find and the power difference with use of automatic tilting mechanism used for every day with respect to sun rays, but the direction of sun rays changes drastically from a season to season, and hence, the seasonal tilting system is sufficient for the agriculture water pumping using solar systems.

Mihailidis et al. [4] analyzed different solar panel support structures, classified the support structures as fixed and adjustable designs, and discussed the methods to estimate the wind loads. Cao et al. [5] conducted the wind tunnel experiments to determine the wind loads and the structural deformations. Bitsuamlaka et al. [6] studied the effect of boundary layer on the solar panel loads. Sarode et al. [7] build a three-dimensional model and analyzed the structural integrity using the FEM-based software package (Fig. 1).

2 Agricultural Solar Panel—Design Requirements

To meet the power requirements for the different pumping system and number of solar panels, the overall system approximate weight is listed in Table 1.

The solar system must be employed with the tilting mechanism for seasonal tilting as well as for everyday tilting with respect to the position of the sun. It is found from the available literature that the sun radiation direction changes with respect to day time and seasons of the year. To achieve maximum efficiency of power generation, the tilt angles are fixed for day tilting and seasonal tilting.

Fig. 1 Eight panels—solar panel supporting structure

Table 1 Estimated weight of solar system

S. No.	Solar pumping system motor capacity (HP)	Number of solar panel required	Approximate weight (kg)
1	3	10	260
2	5	16	410
3	7.5	24	615

2.1 *Problem Description*

An eight-panel solar support system was designed and analyzed for the structural integrity with the help of FEA package Nastran at the wind speed of 180 km/h with 15° seasonal tilting shown in Figs. 2 and 3 which shows the 30°-day tilting of the entire panel system and optimizes the weight for the design requirements.

3 Numerical Formulation

3.1 *Mesh Generation*

The total weight of the eight panels is estimated as 200 kg and applied as the lumped mass on the center of the panel system and the support system meshed using the 1-D beam elements, and bolts are modeled by using the CBUSH and RBE2 elements. The mesh quality is reported in Table 2.

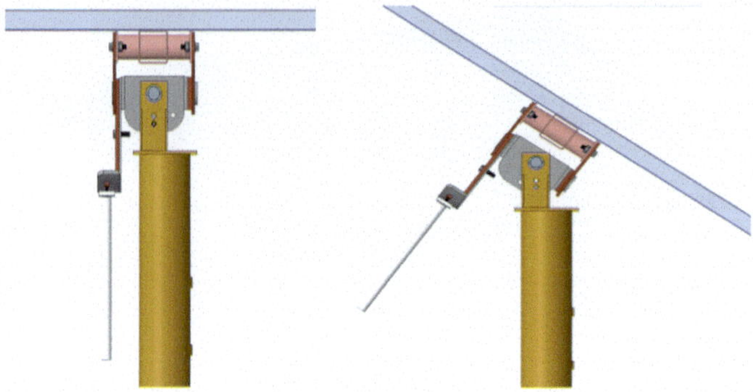

Fig. 2 Seasonal tilting (15 + 15 = 30° from vertical axis)

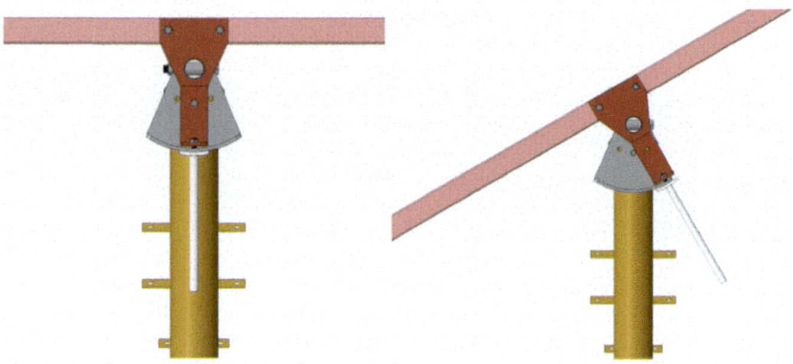

Fig. 3 Day tilting (45 + 45 = 90°)

Table 2 FE mesh quality

Parameters		Value
Type of elements	QUAD4	9890
	TRIA	86[a]
	CBUSH	52
	RBE2	29
Element quality check		
Aspect ratio		≦4.5
Jacobian		≧0.7
Warpage		≦0.0194
Avg. element size		10 mm

[a]TRIA3/QUAD4 = 0.87%

CBUSH elements are assigned with the stiffness of 10^6 and 10^8 MPa, and the panel lumped mass is applied using the CONM2 element. To ensure the accuracy of the results, the ratio between the number of quadrilateral element and triangular element is kept as low as possible in the analysis (Figs. 4 and 5).

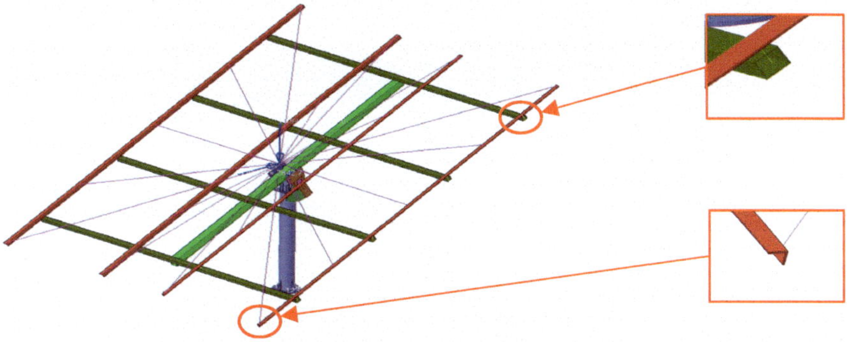

Fig. 4 3D visualization of the 1D beam elements

Fig. 5 Bolt modeled using CBUSH and RBE2 elements

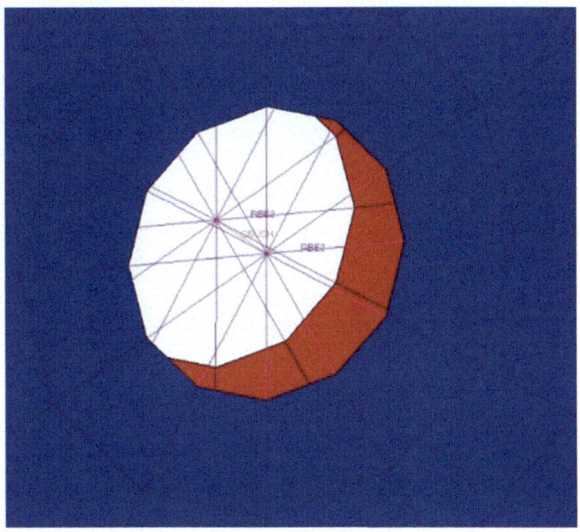

3.2 Load Estimation

Based on design code for wind loads on building and structures IS 875 for the solar panel structure, C_L and C_D are found to be 0.4 and 0.15, respectively, and the load due to the wind force is calculated for different angle attack and found that the load is maximum at 30°-day tilting and 45° seasonal tilting.

$$L = C_L.\frac{1}{2}\rho.V^2.S \tag{1}$$

$$D = C_D.\frac{1}{2}\rho.V^2.S \tag{2}$$

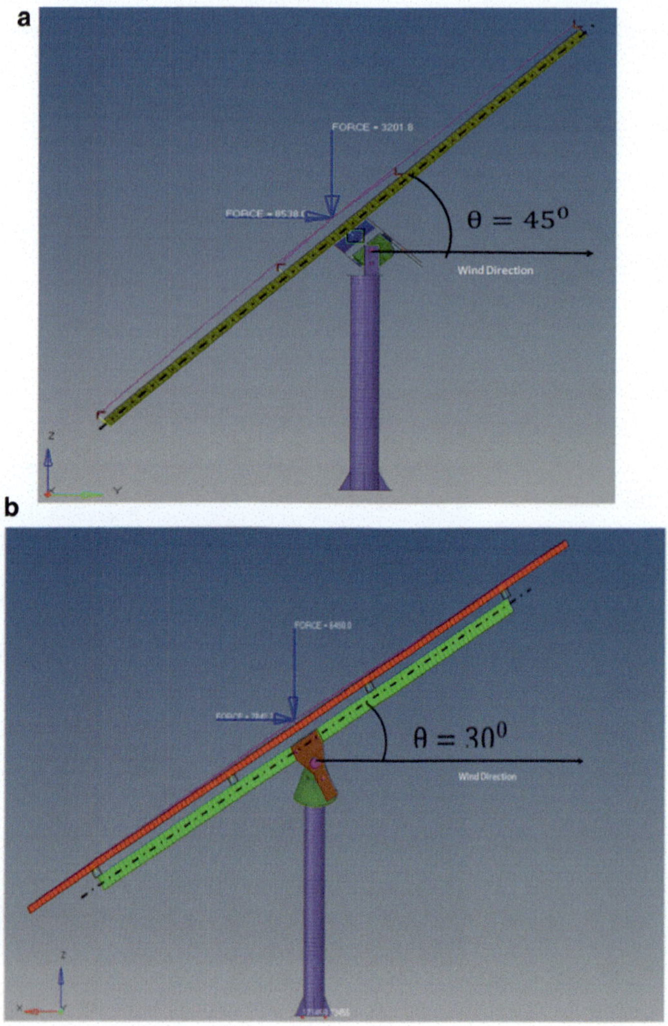

Fig. 6 **a** Wind loads—day tilting. **b** Wind loads—seasonal tilting

Using the above equations, the load on the solar panel is estimated for the different tilting positions. The base of the solar panel support system was arrested for all six degrees of freedom (Fig. 6).

4 Results and Discussion

The entire structure is made of steel having the properties as listed in Table 3.

Table 3 Material properties

Properties	Description
Young's modulus	205 GPa
Poisson ratio	0.3
Density	7870 kg/m^3
Tensile yield strength	370 MPa

Table 4 Analysis results

Condition	Max. stress (MPa)	Max. deflection (mm)	Reserve factor
0° Tilt	21	1.45	17.62
30° Tilt	285	36.17	1.30
45° Tilt	322	45.27	1.15

The static analysis results are within the allowable limits of the design, and the displacement and stress levels are within the yield limit as indicated in Table 4 (Figs. 7, 8, and 9).

5 Conclusions

Finite element method is an efficient tool for determining structural stability of the engineering structures. In this paper, the solar panel support system is analyzed to

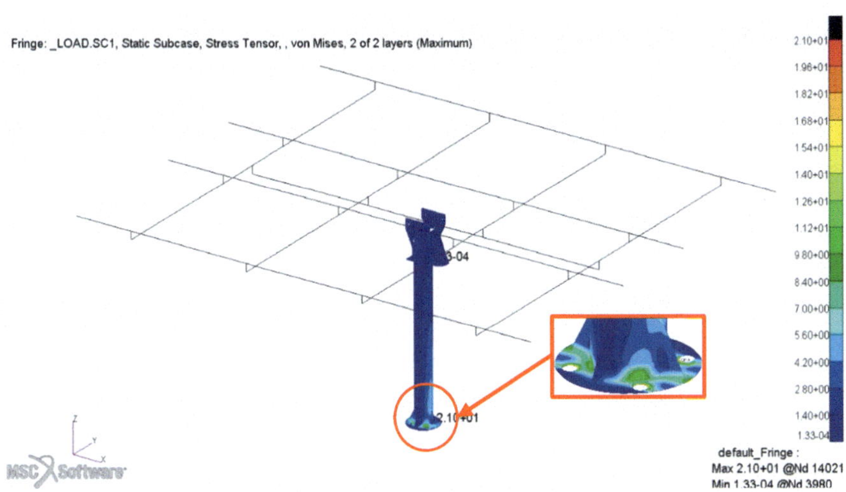

Fig. 7 Von Mises stress distribution for 0° tilt angle

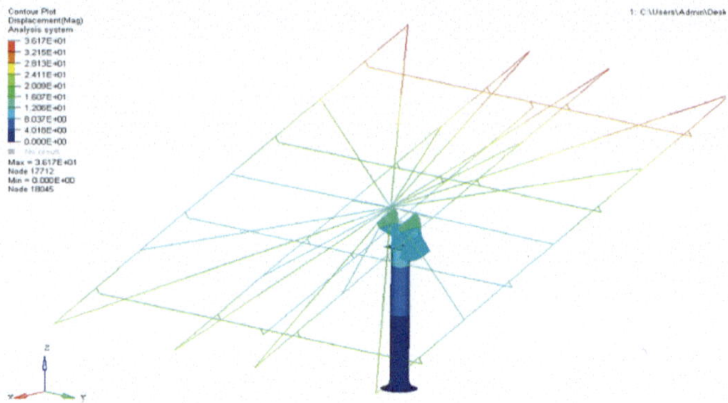

Fig. 8 Static deflection at a titling angle of 30°

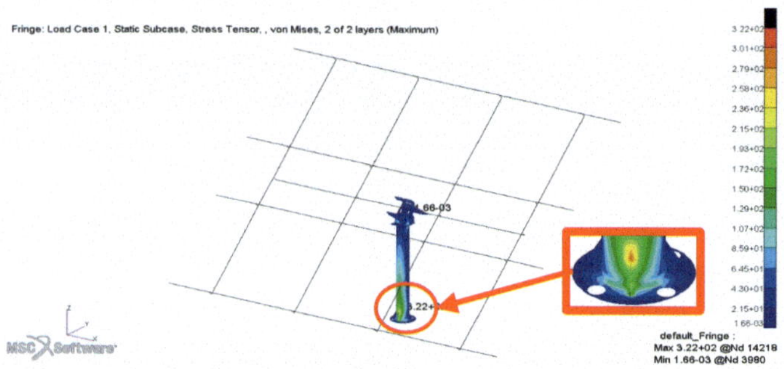

Fig. 9 Von Mises stress distribution for 45° tilt angle

determine the structural behavior for the wind load condition with different tilting angles. It is found that the reserve factors are gradually decreasing as the tilt angle is increasing, and hence, the strength of the supporting structure is highly depending on the tilt angle, and it is safe for 0° and 30° positions and slightly critical at 45° position as the reserve factor is 1.15, i.e., the margin of safety is only 15%. The analysis confirms that the stress levels are within limit. As per Table 4, it clearly indicates the material strength (allowable yield strength) is more than the analyzed results.

References

1. MSC Nastran Quick Reference Guide
2. Indian Standard IS: 875(Part 3) (1987)
3. GadhaviAkash G, Kundaliya DD (2015) Design and analysis of solar panel support structure—a review Paper. Int J Adv Res Eng Sci Technol 2

4. Mihailidis A, Panagiotidis K, Agouridas K (2011) Analysis of solar panel support structures. In: 3rd ANSA &μETA international conference
5. Cao J, Yoshida A, Saha PK, Tamura Y (2013) Wind loading characteristics of solar arrays mounted on flat roofs. J Wind Eng Ind Aerodyn 123:214–225
6. Bitsuamlak GT, Dagnew AK, Erwin J (2010) Evaluation of wind loads on solar panel modules using CFD. In: The fifth international symposium on computational wind engineering (CWE2010) Chapel Hill, North Carolina, USA (2010)
7. Sarode VB, Ulhe PN (2014) Design & optimization of steel structure for solar electrical panel. Int J Res Advent Technol 2:388–394

Design and Fabrication of Ocean Water Pumping and Storage System

M. Muthukannan, S. Anantha Krishnan, and Ajay Pratap Kushwaha

Abstract Existence of fossil fuels in the near future is not promising because of their depletion at a faster rate and their limited availability. Further, owing to the growing concern for the environmental degradation has led to the world's interest in renewable energy resources. In such a scenario, wave power, which is a potential source of energy, can cater to the power needs of future generation and is clean. Our objective is to develop a model which can be used for efficient conversion of wave energy into electrical power. In the present study, the concept of buoyancy has been utilized to pump the ocean water and storing it at a higher elevation. From this elevation, the potential energy of water can be converted into kinetic energy for power generation. From the study undertaken, it is observed that the possibility of electricity generation by using this method and on conducting the experiment, it is observed that for a wave power of 150 W, only 10% of the wave power has been converted and stored in the form of potential energy of water and the remaining unutilized wave power shows that there still exists scope for research work for improving the efficiency of extraction.

Keywords Buoyancy · Efficiency · Potential energy · Wave energy

---

M. Muthukannan (✉) · S. Anantha Krishnan · A. P. Kushwaha
SSM Institute of Engineering and Technology, Dindigul, Tamil Nadu 624002, India
e-mail: mmk.mech59@gmail.com

S. Anantha Krishnan
e-mail: sakrish007@gmail.com

A. P. Kushwaha
e-mail: kushwahapratap0719@gmail.com

L.-J. Yang et al. (eds.), *Proceedings of ICDMC 2019*,
Lecture Notes in Mechanical Engineering,
https://doi.org/10.1007/978-981-15-3631-1_22

1 Introduction

1.1 Need for Wave Energy

Owing to the commitment to reduce the emission of greenhouse gases and to provide adequate energy, efforts are being made to supplement the energy base with renewable resources. Several countries have formulated policy frameworks to ensure that renewable splay an important role in the future scenario. Wave energy is an economically competitive renewable source. It plays a major role in meeting energy demand. With the increasing thrust on renewables, the growth of wave energy will continue in the years to come.

1.2 Wave Energy Basics

Wave energy comes from the interaction between the winds and surfaces of oceans. Ocean wave is due to the periodic to and fro, up and down motion of water particles in the form of progressive waves. The waves originate in different parts of the ocean and travel toward the shore. The energy available varies with the size and frequency of waves. Ocean waves possess both potential energy and kinetic energy. It is estimated that about 50 KW power is available for every meter width of true wave front. The height of the waves depends upon the wind velocities, depth of the ocean, and contour of the shore [1].

1.3 Advantages of Wave Energy

- It is a clean, renewable energy source.
- Wave power devices do not require large land masses.
- The action of large waves minimizes erosion.
- Relatively pollution free.

1.4 Limitation of Wave Energy

- Lack of dependability.
- Relative scarcity of accessible sites of large wave activity.
- Construction of conversion devices is complicated.
- Devices have to withstand enormous power of stormy seas.
- There are unfavorable economic factors such as large capital investment and costs of repair, replacement, and maintenance.

Table 1 Wave energy potential on Indian coasts

No.	Location	Northeast monsoon			Southwest monsoon		
		Mean wave height (m)	Mean wave period (s)	Wave power (kW/m)	Mean wave height (m)	Mean wave periods (s)	Wave power (kW/m)
1	Near Calcutta	1.33	8.00	13.85	1.95	7.65	28.80
2	Near Vizag	1.60	6.25	15.70	2.05	8.25	33.65
3	Near Chennai	1.55	5.85	13.45	1.70	5.80	16.60
4	Near Mumbai	1.00	5.00	4.90	2.65	6.95	47.00

2 Literature Review

2.1 Wave Energy Sites in India

The most attractive locations are the Gulf of Kutch and the Gulf of The Khambhat on the west coast where the maximum tidal range is 11 m and 8 m with average tidal range of 6.77 m and 5.23 m, respectively. The Ganges Delta in the Sunderban in West Bengal also has good locations for small-scale tidal power development. The maximum tidal range in Sunderban is approximately 5 m with an average tidal range of 2.97 m. The identified economic tidal power potential in India is of the order of 8000–9000 MW with about 7000 MW in the Gulf of Khambhat about 1200 MW in the Gulf of Kutch and less than 100 MW in Sundarbans. Wave energy potential on Indian coasts is shown in Table 1 [2].

3 Existing Wave Energy Projects

3.1 Wave Barrages

Wave barrages work like a hydro-electric scheme, but the dam is much bigger. A huge dam or a barrage is built across a river estuary. When the wave comes in and out, water flows through tunnels in the dam. As the wave comes in, the dam allows the sea water to pass through the holding basin. As soon as the wave is about to go down, dam is closed. The water held back in this way will be used to feed turbines at fewer waves. The ebb and flow of the tide can be used to turn a turbine, or it can be used to push air through a pipe, which then turns the turbine [4]. The concept of wave barrages is shown in Fig. 1.

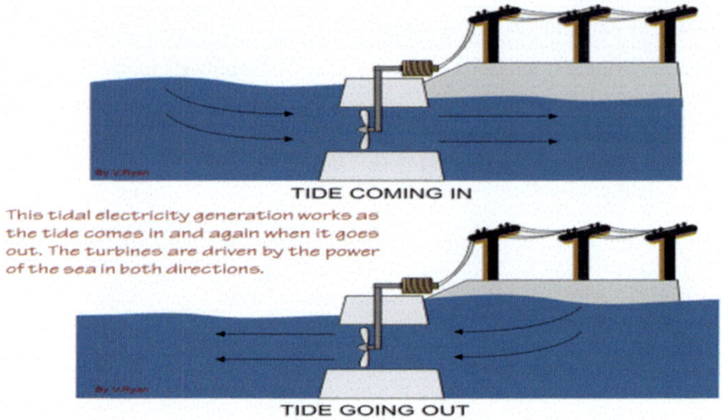

Fig. 1 Concept of tidal power plant

3.2 Marine Propeller

Marine current turbines generate electricity by utilizing wave current flows. The turbine is fixed on a structure and is driven by the flow of the waves. This technology effectively is similar to that of a wind turbine with the rotor blades driven not by wind power but by wave currents. Water has an energy density of more than 800 times that of wind. Twin rotors rotate with the movement of the tidal flow and pitch through 180 degrees to optimally track wave current direction and speed. The marine propeller is shown in Fig. 2.

Fig. 2 Marine propeller

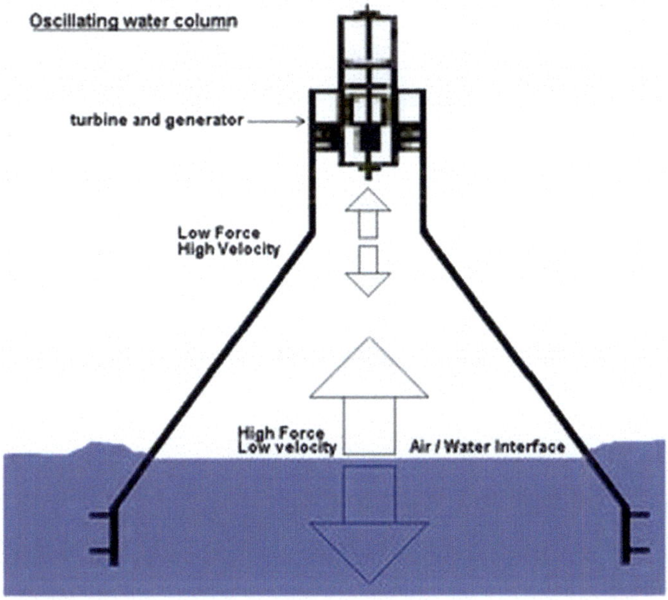

Fig. 3 Concept of OWC

3.3 Oscillating Water Column

An oscillating water column (OWC) is a wave energy converting technology that can be installed onshore preferably on rocky shores. Near shore in up to 10 m of water or offshore in 40–80 m deep water can be used. The device consists of a large wave capture chamber, a platform for an air turbine, a lip, wing walls, and an air chamber. When waves approach the device, they enter under the partially submerged lip that traps air in a piston-type system, forcing the air upwards through the air turbine. This pressure forces the turbine to spin, which is how the energy is harnessed from the waves. As the waves retreat, air enters back into the air chamber from the other side of the turbine. This principle is used in Vizhinjam power plant in Kerala. The concept of OWC is shown in Fig. 3.

4 Factors Affecting Wave Energy

4.1 Wind Speed

With the increase in wind speed, there is an increase in wave energy. The amplitude of the waves depends on wind speed.

4.2 *Depth of Ocean Water*

The greater the depth of ocean waters the higher the wave velocity, very large energy fluxes are available in deep ocean waves which have unlimited power and energy.

4.3 *Effective Pitch Value*

It is the uninterrupted distance on the ocean over which the wind can blow before reaching the point of reference. The larger the distance higher the wave energy. This distance may vary from 5 to 45 km.

5 Project Concept and Working

5.1 *Introduction*

The present project work of ocean water pumping and storage utilizes the concept of buoyancy force of the waves to pump water to higher level for storage.

5.2 *Buoyancy*

Buoyancy is an upward force exerted by a fluid, which opposes the weight of an immersed object. In a column of fluid, pressure increases with depth as a result of the weight of the overlying fluid. Thus, a column of fluid, or an object submerged in the fluid, experiences greater pressure at the bottom of the column than at the top. This difference in pressure results in a net force that tends to accelerate an object upwards. This force is called buoyant force. The magnitude of this force is proportional to the difference in the pressure between the top and the bottom of the column and also equivalent to the weight of the fluid that would otherwise occupy the column, i.e., the displaced fluid. The object should be either less dense than the liquid or is shaped appropriately (as in a boat), so that the force can keep the object afloat. This phenomenon is called buoyancy [3].

5.3 *Reason for Utilization of Buoyancy*

The concept of wave barrages, marine propellers, and oscillating water column had several negative issues. In case of wave barrages and marine propellers, aligning the turbine blades according to the direction of wave current requires an extra yaw mechanism and they have negative effect on marine life. Trapping atmospheric air becomes difficult in oscillating water column if sea level increases. Thus, buoyancy concept was found to eliminate such negative issues and comparatively efficient.

5.4 *Working of Ocean Water Pumping*

The design of assembled model is shown in Fig. 4. Due to the rise and fall of the waves, the buoy will move up and down. The buoy is connected to the piston of a hydraulic pump. The pump has an inlet valve and an outlet valve. During the rise of the waves, the piston moves up and creates a low-pressure area below it in the cylinder. The pressure of water surrounding the cylinder is higher than the pressure inside the cylinder and water flows from high pressure area to low-pressure area. During the fall of the waves, the piston comes down and the total force of the buoy acts on the water trapped in the cylinder. The exit of water through the inlet valve is arrested by using a non-return valve. The water is pushed out through the outlet valve with great pressure. This pressure causes the water to move upwards against the gravity, up to a height of 5.12 m. At this height, a storage tank is kept and the pumped

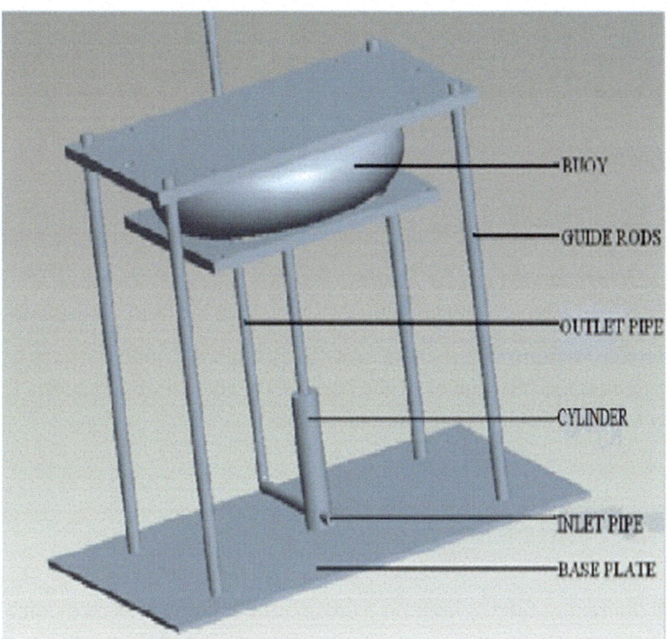

Fig. 4 Pro e-design of assembled model

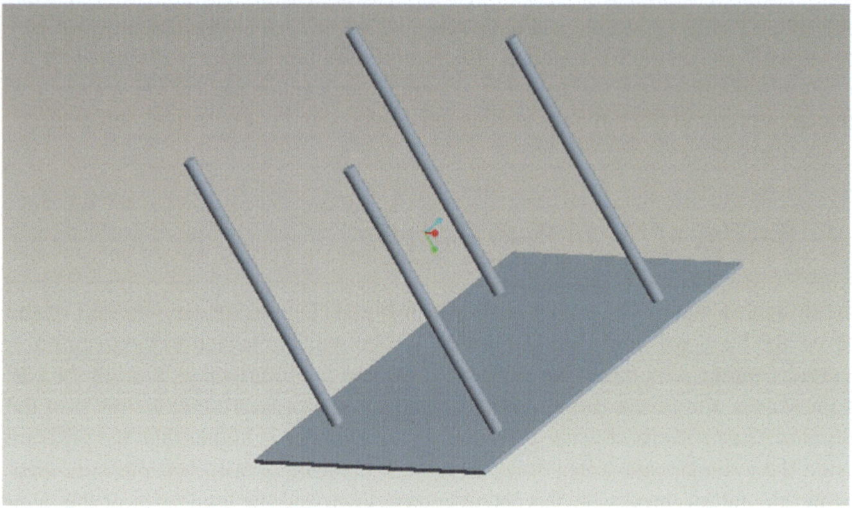

Fig. 5 Pro e-design of base plate and guide rods

ocean water is stored. The return of pumped water into the cylinder is arrested by using another non-return valve in the outlet pipe.

6 Design and Calculations

The design of individual components is as follows.

6.1 Base Plate and Guide Rods

The base plate is a rectangular plate of size 100 × 80 cm. Guide rods are fixed to the base plate to guide the movement of the buoy in up and down direction. The design of the base plate and guide rods is shown in Fig. 5.

6.2 Cylinder and Piston

The internal diameter of the cylinder is 5 cm, external diameter is 6 cm, and its height is 20 cm. The diameter of piston is 5 cm, the diameter of piston rod is 3 cm, and its height is 60 cm. The design of cylinder and piston is shown in Fig. 6.

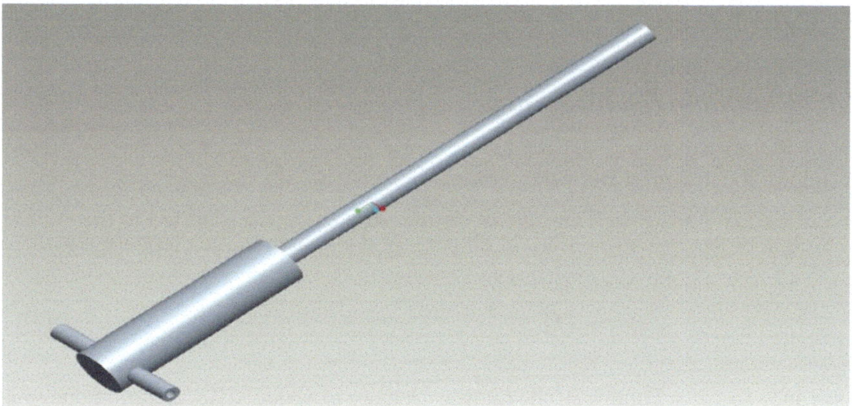

Fig. 6 Pro e-design of cylinder

6.3 *Buoy (Float)*

The buoy consists of two plates with a rubber tube in between. The top plate is of size 77.5 × 55 cm and bottom plate is of 55 × 55 cm. The diameter of the tube is 57.5 cm. The design of buoy is shown in Fig. 7.

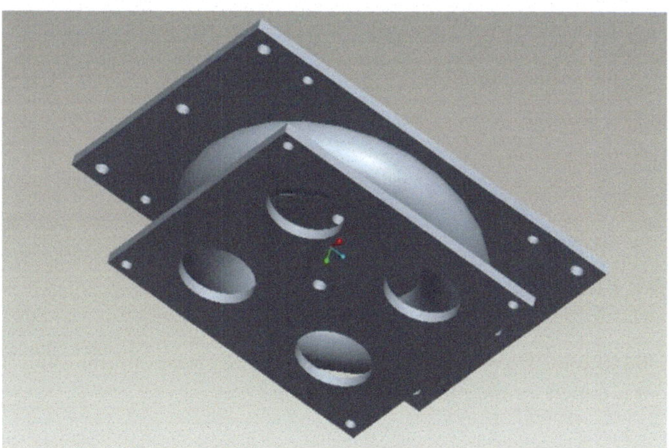

Fig. 7 Pro e-design of buoy

7 Calculations

7.1 Condition for Buoyancy

Weight of the system < Buoyancy force

$$\begin{aligned}\text{Weight of the system} &= \text{Mass of all parts of the system} \times \text{Acceleration due to} \\ &= 30 \times 9.81 \\ &= 294.3\ \text{N}. \end{aligned} \tag{7.1}$$

$$\begin{aligned}\text{Buoyancy force} &= \text{Volume of the buoy} \times \text{Density of water} \times \text{Acceleration due to gravity} \\ &= \left(\pi \times 0.4^2 \times 0.3\right) \times 1000 \times 9.81 \\ &= 1479.3\,\text{N}. \end{aligned} \tag{7.2}$$

Therefore, the weight of the system is less than the buoyancy force.

7.2 Wave Fume Parameters

The Proposed model is going to be tested in the wave flume facility in Anna University.

1. Wave velocity = 1.5 m/s
2. Wave height = 8 cm
3. Time period of consecutive waves = 2 s.

7.3 Water Pumping Calculation

Since the maximum wave height is 8 cm in the wave flume, the maximum displacement attained by the piston will be 8 cm.

$$\begin{aligned}\text{Working area of the piston} &= \pi \times r^2 \\ &= \pi \times .025^2 \\ &= 0.0019625\,\text{m}^2 \end{aligned} \tag{7.3}$$

$$\begin{aligned}\text{Pressure inside the cylinder} &= \text{force/area} \\ &= 300/0.00196 \\ &= 1.52\ \text{bar}. \end{aligned} \tag{7.4}$$

$$\begin{aligned}\text{Volume of piston displacement} &= \pi \times r^2 \times h \\ &= \pi \times 0.025^2 \times 0.08 \\ &= 0.0001568\,\text{m}^3.\end{aligned}$$

$$\begin{aligned}\text{Volume of water displacement} &= 0.0001568/2 \\ &= 0.0000784\ \text{m}^3/\text{s}\end{aligned}$$

Inner diameter of connecting pipe = 1 cm.

Cross sectional area of pipe = $\pi \times 0.005^2 = 0.000078\,\text{m}^2$

Quantity of discharge = area × velocity

$\therefore 0.0000784$

$= 0.000078 \times$ velocity

Velocity at inlet of the connecting pipe = 1 m/s.

By Bernoulli's equation, [5]

$$\begin{aligned}P_1/\rho g + v_1^2/2g + z_1 &= P_2/\rho g + v_2^2/2g + z_2 \\ &\times \frac{1.52}{9810} + \frac{1^2}{19.62} + 0 = \frac{1.03}{9810} + 0 + z_2 \\ &\times \text{(since exit velocity and head at inlet is zero)} \\ &\times 1.55 \times 10^{-4} + 0.05 = 1.15 \times 10^{-4} + z_2 \\ \text{Hence } z_2 &= 5.05\ \text{m}. \end{aligned} \tag{4.8}$$

From the data, it is showed that water can be raised to the height of 5.05 m.
Efficiency of utilization from wave power:
Available power on the generated wave,

$$P_{\text{available}} = \frac{1}{64\Pi}\rho g^2 H^2 T$$

where ρ is the density of water, g is the acceleration due to gravity, H is the wave height from trough to crest, and T is the time period between two consecutive waves.

Therefore, available power,

$$P = \frac{1}{64\Pi} \times 1000 \times 9.812 \times 0.082 \times 2 = 6.13\ \text{W}$$

Power generated on the experimental setup,

$P = 0.33$W Potential energy collected on container per second

$$= (\text{mass of water delivered per second in the container} \times \text{acceleration due to gravity} \times \text{height of the container from ground})$$
$$= 0.014 \times 9.81 \times 2.4$$
$$= 0.33\ \text{W}$$

Therefore, efficiency of newly developed device,

$$\eta = \left(P_{\text{generated}}/P_{\text{available}}\right) \times 100\%$$
$$= (0.33\ \text{W}/6.13\ \text{W}) \times 100$$
$$= 5.38\%$$

If water can be stored at the maximum possible height of 5.05 m, then the efficiency of the device becomes 11.3%.

8 Fabrication

8.1 Material Selection

The first step in the fabrication is to select the materials for individual components. The foundation for the complete model is the base plate. It should withstand the dead load and working pressures of the model. The cylinder should withstand the pressure exerted by the piston. As the complete model is in contact with water, all the components must be corrosion resistant. So the following materials were found to efficiently suit our requirements.

1. Base plate—Mild steel
2. Guide rods—Stainless steel
3. Cylinder—Acrylic tube
4. Piston—PVC
5. Piston rod—Stainless steel
6. Buoy—Acrylic plates and rubber tube
7. Bolts, nuts, and clamps—Stainless steel.

8.2 Procedure

Raw base plate was procured and cut to the required size (100 × 80 cm) with the help of shearing machine. Markings were made on the base plate for guide rods and

Fig. 8 Assembled view of base plate and guide rods

cylinder position. The base for the guide rods was made by turning, drilling, and internal threading operations. The base for guide rods was fixed to the base plate by welding. The guide rods were inserted into their bases and bolted firmly. The bolt and the base were welded. An extra bolt was provided to increase the rigidity. Assembled view of base plate and guide rods is shown in Fig. 8.

For the cylinder, the standard 5 cm inner diameter acrylic tube was procured and cut to required length with cutter. Two holes were drilled in the cylinder at the bottom with internal threads for inlet and outlet valves. For valves, the standard 1 cm inner diameter acrylic tubes were procured and threaded both sides.

The valves were affixed to cylinder and non-return valves (NRV) are attached. Cylinder was fixed to the base plate with flange and bolts. Araldite was used to increase the firmness. Piston was made of PVC. PVC is machined by turning operations to reach an outer diameter of 5 cm. Internal threading was made at the top of the piston by tapping. Piston rod was inserted into the threaded hole on the top of the piston and bolted. Assembled view of cylinder and NRV is shown in Fig. 9.

For buoy, a large acrylic sheet was procured. It was cut into required size by cutting machine.

Markings were done on the acrylic sheet for drilling holes. By manual drilling, holes were drilled on the acrylic sheet for insert in guide rods and bolting.

A PVC bush was inserted into the guide holes and bolted. A rubber tube was inflated and kept between the acrylic plates and bolted. Assembled view of buoy is shown in Fig. 10.

Fig. 9 Assembled view of cylinder and NRV

Fig. 10 Assembled view of buoy

1. Finally, the PVC tube for raising the water was attached to the outlet valve and clamped for support. Assembled view of model is shown in Fig. 11.

9 Conclusions

The ocean water pumping system has been designed and fabricated. The project utilizes the ocean waves which has unlimited energy and also available in abundance. It does not need any fuel which reduces the fuel cost. Thus, the project will help to meet the growing concern for the increasing energy demand by harnessing wave energy.

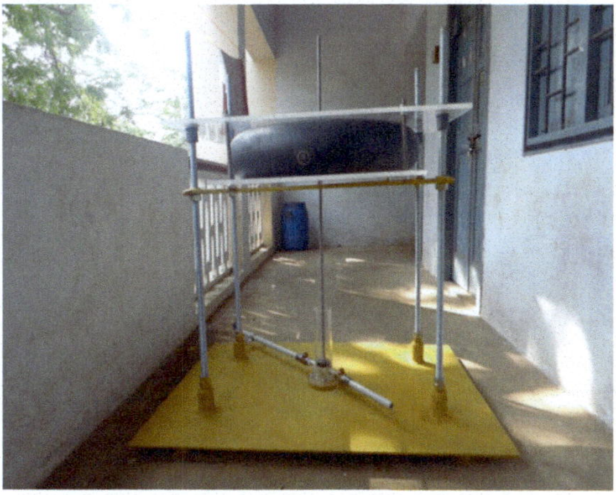

Fig. 11 Assembled view of model

In the future, the optimization will be done in buoy and other parts of the system in order to increase the efficiency even better and also it further requires research on the installation methods and designing based upon the varying wave parameters.

References

1. Kothari DP (2008) Renewable energy sources & emerging technologies. PHI Learning Private Limited, pp 101–128
2. Elsevier (2003) Elsevier ocean engineering series. Elsevier Publications Ltd., pp 41–78
3. McCormick ME (2003) Ocean wave energy conversion. McCormick Publications, pp 54–81
4. http://www.inhabitat.com/wavebarrages-generates-electricity-from-waves
5. http://www.engineeringtoolbox.com/pumppower

Influence of Cylindrical Threaded Tool Pin Profile on the Mechanical and Metallurgical Properties of FSW of ZE42 Magnesium Alloy

A. K. Darwins, M. Satheesh, and G. Ramanan

Abstract Advanced ZE42 series of magnesium alloys are constantly developed due to its outstanding properties in terms of any occasional earth components. The ZE42 magnesium alloy is friction stir welded with threaded cylindrical tool pin profile at various rotational and welding speeds, and the influence of the process parameters over the responses is studied. In the study of observing the quality of weld parameters through the properties of welded joints are analysed using the mechanical tests, including microhardness, tensile strength and metallurgical testing such as X-Ray diffraction, FTIR and thermogravimetric analysis. The welded joints possess enhanced strength and consistent quality of surface that represent a solidity of the process parameters that were considered. The results depicted that the existence of Cu particles in stir zone was due to tool wear, and the development of a ferrite affects the tensile properties at the rotational speed of about 850 rpm, where reduction of strength is due to the HAZ softening. The tensile samples broke at base material, which proved that the joints obtained higher strength. The microhardness peaks of 78 HV were rectified in all the joints at 1000–1200 rpm. Microstructure images clearly show the tool contamination in the weld region. FTIR results help in distinguishing strong absorption appear in the weld zone region which is easily recognized by bands.

Keywords Magnesium alloy · Friction stir welding · Microstructure · Hardness · X-ray diffraction

A. K. Darwins (✉) · M. Satheesh
Department of Mechanical Engineering, Noorul Islam Centre for Higher Education, Kumaracoil, India
e-mail: akdarwins@gmail.com

M. Satheesh
e-mail: satheeshudhaya@gmail.com

G. Ramanan
Department of Aeronautical Engineering, ACS College of Engineering, Bengaluru, India
e-mail: gramanan1987@gmail.com

L.-J. Yang et al. (eds.), *Proceedings of ICDMC 2019*,
Lecture Notes in Mechanical Engineering,
https://doi.org/10.1007/978-981-15-3631-1_23

1 Introduction

The magnesium alloys of series ZE42 attracted the interest due to its enhanced mechanical properties comparable as compared to various other unusual earth components that are used in commercial fields, such as aerospace, automobile and transport [1]. The microstructure and thermal evaluation of the sheet particles of Mg alloy concluded that the alloy exhibits enhanced tensile properties even at varying temperatures [2, 3]. The optimization of ZE42 alloys in terms of composition finds its application in greater age-hardening response, the behaviour of discharge and properties associated with its electrochemical nature of the anode of Mg–Al–Sn alloy. Though, comparatively diminutive consideration was rewarded to the process of welding of alloys of Mg–Al–Sn [4]. Moreover, the dependable process of welding is crucial in case of commercial area of applications of Mg–Al–Sn. FSW is a technology of green solid-state joining that tends to get rid of the issues associated with welding usually related to traditional fusion welding technologies [5]. The tapered tool pin profile effect on mechanical and the microstructure characteristics of FSW ZE42 magnesium alloy are detailed in [6]. Author exposed about effect of micro-structural variations and tempted residual stresses on AZ31 magnesium alloy joints tensile properties after FSW [7]. The multi-response parameter optimization for the process of FSW AM20 Mg alloy is performed using Taguchi grey relational analysis. In addition, the rare-earth consisting of ZEK120 sheets of Mg alloy have been effectively combined by means of friction stir spot welding [8]. Although only few restricted research on FSW Mg-ZE42 alloys were performed, and in the preceding research, we described that the alloys of Mg have been effectively welded by FSW [9]. Though, there exists no other investigation exists in revealing the micro-structural change, DSC, FTIR and mechanical properties of ZE42 alloys later to FSW at numerous welding speeds for threaded tool pin profile. Consequently, this research targets principally on the special influence of FSW over XRD, FTIR [10, 11], microstructure, mechanical and the thermogravimetric characteristics of Mg-ZE42 alloy for dissimilar speeds of welding involving specific attention in the analysis of association between mechanical and microstructure properties of Mg–5Al–3Sn after FSW.

2 Materials and Methods

The considered materials are the base known as Mg-ZE42 plates with the dimensions of 120 mm (length) × 50 mm (width) × 6 mm (thickness). A simple tool of FSW made up of H13 steel has been used with the combination of a tapered tool pin of 3.7 mm length and 5 mm diameter. While holding base material rigidly the positioning over the anvil and the FSW process has been performed over the direction of extruding at the shoulder plunge with depth of about 0.15 mm that is set with the FSW machine at 2.5 tilt angle. With the involvement of greater choice of numerous pre-experiments, it has been finalized that the rate of rotation has to be approximately 1050 rpm in

Fig. 1 Vertical milling machine used in FSW

order to attain less defect joints of weld holding outstanding mechanical and exterior properties with FSW, confident range of speed of welding with microstructure study [10]. Thus, the constant rate of rotation was 1050 rpm. For the examination of the properties and microstructure of Mg-ZE42 after the completion of FSW at numerous speeds of welding, three dissimilar speeds of welding have been utilized in this research. The vertical milling machine setup used for FSW is shown in Fig. 1.

The joints after FSW is analysed using the X-ray radiography and thermogravimetric analysis have been accomplished over the welds cross-section perpendicular to the direction of welding [12]. Later on grinding mechanically deprived of polishing, the chemical etching of weld cross-sections using a combination of acetic acid of 5 ml, ethanol of 120 ml, picric acid of 5 g and distilled water of 10 ml is done. Joints microstructures have been perceived by an optical microscope and scanning electron microscope (SEM) at about 20 kV with the INCA Energy 350 energy-dispersive X-ray spectroscopy (EDS) system. Once fragmented to little pieces and place it in a crucible, the sample's thermal analysis has been performed with the temperature of 50 ml/min and 10 °C/min argon flow rate by utilizing the differential scanning calorimetric. On successful completion after grinding of samples utilizing papers of silicon, crystallographic texture distribution and phase composition from top surface of surface have been analysed using X-ray diffractometer (XRD) with the Cu Kα radiation of about 30 mA and 60 kV with 10–90° of tilt angle. In transverse direction, hardness has been measured over the centre of the joint transversely over the mid-thickness for each 1 mm positioning of a load of 120 g for about 10 s on the MH-3 Vickers hardness tester. Transverse tensile test samples have been perpendicularly cut to extrusion direction for weld joints and BM before performing annealing and after performing annealing and the area that has been welded was put in centre within the gauge length of 50 mm, satisfying the standard of ASTM [13]. Tensile tests have been performed over a mechanical tester, namely CMT-5105 with the crosshead speed of about 0.5 mm/min at the room temperature. Outcomes from the tensile analysis have been grabbed off from specimens that count three in safeguarding investigational outcomes accuracy and reliability.

3 Results and Discussion

3.1 Mechanical Properties

Distribution in hardness of Vickers through joint centre that is measured laterally with mid-thickness is illustrated in Fig. 2. The BM hardness tends to be in choice of 60–73 HV. It has been identified that the hardness distribution profile showed subsequently to FSW. The profile hardness exposes minor disparity with collective welding speed. Hardness values at lower range have been experimental in HAZ at irrespective speed of welding [12]. The yield strength (YS), BM elongation and the ultimate tensile strength (UTS) tend to resolute at about 180, 150 MPa and 8% correspondingly is illustrated in Fig. 3. Notification of the elastic characteristics known as the Young's modulus of BM compared to weld joints has been dissimilar [8]. After FSW, the decrease in elongation and strength deprived of concern to the speed of welding, particularly in case of elongation. Both YS and UTS decreases with the increase in speed of welding has been noticed, though the cumulative degree of UTS was

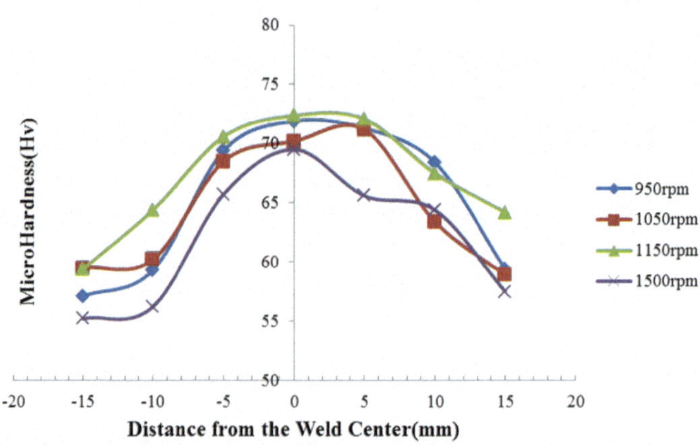

Fig. 2 Properties of hardness of ZE42 Mg alloy processed with FSW

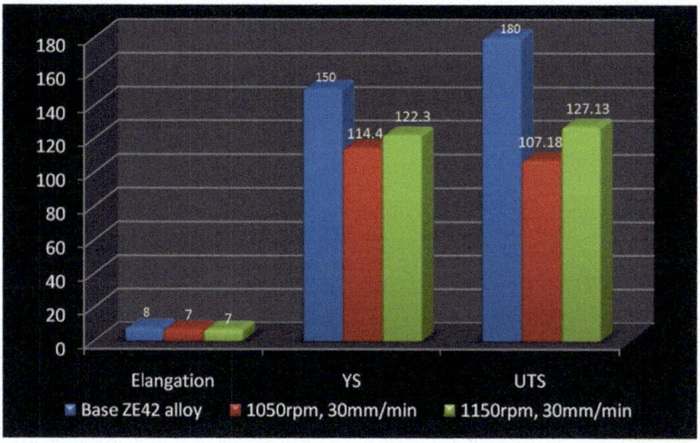

Fig. 3 Histogram image of the tensile properties corresponding to ZE42 magnesium alloy

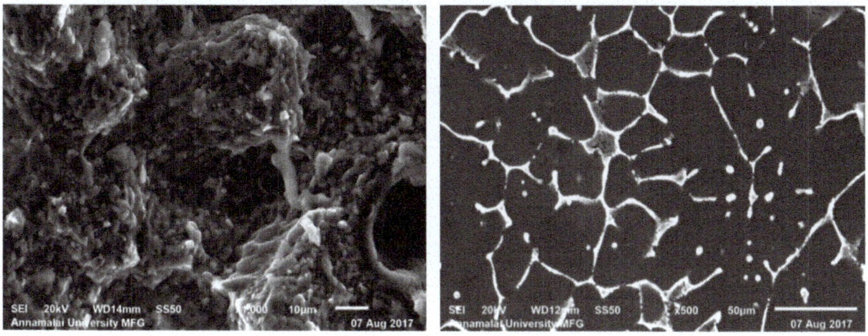

Fig. 4 SEM images of weld zone of FSW and tensile fractography of ZE42 alloy

not much sure to YS. Joints elongation after FSW had unnoticeable variations with rising welding speed. The SEM characteristic micrographs foe the tensile fracture surfaces of BM with 120 mm/min joint are illustrated in Fig. 4. In case of BM, the inclusions of elongated dimples in the tensile fracture escorted with certain tear ridges, as illustrated in Fig. 4. Also, dimple-like features is illustrated at the fracture surface of weld joints after the completion of FSW.

3.2 *Microstructure*

The microstructures of SEM at dissimilar zones of cross-section for distinctive joints are illustrated in Fig. 5. The XRD and EDAX analysis for Mg alloy is depicted in Figs. 6 and 7 correspondingly. Figure 6 reveals the XRD for NZ in joints after the completion of FSW [09]. Great range of peaks is identified with the rapid rise of

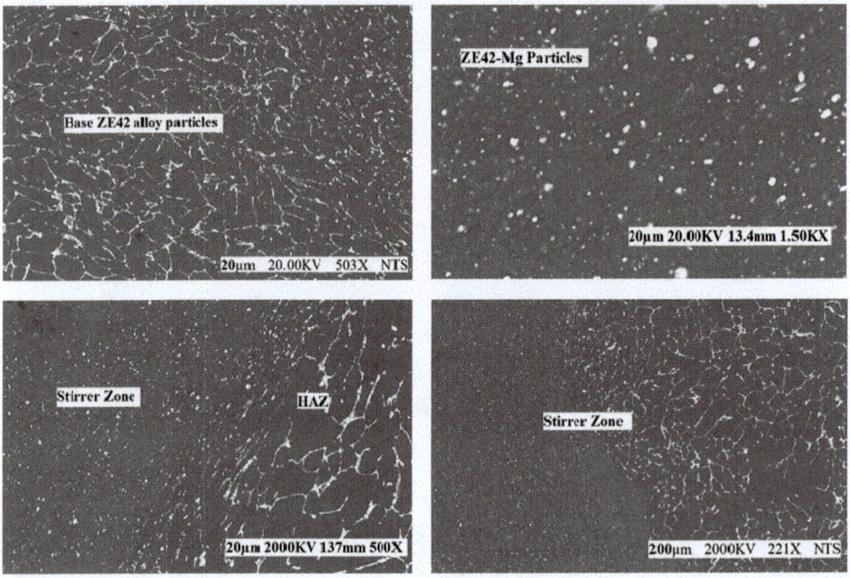

Fig. 5 Microstructure of FSW processed ZE42 alloy at three zones

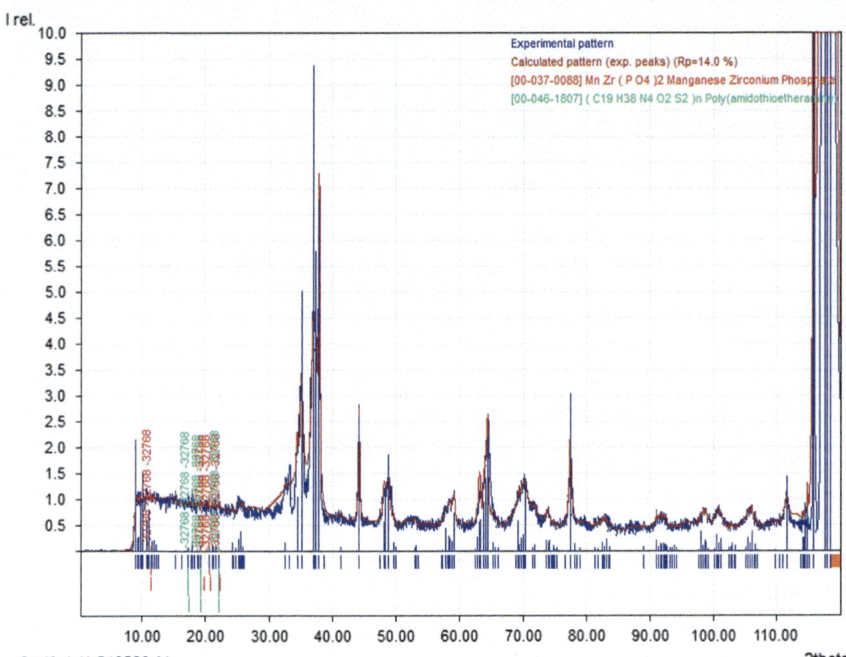

Fig. 6 XRD analysis of FSW processed Mg-ZE42 alloy

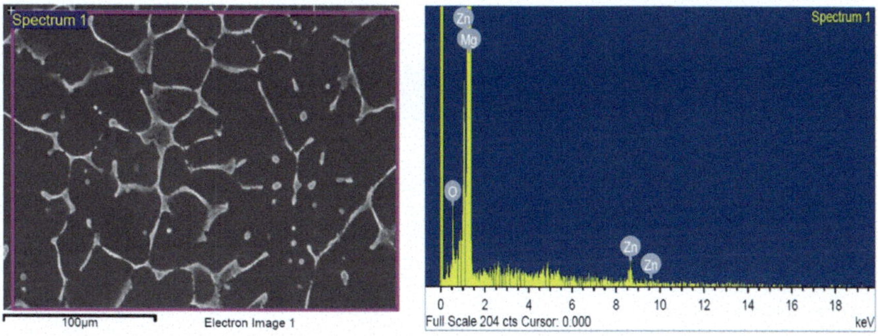

Fig. 7 EDS analysis of ZE42 alloy joints after FSW

speed of welding at a continuous rate of rotation of 1050 rpm. The size of grain of Mg matrix diminished, though few Zn, which acts as the intermetallic particles that are scratched in the extraordinary speed of welding of 120 mm/min. It discloses that ZE42 Mg phase has been softened, whereas base alloy phase persisted in NZ of welds after the completion of FSW has been proved reliable in SEM and XRD analysis [12].

3.3 *Thermal Analysis*

It is identified that the temperature associated to the peak attained at the time of processing of friction stir that ranges from 600 to 650 °C with constant rotational speed of 1050 rpm and translational speeds of 20–30 in/min and the peak temperature increases with the increase in rate of rotation. The temperature during welding of the commercial ZE42 Mg alloy when FSW was in the range of about 648 °C at 200 mm/min and 950 rpm is depicted in Fig. 8. Consequently, numerous reasons for believing that the temperature at maximum range while FSW does not overdo 600 °C for the parameters of process involved in the current research. The phase of Mg–Zn is provided in the presence of high thermal stability, and the point of dissolution is abundantly greater than the temperature of about 600 °C [13]. Thus, the Mg–Zn phase with improved thermal stability is used in NZ after FSW. The warm strength of the base metal is less than that of the Mg-ZE42. Henceforth, with the increase in percentage of alloy in the composition its primer of initial temperature additions or fortified [12]. This takes place due to the adequate heating of HAZ throughout FSW in the absence of plastic deformation of novel grains using cold work recovery and precipitates coarsening. The grains that are present in the TMAZ are changed with convinced metal-flow direction orientation due to stirring for the period of FSW.

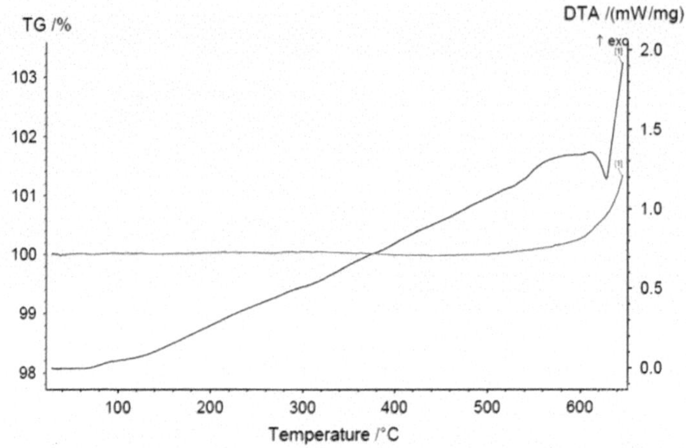

Fig. 8 DSC heating temperature curves of FSW processed ZE42 alloy

3.4 *Fourier-Transform Infrared Spectrum*

Fourier-Transform Infrared is just passing infrared radiation bound to provide a rapid and qualitative specification with the consideration of modification in chemical structure. ZE42 FSW spectrum has been managed are illustrated in Fig. 9. Broad adsorption band in area of 1120 cm^{-1} are alkenes residues, which are noticed using O–H deformation absorptions [11] and C–H stretching. The stretching absorption bands of C–O are available just below 1600 cm^{-1} one towards asymmetrical vibration and other towards symmetrical vibration [14].

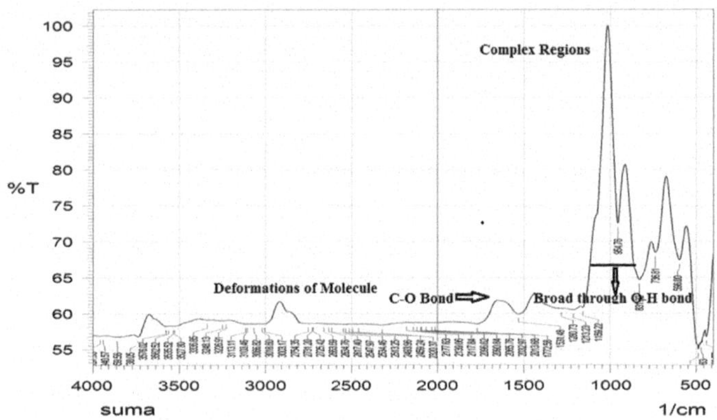

Fig. 9 FTIR result of tapper tool FSW processed ZE42 alloy

4 Conclusions

In this research, friction stir welding (FSW) is done in the ZE42 magnesium alloy involving welding speeds between the range of 120 and 120 mm/min with constant rotational speed of 1050 rpm. In addition, the characteristics of FSW on XRD, microstructure, EDAX, DSC, FTIR, tensile properties, and the hardness of magnesium alloy Mg-ZE42 have been examined. Predominant decisions are stated as,

(1) The magnesium alloy Mg-ZE42 has less joint imperfections at numerous speeds of welding with FSW process. Joint which is welded at 120 mm/min provides higher ultimate tensile strength (UTS) of 128 MPa. In every scenario, the tensile strength of joints was less compared to the BM strength after FSW that is due to the phase dissolution of nugget zone (NZ). The NZ is the region that is softened in the heat-affected zone (HAZ) with the development of grain, residual stress and the TMAZ content of dislocation.
(2) On increasing the choice of speed of welding, the aggregate degree of the UTS has not been noticeable similar to the yield strength (YS) of the joints after performing FSW, as the influence of welding speed on YS has been involved deeper compared to the UTS.
(3) XRD and EDS results confirm the existence of the alloy particles of Zn and Mg in the peak, with the compounds in the alloy. SEM images indicate that the speed zone elements of welding are consistently disseminated and the interfaces among the particles are calculated.
(4) TGA/DTA illustrates that with the increase in percentage of alloy the rate of ignition also increases. Intensity of FTIR is concerning 3120.42 cm^{-1} band indicating greater attraction absorption specifies that within chemical structure of zones that are with heat at maximum speed of rotation.

References

1. Luo AA (2013) Magnesium casting technology for structural applications. J Magnes Alloys 1(1):2–22
2. Diju Samuel G, Edwin Raja Dhas J, Ramanan G, Ramachandran M (2017) Production and microstructure characterization of AA6061 matrix activated carbon particulate reinforced composite by friction stir casting method. Rasayan J Chem 10(3):784–789
3. Schumann S, Friedrich H (2003) Current and future use of magnesium in the automobile industry. Mater Sci Forum 419:51–56
4. She J, Pan F, Zhang J, Tang A, Luo S, Yu Z, Rashad M (2016) Microstructure and mechanical properties of Mg–Al–Sn extruded alloys. J Alloy Compd 657:893–905
5. Motalleb-Nejad P, Saeid T, Heidarzadeh A, Darzi K, Ashjari M (2014) Effect of tool pin profile on microstructure and mechanical properties of friction stir welded AZ31B magnesium alloy. Mater Des 59:221–226

6. Darwins AK, Satheesh M, Ramanan G (2018) Modeling and optimization of friction stir welding parameters of Mg-ZE42 alloy using grey relational analysis with entropy measurement. IOP Conf Ser: Mater Sci Eng 402:012162
7. Commin L, Dumont M, Rotinat R, Pierron F, Masse JE, Barrallier L (2012) Influence of the microstructural changes and induced residual stresses on tensile properties of wrought magnesium alloy friction stir welds. Mater Sci Eng, A 551:288–292
8. Ramanan G, Dhas JER (2018) Multi objective optimization of wire EDM machining parameters for AA7075-PAC composite using grey-fuzzy technique. Mater Today: Proc 5(2):8280–8289
9. Zhou X, Liu RR, Zhou HT, Jiang WX (2017) A revisited study of the processing map and optimized workability of AZ61 magnesium alloy. J Mater Eng Perform 26(5):2423–2429
10. Zhou L, Nakata K, Liao J, Tsumura T (2012) Microstructural characteristics and mechanical properties of non-combustive Mg–9Al–Zn–Ca magnesium alloy friction stir welded joints. Mater Des 42:505–512
11. Darwins AK, Satheesh M, Ramanan G (2018) Effect of thermo gravimetric and FTIR analysis on friction stir processed Mg–ZE42 alloy. Rasayan J Chem 11(1):365–371
12. Pan F, Xu A, Deng D, Ye J, Jiang X, Tang A, Ran Y (2016) Effects of friction stir welding on microstructure and mechanical properties of magnesium alloy Mg–5Al–3Sn. Mater Des 110:266–274
13. Rao HM, Rodriguez RI, Jordon JB, Barkey ME, Guo YB, Badarinarayan H, Yuan W (2014) Friction stir spot welding of rare-earth containing ZEK100 magnesium alloy sheets. Mater Des 56:750–754
14. Ramaman G, Edwin Raja Dhas J, Ramachandran M (2017) Influence of activated carbon particles on microstructure and thermal analysis of AA7075/PAC metal matrix composites. Rasayan J Chem 10(3):375–384

Free Vibration and Buckling Analysis of FG-CNT Plates

Lenin Nagarajan and I. Mohammed Irfan

Abstract This work deals with free vibration and buckling analysis of Functionally Graded single walled Carbon nanotube reinforced composite plate. The material properties were computed by the extended rule of mixtures. The non-dimensional frequency parameter was found by generating eigenvalues and eigenvectors solved by using ABAQUS. The subspace method was used for generating Eigenvalues of buckling for the plate. This work was intended to study the effects of varying temperature, volume fractions of CNT and CNT grading on non-dimensional natural frequencies and buckling parameters. The results of this study were validated from the results available in literature. It was found that the CNT volume fraction and CNT grading type has significant effect on non-dimensional frequency parameters. Similarly, ample effect of these parameters were observed for buckling of plate.

Keywords Buckling · Carbon nanotubes · Functionally graded · Vibration

1 Introduction

A major contribution towards research on free vibration analysis of CNTs is done by Zhang et al. where he uses theories such as extended rule of mixture, first-order shear deformation plate theory and element-free kp-Ritz method to observe deflection decrease with enhancing V-CNT, the central deflection for the FG-O being highest, FG-X the lowest and the central deflection decrease with increasing breath/thickness ratio (b/t) and also observed that CFFF has the highest value of central deflection and CCCC (Clamped-Clamped-Clamped-Clamped) has the lowest the normal stress increase with the increase of lamination angle. The work done by Mohammadimehr et al. [1] reveal that natural vibration enhances with increase in material length scale parameters. Elastic foundation, Van der Waals constant and magnetic field enhance natural frequency which reduces with increasing structural damping and foundation

L. Nagarajan · I. Mohammed Irfan (✉)
Department of Mechnical Engineering, Vel Tech Rangarajan Dr. Sagunthala R&D Institute of Science and Technology, Chennai, India
e-mail: irfan2390@gmail.com

L.-J. Yang et al. (eds.), *Proceedings of ICDMC 2019*,
Lecture Notes in Mechanical Engineering,
https://doi.org/10.1007/978-981-15-3631-1_24

damping coefficients. Natural frequency decreases with an increase in temperature change, humidity and an enhances in volume fraction of SWCNT (Single Walled carbon Nano Tube) which gives rise to increase in stiffness. García-Macías et al. [2] investigated that frequency parameters increase for higher values of skew angle and that higher volume fractions of CNTs causes greater values of frequency parameters. Zghal et al. [3] found that frequency is higher for clamped condition than for simply supported and that increasing volume fraction of CNT, increases the frequency along with greater concentration of CNT towards the top. Kiani [4] proved that the frequency of a closed circuit plate is always greater than a plate with open circuit boundary conditions and that FG-X CNTRC plate has the greatest frequency. Daneshmehr et al. [5] researched that on increasing the length of a nano plate, the frequency ratio moves towards 1 and that thickness does not have any effect on size changes. Setoodeh et al. [6] proved that as the side angle increases, the effect of nonlinearity increases. Nguyen Dinh Duca et al. investigated that the central deflection of FG-X plate was lowest, while the FG-O plate had higher deflection and the UD plate lied between FG-X and FG-O. Natarajan et al. [7] have shown that with enhancement in the volume fraction of CNT distribution in the facesheet, deflection reduces. Further work by Yas et al. [8] reveals that the natural frequency reduces with greater L/R ratio for short cylindrical panels and then remains unchanged for long cylindrical panels. Increasing the V-CNT fraction, enhances the normalized natural frequency and FG-X has highest natural frequency. The research by Wang et al. [9] reveals that the CCCC, FG-X CNT grading, low aspect ratio, and greater CNT volume fraction are effective to enhance the fundamental frequencies and critical buckling loads. Yas and Samadi [8] discuss that it is important to know that the natural frequencies and critical buckling loads reduces with increasing the slenderness ratio. They also unveil that CeC beam has the greatest natural frequency and FG-X distribution has higher fundamental frequency. Zeyu Shia et al. use Fourier series method, first-order beam theory and Rayleigh–Ritz procedure to derive the values of FGV, FGX, FGA, UD and FGO frequencies for future reference purposes. Yas et al. [10] investigated the numerical results which show that the agglomeration parameters μ and η can greatly affect the natural frequencies, but does not effect the wave number of the respective fundamental frequencies. Panyatong et al. [1] have researched that small-scale effect started from the nonlocal elasticity leads in a reduction in the natural frequency of the FG nanoplate, however an elastic medium supporting the FG nanoplate leads to an enhanced natural frequency. Mirzaei and Kiani [11] have studied that frequencies of the panel are dependent on both, volume fraction of carbon nanotubes and their distribution pattern across the thickness and that on increasing the volume fraction of carbon nanotubes frequencies of the panel increases. Malekzadeh n and Zarei [12] discuss that for both thin and moderately thick plates, the FG-X distribution of CNTs through the thickness of the plate gives stiffer composite plates than other types of distributions. Li and Hu [13] researched on approaches to demonstrate efficient methods to measure the uncertainty relation on the vibrational characteristics of FG-CNTRC plates. They found the importance of power-law index (k) on bending stiffness of FG-CNTRC. Another outcome obtained by them was that any uncertainty in the material parameters set up at mid-plane level is propagated along the thickness of the plates by means of the index k. The work done by Reza Ansari et al. [14] reveals that functionally grading of CNTs through the thickness direction can considerably improve the vibrational characteristics of FG-CNTRC elliptical plates. Mohammadzadeh-Keleshteri et al. [15] unveil that the responses of the CNTRC sector plate can be optimized by playing with values of the CNT volume fraction and distribution. Their numerical results indicate that the increment in the nonlinear frequency decreases at certain vibration amplitude due to vibration redistribution. Mehar and Panda [16] in their paper, compute numerically using the direct iteration method in conjunction with the finite element method and validate the convergence behavior of the present nonlinear model. Moussa Abualnoura et al. Mahmoud in their paper conclude that the present theory is not only accurate but also simple in predicting the natural frequencies of functionally graded plates with stretching effect.

2 Methodology and Formulations

2.1 *Modelling of Material Properties*

It is supposed that the FG-CNTRC (Functional Graded Carbon Nano Tube) plate is made from a blend of isotropic matrix (epoxy resin) and fibers (CNTs) and the material properties are presumed to be graded along the thickness direction through the linear distributions (UD, FG-X and FG-O) of the volume fraction of carbon nanotubes.

2.1.1 Linear Material Distribution

In order to evaluate the affect of varying distributions of the CNT on the behavior of the FG-CNTRC plate, uniform distribution (UD) and functionally graded distributions (FG-X, and FG-O) of carbon nanotubes along the thickness direction of the nanocomposite plate are considered. The CNT volume fractions V_{cnt} of different types of FG-CNTRC plate can be shown as:

$$V_{\text{cnt}} = \begin{cases} V_{\text{cnt}}^{*} \quad \text{(UD CNTRC)}, \\ \frac{4|z|}{h} V_{\text{cnt}}^{*} \quad \text{(FG-X CNTRC)}, \\ 2\left(1 - \frac{2|z|}{h}\right) V_{\text{cnt}}^{*} \quad \text{(FG-O CNTRC)}, \end{cases} \tag{1}$$

Here, V_{cnt}^{*} depends on CNT percentage, mass fraction and density.

$$V_{\text{cnt}}^{*} = \frac{w_{\text{cnt}}}{w_{\text{cnt}} + \frac{\rho_{\text{cnt}}}{\rho_m} - \left(\frac{\rho_{\text{cnt}}}{\rho_m}\right) w_{\text{cnt}}} \tag{2}$$

where w_{cnt} is the mass fraction of nanotube, ρ_{cnt} and ρ_m are the densities of carbon nanotube and matrix respectively. For this linear material property variation, material properties can be obtained by replacing the value of V_{cnt} in Eq. (1).

2.1.2 Extended Rule of Mixture

To determine the effective material properties of CNTRC shell extended rule of mixture is employed, which can be expressed as [17]

$$\begin{aligned} E_{11} &= \eta_1 V_{\text{cnt}} E_{11}^{\text{cnt}} + V_m E_m, \\ \frac{\eta_2}{E_{22}} &= \frac{V_{\text{cnt}}}{E_{22}^{\text{cnt}}} + \frac{V_m}{E_m}, \\ \frac{\eta_3}{G_{12}} &= \frac{V_{\text{cnt}}}{G_{12}^{\text{cnt}}} + \frac{V_m}{G_m} \end{aligned} \tag{3}$$

Since Poisson's ratio depends weakly on the position, we assume υ_{12} to be:

$$\upsilon_{12} = V_{\text{cnt}}^{*} \upsilon_{12}^{\text{cnt}} + V_m \upsilon_m$$

Figure 1 explains the different orientation of carbon Nanotubes. Figure 1a shows the UD (uniformly distributed) grading where the carbon nanotubes percentage/fraction is constant throughout. Figure 1b shows the *X* type grading in which fraction is highest on the top and bottom surfaces while least at the middle. Figure 1c shows the exact opposite of *X* type that is the *O* type where in the nanotubes are abundantly present in the middle but sparsely at the top regions.

2.2 *Free Vibration Analysis*

After determining the effective material properties of proposed FG-CNTRC plate FEM modeling and analysis has been carried out using ABAQUS. Finally, the free vibration analysis is carried out using the following equation of motion:

$$[M]\{\ddot{q}\} + [K]\{q\} = \{F\} \tag{4}$$

where $[M]$ denotes the mass matrix, $[K]$ denotes the stiffness matrix.

In the present analysis, the natural frequency of the FG-CNTRC plate is non-dimensionalized using the formula:

$$\overline{\omega} = \omega\left(\frac{a^2}{h}\right)\sqrt{\frac{\rho_m}{E_m}}$$

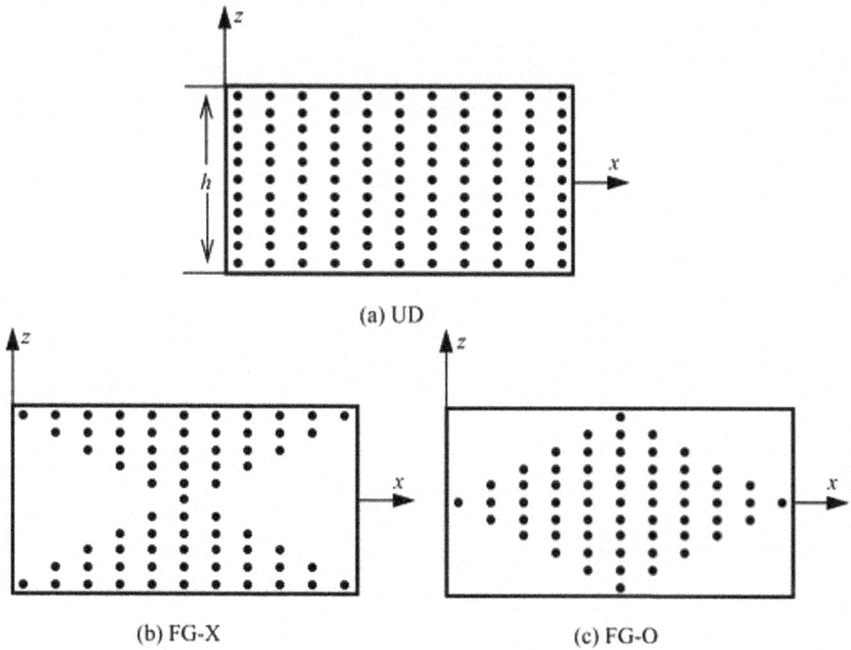

Fig. 1 Orientation of carbon nano tubes

where E_m is the Young's modulus of the matrix at the respective temperature.

2.3 *Buckling Analysis*

The buckling analysis of FG-CNTRC plate considers the solution of the Eigen problem:

$$[K - \lambda K_g]X = 0$$

where λ are the critical loads and X the buckling modes. The geometric stiffness matrix K_g is obtained by

$$K_g = \int_0^L \left[\frac{\mathrm{d}N_w}{\mathrm{d}x}\right]^{\mathrm{T}} P \left[\frac{\mathrm{d}N_w}{\mathrm{d}x}\right] \mathrm{d}x$$

In the present work a non-dimensional buckling parameter is used as given by:

$$\overline{N}_{\text{cr}} = \frac{N_{\text{cr}} b^2}{E_m h^3}$$

2.4 Abaqus Model and Stacking Sequence

Model with composite layup showing orientation angle (zero degrees), and number of plies (10).

The above snapshots of the model are taken to explain the modelling, orientation and boundary condition at which the analysis has been performed (Fig. 2).

The stacking sequence as clearly visible, shows 10 plies stacked together of equal thickness (relative thickness of each ply is thus $t = 0.1$ or 10%). the horizontal red lines signify the angle at which plies have been with respect to each other. since all lines are parallel, the orientation angle was thus 0/0/0/0/0/0/0/0/0/0.

The boundary condition of CCCC that all four sides of the plate is clamped was applied while doing the analysis. Although this is usually not the case in real life scenarios but the materials while testing are some time or another sustained to such kind of a boundary condition and their frequency can be crucial.

In this analysis 3 types of grading have been used for the CNT plates. A 2-Dimensional cut section view is provided in the photo below for different types of grading. The grading is assumed to be through the thickness which is h ($+h/2$ to $-h/2$). The length of plate is 'a' and width is 'b', since square plates are in consideration, length is equal to width which means $a = b$.

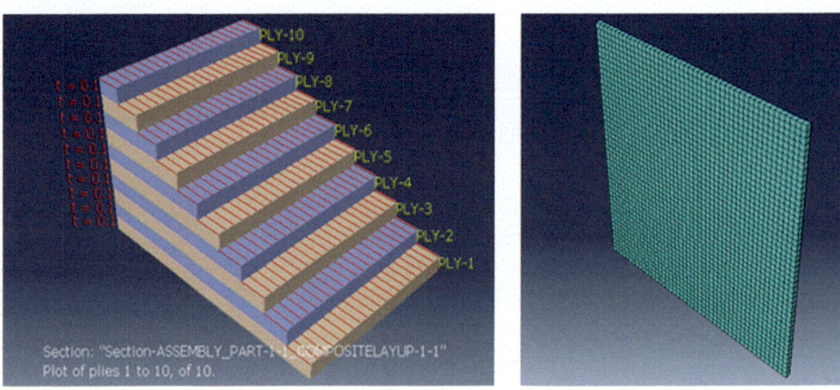

Fig. 2 Stacking sequence and Abaqus model

Table 1 Validation table for free vibration at 300 K for $a/h = 50$ for uniformly distributed CNT

Volume percentage of CNT	Method	Node 1	Node 2	Node 3	Node 4
11	Present	43.54778	49.3173	63.94229	88.97827
	Reference [16]	40.09760	45.64759	63.77431	100.083
	Reference [19]	39.730	43.876	54.768	74.488
14	Present	47.93973	53.0635	67.04377	92.21469
	Reference [16]	43.8826	49.14055	66.81946	107.979
	Reference [19]	43.583	47.479	57.968	77.395
17	Present	51.6191	56.41688	69.91541	94.7705
	Reference [16]	49.53276	56.54438	79.42928	123.9226
	Reference [19]	49.074	54.324	68.069	92.968

3 Results and Discussion

In the following table, validation and convergence behaviour of free vibration responses on the FG-CNT (Functionally Graded-Carbon Nano Tube) plate is computed for the CCCC condition, volume fraction of CNT 11, 14 and 17%. The present values of non-dimensional frequency parameter agree well with reference [18] and reference [19] (Tables 1, 2, 3, 4 and 5; Figs. 3, 4, 5, 6 and 7).

4 Conclusions

Free vibration analysis of a square plate of SWNT with different grading type (UD, FG-X, FG-V, FG-O) were done at varying temperatures (300, 500, 700 K) and for different volume fractions of CNT (11, 14 and 17%) with constant a/h ration (=50) and boundary conditions (CCCC). Extended rule of mixtures was employed for finding out the material properties FG-CNT. The eigenvalues were extracted using the lanscoz solver in an FEA software—ABAQUS. Various results were found:

(1) At any percentage of volume fraction of reinforcements and at any temperature, FG-X yields the highest values of non-dimensional frequency parameter followed by uniformly distributed and FG-O.
(2) At a constant temperature and a/h ratio, as the volume fraction of CNT increases, the non-dimensional frequency also increases due to increasing stiffness.
(3) At constant volume fraction of CNT and for the same grading type, the frequency reduces with increasing temperature.
(4) The effect of meshing size on non-dimensional frequency parameter isn't very considerable but yields in reduction in values as mesh size reduces and number of mesh increases.

Table 2 For grading type UD and FG-V, *a*/*h* = 50, varying temperatures and volume percentages of CNT

Temperature	CNT (%)	Grading	1	2	3	4	5	6
300	11	UD	43.5477	49.3173	63.9422	88.9782	109.3622	111.8495
		FGO	27.5456	36.1878	54.7398	69.8637	74.4846	83.2418
	14	UD	47.9397	53.0635	67.0437	92.2146	119.3927	120.8158
		FGO	27.9634	36.5193	54.9882	71.0798	75.6108	83.4471
	17	UD	51.6191	56.4168	69.9154	94.7705	127.3408	128.4589
		FGO	29.5978	37.7396	55.7786	75.2482	79.3759	83.9796
500	11	UD	43.1988	49.0250	63.7390	88.8579	108.6022	111.1352
		FGO	23.7889	33.5222	52.8206	59.5295	65.6316	79.1813
	14	UD	47.4716	52.8555	66.9276	91.5994	118.1802	120.3352
		FGO	27.3233	36.0831	54.4977	69.0868	74.1813	82.3292
	17	UD	51.1831	56.2611	69.8829	94.2641	126.2951	128.1847
		FGO	29.3516	37.6222	55.5386	74.4185	79.0429	83.0732
700	11	UD	43.0159	48.8635	63.6112	88.7586	108.1656	110.7148
		FGO	24.9857	34.3792	53.3902	62.7930	68.5275	81.5170
	14	UD	47.2816	52.6903	66.8038	91.5155	117.7697	119.944
		FGO	27.2208	36.0086	54.4518	68.8158	73.9373	82.3024
	17	UD	50.9814	56.0835	69.7470	94.1705	125.875	127.782
		FGO	29.2383	37.5372	55.4847	74.1263	78.7769	83.0409

Table 3 At 300 K, *a*/*h* = 50, for different mesh sizes, grading types and volume fractions

CNT (%)	Grading	8 × 8	16 × 16	24 × 24	32 × 32
11	UD	43.63159	43.54778	43.4659	43.458
	FGX	50.6053	50.51882	50.4995	50.4933
	FGO	28.1	27.96348	27.932	27.9237
14	UD	48.052	47.93973	47.9138	47.9093
	FGX	55.2792	55.23794	55.2298	55.2241
	FGO	27.6838	27.54566	27.5152	27.50731
17	UD	51.7019	51.6191	51.6006	51.5946
	FGX	59.2274	59.24614	59.2464	59.2464
	FGO	29.7428	29.59782	29.5656	29.5579

Table 4 Validation of buckling parameter at 500 K, *a*/*h* = 10 and for 11% volume fraction of CNT which is uniformly graded under uniaxial compression

	Node 1	Node 2	Node 3	Node 4
Present	24.23	32.976	41.988	47.198
Reference value [20]	24.95	33.4917	47.8115	50.0545

Table 5 Buckling parameter at different temperatures, $a/h = 10$ and for varying volume fraction of CNT, grading type under uniaxial compression

Temperature	CNT (%)	Grading	Mode 1	Mode 2	Mode 3	Mode 4
300	11	UD	24.55	33.262	42.194	47.33
		FGO	10.2456	10.51144	16.31544	23.6916
	14	UD	28.8906	37.388	45.698	50.2386
		FGO	10.45622	10.9858	17.5339	25.0088
	17	UD	32.798	40.926	48.696	52.798
		FGO	10.9142	11.66848	17.16312	24.7188
500	11	UD	24.23	32.976	41.988	47.198
		FGO	9.1778	9.4414	15.2746	22.433
	14	UD	28.4606	37.0382	45.4612	50.0876
		FGO	10.1858	10.4032	16.244	23.614
	17	UD	32.407	40.6776	48.6038	52.8086
		FGO	10.851	11.5476	17.091	24.6426
700	11	UD	24.058	32.814	41.854	47.094
		FGO	9.132	9.414	15.2408	22.395
	14	UD	28.2832	36.8924	45.3634	50.0274
		FGO	10.156	10.3502	16.3406	23.5764
	17	UD	32.2116	40.5292	48.5096	52.752
		FGO	10.81994	11.48834	17.05546	24.607

(5) Buckling analysis was done on the same plate with a/h ratio of 10. The initial results were validated by Reference [20] and the following results were noted:

(a) As the temperature increases the buckling parameter value for a given CNT percentage decreases.
(b) For a given temperature, grading as the CNT percentage increases the buckling parameter value increases.
(c) The buckling parameter value of uniformly distributed grading at a given temperature and CNT percentage is much higher than FG-O.

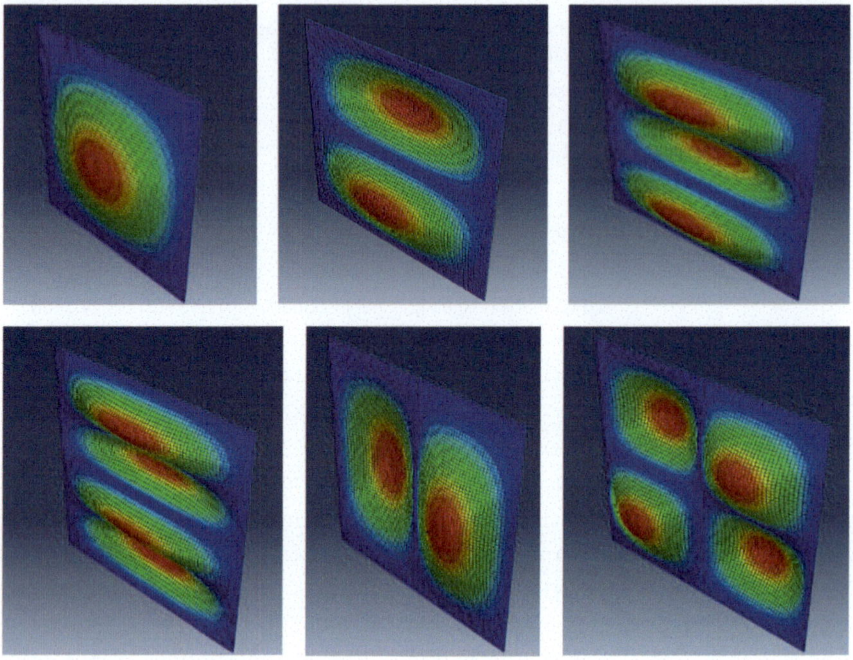

Fig. 3 The modal shapes have been plotted for the first 6 fundamental natural frequencies

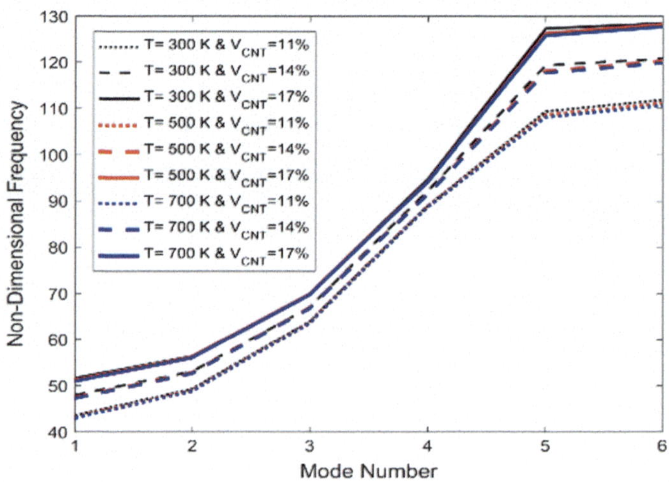

Fig. 4 Non-dimensional frequency versus mode number of UD at different temperatures and volume fractions

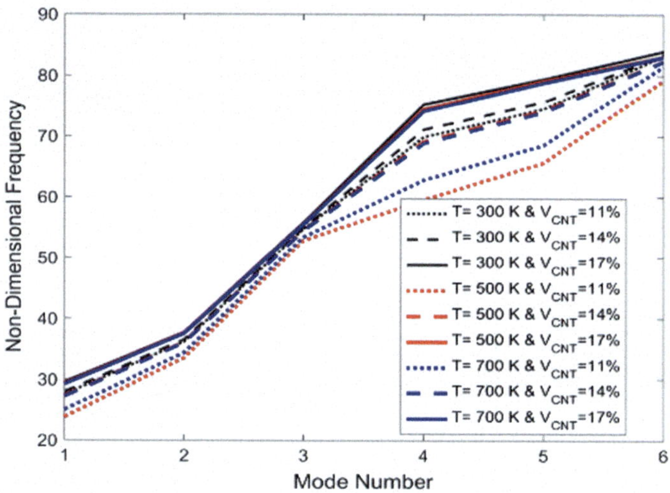

Fig. 5 Non-dimensional frequency versus mode number of FGO at different temperatures and volume fractions

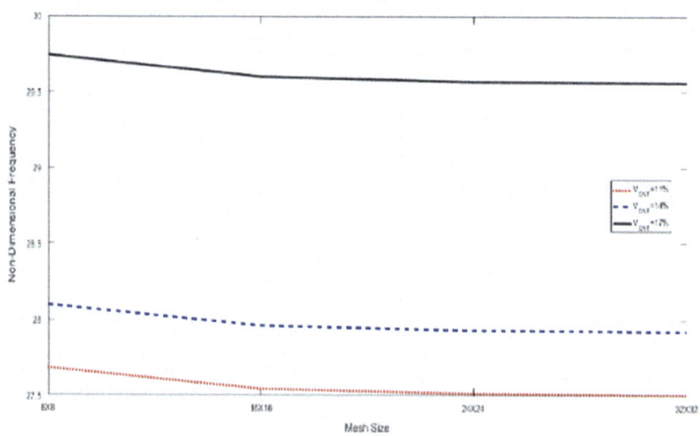

Fig. 6 Non-dimensional frequency versus mesh size at 700 K

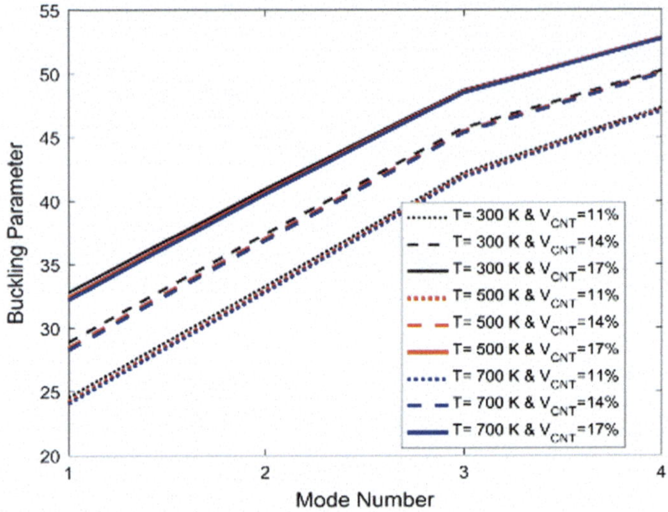

Fig. 7 Buckling parameter versus mode number of *UD*

References

1. Mohammadimehr M, Rousta Navi B, Ghorbanpour Arani A (2015) Free vibration of viscoelastic double-bonded polymeric nanocomposite plates reinforced by FG-SWCNTs using MSGT, sinusoidal shear deformation theory and meshless method. Compos Struct 131:654–671
2. García-Macías E, Castro-Triguero R, Friswell MI, Adhikari S, Sáez A (2016) Metamodel-based approach for stochastic free vibration analysis of functionally graded carbon nanotube reinforced plates. Compos Struct 152:183–198
3. Zghal S, Frikha A, Dammak F (2018) Free vibration analysis of carbon nanotube-reinforced functionally graded composite shell structures. Appl Math Model 53:132–155
4. Kiani Y (2016) Free vibration of FG-CNT reinforced composite skew plates. Aerosp Sci Technol 58:178–188
5. Daneshmehr A, Rajabpoor A, Hadi A (2015) Size dependent free vibration analysis of nanoplates made of functionally graded materials based on nonlocal elasticity theory with high order theories. Int J Eng Sci 95:23–35
6. Setoodeh AR, Shojaee M (2016) Application of TW-DQ method to nonlinear free vibration analysis of FG carbon nanotube-reinforced composite quadrilateral plates. Thin-Walled Structures 108:1–11
7. Natarajan S, Haboussi M, Manickam G (2014) Application of higher-order structural theory to bending and free vibration analysis of sandwich plates with CNT reinforced composite facesheets. Compos Struct 113(1):197–207
8. Kamarian S, Salim M, Dimitri R, Tornabene F (2016) Free vibration analysis of conical shells reinforced with agglomerated carbon nanotubes. Int J Mech Sci 108–109:157–165
9. Wang M, Li ZM, Qiao P (2016) Semi-analytical solutions to buckling and free vibration analysis of carbon nanotube-reinforced composite thin plates. Compos Struct 144:33–43
10. Yas MH, Pourasghar A, Kamarian S, Heshmati M (2013) Three-dimensional free vibration analysis of functionally graded nanocomposite cylindrical panels reinforced by carbon nanotube. Mater Des 49:583–590
11. Mirzaei M, Kiani Y (2016) Free vibration of functionally graded carbon nanotube reinforced composite cylindrical panels. Compos Struct 142:45–56

12. Malekzadeh P, Zarei AR (2014) Free vibration of quadrilateral laminated plates with carbon nanotube reinforced composite layers. Thin-Walled Structures 82:221–232
13. Li L, Hu Y (2016) Nonlinear bending and free vibration analyses of nonlocal strain gradient beams made of functionally graded material. Int J Eng Sci 107:77–97
14. Fantuzzi N, Tornabene F, Bacciocchi M, Dimitri R (2017) Free vibration analysis of arbitrarily shaped Functionally Graded Carbon Nanotube-reinforced plates. Compos B Eng 115:384–408
15. Mohammadzadeh-Keleshteri M, Asadi H, Aghdam MM (2017) Geometrical nonlinear free vibration responses of FG-CNT reinforced composite annular sector plates integrated with piezoelectric layers. Compos Struct 171:100–112
16. Mehar K, Panda SK (2016) Geometrical nonlinear free vibration analysis of FG-CNT reinforced composite flat panel under uniform thermal field. Compos Struct 143:336–346
17. Kiani Y (2016) Free vibration of functionally graded carbon nanotube reinforced composite plates integrated with piezoelectric layers. Comput Math Appl 72(9):2433–2449
18. Wu C-P, Li W-C (2017) Free vibration analysis of embedded single-layered nanoplates and graphene sheets by using the multiple time scale method. Comput Math Appl 73(5):838–854
19. Zhu P, Lei ZX, Liew KM (2012) Static and free vibration analyses of carbon nanotube-reinforced composite plates using finite element method with first order shear deformation plate theory. Compos Struct 94(4):1450–1460
20. Lei ZX, Liew KM, Yu JL (2013) Buckling analysis of functionally graded carbon nanotube-reinforced composite plates using the element-free kp-Ritz method. Compos Struct 98:160–168

The Role of Processing Temperature in Equal Channel Angular Extrusion: Microstructure Mechanical Properties and Corrosion Resistance

Gajanan M. Naik, S. Narendranath, and S. S. Satheesh Kumar

Abstract Equal channel angular extrusion, patented in Russia by V. M. Segal in 1977, has become a promising technique to enhance tensile strength and corrosion resistance of Mg alloys. It is believed that the processing temperature ensures the production of ECAE-processed billet without surface defects. Indeed, ECAE processing temperature affects microstructure, tensile behavior, and corrosion resistance of the material. Therefore, this chapter investigates the impact of ECAE pressing temperature on microstructure, mechanical behavior, and corrosion resistance of AZ80 Mg alloys. The processing temperature of 533 and 663 K was selected based on the recrystallization temperature of Mg alloys. As a result, the processing temperature has a substantial impact on material properties. The axial tensile strength and hardness decrease by 25.45% and 6.56%, respectively, due to thermal softening of materials. The corrosion resistance increases by 84% due to grain size reduction and distribution of secondary phases, when the ECAP-4P processing temperature is increased from 533 K to 663 K.

Keywords AZ80 · ECAE · Polarization · Micro-hardness · Corrosion

1 Introduction

Wrought magnesium alloys are lightest engineering material, and it has quite special properties which lead to particular applications. In specific, their highest strength to

G. M. Naik (✉) · S. Narendranath
Corrosion Engineering Laboratory, Department of Mechanical Engineering, National Institute of Technology Karnataka, Surathkal, Mangaluru, Karnataka 575025, India
e-mail: gajamnaik@gmail.com

G. M. Naik
Department Mechanical Engineering, Mangalore Institute of Technology and Engineering, Moodbidri, Mangaluru, Karnataka 574225, India

S. S. Satheesh Kumar
Defence Metallurgical Research Laboratory, Kanchanbagh, Hyderabad, Telangana 500058, India

L.-J. Yang et al. (eds.), *Proceedings of ICDMC 2019*,
Lecture Notes in Mechanical Engineering,
https://doi.org/10.1007/978-981-15-3631-1_25

weight ratio, good castability, and high damping capability make Mg alloys tremendously attractive in aerospace, electronics, marine, and automobile industries [1, 2]. Indeed, magnesium alloys have poor tensile strength, ductility, and corrosion resistance properties associated with other engineering materials like aluminum alloys, steels, etc. Wu et al. [3] examine the effect on mechanical and corrosion improvement of AZ91 Mg alloys after adding Ca and rare earth (RE), developed high-strength and corrosion-resistant alloy, the ultimate tensile strength (UTS) improved by 16% and the rate of corrosion reduced to 0.087 mg/cm^2/day owing to the presence of the Al_2Ca second phases. But, as per the literature alloying is not giving a significant improvement. Therefore, in order to increase the mechanical and corrosion resistance of Mg alloys researchers are using severe plastic deformation techniques. Chen et al. [4] have studied the ECA pressing of AZ91 Mg alloy and its effects on grain refinement and mechanical behavior; as a result, the yield strength, UTS, and ductility of AZ91 Mg alloys are remarkably achieved 290 MPa, 417 MPa, and 8.45%, respectively, after two-step ECAP-4P at 225 °C and two passes at 180 °C; this is mainly due to reduced grain and to the precipitation of $Mg_{17}Al_{12}$ secondary phase during ECAP. A similar remark was made by Kim et al. [5] on AZ61 Mg alloy. Shaeri et al. [6] examined the impact of ECA processing temperatures of room temperature 120, 150, and 180 °C on grain refinement and mechanical behavior of Al-7075 heat-treated alloy, above study shown that the enhancement of mechanical properties was largely attributed to ECA processing temperature, grain boundary strengthening, and dislocation densities. Therefore, in the current study, an attempt is made to investigate the impact of processing temperature on microstructure, mechanical behavior, and corrosion performance of AZ80 Mg alloy during ECAE.

2 An Overview of Equal Channel Angular Extrusion

The principle of ECAE is explained in Fig. 1. The ECAE die involves two channels of equal in cross section, intersecting at an angle ϕ: channel angle and ψ: corner angle. The specimens in the form of a bar or rod are machined to suitable within the die channel, and the material is pressed through the ECAP die using a punch/RAM. The imposed distortion is simple pure shear, which induces as the sample passes through the ECAE die without changing the shape of the specimen. Thereby, samples are permitting repeated press in order to introduce large strains [7]. The induced accumulated strain can be calculated through Eq. (1).

Also, the significant changes in the microstructure and texture can be attained through ECAE processing routes such as Route-A: Billets are extruded without any rotations. Route-B_A: Sample is rotated by 90° in a CW direction between successive passes. Route-B_c: Sample is rotated by 90° in CCW direction between successive passes. Route-C: Billet is rotated by 180° between the passes. Route-R: Here, sample is inverted from the initial position in each ECAE passes, out of this route-B_c and route-R are considered as effective routes to achieve ultra-fine grain structures with minimum number of ECAE passes [8, 9].

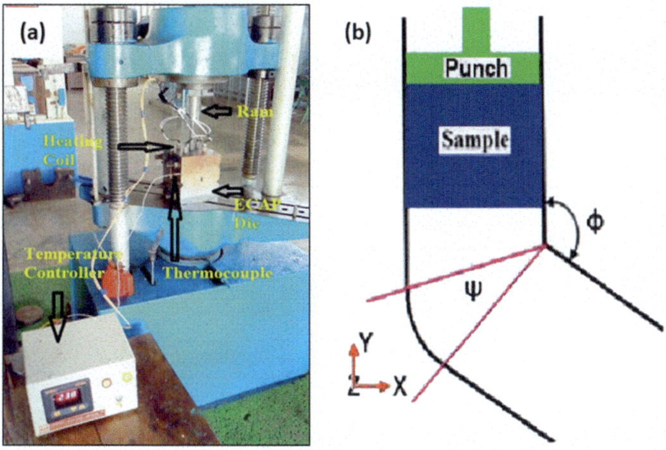

Fig. 1 Equal channel angular extrusion, **a** experimental setup, **b** schematic [2]

$$\varepsilon_{eq} = \frac{N}{\sqrt{3}}\left(2\cot\left(\frac{\phi + \Psi}{2}\right) + \Psi \operatorname{cosec}\left(\frac{\phi + \Psi}{2}\right)\right) \tag{1}$$

γ = shear strain; ε_{eq} equivalent strain.

3 Experimental Work

3.1 Materials and Equal Channel Angular Extrusion Process

An AZ80 alloy having a material composition of Al-8, Zn-0.5 Mg-Bal. (wt%) was received from Exclusive Magnesium Pvt. Ltd. Hyderabad, India, and it was cut into billets with a diameter of 16 mm and length of 80 mm, prior to ECAE. Equal channel angular pressing was accomplished at 533 K and 663 K processing temperature with 1 mm s^{-1} speed through the 110° intersecting channel angle and 30° corner angle by route-R; here, the sample is inverted from its original position during each ECAP passes. Four successive passes were given to the sample. Prior to pressing, the alloy undergone homogenization at 673 K for 24 h followed by furnace cooling. ECAP sample preparation of AZ80 alloys under this practice has been described scientifically in Fig. 1 and Ref. [10].

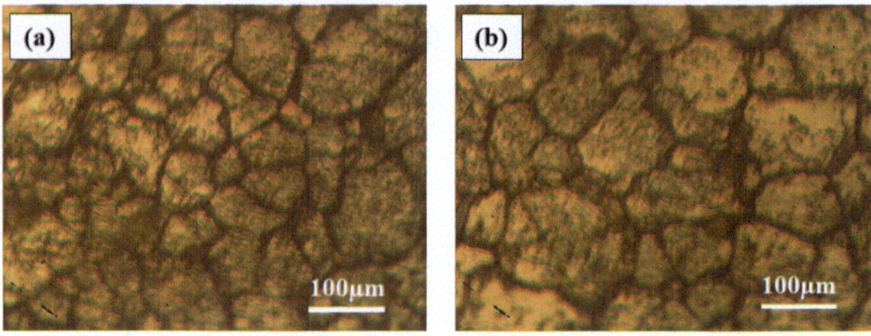

Fig. 2 Optical images of AZ80 Mg alloy, **a** as-received, **b** homogenized at 673 K-24 h

3.2 Microstructure Characterization and Mechanical Test

Microstructures of the processed and unprocessed samples were observed by OM (BIOVIS material plus) after mechanical polishing and etching at room temperature using 4.2 g picric acid, 10 ml distilled water, 10 ml acetic acid, and 70 ml ethanol for 3–5 s. The observed microstructure of as-received and solution-treated samples is presented in Fig. 2.

Tensile specimens (ASTM E8) with 16 mm gauge length (GL), 4 mm diameter, and 2 mm shoulder radius were extracted from the ECAP-processed materials by conventional turning. All the tensile samples, including as-received material, were obtained at room temperature at strain rate 0.5 mm/s. The tensile test was conducted thrice on each processed and unprocessed sample under similar testing conditions. Measurements of the micro-hardness were carried out at an indenter load of 100 g and dwell period of 13 s.

3.3 Corrosion Test

Electrochemical corrosion experimentations were carried out in 3.5 wt% NaCl solution using potentiodynamic polarization technique. All experiments were performed in a corrosion cell kit at a room temperature of 25 °C. The cell kit setup was fitted with a three-electrode system such as saturated calomel electrode used as a reference electrode, a graphite rod as an auxiliary electrode, and the sample (AZ80) as the working electrode. Before potentiodynamic polarization test, the working electrodes (AZ80) were immersed in NaCl corrosion media for 30 min to maintain open-circuit potential (OCP). The polarization experiments are carried out with −250 to +250 mV and scan rate of 2 mV/s.

4 Results and Discussion

4.1 Effect of Processing Temperature on Microstructure

Figure 2 displays the optical images of as-received and homogenized sample; those ECAE-processed specimens at two different processing temperatures are presented in Fig. 3; and the average grain size was calculated by using a lineal intercept method.

The mean grain size of the as-received and homogenized AZ80 Mg alloys is typically non-uniform, and the mean grain size is about 50.20 μm and 50.75 μm, respectively, as shown in Fig. 2a, b. This slight increase of grain size during the homogenization of Mg alloy is due to grain growth effect [13]. The grain size after 2P-ECAE at 533 K and 4P gradually decreases to 26.34 μm and 5.12μm, respectively, which is depicted in Fig. 3a, b. Likewise, Fig. 3c, d presents the microstructure of sample ECAE extruded at 663 K 2P and 4P of grain size 37.26 μm and 10.54 μm, respectively. From the results, it is deduced that the average grain size was increased at higher ECAE processing temperature (663 K) compared to sample processed at 533 K; this is due to the grain growth effect at higher processing temperature.

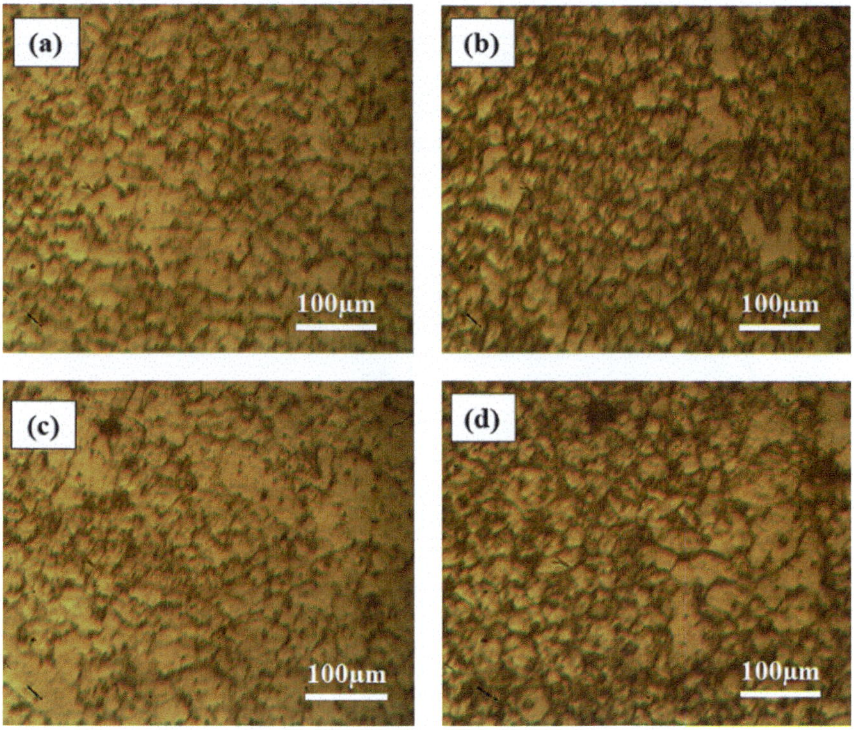

Fig. 3 Microstructure of AZ80 Mg alloys, **a** ECAP-2P at 533 K, **b** ECAP-4P at 533 K, **c** ECAP-2P at 663 K, **d** ECAP-4P at 663 K

Although grains were not uniformed at second pass of ECAE, it is observed bimodal grains, and then, grains are significantly refined after the fourth pass of ECAE represented in Fig. 3. Similar observations are made on AZ80 alloys by Avvari et al. [11].

4.2 *Effect of Processing Temperature on Mechanical Properties*

The engineering stress–strain curves of as-received (AZ80) and equal channel angular extruded AZ80 magnesium alloys at two different processing temperatures are shown in Fig. 4. The engineering stress–strain curves display ultimate tensile strength and ductility of as-received and ECAE-processed specimens at 533 K and 663 K. From the obtained results, it is observed that sample exhibited improved ultimate tensile strength and ductility after ECAP at 533 K and 663 K compared to as-received alloy. Moreover, from the study it is also observed that ECAE-processed sample at 533 K possessed the utmost ultimate tensile strength and ductility compared to sample processed at 663 K as shown in Fig. 4a. The experimental results of Vickers micro-hardness test are shown in Fig. 4b. The figure specifies that the mean micro-hardness of the as-received and ECAE of Mg alloys processed at 533 K and 663 K showed improved micro-hardness from 0P to 4P. Also, Mg alloy processed at 663 K exhibited lower micro-hardness values associated with Mg alloy processed at 533 K and this is due to thermal softening of the alloy during high-temperature ECAE [6, 12].

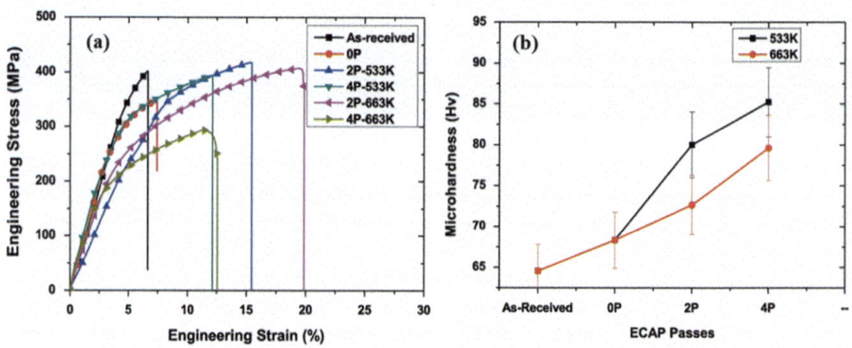

Fig. 4 Mechanical behavior of ECA extruded AZ80 Mg alloy, **a** tensile behavior, **b** micro-hardness

4.3 *Effect of Processing Temperature on Electrochemical Corrosion*

As-received and ECAE-processed specimens were subjected to potentiodynamic polarization tests to investigate the impact of processing temperature on corrosion resistance of AZ80 Mg alloys. During study corrosion current densities (Icorr), corrosion potentials (Ecorr) were obtained from electrochemical polarization curves as presented in Fig. 5. From the experimental data, it was observed that equal channel angular pressed Mg alloy showed improved corrosion resistance compared to as-received Mg alloy. From Fig. 5a, it is observed that Mg alloy processed at 663 K exhibited improved pitting corrosion resistance. Remarkably, AZ80 Mg alloy processed at higher temperature (663 K) revealed decreased corrosion rate compared with the Mg alloy processed at 533 K shown in Fig. 5b. From study, the enhanced corrosion resistance of the AZ80 alloy was stated owing to better distribution of $Mg_{17}Al_{12}$ secondary phases at higher processing temperature (663 K) [11, 13–18]. Figure 6a–c shows the corrosion morphologies for as-received, ECAP-2P at 533 K, and ECAP-2P at 663 K, respectively. The surface of as-received AZ80 alloys has exposed abundant corrosion attack. In addition, in comparison with Fig. 6a, the surface area of corrosion pits was reduced in as-processed alloy as shown in Fig. 6b, c. Specifically, AZ80 Mg alloy processed at 663 K after ECAP-2P exhibited less

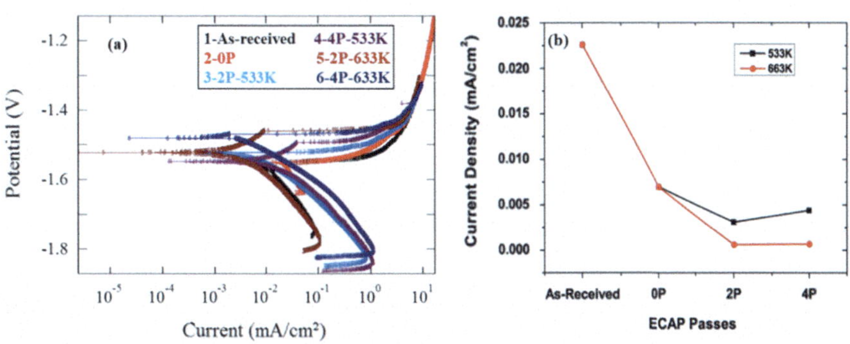

Fig. 5 **a** Polarization curves, **b** corrosion current density versus number of ECAP passes

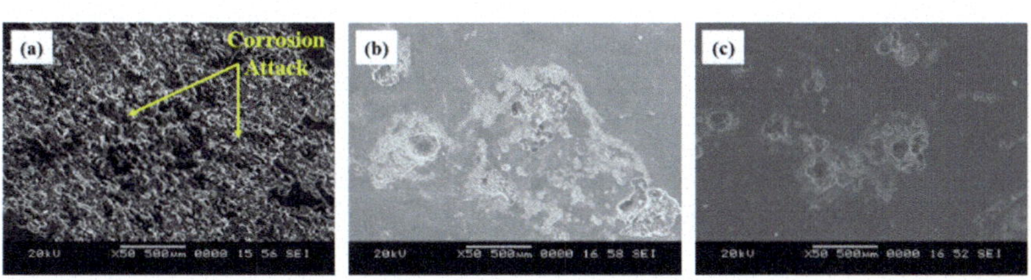

Fig. 6 Corrosion morphology, **a** as-received, **b** 533 K-2P, **c** 663 K-2P

corrosion attack on the surface. This indicates that secondary phases (β-$Mg_{17}Al_{12}$) are distributed after processing, which reduces the number of cathodic sites and, as a result, improves the corrosion resistance.

5 Conclusions

The impact of ECA extrusion temperature on microstructure, mechanical, and corrosion behavior of AZ80 Mg alloys was investigated and reported in this paper. Based on the observation, the following conclusions are drawn.

1. According to the microstructural observations, the initial mean grain size of about 50.20 μm was reduced to less than 5.12μm at 533 K and 10.54 μm at 663 K after the fourth pass of ECAE process. Optical microstructural studies establish that increasing ECAE processing temperature leads to an increase in the grain size due to grain growth effect.
2. Mechanical properties such as ultimate tensile strength, ductility, and microhardness of the AZ80 mg alloy are directly proportional to the ECAE passes but it has an inverse effect on ECAE processing temperature.
3. The electrochemical corrosion of Mg alloys exhibited better corrosion resistance at higher processing temperature of 663 K compared to lower processing temperature 533 K; this revealed an 84% improved corrosion resistance.

Acknowledgements This work was supported by DRDO-NRB, Government of India, under grant number NRB/4003/PG/366.

References

1. Song GL, Atrens A (1999) Corrosion mechanisms of magnesium alloys. Adv Eng Mater 1(1):11–33
2. Kulekci MK (2008) Magnesium and its alloys applications in automotive industry. Int J Adv Manuf Technol 39(9–10):851–865
3. Wu G, Fan Y, Gao H, Zhai C, Zhu YP (2005) The effect of Ca and rare earth elements on the microstructure, mechanical properties and corrosion behavior of AZ91D. Mater Sci Eng, A 408(1–2):255–263
4. Chen B, Lin DL, Jin L, Zeng XQ, Lu C (2008) Equal-channel angular pressing of magnesium alloy AZ91 and its effects on microstructure and mechanical properties. Mater Sci Eng, A 483:113–116
5. Kim WJ, Hong SI, Kim YS, Min SH, Jeong HT, Lee JD (2003) Texture development and its effect on mechanical properties of an AZ61 Mg alloy fabricated by equal channel angular pressing. Acta Mater 51(11):3293–3307
6. Shaeri MH, Shaeri M, Ebrahimi M, Salehi MT, Seyyedein SH (2016) Effect of ECAP temperature on microstructure and mechanical properties of Al–Zn–Mg–Cu alloy. Prog Nat Sci: Mater Int 26(2):182–191

7. Valiev RZ, Langdon TG (2006) Principles of equal-channel angular pressing as a processing tool for grain refinement. Prog Mater Sci 51(7):881–981
8. Estrin Y, Vinogradov A (2013) Extreme grain refinement by severe plastic deformation: a wealth of challenging science. Acta Mater 61(3):782–817
9. Avvari M, Narendranath S (2014) Influence of Route-R on wrought magnesium AZ61 alloy mechanical properties through equal channel angular pressing. J Magnes Alloys 2(2):159–164
10. Sklenicka V, Dvorak J, Svoboda M, Kral P, Kvapilova M (2012) Equal-channel angular pressing and creep in ultrafine-grained aluminium and its alloys. In: Aluminium alloys-new trends in fabrication and applications (InTech)
11. Avvari M, Narendranath S (2018) Effect of secondary Mg 17 Al 12 phase on AZ80 alloy processed by equal channel angular pressing (ECAP). Silicon 10(1):39–47
12. Parshikov RA, Rudskoy AI, Zolotov AM, Tolochko OV (2013) Technological problems of equal channel angular pressing. Rev Adv Mater Sci 34:26–36
13. Naik GM, Gote GD, Narendranath S, Kumar SS (2018) The impact of homogenization treatment on microstructure microhardness and corrosion behavior of wrought AZ80 magnesium alloys in 3.5 wt% NaCl solution. Mater Res Express 5(8):5
14. Naik GM, Gote GD, Narendranath S (2018) Microstructural and Hardness evolution of AZ80 alloy after ECAP and post-ECAP processes. Mater Today: Proc 5(9):17763–17768
15. Naik GM, Narendranath S, Kumar SS (2019) Effect of ECAP Die Angles on Microstructure Mechanical Properties and Corrosion Behavior of AZ80 Mg Alloy. J Mater Eng Perform 28(5):2610–2619
16. Naik GM, Gote GD, Narendranath S, Kumar SS (2018) Effect of grain refinement on the performance of AZ80 Mg alloys during wear and corrosion. Adv Mater Res 7(2):105
17. Naik GM, Narendranath S, Kumar SS, Sahu S (2019) Effect of Annealing and Aging Treatment on Pitting Corrosion Resistance of Fine-Grained Mg-8% Al-0.5% Zn Alloy. JOM 71(12):4758–4768
18. Gajanan MN, Narendranath S, Satheesh Kumar SS (2019) Influence of ECAP processing routes on microstructure mechanical properties and corrosion behavior of AZ80 Mg alloy. In: AIP Conference Proceedings, vol 2082(1), pp 030016

Reciprocating Wear Studies of Inconel 718 and Mod.9Cr–1Mo Ferritic Steel by Surface Profilometric Characterization

K. Adityan, N. L. Parthasarathi, P. Ashoka Varthanan, R. Priya, and Utpal Borah

Abstract The Indian prototype fast breeder reactor (PFBR) is a pool-type reactor in which the steam generator employs liquid sodium as coolant. Modified 9Cr–1Mo is the construction material for steam generator, owing to its excellent high-temperature mechanical properties. Nickel alloy 718 has been used for the tube bundle support structures due to its excellent high-temperature properties and its compatibility with liquid sodium. The coolant sodium is responsible for the flow-induced vibration in the steam generator tubes which in turn leads to the reciprocating fretting wear in the components. This wear is undesirable for both life perspective and functionality. It is important to study the wear of the components, such as the heat exchanger tubes of Cr–Mo steel and the support structures by simulating the industrial operating atmosphere. As a groundwork, reciprocating wear tests are planned and the test matrix has been formulated to study the wear characteristics of Inconel 718 against Al_2O_3 balls and 9Cr–1Mo. The tests were carried out in frequencies of 3, 6 and 9 Hz and varying temperatures such as room temperature (RT), 100 and 150 °C. Optical microscopy is carried on the tested samples to study the metallography. Surface profilometer is employed to characterize the worn scars of the tested specimens and thereby concluding the mode of material removal during the reciprocating wear. Different surface roughness parameters on the scar were studied and compared with the parameters taken for study.

Keywords Reciprocating wear · Inconel 718 · 9Cr–1Mo · Profilometry

K. Adityan · P. A. Varthanan
Department of Mechanical Engineering, Sri Krishna College of Engineering and Technology, Coimbatore, Tamil Nadu, India

N. L. Parthasarathi (✉) · U. Borah
Materials Development and Technology Division, Indira Gandhi Centre for Atomic Research, Kalpakkam, Tamil Nadu, India
e-mail: nlpartha@gmail.com

R. Priya
Corrosion Science and Technology Division, Indira Gandhi Centre for Atomic Research, Kalpakkam, Tamil Nadu, India

L.-J. Yang et al. (eds.), *Proceedings of ICDMC 2019*,
Lecture Notes in Mechanical Engineering,
https://doi.org/10.1007/978-981-15-3631-1_26

1 Introduction

Modified 9Cr–1Mo ferromagnetic steel (Grade 91) is being used as the material for fabrication of steam generators in the prototype fast breeder reactor in Kalpakkam, India. It has been selected as the major construction material, owing to its higher fatigue and increased creep resistance [1]. On the other hand, Inconel 718, an precipitation-hardened nickel-based superalloy, strengthened more often by face centred cubic (FCC) γ'-Ni3 (Al, Ti) and ordered body centred tetragonal (BCT) γ''-Ni_3Nb has been incorporated as the material for tube bundle support structures. Inconel 718 is extensively used in gas mills, rocket motors, spacecraft, nuclear reactors and pumps, due to its high-temperature strength, weldability, corrosion resistance and microstructure stability at elevated temperature (650 °C) [2–4]. For turbine generators, the pressurized water reactor uses the massive heat exchanger which incorporates the heat that is drawn from the primary one, to generate steam with in the secondary aspect. Only one fluid (water) is passed inside the tubes by which the thermal heat transfer occurs, on the other hand, remaining fluid which is the liquid sodium passed through the outer side of the same. Due to the combined effect of massive flow rate that is being offered by the combined effect of water as well as sodium and a little clearance in the middle of the tubes and its support structures, the flow-induced vibration occurs. Fretting wear resulted in noticeable wear rate of the component. This mode of wear is predominant and its attributes for the loss of material in this component compared to other modes of wear [5, 6]. During operations involving continuous higher velocity and longer operating conditions, the amplitude of frequency gets consolidated to get amplified and simultaneously it causes the failure in form of wear caused by fatigue [7, 8]. The wear characteristics of Inconel 690 were studied by fatigue test and it was reported that the fatigue limit was reduced up to 43% by the fretting environment [9]. Tribological properties of the Inconel 690 were investigated at different thermal ageing conditions and reported that due to softening by ageing heat treatment wear loss was more pronounced. Load was pointed out as the most influencing parameter [10]. Several surface modification experiments were done for enhancing the mechanical properties of the Inconel 718. Boronizing Inconel produced twofold increase in depth and the solution annealed borided Inconel offers tremendous wear resistance [11]. The effect of polishing in Inconel 718 using two-dimensional ultrasonic assisted polishing and reported that the polishing reduced the dispersion degree of the machining surface roughness. Further, the polishing showed an adverse impact in improving the material removing rate of Inconel 718 [12]. A detailed review of the wear characteristics of Inconel alloys concluded that they exhibited better wear resistance. Particularly, Inconel 600 exhibited higher wear resistance than Inconel 690. In general, Inconel 718 showed overall better wear quality characteristics [13]. For better understanding of the fretting wear at higher frequencies, it is mandated to study the wear characteristics of 9Cr–1Mo and Inconel 718 in lower frequencies. In PFBR, the steam generator bundle tubes encounter a moderate fretting wear with the steam generator support structures, which is made up of Inconel 718. The wear is to be simulated by a reciprocating sliding wear test set up in laboratory scale.

The objective of this work is to study the reciprocating wear characteristics of 9Cr–1Mo ferritic steel and Inconel 718 in a different test conditions which includes varying temperatures and frequencies as specified. Optical micrographs were taken using Confocal Laser microscope. Surface profilometry of the worn scars was studied and reported. At an outset, the Inconel 718 material needs a surface modification in order to withstand the fretting wear and to resists the wear due to flow-induced vibrations.

2 Experimental Methods

2.1 *Sample Preparation*

Inconel 718 discs of dimensions Ø 25 mm × 6 mm thickness were fabricated by electric discharge machine. The counter bodies used are alumina balls of 6 mm diameter as a reference material procured from M/S Ducom, Bangalore and modified 9Cr–1Mo ferritic steel pins for simulating the steam generator tubes. Alumina balls are lighter and more corrosion/wear resistant than steel balls and can withstand high temperatures up to 1750 °C. The another counter body 9Cr–1Mo is fabricated in a cylindrical pin form with an end diameter of 6 mm. The Inconel 718 discs needed for the sliding wear tests were polished to mirror finish with R_a ~1 µm. The chemical compositions of the both Inconel 718 and 9Cr–1Mo are shown in Tables 1 and 2, respectively.

2.2 *Metallography*

Metallographic samples were prepared for Inconel 718. Colloidal silica polishing was done after achieving the diamond finish and marble's reagent is used as an etchant. Laser confocal microscope was used to capture the optical micrographs. Figure 1 presents the microstructure for Inconel 718 observed at 1000x magnification. The microstructure has fine-grained microstructure with visible precipitates. Figure 2 shows the microstructure of the mod 9Cr–1Mo ferritic steel. Metallographic sample was prepared by sequential polishing and etching with Picral followed by few drops Nitric acid. The microstructure revealed the presence of predominant ferritic phase as expected.

Table 1 Chemical composition of Inconel 718

Elements	C	Mn	Si	Ti	Al	Co	Mb	Cb	Fe	Cr	Ni
Percentage	0.08	0.35	0.35	0.6	0.8	1.0	3.0	5.0	17.0	19.0	52.82

Table 2 Chemical composition of Modified 9Cr–1Mo

Elements	C	Mn	Si	N_2	Al	P	Mb	Nb	V	Cr	Ni
Percentage	0.08–0.12	0.30–0.60	0.00–0.50	0.030–0.070	0.04	0.020	0.85–1.05	0.06–0.10	0.18–0.25	8.00–9.50	0.40

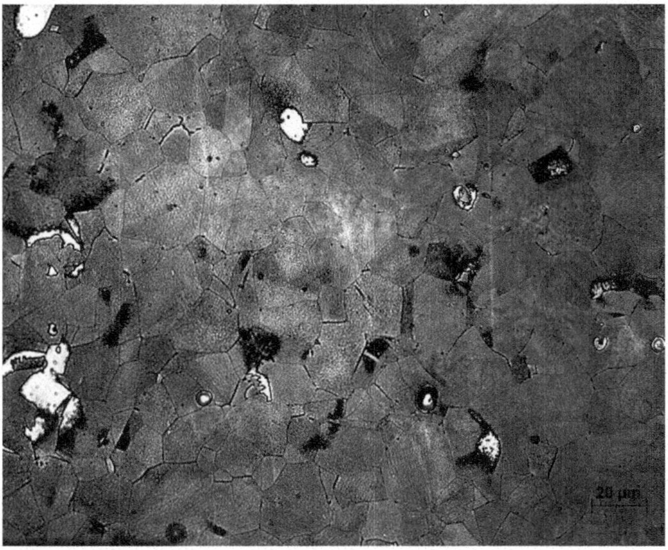

Fig. 1 Microstructure of Inconel 718 specimen

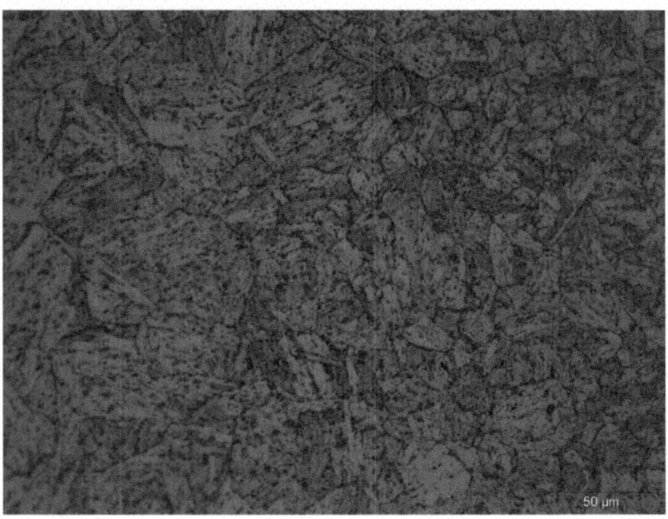

Fig. 2 Microstructure of mod 9Cr–1Mo

2.3 Dry Reciprocating Wear Test

The linear reciprocating test (LRT) was conducted using the LRT test setup model DUCOM-CM9062 as shown in Fig. 3. The tests were carried out in three different temperatures say room temperature (RT), 100 °C, 150 °C. For each of the above

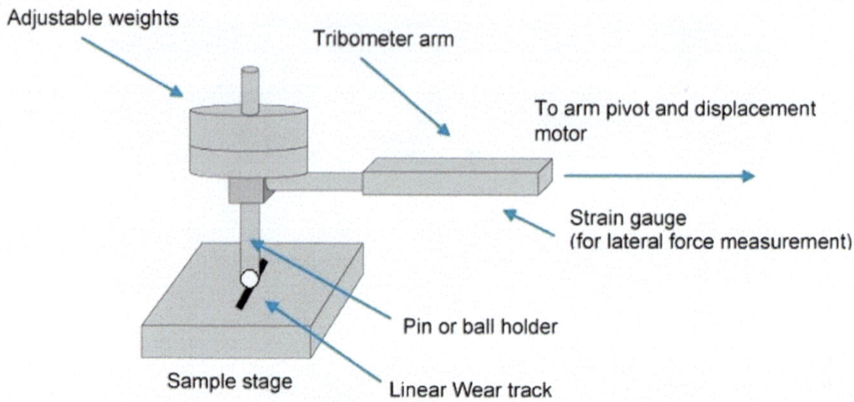

Fig. 3 Linear reciprocating test setup

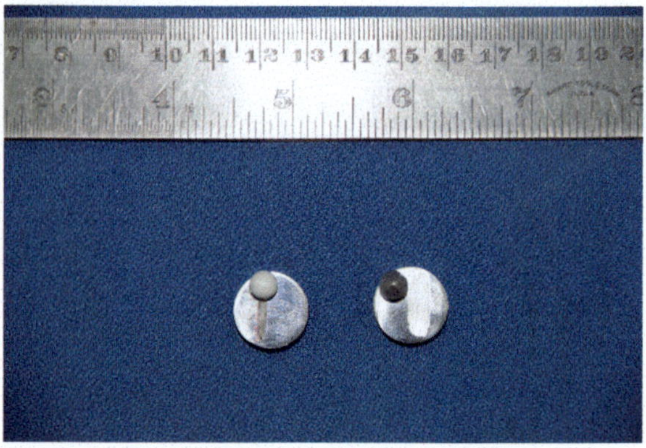

Fig. 4 Reciprocating wear test disc and pin

said temperatures, the frequencies have been varied as 3 Hz, 6 Hz and 9 Hz. By incorporating the above said parameters the test has been conducted for Inconel 718 against the counter surface of a Ø 6 mm alumina balls as well as the 9Cr–1Mo ferritic steel cylindrical pins with of 6 mm end diameter and 7.25 mm height. A uniform weight of 20 N is applied for all the test conditions. The tests were completed and wear loss was calculated using the weight loss method. Figure 4 shows the Inconel 718 specimen with its counter bodies Al_2O_3 ball and mod 9Cr–1Mo pin.

2.4 Surface Profilometry

Surface profilometry studies of the worn scar of the Inconel 718 specimens were carried out using the inductive gauge that possesses diamond indenter over its tip, by Talysurf CLI 1000. The roughness parameters were determined by the profilometer and it includes R_a (Arithmetic mean deviation of roughness profile), R_q (Root mean square deviation of roughness profile), R_p (Maximum peak height of the roughness profile) and R_v (Maximum valley depth of the roughness profile). Table 1 shows the various values obtained by the Talysurf CLI 100 (Taylor Hobson) surface profilometer. TalyMap Platinum software version 4.1 is used to analyse the roughness profiles.

3 Results and Discussions

3.1 *Inconel 718 Versus Al_2O_3 Balls*

Figures 5, 6 and 7 show the plots between coefficient of friction and time of the chosen material pair, namely Inconel 718 and alumina ball. The figures further present the consolidated view of the three operating frequencies, namely 3, 6 and 9 Hz for three different test conditions, namely RT, 100 °C and 150 °C, respectively. In all testing conditions (RT, 100 and 150 °C), the coefficient of friction is found to be more for 9 Hz operating frequency compared to the rest. The higher value of COF is attributed by the higher operating frequency (9 Hz). In addition, when an Inconel 718 with a hardness value 332 Hv is made to mate with a harder reference material, alumina (Al_2O_3) with a hardness of 1200 Hv, in a higher frequency condition, the relatively less hard test material degrades and it suffers weight loss in the form of

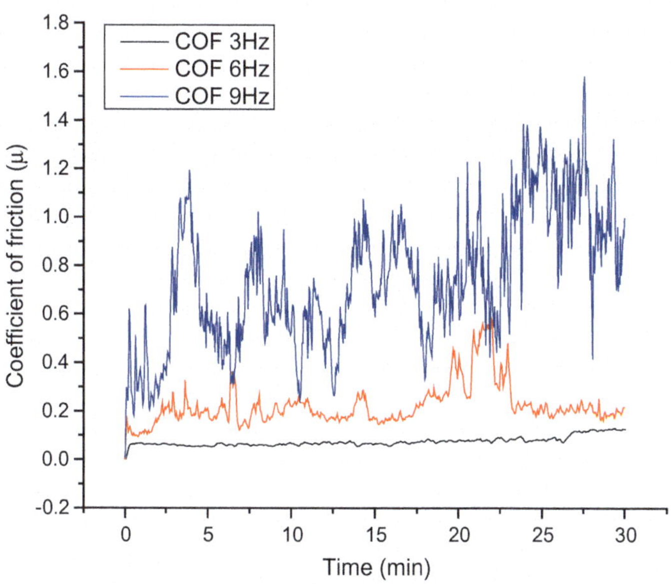

Fig. 5 Time versus Coefficient of friction (Room temperature)

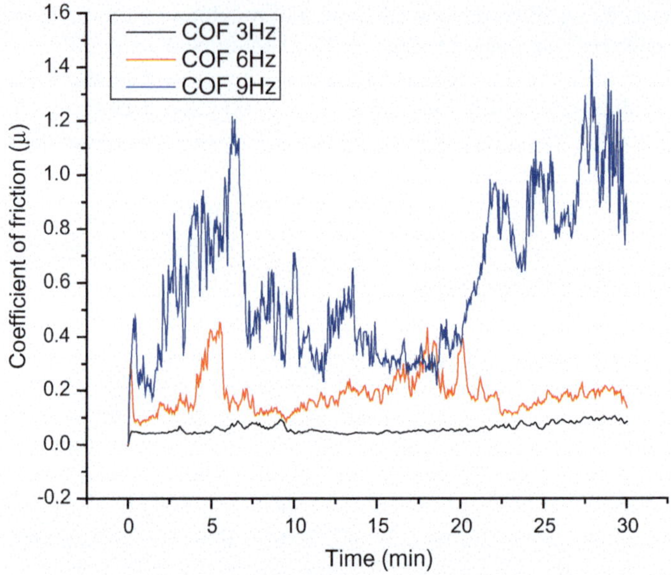

Fig. 6 Time versus coefficient of friction (100 °C)

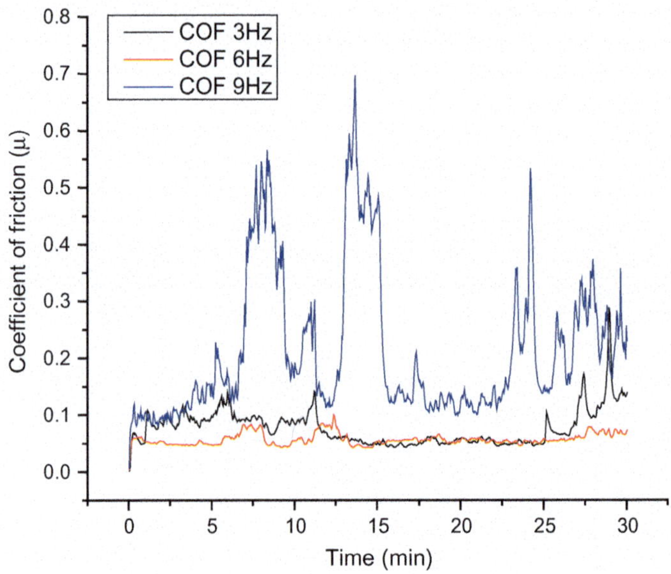

Fig. 7 Time versus coefficient of friction (150 °C)

wear, due to the reciprocating motion of the counter body. In addition, alumina ball, which is a reference material for reciprocating wear testing, is a hard ceramic and it can withstand higher operating temperatures. Alumina hardly wears and there is no surface degradation. At 9 Hz operating condition, the consistent hard surface of the Al_2O_3 ball retained the spherical nature and thereby the point in the alumina ball establishes a line contact with Inconel 718 during reciprocating motion. It is justified with a relatively less wide but deeper worn scar. The higher operating frequency (9 Hz) further justifies the encountered higher COF (1.2) in this pair. The COF curves are distinctly demarcated for different operating temperatures and frequencies.

3.2 *Inconel 718 Versus 9Cr–1Mo*

Figures 8, 9 and 10 depict the variation of coefficient of friction with respect to time. It further showed the consolidated view of three operating frequencies, namely 3, 6 and 9 Hz. Here, when compared to the COF encountered by the Inconel and alumina pair, the coefficient of friction of the Inconel and 9Cr–1Mo pair is considerably low. This is due to the reason that the hardness of the 9Cr–1Mo (220 Hv) is marginally lesser than the hardness of the Inconel 718 material (332 Hv). Modified 9Cr–1Mo cylindrical steel pins are used as a counter body to simulate the steam generators which tend to vibrate with the steam bundle support structures made up of Inconel

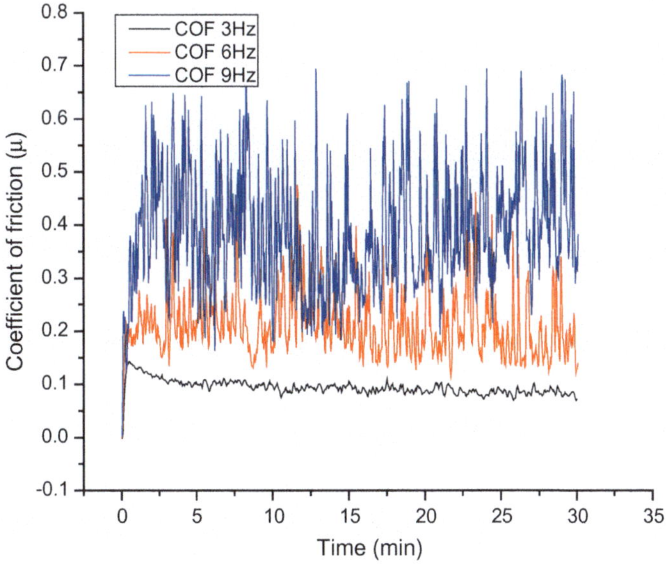

Fig. 8 Time versus coefficient of friction (room temperature)

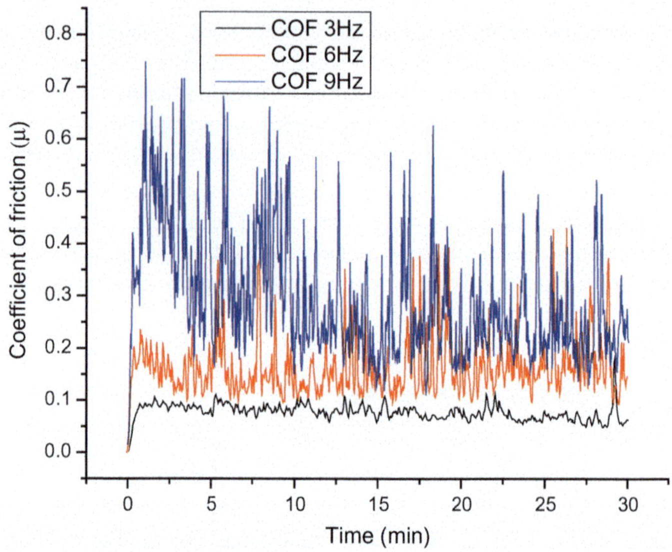

Fig. 9 Time versus coefficient of friction (100 °C)

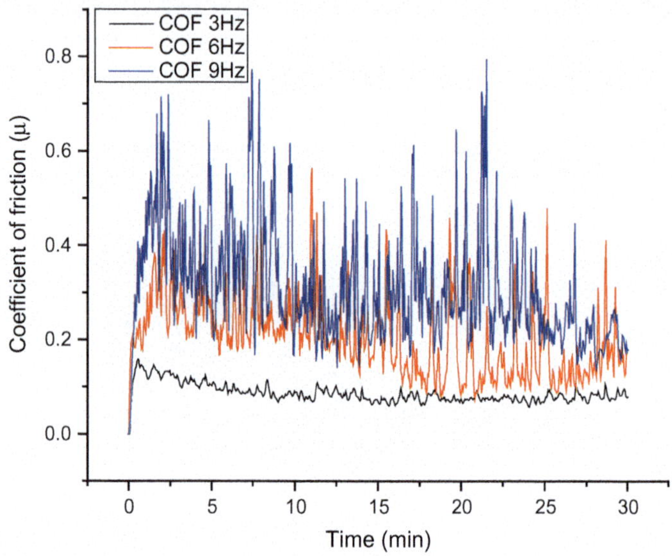

Fig. 10 Time versus coefficient of friction (150 °C)

718. The COF values of mod 9Cr–1Mo mated Inconel 718 are in the range of 0.2–0.8, on the other hand, for alumina Inconel 718 mating pair COF varies between 0.2 and 1.2. The rise in COF values is due to the difference in the hardness values of the matting pairs as discussed earlier. Further, the lesser hardness of mod 9Cr–1Mo steel makes the material to lose its structural integrity during operation at higher temperatures. Further, the cylindrical pin with a hemispherical edge disintegrates to a flattened surface. The point contact between the mating pair becomes an area contact and thereby a wider scar with shallow depth is presented in the worn scar of Inconel 718. The smoothening of the mod 9Cr–1Mo steel pin into a flat face becomes a potential cause for the decreased COF (0.2) relatively at 9 Hz and 150 °C operating frequency. This COF variation pattern and its lesser degree of severity of wear of Inconel 718 against mod 9Cr–1Mo are in well agreement and very well correlated with the worn scars.

Figures 11 and 12 portray the weight loss of Inconel 718 when paired with Al_2O_3 and 9Cr–1Mo ferritic steel respectively. From the figure, it is very clear that the weight loss for Inconel 718 is higher when paired with alumina balls than with the 9Cr–1Mo steel cylindrical pins. This is due to the higher coefficient of friction resulting from the higher hardness of the alumina ball at higher operating frequencies and temperature conditions. At 150 °C and 9 Hz frequency, the weight loss of Inconel 718 against alumina balls as well as 9Cr–1Mo seems to be more, compared to the rest of the operating conditions.

From Fig. 13, we can easily find that the 9Cr–1Mo steel cylindrical pins, which are relatively less hard, suffer a noticeable weight loss, when mated against the relatively

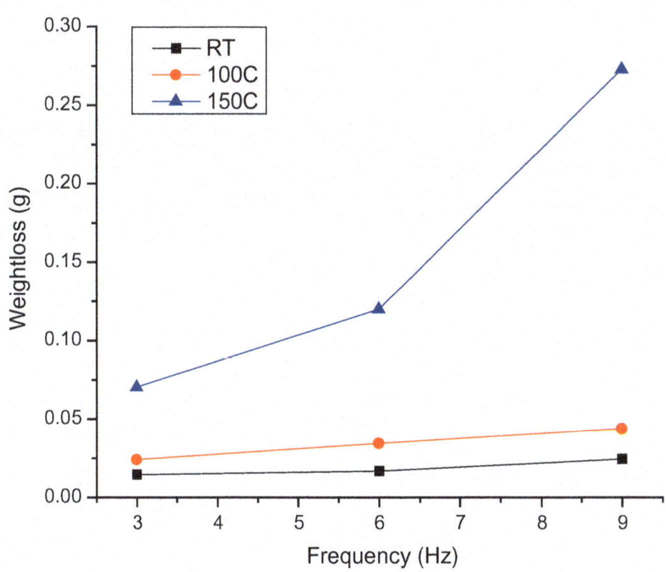

Fig. 11 Weight loss for Inconel 718, paired with Al_2O_3

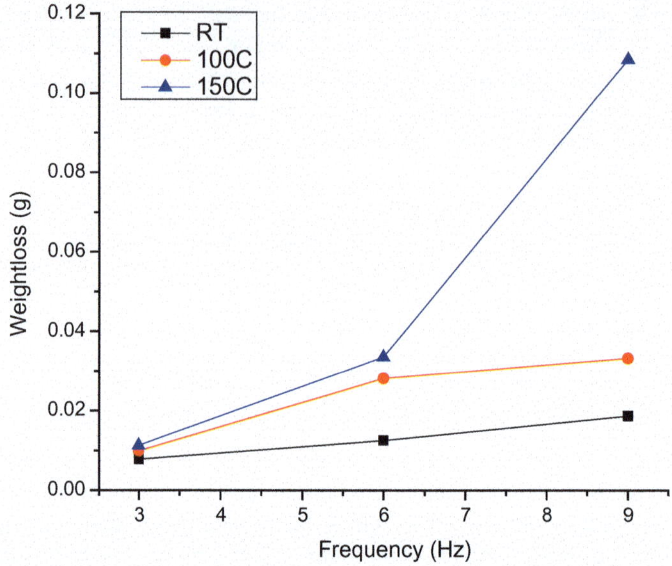

Fig. 12 Weight loss for Inconel 718, paired with 9 Cr–1Mo

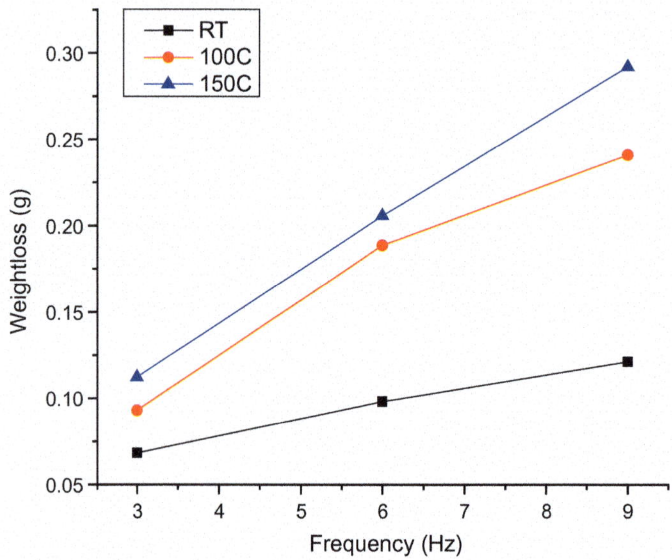

Fig. 13 Weight loss for 9Cr–1Mo when paired with Inconel 718

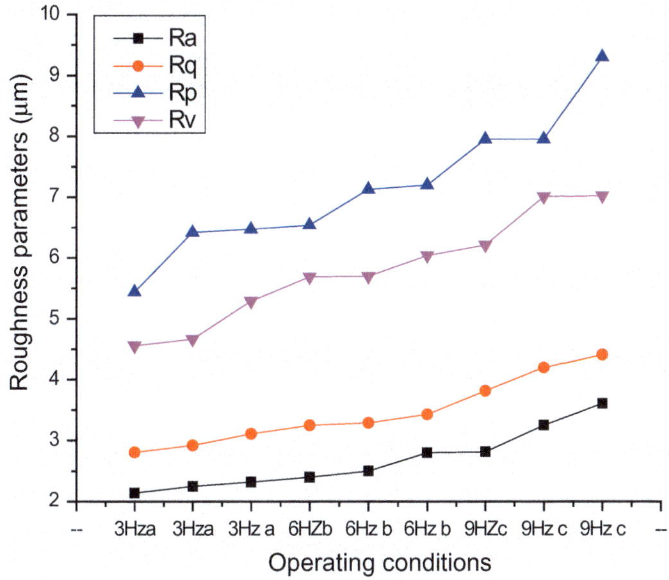

Fig. 14 Roughness parameters for Inconel 718 versus Al_2O_3 balls

harder Inconel 718. Similarly, it is also visible that the weight loss is higher at the maximum frequency (9 Hz), irrespective of the temperature conditions.

Figure 14 shows the variation of the roughness parameters for Inconel 718 versus Al_2O_3 balls. The roughness parameters, namely R_a, R_q, R_p and R_v, showed a rising trend with operating frequencies as well as temperatures. In Fig. 12, a denotes room temperature, b denotes 100 °C and c denotes 150 °C. Figure 15 depicts the variation of roughness parameters for Inconel 718 specimens against 9Cr–1Mo steel counter cylindrical pins. Similar to the previous graph, the results showed a gradual increase in values with rise in operating temperatures as well as frequencies.

4 Conclusions

Reciprocating wear test experiments were carried out at different operating temperatures and frequencies to study the wear of Inconel 718 against Al_2O_3 and mod 9Cr–1Mo steels and the following points were concluded,

- The variation of COF versus time for Inconel 718 versus Al_2O_3 is presented with clear demarcation for different operating temperatures and frequencies but when it is mated with mod 9Cr–1Mo overlapping is present in higher operating temperatures and frequencies.

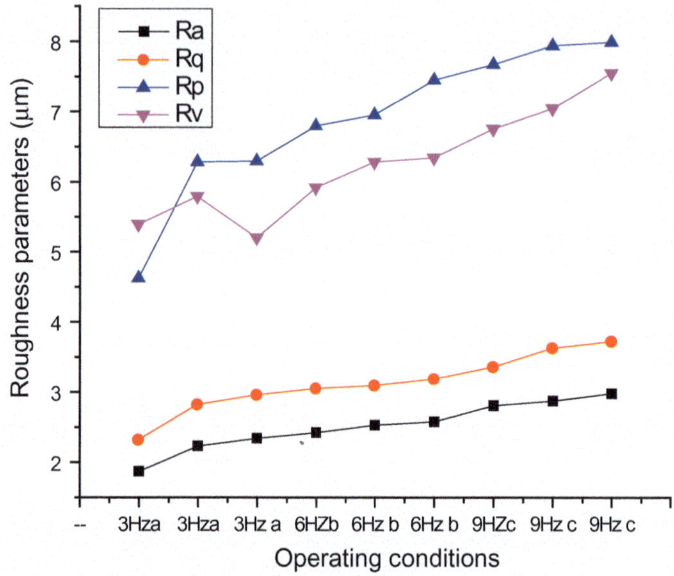

Fig. 15 Roughness parameters for Inconel 718 versus 9Cr–1Mo steel

- The higher COF (1.2) between Inconel 718 and Al_2O_3 is due to the higher hardness, structural integrity at higher operating temperature and the lesser coefficient of friction between Inconel 718 and mod 9Cr–1Mo (0.2) is attributed to the relatively lesser hardness of the counter body compared to the alumina ball.
- In Inconel 718, the wear encountered in terms of mass loss is considerably more when mated with Al_2O_3, whereas when the same is mated with mod 9Cr–1Mo, the counter body experiences more weight loss during reciprocating wear.
- The roughness parameters, namely R_a, R_q, R_p and R_v, showed a rising trend with operating frequencies as well as temperatures, making us to understand that the surface roughens during higher operating temperatures and frequencies.
- A surface modification is prescribed to lower the COF between Inconel 718 and mod 9Cr–1Mo and thereby minimize the wear loss in terms of mass during reciprocating wear.

References

1. Sharatchandra Singh W, Purnachandra Rao B, Thirunavukkarasu S, Mahadevan S, Mukhopadhyay CK, Jayakumar T (2015) Development of magnetic flux leakage technique for examination of steam generator tubes of prototype fast breeder reactor. Ann Nucl Energy 83:57–64
2. Apparao G, Kumar M, Srinivas M, Sarma DS (2003) Effect of standard heat treatment on the microstructure and mechanical properties of hot isostatically pressed superalloy Inconel 718. Mater Sci Eng A 355(1–2):114–125

3. Ping DH, Guy F, Cuic Y, Harada H (2007) Grain boundary segregation in a Ni–Fe based (alloy 718) superalloy. Mater Sci Eng A 456(1–2):99–102
4. Rahman M, Seah WKH, Teo TT (1997) The machinability of Inconel 718. J Mater Process Technol 63(1–3):199–204
5. Fisher NJ, Chow AB, Weckwerth MK (1995) Experimental fretting-wear studies of steam generator materials. J Press Vessel Technol 117:312–320
6. Guerout FM, Fisher NJ, Grandison DA, Weckwerth MK (1996) Effect of temperature on steam generator fretting-wear. In: ASME PVP, vol 328, Flow-induced vibration, pp 233–246
7. Ko PL (1985) Heat exchanger tube fretting wear: review and application to design. J Tribol 107:149–156
8. Taylor CE, Pettigrew MJ, Dickinson TJ, Currie IG, Vidalou P (1998) Vibration damping in multispan heat exchanger tube. J Press Vessel Technol 120(3):283–289
9. JD Kwan, YS Chani, YT Bae, J Choin (2006) A study on fretting wear behavior in room temperature air for Inconel alloy 690. Int J Mod Phys B 20, 25, 26, 27:4303–4308
10. Concessao JI, Quadros JD, Vaishak NL (2016) Effect of ageing on the tribological behavior of Inconel 690 using Taguchi's Method. Am J Mater Sci 6(4A):25–30
11. Deng D, Wang C, Liu Q, Niu T (2015) Effect of standard heat treatment on microstructure and properties of borided Inconel 718. Trans Mater Soc China 25:437–443
12. Yu T, Yang X, An J, Yu X, Zhao J (2018) Material removal mechanism of two dimensional ultrasonic vibration assisted polishing Inconel 718 nickel- based alloy. Int J Adv Manuf Technol 96(1–4):657–667
13. Banker VJ, Mistry JM (2015) Wear mode in Inconel alloys—a literature study. In: Advances in materials and product design (AMPD)—2015. Sardar Vallabhbhai National Institute of Technology (SVNIT), Surat. ISBN No: 978-93-5196-956-3

High-Temperature Sliding Wear Characterization Studies of AISI 316 L(N) by Surface Profilometry

N. Aruldev, N. L. Parthasarathi, B. Rajasekaran, and Utpal Borah

Abstract The major construction material in the Prototype Fast Breeder Reactor (PFBR) is AISI-type 316 L(N) austenitic stainless steel due to its good mechanical properties and compatibility with liquid sodium. Sliding wear experiments were carried out at various temperatures up to 550 °C at constant load (20 N) and sliding speed (0.8 m/s) using a pin-on-disc test rig as per the ASTM standard G99-05. Analysis of the test results presented that the wear increased considerably with the temperature. The characterization of worn surface topography is done by a complete profilometry study using Talysurf CLI 1000 surface profilometer. The 3D surfaces were captured both by induction mode by diamond stylus as well as non-contact high-resolution confocal point gauge having range of 3000 μm with 0.25 nm resolution was used for surface profiling. The cold-welded surfaces were analysed by the profilometer, and the geometry of the deposit on the wear track was analysed by the profilometer. The roughness parameters were correlated with the amount of wear data obtained from the experiments at various testing temperatures. As the temperature increases during the sliding wear, the material loss is presented with more furrows resulting in enhanced surface roughness values.

Keywords AISI 316 L(N) · High temperature · Sliding wear · Profilometry

1 Introduction

Profilometry has a major role in the characterization of surface profiles of materials. In addition, this technique is predominantly used to understand the topological data from surface. The characterization can be done by analysing a single point, a line (two dimensional) or even an area (three dimensional). The method of capturing is

N. Aruldev · B. Rajasekaran
Department of Metallurgical and Materials Engineering, NIT-K, Surathkal, India

N. L. Parthasarathi (✉) · U. Borah
Materials Development and Technology Division, Indira Gandhi Centre for Atomic Research, Kalpakkam, Tamil Nadu, India
e-mail: nlpartha@gmail.com

L.-J. Yang et al. (eds.), *Proceedings of ICDMC 2019*,
Lecture Notes in Mechanical Engineering,
https://doi.org/10.1007/978-981-15-3631-1_27

as simple as that of giving x and y coordinates to specify an area of interest. The main objectives of the profilometry are to get surface roughness, surface morphological parameters and step height measurement. Here, the measurement is done by two modes, viz. by a physical probe (stylus) contact-type and non-contact-type (white light interferometry principle) light.

In case of contact surface profilometry, there is physical contact of moving probe (diamond stylus) along the surface, in case of non-contact profilometry there is no physical contact to the surface [1]. Here, we used diamond stylus in case of contact profilometry and light in case of non-contact profilometry. The surface characterization of surface profile is a challenging task, because the values of surface parameters should have accuracy as well as precision, which mandates repeatability and consistency. A comparison of the results for roughness parameters is obtained with the help of two methods of evaluating the quality of the finished surfaces, namely 2D profilometry and 3D profilometry [2]. The basics of surface texture parameters and its measurement of surface profiles were explained for better understanding [3, 4]. A method was developed for measuring and finding the volumetric wear expected in ceramic specimens using non-contacting optical methods [5].

For the Prototype Fast Breeder Reactor (PFBR), AISI 316 L(N) is the major construction material designed by Indira Gandhi Centre for Atomic Research (IGCAR), Kalpakkam; the remarkable properties of AISI 316 L(N) austenitic steel includes good high-temperature strength, toughness, creep strength, low cycle fatigue strength and compatibility with liquid sodium. 316 L(N) and contact pairs undergo sliding wear at different temperatures (550 °C). As the temperature increases during the sliding wear, the material loss is presented with more undulations resulting in higher surface roughness values [6]. Surface profiles across the grooves in ultrahigh molecular weight polyethylene were determined using a Taylor Hobson surface profilometer. Four profilometer measurements were obtained across the different wear tracks [7]. In managing tool steel materials, the tribological behaviour was compared by using different methods for worn surface analysis. Worn area of the surface was determined by the surface profiles of 2D measurements, and the wear volume was calculated by using 3D images captured by the microscope [8].

The diamond stylus and white light profilometry were employed to measure the surface profiles of the worn tracks according to their corresponding temperature. The objective of this work is to find the suitability of the profiling technique for the better characterization of worn surface profile and its effective characterization. This paper attempts to report the results from the surface profile analysis and to differentiate the values of surface parameters obtained by diamond stylus and optical measurement.

2 Experimental Procedure

2.1 *Material*

AISI 316 L(N) is the most common type of austenitic steel used in nuclear reactor. The carbon content of alloy is less than 0.03% to reduce the sensitization effect caused by high temperature during welding process and the nitrogen expanded austenite is also present. L denotes the carbon content of alloy, and N denotes the presence of nitrogen. Compared to 316 L, it has more corrosion resistance, even if it is continuously exposed to the temperature ranging between 900 and 1500 °C. The presence of nitrogen adds more resistance to sensitization, which increases the yield strength and provides some solid solution hardening. Compared to 316 L, 316 L(N) has more resistance against corrosion as well as crevice or pitting corrosion. Table 1 shows the chemical composition of AISI 316 L(N).

2.2 *Wear Testing Procedure*

The sliding wear experiments were carried out in DUCOM to make TR-20-M12EV high-temperature pin-on-disc tribometer as per ASTM: G99-05 standard test method for wear testing. AISI 316 L(N) discs of 130 mm diameter and 10 mm thick were machined out from a block. AISI 316 L(N) pins with hemispherical contact, which is capable of establishing point contact, were also machined in required quantity. The nose radius of the pin is 5 mm. A constant load of 20 N was applied, and the tests were carried out for 800 s each at a constant sliding velocity of 0.8 m/s resulting in total sliding distance of 640 m in each test. The wear tests were conducted at various temperatures, namely room temperature (25 °C), 100, 150, 250, 350, 450 and 550 °C. Figure 1 shows the worn tracks by high-temperature sliding wear tests carried out at 250, 350, 450 and 550 °C for representative purpose.

Table 1 Chemical composition of AISI 316 L(N)

Elements	C	Cr	Mo	Ni	Mn	Si	N	S	P	Cu	Fe
Weight (%)	0.024	17.8	2.38	13	1.84	0.25	0.05	0.005	0.03	0.519	Bal

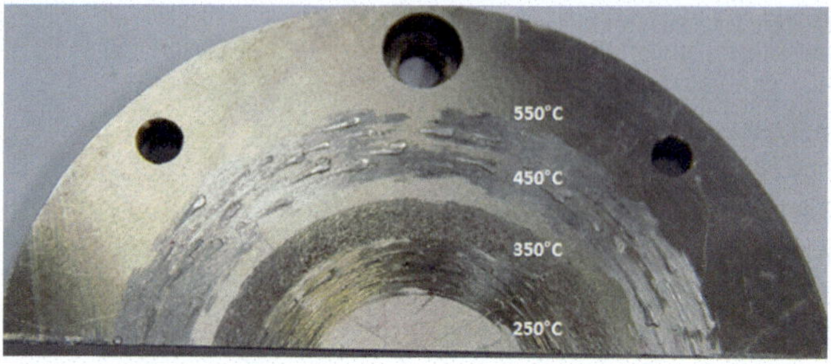

Fig. 1 Worn tracks carried out at 250, 350, 450 and 550 °C

2.3 Surface Profilometer

The characterization of the worn track surfaces is done by TAYLOR HOBSON TALYSURF CLI 1000 profilometer. There are two types of characterization in profilometer: one is a contact type (diamond stylus) and the other one is non-contact (white light interferometry). The inductive gauge consists of a tactile gauge equipped with the diamond stylus tip for tracing surface profile. The working principle of the tactile measuring is shown in Fig. 2. When the sample is tracing the up and down movements of the surface profile, stylus movements change the inductance of electrical circuit. Further, the inductance change will be converted to the surface profile plots. The variation of up and down movement of stylus arm aids to calculate the

Fig. 2 Diamond stylus principle

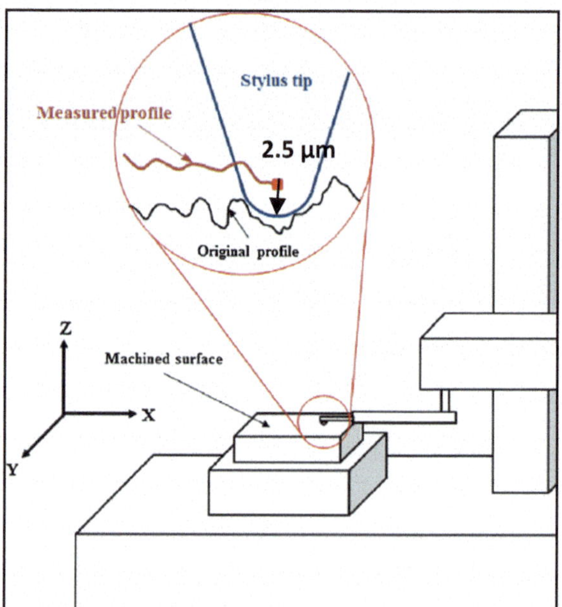

height variations of peaks and valleys in z axis. The diamond stylus has the active contact between the instrument and surface, so the variations of stylus movement directly affect the traced profile. Inductive Gauge has a working range 2.5 mm, vertical resolution 40 nm, accuracy of 0.3 μm and the diameter of diamond tip stylus 2.5 μm.

Figure 3 shows the working principle of the confocal surface profiling using the white light interferometry principle. A white light beam is focussed on a surface through a spherical aberration lens (chromatic length aberration). Due to this aberration, the focus point is at different Z-positions for different wavelengths. The reflected light is sent to the spectrometer through an optical pinhole. The spectrometer provides an intensity curve depending on the wavelength. The focussed wavelength is the one corresponding to the maximum of intensity. Chromatic linear aberration (CLA) confocal gauge has a working range 300 μm, working distance of 5 mm, maximum slope 25°, lateral precision of eight micrometers and lateral resolution of one and two micrometers. When we compare both techniques, the resolution is found to be more in optical measurement, one micrometers resolution optimized for step height measurement and two micrometers for roughness measurement. The working range is more in diamond stylus that is 2.5 mm whereas in the other hand for optical measurement it is up to 300 μm, but it has the advantage of lateral precision, resolution, repeatability and linearity.

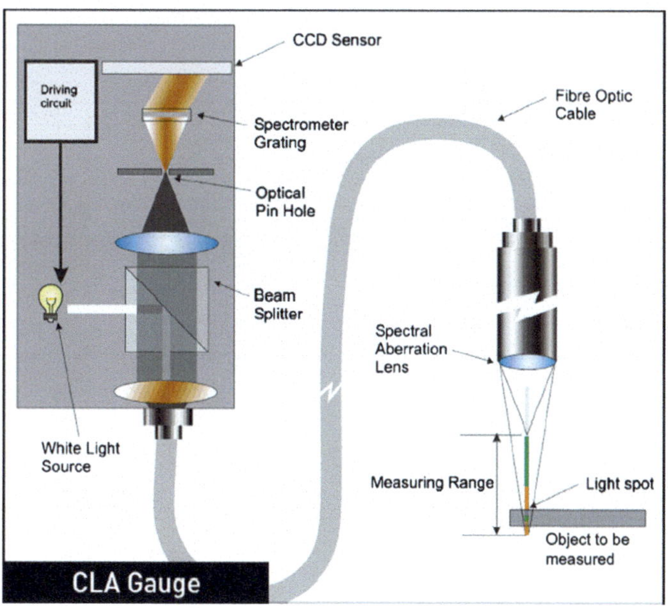

Fig. 3 Confocal surface profiling principle [9]

2.4 Surface Parameters

The following are the surface parameters taken for the study, namely Sa—arithmetic mean deviation of surface profile, Sq—root mean square deviation of surface, Sp—maximum height of summits, Sv—maximum depth of valleys, St—total height of surface, Ssk—skewness of topographic distribution, Sku—kurtosis of topography height distribution and Sz—ten-point height surface.

3 Results and Discussion

3.1 Surface Profile by Diamond Stylus

See Figs. 4, 5, 6 and 7.

3.2 Surface Profile by White Light Interferometry

See Figs. 8, 9, 10 and 11.

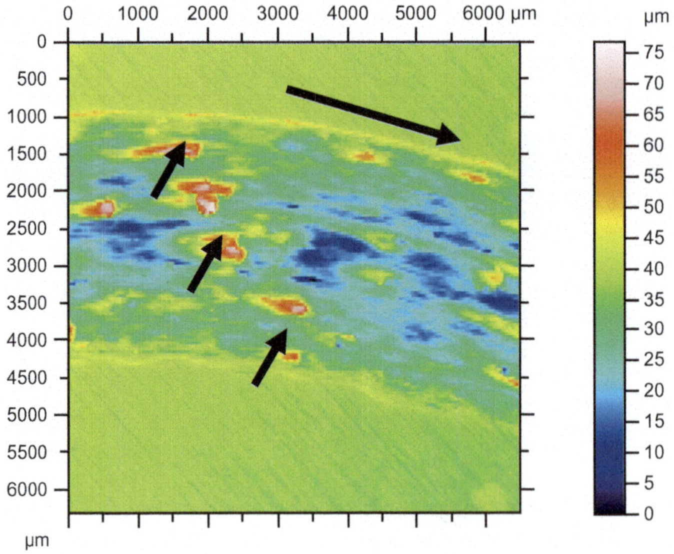

Fig. 4 Worn track at 25 °C

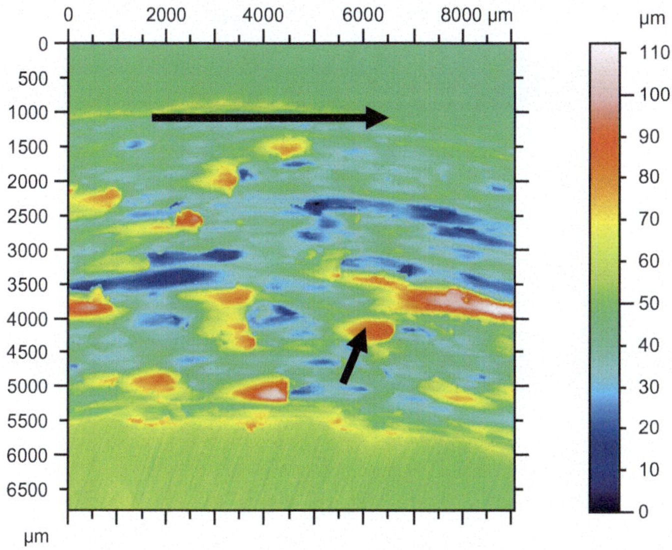

Fig. 5 Worn track at 150 °C

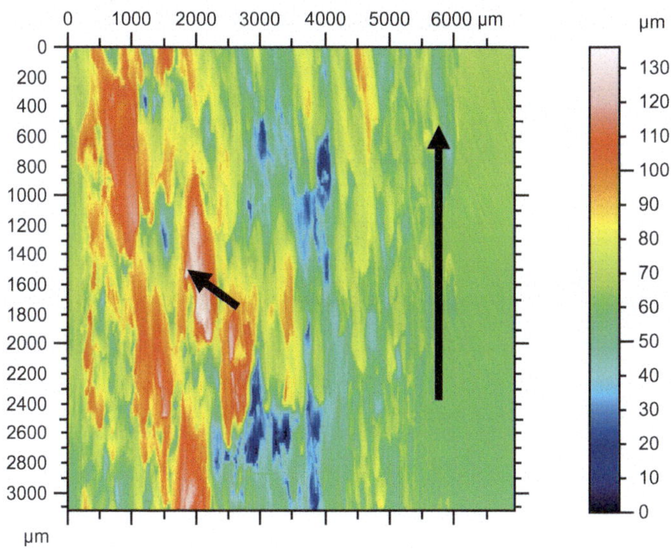

Fig. 6 Worn track at 350 °C

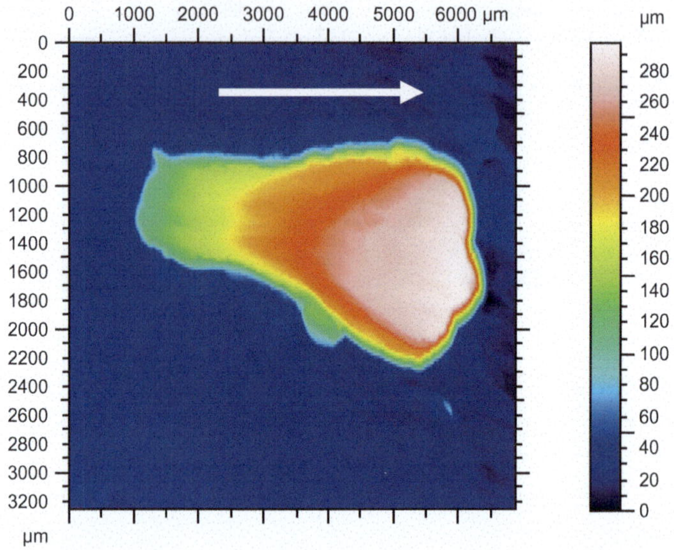

Fig. 7 Worn track at 550 °C

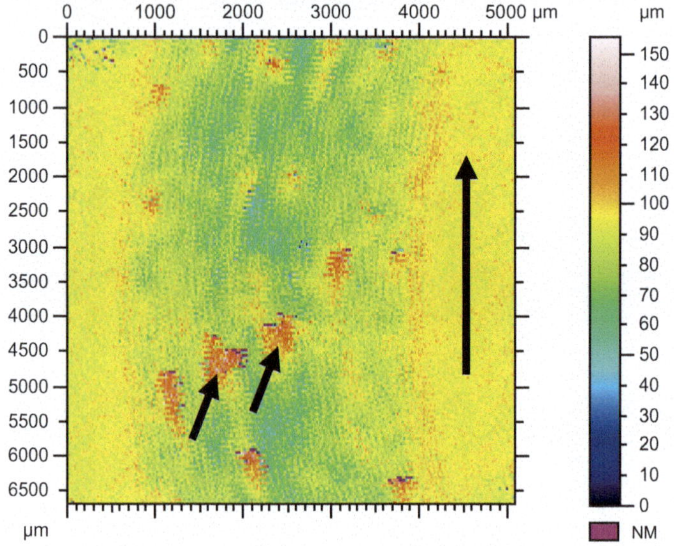

Fig. 8. Worn track at 25 °C

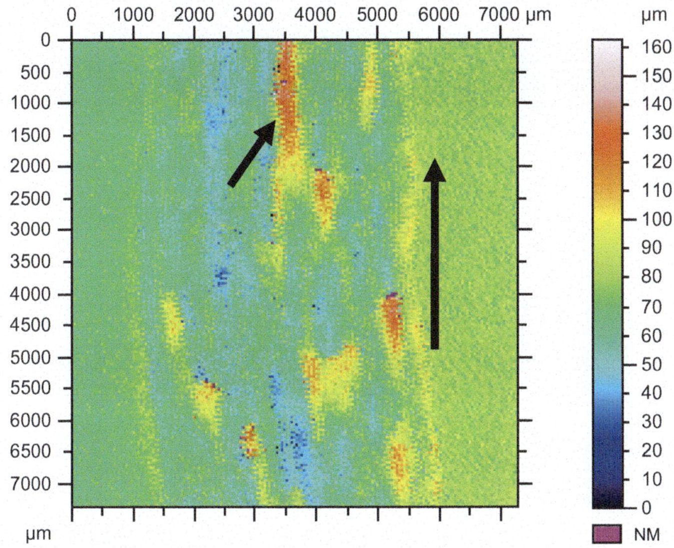

Fig. 9. Worn track at 150 °C

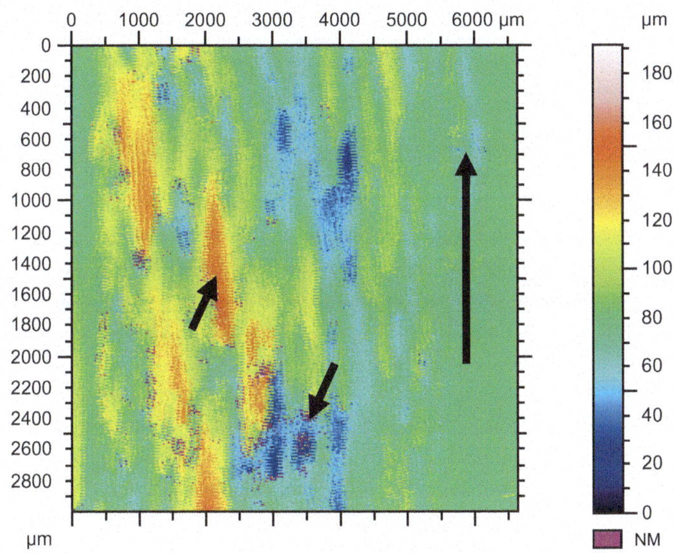

Fig. 10 Worn track at 350 °C

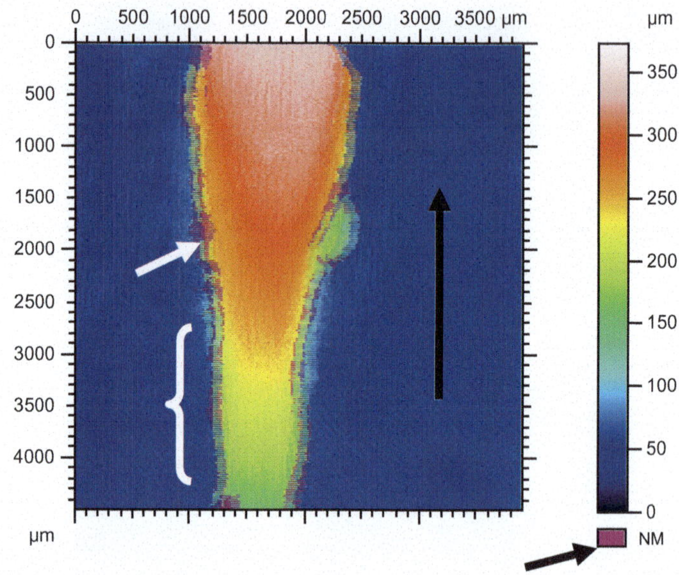

Fig. 11 Worn track at 550 °C

3.3 *Surface Parameter Readings and Comparison*

Figures 4, 5, 6 and 7 show the surface profile worn at temperatures, namely 25, 150, 350 and 550 °C for representative purpose. The colour scale is shown in the right side of the surface profile. The varying colours from black to red indicate the gradual increase in the height of the roughness profile of the worn tracks. In Figs. 4, 5 and 6, smaller arrow marks show the raised points in the worn track due to the material transfer from pin material and the longer arrow marks notify the sliding direction. In Fig. 7, a prominent human foot-like island structure is captured by the surface profile; this feature significantly depicts the material transfer (cold-welded) from the counter body, due to adhesive wear. This zone was chosen for displaying the vulnerability of the high-temperature wear. The profile was presented with several such features in this high operating temperature (550 °C). The bottom-most point of the profile is shown in black colour (lowest point of the valley), and the topmost point is shown in red colour (highest point in the peak). Tables 2 and 3 give the surface parameter values by stylus measurement and optical measurement, respectively.

Figure 12 depicts the variation of arithmetic mean deviation of surface (Sa) by tactile method (diamond stylus) and optical method (white light interferometry) with respect to the corresponding temperatures. Since the values obtained are computed from arithmetic averages, there is no remarkable difference in the values obtained by both the techniques, so Sa values typically vary from 6 to 80 µm. When measured by white light interferometry in higher operating temperatures, namely 450 and 550 °C, Sa values are found to be in the order 50 µm and 79.2 µm, respectively.

Table 2 Surface parameter values by stylus measurement

Temp (°C)	Sa (μm)	Sq (μm)	Sp (μm)	Sv (μm)	St (μm)	Ssk (μm)	Sku (μm)	Sz (μm)	Number of motifs
25	6.2	8.03	42.7	34.2	76.9	−0.316	4.7	76.9	36
100	2.44	3.46	23	16.2	39.2	−0.023	6.05	39.2	102
150	9.19	12.6	65.4	46.9	112	0.843	5.58	112	54
250	19.2	35.9	172	62.5	234	3.28	13.3	234	18
350	13.8	18.4	67.8	68.5	136	0.498	3.8	136	91
450	52.5	75.3	257	60.8	318	2.2	6.48	318	3
550	70.2	86.4	227	71	298	1.47	3.53	298	3

Table 3 Surface parameter values by optical measurement

Temp (°C)	Sa (μm)	Sq (μm)	Sp (μm)	Sv (μm)	St (μm)	Ssk (μm)	Sku (μm)	Sz (μm)	Number of motifs
25	8.88	11.3	68.2	87.3	155	−0.316	4.56	155	101
100	4.64	6.58	59	88.1	147	−1.25	13.6	147	165
150	10.5	14.1	91	71.6	163	0.719	5.49	163	61
250	18.8	35.6	211	142	354	3.3	13.9	354	199
350	15.4	20.8	108	83.8	192	0.442	3.81	192	141
450	50	75.3	293	116	409	2.36	7.1	409	43
550	79.2	94.5	261	111	373	1.33	3.06	373	98

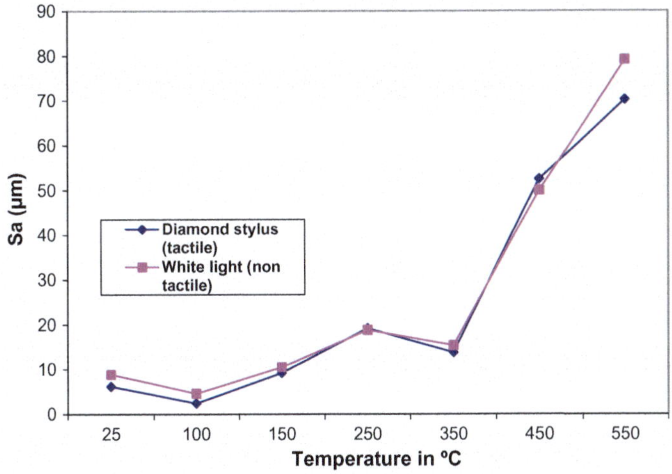

Fig. 12 Variation of Sa values by diamond stylus and optical method with respect to corresponding temperatures

Figure 13 depicts the variation of root mean square deviation of surface (Sq) by tactile method and optical method with respect to the corresponding temperatures. The trend was found to be similar to the variation of Sa values since it is derived from average values.

Figure 14 portrays the variation of maximum height of peak or summit (Sp) values by diamond stylus and optical method with respect to corresponding temperatures. The Sp values measured by non-tactile method showed increased values of 20 µm, uniformly in all the operating temperatures. Optical methods work in the principle of

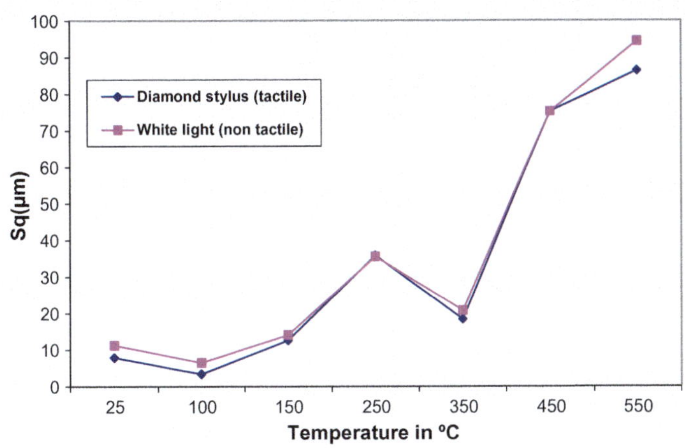

Fig. 13 Variation of Sq values by diamond stylus and optical method with respect to corresponding temperatures

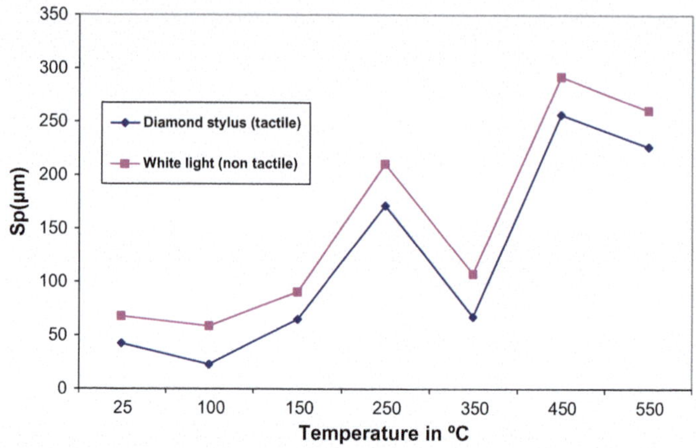

Fig. 14 Variation of Sp values by diamond stylus and optical method with respect to corresponding temperatures

interferometry, which the prominent peaks reflect light with shining surfaces, which enables the exact capturing of the peak values. In 450 °C operating temperature, Sp value is found be maximum (293 µm).

Figure 15 shows the variation of maximum height of valley (Sv) values by diamond stylus and optical method with respect to corresponding temperatures. The Sv values clearly show the demerits of using diamond stylus (tactile method). Certain deep valley points were missed by the diamond stylus because of its small nose diameter (2.5 µm), which results in skidding, which results in the unobserved valleys. In 250 °C operating condition, it is observed that Sv value by optical method is 142 µm,

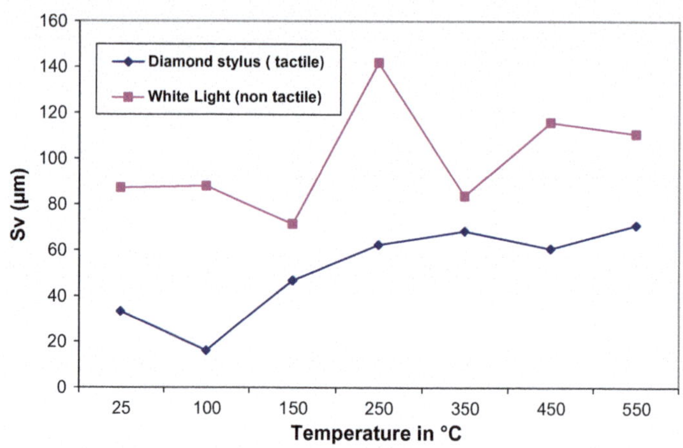

Fig. 15 Variation of Sv values by diamond stylus and optical method with respect to corresponding temperatures

which is greater than the value measured by the tactile diamond stylus (62.5 μm). In addition, the maximum slope value which can be sensed by optical method is 25°, which further enables the method to measure the deepest valley points without getting unobserved.

Another effect of the geometry of the stylus is "flanking". If the slope of the roughness exceeds the angle of the conisphere, the stylus will no longer contact on the spherical tip portion, but on the straight edge. This is most clearly evident when measuring a step, as illustrated in Fig. 16. Typically, the profile will exhibit a radius followed by an angle that corresponds to the stylus "flank" angle. This is a major disadvantage of the diamond stylus.

Figure 17 shows the variation of total height of the surface profile (St) by diamond stylus and optical method with respect to corresponding temperatures. The maximum total height (St) of the worn track profile varied from 75 to 300 μm, when measured by diamond stylus and when the same surface profile instances were measured by white light interferometry, it varied between 150 and 380 μm. In case of light profilometry, there is one more colour (purple) in the surface profile that indicates the

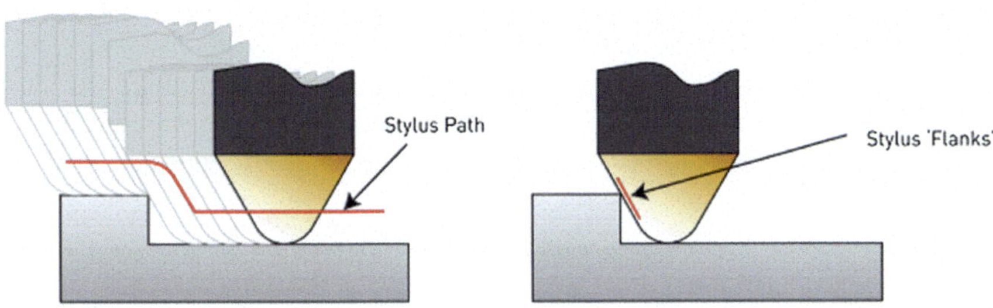

Fig. 16 Stylus flanking [11]

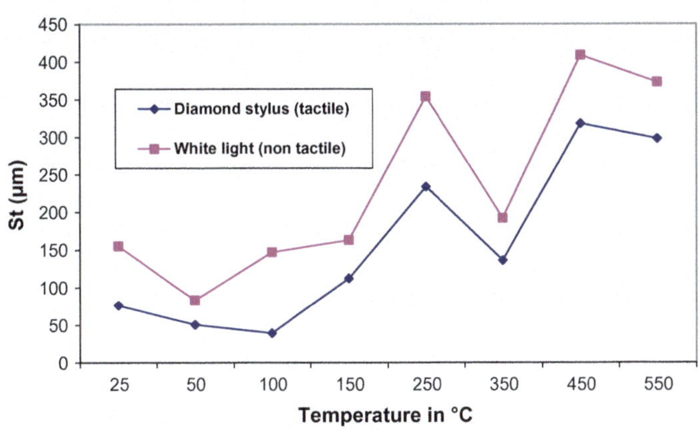

Fig. 17 Variation of St values by diamond stylus and optical method with respect to corresponding temperatures

Non-measured points on the surface that is shown below the colour scale. The stylus measurement does not have any non-measured points in the evaluation length, but in case of optical measurement, in every temperature non-measured points are present. It is due to the insufficient reflection from relatively dark valley surfaces. In Figs. 8, 9, 10 and 11, few non-measured points are denoted by small arrow marks for better understanding.

3.4 Motif Analysis

Motif is a representation of either hills or dales. A dale is surrounded by a ridge line and its minimum point is the pit, a hill is surrounded by a course line and its maximum point is peak. The intersection of course line and ridge line is called saddles points; this is also presented in the profile. The white cross indicates the location of peak on each motif; it shows how this will affect the lubrication, wear and nature of contact; and it gives the appearance of surface texture. The representation of hill and dales is inferred by segmentation. Figure 18 shows the motif analysis of a worn track by diamond stylus and optical method at 250 °C. Using diamond stylus measurement (a) and by white light interferometry (b), 18 and 199 are the total number of motifs present in the surface profile by diamond stylus measurement and white light interferometry, respectively. The profiles measured by optical method are

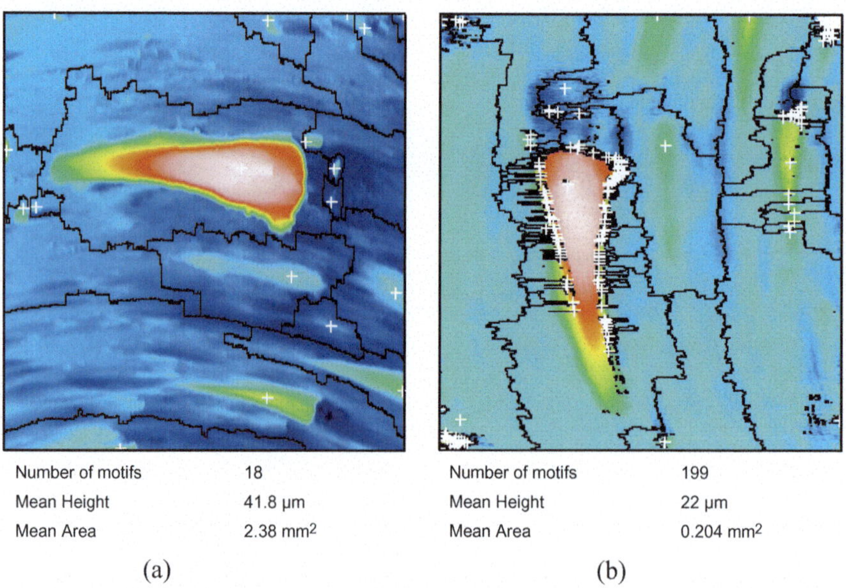

Fig. 18 Worn track at 250 °C **a** using diamond stylus **b** using optical measurement

presented with more number of motifs compared to the stylus making us to conclude optical method is superior for analysing the worn surface.

4 Conclusion

A detailed experimental study was conducted to understand the effect of stylus and optical measurement in observing the surface roughness parameters. From the experimental data, we can conclude the following.

- Worn track with highest operating temperature (550 °C) was presented with higher average roughness value parameters, namely Sa, Sq, etc.
- In the optical method, Sp values showed a uniform increase of 20 μm, in all the operating temperatures compared to the stylus method. The prominent peaks, which were reflected by light, enabled the exact capturing of the Sp values by optical method.
- The maximum slope value, which can be sensed by optical method is 25°, which enables to measure the deepest valley points in measuring the Sv values.
- More number of motifs are captured by optical method (199 nos) compared to the stylus method (18 nos), ensures the effectiveness of the optical method in the worn surface characterization.
- For high-temperature worn surface characterization, optical method is more accurate than the stylus method, as far as the worn surfaces are presented with bright and reflecting features.
- In contrast, the non-measured points are almost negligible (except deep valleys) in the diamond stylus. For non-reflecting, dull finished surfaces, the deployment of the tactile diamond stylus becomes inevitable and more appropriate.

References

1. Arvinth Davinci M, Parthasarathi NL, Borah U, Albert SK (2014) Effect of the tracing speed and span on roughness parameters determined by stylus type equipment. Measurement 48:368–377
2. Deleanu L, Georgescu C, Suciu C (2012) A comparison between 2D and 3D surface parameters for evaluating the quality of surfaces. The Annals of “dunărea de jos” University of Galati, Fascicle v, Technologies in Machine Building. ISSN 1221-4566
3. TALYSURF CLI operator’s handbook
4. Leach R (ed) Characterization of Areal Surface Texture, ISO 25178 field parameters in Chapter 2
5. Parthasarathi NL, Borah U, Albert SK (2013) Effect of temperature on sliding wear of AISI 316 L(N) stainless steel—analysis of measured wear and surface roughness of wear tracks. Mater Des 51:676–682
6. Green N, Bowen J, Hukins DS (2015) Assessment of non-contacting optical methods to measure wear and surface roughness in ceramic total disc replacements. J Eng Med 229(3):245–254

7. De Bona J, Laino S, Pettarin V, Broitman E, Dommarco R, Frontini P (2012) Differences in the sliding wear track patterns between UHMWPE/steel and UHMWPE/CNx pairs. Procedia Mater Sci 1:329–336
8. Hatos I, Hargitai H, Solecki L (2015) Study of 2D and 3D methods for worn surface. Anal Tool Mater 8(2):165–178
9. Phillips Chris (2005) Taylor Hobson Talysurf CLI systems user training manual
10. Lee D-H, Cho NG (2012) Assessment of surface profile data acquired by a stylus profilometer. Measur Sci Technol 23:12. https://doi.org/10.1088/0957-0233/23/10/105601
11. Exploring Surface Texture, A fundamental guide to the measurement of surface finish, 7th edn. Taylor Hobson operator's handbook

A Review on Mechanical Properties of Medium Density Fiberboard Prepared from Different Fiber Materials

N. Pugazhenthi and P. Anand

Abstract In recent times, the utility of medium density fiberboard plays an important role in household application and interior design purpose. The quality of this board was increased by increasing the properties in recent decade for various applications. For making this fiberboard, the thermosetting resin (urea-formaldehyde, phenol-formaldehyde, polyurethane, melamine) was used to increase the properties, and selection of fibers was made on availability and protein properties. In addition to this, resin, wax and ammonium chloride were also used. The MDF boards were made from different fibers like rubber wood, empty fruit bunch, pineapple leaf, bagasse (sugarcane), bananas stem and midrib, canola straw, bamboo and rice straw, etc. Here, we have discussed about the properties of various MDF panels prepared from different fiber materials. Through this, the further research on MDF will helps the authors to analyze the mechanical properties for various fiber boards.

Keywords Medium density fiberboard · Resin · Modulus of rupture · Modulus of elasticity

1 Introduction

The medium density fiberboard was introduced in the late 1980s and gradually its application was increased more nowadays. The board was prepared from fiber materials and resin content with hot pressing [1–3]. Plywood and hardboard were replaced by the MDF in molding and door production [4]. This was used in factory for making furniture, counter tops, underlayment, etc. Cellulosic fiber has great importance in mechanical properties, when reinforcing [5, 6]. This has some special characteristics like low density, biodegradeable, sustainable and easy available [7–9]. This can be obtained from wood species or agro-based fiber [10]. Recently, the agricultural residues were used to produce the MDF panel than the wood raw materials [11–18].

N. Pugazhenthi (✉) · P. Anand
Department of Mechanical Engineering, Vel Tech Rangarajan Dr. Sakunthala R&D Institute of Science and Technology, Avadi, Chennai 600062, India
e-mail: pugal24@gmail.com

L.-J. Yang et al. (eds.), *Proceedings of ICDMC 2019*,
Lecture Notes in Mechanical Engineering,
https://doi.org/10.1007/978-981-15-3631-1_28

The MDF board has great importance in structural and non-structural purpose. The board was classified into three types, based on its application and density, which were light plate, moisture resistant plate and fire retardant [19–25]. The authors have considered this type of application and manufactured different types of MDF by considering some parameters. Flame-retardant chemicals like sodium aluminate, antimony trioxide, magnesium oxide, zinc borate, aluminum trihydrate and ammonium polyphosphate were used to reduce the weight loss and internal bonding and increased thermal property [26–32]. Similarly, the moisture resistance can also be changed by adding chicken feather and by increasing hot pressing time [33–35]. The wood boards increase the decay resistance and susceptible to microorganism [36]. It was important to understand the mechanical strength of wood panels with the impact on biological and thermal degradation [37–39]. The bio-resistance can also be improved through heat treatment, and through this, the chemical modification takes place [40–49]. The mechanical properties were studied periodically and compared with outdoor exposure. The review on preparation of medium density fiberboard was investigated, and through this, different materials were found for preparing the MDF [50]. In this work, the properties of different MDF were studied and described clearly, and this will help the researchers, in further to analyze the properties of MDF.

2 Selection of Different Materials

All types of wood were made of cellular structure with the composition of cellulose, lignin, hemicelluloses and minor amounts (usually less than 10%) of extraneous materials. Based on this, composition and structure, the characteristics of wood will be hard or soft, then heavy or light, then stiff or flexible. Medium density fiberboard (MDF) is a technical term used for a panel primarily with the composition of fibers and compact density throughout the panel. Here, the authors had selected the material based on the lignin properties, based on the availability of the materials and some peculiar advantage of the fiber.

3 Properties of Medium Density Fiberboard

In India, the MDF was basically classified into two types, namely standard board grade I and standard board grade II, which was formulated by BIS (Bureau of Indian Standards). This committee described the specification and property value of MDF.

In formulating the quality of MDF, the weightage and practices prevailed in different countries and also the climatic conditions were considered. In this revision, properties have been incorporated to facilitate the proper application of products, and properties like modulus of elasticity and the values indicating minimum and average have been included.

The below-mentioned value shows the reference value for the properties of MDF (Table 1) [51].

Table 1 Mechanical properties and values of MDF

S. No.	Properties	Range	
		SBG I	SBG II
1	**Density, kg/m^3**	600–900	600–900
2	**Modulus of rupture, N/mm^2**		
	Up to 20 mm thickness		
	Average value	28	28
	Minimum value	25	25
	Above 20 mm thickness		
	Average value	25	25
	Minimum value	22	22
3	**Modulus of elasticity, N/mm^2**		
	Up to 20 mm thickness		
	Average value	2800	2800
	Minimum value	2500	2500
	Above 20 mm thickness		
	Average value	2500	2500
	Minimum value	2300	2300
4	**Internal bonding, N/mm^2**		
	Up to 20 mm thickness		
	Average value	0.8	0.9
	Minimum value	0.7	0.8
	Above 20 mm thickness		
	Average value	0.7	0.8
	Minimum value	0.6	0.7
5	**Screw withdrawal strength, N**		
	Face	1500	1500
	Edge	1250	1250

4 Properties of MDF Prepared with Different Fiber Materials

The fibers were selected and cleaned with water for removing any impurities. Then the fiber was chopped into small pieces. The chips were defibrated using the machine and streamed to get better quality. The resin was added as a adhesive into the chopped fiber and the mixture was poured into the mold with the standard dimensions. Finally, the mat was produced and cutted into standard size. This pieces were conducted mechanical test to measure the property value.

4.1 Arash Chavooshi et al. [52]

In this, the author has taken different composition of materials to test the mechanical properties. The material used was MDF dust in the range of 40–60% and aluminum powder of 0–15%. The matrix material used was polypropylene in a range of 21–56% and coupling agent of 4% as constant. In this, the different compositions were formulated and named in a codes of A, B, and C and each code has four different compositions. He has conducted bending strength, through which the maximum value achieved for modulus of elasticity and modulus of rupture for A1 [40% MDF, 56% PP, 0% AlP and 4% MAPP] with the value of 2433 MPa and 25.94 MPa, respectively. The reason behind was, when increasing the content of MDF, the adhesive bonding was less with other compositions. Similarly, the tensile property of this composition also has a maximum in A1 material with the value of 3343.67 MPa. The energy was decreased by increasing the fiber content. The withdrawal strength of fasteners was increased by decreasing MDF dust and aluminum process.

4.2 Evan D. Sitz et al. [53]

The study was made on different MDF materials like soybean straw, wheat straw and combination of both. The mechanical properties were tested with this MDF board. Through this, the bending strength like MOE and MOR of soybean was 1420 MPa and 7.84 MPa, respectively. This value has max when compared with wheat straw and combination of both. Similarly, the internal bonding strength for soybean straw has a maximum value of 203 MPa. But the screw withdrawal load of wheat straw shows 309 N which was 88% improvement over soybean straw.

4.3 H. P. S. Abdul Khalil et al. [54]

Here, the author has prepared a new hybrid MDF board from rubber wood [RW] and empty fruit bunch [EFB]. He has conducted mechanical properties like bending strength and internal bonding for different composition of boards and compared each other. The four different types of boards were made to compare the properties. They were 20% EFB, 80% RW and 50% EPB, 50% RW with this 65 and 95% relative humidity for each. After testing the properties, when considering without relative humidity, it was found that 20:80 composition board shows maximum flexural strength when compared to 50:50 composition board with the value of 32.21 MPa and flexural modulus also a maximum for 20:80 composition with the value of 2634 MPa. This was happening, due to the stress transformation from fiber. Similarly, the flexural strength and flexural modulus also have increased the value for 20:80 compositions with 65% relative humidity when compared with 95% RH composition boards. This was due to higher humidity may increase the degradation of the board. In this, the board value was measured with and without relative humidity (RH), in which 20:80 composition board has max internal bonding strength for without RH. Then for with RH, 50:50 composition board has max IB strength.

4.4 R. Hashim et al. [30]

Here, the author has made flame-retardant urea formaldehyde MDF and conducted mechanical properties like bending strength and internal bonding strength. The fiber selected was rubber wood, and flame-retardant chemicals were taken to prepare the MDF. The boards were prepared with different chemicals with each 0, 10, 15, 20 and 30%. Through this, the modulus of rupture was more by having 10% flame-retardant chemicals in all types of MDF with the value of 25.67, 20.99 and 32.05 MPa. When the percentage of chemical was increased, the strength of MOR will be decreased due to the deposition of chemicals in the fiber of the board. This was similar to the property of internal bonding. So, the MDF of 10% aluminum trihydrate has maximum value when compared to 0% chemical MDF.

4.5 Umit Buyuksari et al. [55]

This study helped to found the mechanical property like bending strength (MOR and MOE) of MDF panels with different pressure and temperature of the compressed veneer sheet. Here, different types of panels were prepared with different thickness and laminated with veneer sheet. The panels also compared with sandwich panel. Finally, the panels were named from A to H, each has different conditions. The results were analyzed through ANOVA and Duncan's mean separation through which the MOE and MOR has a maximum value of 8.89 GPa and 97.5 MPa, respectively, for the H-type panel when compared to other panels. This type of panel has a condition of temperature of 200 °C and with a pressure of 6 MPa. This shows that the surface quality, increased by having high temperature and pressure and also linearly proportional to mechanical properties.

4.6 M. A. Norul Izani et al. (2013) [56]

In this, the author has prepared MDF using an empty fruit bunch of oil palm [EFB] and with 8, 10 and 12% of phenol formaldehyde resin. He has made different types of panels by treating and without treating. He has taken four types of panels, one without treatment, then with boiled water, and then soaked with sodium hydroxide (NaOH) and a combination of both treatments. From these panels, he has compared the mechanical properties like bending strength and internal bonding strength. The board with 12% resin content with NaOH treated, produced max value in modulus of rupture and modulus of elasticity through ANOVA analysis. The internal bonding strength also has a maximum value of 0.67 MPa for the same panel when compared with other panels. This was due to the treatment of alkalization, shows the improvement bonding between fiber and resin.

4.7 *Md Mamunur Rashid et al. (2014) [57]*

This study tells about the mechanical properties of MDF made of banana stem and midrib. Here, MDF board was made with these fibers and with 20% of urea formaldehyde as binding agent. The author has conducted modulus of rupture and modulus of elasticity for both boards, and compared with commercial MDF available. Through this, MOR and MOE has a maximum value of 50.91 MPa and 3939.25 MPa, respectively, for banana stem MDF when compared with midrib and commercial available MDF. The value was increased due to increase in density of the board.

4.8 *Alireza Ashori et al. (2009) [58]*

Here, the author has considered the bagasse fiber (sugar cane) for preparing MDF with urea formaldehyde. The board was made and six panels (A–F) were created to test the mechanical properties like MOR, MOE and internal bonding strength. First three panels were treated with 10% maleic anhydride (MA) and then hot pressed at 180, 190 and 200 °C. In these, the panel B treated with 10% MA and hot pressed at 190 °C shows maximum values of 28.6 MPa and 2582 MPa for MOR and MOE, respectively. But when the temperature increased to 200 °C, the panel shows less value and this was due to degradation of fiber. But when considering the internal bonding strength, the panel A treated with 10% MA and hot pressed at 180 °C shows maximum value of 0.32 MPa.

4.9 *Zawawi Ibrahim et al. (2013) [59]*

In this, MDF was prepared using oil palm trunk (OPT) and urea formaldehyde and conducted different mechanical properties like MOR, MOE and internal bonding strength. This MDF was made into different panels with varying conditions, in which the refined pressure of 2, 4, 6 and 8 bar and each has 100–400 s for preheating. From these, the total of 16 panels were prepared and conducted the mechanical test. The result was analyzed through ANOVA method showing maximum values of 39.89 MPa for modulus of rupture for the panel, which has 6 bar refined pressure and preheating time of 200 s. The modulus of elasticity also has a maximum value in 6 bar pressure panel, but in 300 s preheating time. The internal bonding strength also shows maximum values in 6 bar refined pressure. The property value increases, by increasing the pressure and this was due to better arrangement and orientation of fiber. But when the pressure was increased to 8 bars, the property value decreases, and this may be damage to fiber wall.

4.10 H. A. Aisyah et al. [21]

Here, the author has conducted three mechanical properties like MOR, MOE and internal bonding strength for MDF material. This was prepared from kenaf core with urea formaldehyde resin. He prepared different types of boards from kenaf, in which he has selected three different pressures (3, 5 and 7 bar) and two different heating time (3 and 5 min). The result was analyzed through ANOVA method, which has a maximum value of 32.6 MPa and 3735 MPa for MOR and MOE, respectively. The maximum value was got for material with 3 bar pressure and 3 min heating time, but for 5 min heating time material has reduced and similarly for remaining materials also has less value when compared with 3 bars and 3 min material. The internal bonding has a maximum value for 5 bar pressure and 3 min heating time material. The result shows that the bending strength has increased value, when the longer fiber has strong network within itself. But for IB, it was vise versa.

4.11 A. M. El-Kassas et al. (2013) [60]

In this, the author has considered rice straw fiber for preparing MDF with urea formaldehyde resin as a binder. The material was tested to get mechanical properties like MOR, MOE and IB. He has considered different parameters of resin content (16, 18 and 19%) and material thickness like 7, 9 and 16 mm. The materials were grouped from 1 to 5 and each has 4 samples with same parameter. In this, the maximum value of MOR, MOE and IB is 42.17 MPa, 4443 MPa and 1.13 MPa, respectively, for group 5 with the parameter of 9 mm thickness and 19% resin content.

4.12 Hudaverdi Eroglu et al. (2000) [61]

In this, the source was taken as wheat straw (WS), wood fiber (WF) and a combination of both and made MDF with different parameters like adhesive (8, 10 and 12%), pressing time (5, 6 and 7) and hardening agent as 1%. The MDF was made by WS, and WF was prepared initially with 0.7 g/cm^3 and 0.8 g/cm^3 density and conducted mechanical properties. Then, the results were compared with combination MDF. The combination of both sources was made with 70–30, 50–50 and 30–70% proposition with 0.8 g/cm^3. Through this, the maximum value of bending strength and internal bonding strength shows for 50–50% straw and wood fiber MDF with the condition of 12% adhesive, pressing time of 6 min.

4.13 Xianjun Li [62]

Here, the rice straw fiber was pretreated with oxalic acid (OA) and steam and prepared MDF. This was compared with raw rice straw MDF to check the maximum mechanical properties. The pretreatment condition was time (5 and 10 min) and temperature (100, 120, 140 and 160 °C). The internal bonding strength shows 9.6% increment for steam treated MDF and 13.4% increment for OA treated MDF but MOE and MOR increased for steam treated MDF with 6% and 13.9%, respectively. This was due to bonding ability and physical change of the MDF.

4.14 Xin Li et al. [63]

In this work, the soybean protein was used as an adhesive instead of commercial resin with wood fiber for making MDF. This adhesive was coated on the surface of wood fiber and conducted mechanical properties with processing parameters. The parameters considered to be initial moisture content (10–35%), press time (5–20 min) and temperature (130–205 °C). Through this, the maximum value shows 14.45 MPa for tensile strength, 0.86 MPa for internal bonding strength, 29.65 MPa for modulus of rupture and 2992.89 MPa for modulus of elasticity. This was due to an interaction between fiber and soybean protein by long pressing the board.

4.15 Hossein Yousefi [64]

The author has considered different parameters to prepare the MDF material and compared mechanical properties like MOR, MOE and internal bonding strength. The parameters were used with canola straw in MDF preparation like different steaming time, addition of resin and press time. The steam time was 2, 5 and 8 min and the resin content was 9 and 11%, then the press time was 4 and 6 min. The mechanical properties were tested and show the maximum value of 20.51 MPa for MOR, 2118 MPa for MOE and 0.512 MPa for IB. This value shows for MDF with a 8 min stream time, then 11% resin content and 4 min press time. This result shows that the value was increased, due to the bonding between the fiber and resin and max stream time.

4.16 *X. Philip Ye et al. [65]*

In this, MDF was prepared for wood straw, wheat straw (WS) and soybean straw (SS) separately with urea formaldehyde in a proportion of 6, 9 and 12%. The straw was made into three compositions like 100% wood straw, 100% ag-fiber (WS and SS) and a combination of both. The author has made different types of boards with different composition and conducted mechanical properties like MOR, MOE and IB. Through this, the internal bonding strength shows maximum value of 0.68 N/mm^2 for 50% Ag- fiber composition material. The MOE and MOR shows the maximum value of 2755 N/mm^2 and 25.23 N/mm^2 for 12% adhesive level of 100% wood straw, respectively. This was due to increase in adhesive level and bonding between the fibers.

4.17 *Ayfer Donmez Cavdar et al. [66]*

Here, the author has prepared thin MDF by treating with sunflower waste oil vapor and heat with the temperature of 200 °C and the duration of 10 and 20 min. The panels were tested the bending properties of MOR and MOE, in which MOR shows the reduction value in the range of 3–11% and 4–12% for oil vapor and heat treatment panels with 10 min press, respectively. For 20 min panels, the MOR shows reduction value in the range of 6–20% and 6–21% for oil vapor and heat treated panels, respectively. The MOE shows the reduction value in the range of 1–6% and 3–7% for oil vapor and heat treatment panels with 10 min press, respectively. For 20 min panels, the MOE shows reduction value in the range of 8–18% and 12–18% for oil vapor and heat treated panels, respectively. This reduction is due to treatment of panels with some chemical modification in the fibers.

5 Conclusion and Future Scope

I would like to conclude that this review explored the properties of MDF which described about the mechanical properties like tensile strength, bending strength and internal bonding strength. The bending strength reduces, when the fiber was heat treated more. The average treatment of fiber shows the maximum value in all type of MDF panel. Generally, the mechanical properties of any MDF material were greatly influenced by chemical treatment of fibers, bonding between the fibers with adhesive level and fiber length. This work will help the researcher in further investigation of medium density fiberboard, because the importance of fiberboard was increasing drastically in India. The availability of natural materials in India was more and the usage of that material was less. With the availability of those materials, the medium density fiberboard can be manufactured and the properties were analyzed. Through this, the import of this engineering product can be reduced.

References

1. Koch P (1972) Utilization of hardwoods growing on southern pine sites. United States Department Of Agriculture, Forest Service, Agriculture Handbook, No 605
2. Manloney T (1993) Modern particleboard and dry process fiberboard manufacturing. Miller Freeman Publishing Inc., San Francisco, CA
3. Saligna A, Kizysik M, Muehl JH, Youngquist AJ, Franco FS (2001) Medium density fiberboard made from eucalyptus. For Prod J 51:10
4. Krzysik AM, Youngquist JA, Franca FS (1999) Medium density fiberboards from plantation-grown Eucalyptus saligna. In: Proceedings of international conference on E&Ective utilization of plantation timber—timber and wood composites for the next century, 1999 May 21–23. Chi-Tou, Taiwan, pp 156–160
5. Xue Y, Veazie DR, Glinsey C, Horstemeyer MF, Rowell RM (2007) Environmental effects on the mechanical and thermomechanical properties of aspen fiber-polypropylene composites. Compos B Eng 38(2):152–158
6. Yan L, Kasal B, Huang L (2016) A review of recent research on the use of cellulosic fibres, their fibre fabric reinforced cementitious, geo-polymer and polymer composites in civil engineering. Compos B Eng 92:94–132
7. Kabir MM, Wang H, Lau KT, Cardona F (2012) Chemical treatments on plant-based natural fibre reinforced polymer composites: an overview. Compos B 43(7):2883–2892
8. Zafeiropoulos NE (2011) Interface engineering of natural fibre composites for maximum performance. Woodhead Publishing Limited, Cambridge
9. Zhou Y, Fan M, Chen L, Zhuang J (2015) Lignocellulosic fibre mediated rubber composites: an overview. Compos B 76:180–191
10. Benthien JT, Thoemen H (2012) Effects of raw materials and process parameters on the physical and mechanical properties of flat pressed WPC panels. Compos A 43:570–576
11. Halvarsson HE (2009) Magnus Norgren: manufacture of non-resin wheat straw fibreboards. Ind Crops Prod 29:437–445
12. Akgu M, TozluoğLu A (2008) Utilizing Peanut Husk (*Arachis Hypogaea* L.) in the manufacture of medium-density fiberboards. Bioresour Technol 99:5590–5594
13. Akgu M, Çamlibel O (2008) Manufacture of medium density fiberboard (MDF) panels from rhododendron (*R. Ponticum* L.) biomass. Build Environ 43:438–443
14. Halvarsson S, Edlund H, Norgren M (2008b) Properties of medium-density fibreboard (MDF) based on wheat straw and melamine modified urea formaldehyde (UMF) resin. Ind Crops Prod 28:37–46
15. Bader TK, Dastoorian F, Ebrahimi G, Unger G, Lahayne O, Hellmich C, Pichler B (2016) Combined ultrasonic-mechanical characterization of orthotropic elastic properties of an unrefined bagasse fiber-polypropylene composite. Compos B 95:96–104
16. Yang HS, Kim DJ, Kim HJ (2003) Rice straw-wood particle composite for sound absorbing wooden construction materials. Bioresour Technol 86(2):117–121
17. Indrayania Y, Setyawati D, Munawar SS, Umemur K, Yoshimur T (2015) Evaluation of termite resistance of medium density fiberboard (MDF) manufacture from agricultural fiber bonded with citric acid. Procedia Environ Sci 28:778–782
18. Zou Y, Huda S, Yang Y (2010) Lightweight composites from long wheat straw and polypropylene web. Bioresour Technol 101:2026–2033
19. Copur Y, Guler C, Tascioglu C, Tozluoglu A (2008) Incorporation of hazelnut shell and husk in MDF production. Bioresour Technol 99(5):7402–7406
20. Halvarsson S, Edlund H, Norgren M (2008a) Properties of medium-density fiberboard (MDF) based on wheat straw and melamine modified urea formaldehyde (UMF) resin. Ind Crops Prod 28:37–46
21. Aisyah HA, Paridah MT, Sahri MH, Anwar UMK, Astimar AA (2013) Properties of medium density fibreboard (MDF) from Kenaf (*Hibiscus Cannabinus* L.) core as function of refining conditions. Compos B 44:592–596

22. Hiziroglu S, Jarusombuti S, Bauchongkol P, Fueangvivat V (2008) Overlaying properties of fiberboard manufactured from bamboo and rice straw. Ind Crops Prod 28:107–111
23. Lee S, Shupe TF, Hse CY (2006) Mechanical and physical properties of agro-based fiberboard. Holz Als Roh-Und Werkstoff 64:74–79
24. Mobarak F, Fahmy Y, Augustin H (1982) Binderless lignocellulose composite from bagasse and mechanism of self-bonding. Holzforschung 36(3):131–135
25. Ye XP, Julson J, Kuo M, Womac A, Myers D (2007) Properties of medium density fiberboards made from renewable biomass. Bioresour Technol 98(5):1077–1084
26. Altuntas E, Narlioglu N, Alma MH, Investigation of the fire, thermal, and mechanical properties of zinc borate and synergic fire retardants on composites produced with PP-MDF wastes. Bioresources. ISSN: 1930-2126
27. Özdemi F, Ayaz A (2017) Investigation of the effect on combustion resistance of ammonium polyphosphate and boric acid chemicals added to surface coating. J For Fac 17(2):290–297
28. Yang H-S, Kim D-J, Kim H-J, Combustion and mechanical properties of fire retardant treated waste paper board for interior finishing material. J Fire Sci 20(6):505–517
29. Sulaiman O, Hashim R, Kumar RN, Tamyez P, Murphy RJ, Ali Z (2008) Effect of incorporation of flame retardants on some of the properties of phenol formaldehyde medium density fiberboard. Int J Agric Res 3:331–339
30. Hashim R, Sulaiman O, Kumar RN, Tamyez PF, Murphy RJ, Ali Z (2009) Physical and mechanical properties of flame retardant urea formaldehyde medium density fiberboard. J Mater Process Technol 209(2):635–640
31. He X, Li X, Zhong Z, Yan Y, Mou Q, Yao C, Wang C (2014) The fabrication and properties characterization of wood-based flame retardant composites. J Nanomater 2014, 6. Article ID 878357
32. Hashim R, How LS, Kumar RN, Sulaiman O (2005) Some of the properties of flame retardant medium density fiberboard made from rubberwood and recycled containers containing aluminum trihydroxide. Bioresour Technol 96:1826–1831
33. Gul W, Sadiq M, Khan A, Shakoor A, Shah J (2017) Improving water resistance property of medium density fibreboard (MDF). Life Sci J 14(3):90–96. ISSN: 1097-8135 (Print)/ISSN: 237-613X (Online)
34. Gul W, Khan A, Shakoor A (2017) Impact of hot pressing temperature on medium density fiberboard (MDF) performance. Adv Mater Sci Eng, 6. Article ID 4056360
35. Winandy JE, Muehl JH, Glaeser JA, Schmidt W (2008) Chicken feather fiber as an additive in MDF composites. J Nat Fibers 4(1):35–48
36. Nami Kartala S, Green III F (2003) Decay and termite resistance of medium density fiberboard (MDF) made from different wood species. Int Biodeterior Biodegradation 51:29–35
37. Muin M, Tsunoda K (2004) Biological performance of wood-based composites treated with a formulation of 3-Iodo-2-propynyl butylcarbamate and silafluofen using supercritical carbon dioxide. J Wood Sci 50(6):535–539
38. Garrote G, Dominiguez H, Parajo JC (1999) Hydrothermal processing of lignocellulosic materials. Holz Roh Werkstoff 57:191–202
39. Winandy JE, Krzysik AM (2007) Thermal degradation of wood fibers during hot-pressing of MDF composites: part I. Relative effects and benefits of thermal exposure. Wood Fiber Sci 39(3):450–461
40. Tjeerdsma B, Boonstra M, Pizzi A, Tekely P, Militz H (1998) Characterisation of thermally modified wood: molecular reasons for wood performance improvement. Holz Roh Werkst 56:149–153
41. Fengel D, Wegener G (1989) Wood—chemistry, ultrastructure, reactions. Walter De Gruyter, Berlin, New York
42. Inoue M, Norimoto M, Tanahashi M, Rowell RM (1993) Steam or heat fixation of compressed wood. Wood Fiber Sci 25(3):224–235
43. Mohebby B, Sanaei I (2005) Influences of the hydro-thermal treatment on physical properties of beech wood. In: Proceeding of IRG/WP 05-40303, 36th Annual meeting, India

44. Sanderman W, Augustin H (1964) Chemical investigations on the thermal decomposition of wood, Part III: chemical investigation on the course of decomposition. Holz Roh Werkst 22:377–385
45. Tjeerdsma BF, Militz H (2005) Chemical changes in hydrothermal treated wood: FTIR analysis of combined hydrothermal and dry heat-treated wood. Holz Roh Werkstoff 63:102–111
46. Yildiz S (2002) Effect of heat treatment on water repellence and anti-swelling efficiency of beech. In: Proceeding of IRG/WP 02-40223. The International research group on wood preservation conference
47. Kartal SN, Imamura Y, Effects Of N′-N-(1, 8-Naphthalyl) hydroxylamine (NHA-Na) and hydroxynaphthalimide (NHA-H) on boron leachability and biological degradation of wood. Eur J Wood Wood Prod. ISSN: 0018-3768 (Print) 1436-736X (Online)
48. Kollmann F, Fengel D (1965) Changes in the chemical composition of wood by thermal treatment. Holz Roh Werkst 23:461–468
49. Hillis WE (1984) High temperature and chemical effects on wood stability. Wood Sci Technol 18:281–293
50. Pugazhenthi N, Anand P (2018) A review on preparation of medium density fiberboard with different materials. Int J Eng Technol 7(3.34):962–965
51. Indian Standard Reaffirmed (2008) Medium density fibre boards for general purpose — specification
52. Chavooshi A, Madhoushi M (2013) Mechanical and physical properties of aluminum powder/MDF dust/polypropylene composites. Constr Build Mater 44:214–220
53. Sitz ED, Bajwa DS (2015) The mechanical properties of soybean straw and wheat straw blended medium density fiberboards made with methylene diphenyl diisocyanate binder. Ind Crops Prod 75:200–205
54. Abdul Khalil HPS, Nur Firdaus MY, Jawaid M, Anis M, Ridzuan R, Mohamed AR (2010) Development and material properties of new hybrid medium density fibreboard from empty fruit bunch and rubberwood. Mater Des 31:4229–4236
55. Buyuksarı U, Hiziroglu S, Akkılıc H, Ayrılmıs N (2012) Mechanical and physical properties of medium density fiberboard panels laminated with thermally compressed veneer. Compos B 43:110–114
56. Izani M, Paridah M, Nor M, Anwar U (2013) Properties of medium-density fiberboard (MDF) made from treated empty fruit bunch of oil palm. J Trop For Sci 25(2):175–183
57. Rashid MM, Das AK, Shams MI, Biswas SK (2014) Physical and mechanical properties of medium density fiber board (MDF) fabricated from banana plant (Musa sapientum) stem and midrib. J Indian Acad Wood Sci 11(1):1–4
58. Ashori A, Nourbakhsh A, Karegarfard A (2009) Properties of medium density fiberboard based on bagasse fibers. J Compos Mater 43(18):1927–1934
59. Zawawi Ibrahim, Astimar Abdul Aziz, Ridzuan Ramli, Anis Mokhtar, SiJoon Lee, (2013) Effect of refining parameters on medium density fibreboard (MDF)pProperties from oil palm trunk. Open JCompos Mater 03(04):127–131
60. Elkassas AS, Mourad AI (2013) Novel fibers preparation technique for manufacturing of rice straw based fiberboards and their characterization
61. Eroglu H, Istek A, Usta M (2000) Medium devsity fireboard (MDF) Manufacturing from wheat straw (Triticum aestivum L.) and straw wood mixture
62. Li X, Yiqiang Wu, Cai Z, Winandy JE (2013) Primary properties of MDF using thermomechanical pulp made from oxalic acid pretreated rice straw particles. Ind Crops Prod 41:414–418
63. Li X, Li Y, Zhong Z, Wang D, Ratto JA, Sheng K (2009) Xiuzhi Susan Sun: mechanical and water soaking properties of medium density fiberboard with wood fiber and soybean protein adhesive. Bioresour Technol 100:3556–3562

64. Yousefi H (2009) Canola straw as a bio-waste resource for medium density fiberboard (MDF) manufacture. Waste Manag 29:2644–2648
65. Philip Ye X, Julson J, Kuo M, Womac A, Myers D (2007) Properties of medium density fiberboards made from renewable biomass. Bioresour Technol 98:1077–1084
66. Cavdar AD, Ertas M, Kalaycıoğlu H, Alma MH (2010) Some properties of thin medium density fiberboard panels treated with sunflower waste oil vapour. Mater Des 31:2561–2567

Investigation of Wear Behavior of Rapid Solidified Al–Si Alloys

N. D. K. Malleswararao and I. N. Niranjan Kumar

Abstract The development of components like, e.g., piston and cylinder liners were normally based on traditional materials but global need more performance, and low-cost alloys have caused to change in the interest of researchers to select alloy materials. Friction and wear are the most common problems in the mechanical systems which may lead to replace the components. Though with the replacement of usual cast iron in engine components with lightweight Al–Si alloys, the friction and wear problem are overthrown. Moreover, Al–Si alloys lead to increase efficiency and less fuel consumption. In this paper, the developments in wear resistance and coefficient of friction performance of Al–Si alloy are taken into consideration based on past research.

Keywords Al–Si alloy · Tribology · Friction · Wear

1 Introduction

Al–Si alloys have excellent strength, low thermal expansion, wear resistance, etc., thereby these are widely used in numerous frictions and wear applications like pistons, engine blocks and cylinder liners [1–3]. Various friction and wear reports have been stated on the properties of Al–Si alloys, though Clarke and Sarkar reported that hypereutectic Al–Si alloys phased have less wear rate than the hypoeutectic phased alloys [4]. Moreover, Shivanath et al. [5] stated that the alloys within the hypereutectic range are having better wear resistance. Warmuzek et al. reported that reinforcement of silicon particles improves the wear performance of Al–Si materials [6]. Though the report of Torabian et al. [7] shows that in the manufacturing process of hypereutectic phased Al–Si alloy materials the additions of Si (silicon) particles improves the wear resistance. Later, Chen et al. [8] performed a series of trails on hypereutectic phased Al–Si alloys and expresses as with the addition of silicon particles hardness of the materials increased. Additionally, the report of Dwivedi et al. [9, 10] explains that the

N. D. K. Malleswararao (✉) · I. N. Niranjan Kumar
Andhra University, Visakhapatnam, Andhra Pradesh 530003, India
e-mail: nottymalli@gmail.com

L.-J. Yang et al. (eds.), *Proceedings of ICDMC 2019*,
Lecture Notes in Mechanical Engineering,
https://doi.org/10.1007/978-981-15-3631-1_29

mixing of silicon particles enhances the load-carrying capacity and wear rate. Wang et al. carried out the experiments on hypereutectic Al–Si alloys (Al–12Si to Al–25Si) by varying load and expressed as increasing load Al–20Si alloy shown better wear performance [11]. Though, Reddy et al. in their investigation noted that wear rate improved at distinct degrees of temperature [12]. Alireza Hekmat et al. fabricated Al–Si alloy by rapid solidification process-T6 condition and reported that the heat treatment process enhances the wear resistance of hypereutectic Al–Si alloys [13]. Although the papers of [14–16], it is shown that the eutectic and primary Si phases are restrained by RSP. Alshmri et al. state that the addition of increasing Si particles leads to abrasive wear [17].

2 Materials and Methods

2.1 Materials

Basically, aluminum–silicon alloys are manufactured by common methods of fabrication process. Though with the presence of Si precipitates these fabrication processes present poor ductility, machinability and wear properties. Therefore, RSP (T6 heat-treated condition) method is used for limiting the coarsening (high cooling rate) of Si particles. In heat-treated condition, the samples were placed under 525 °C for eight hours and 160 °C for eight hours. After heat-treated condition, the rates of solidification improve because of the formation of strained silicon particles which, in turn, improves the mechanical and physical behaviors of the materials. RSP-T6 is a method of improving the strength of materials by changing their medium crystallite size. Hence, the alloys are strengthening due to thin suspended particles in dispersion, which will delay grain boundary movement [14, 15, 18–27]. Some of the hyper-eutectic Al–Si alloys physical properties are listed in Table 1. Whereas properties of lubricant oil grade SAE 15W40 are displayed in Table 2.

Table 1 Physical properties of Al–Si alloys [22]

Alloys	Hardness (HV)	Density (g·cm^3)	Tensile strength (MPa)	Elongation %
Al–Si17	175	2.61	454	0.4
Al–Si20	96	2.86	317	2.4
Al–Si24	189	2.51	466	0.5

Table 2 Properties of lubricant oil grade SAE 15W40 [27]

Lubricant properties	Value
Kinematic viscosity	125 mm^2 s^{-1}at 100 °C
Viscosity index	120
Flash point	220 °C
Pour point	−21 °C

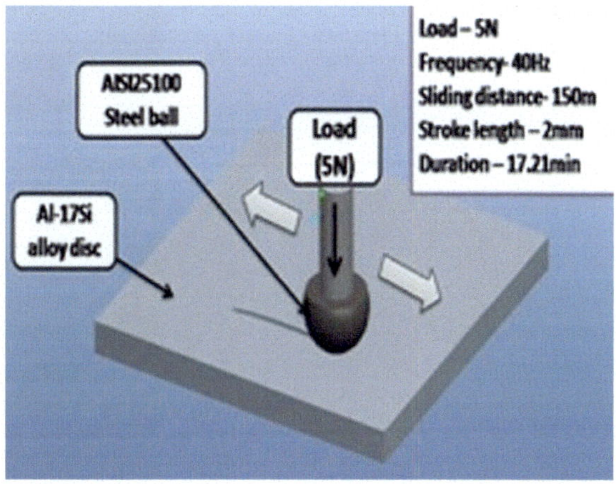

Fig. 1 Experimental set up of linear reciprocating tribometer

2.2 *Methods*

Al–Si alloys surface hardness was determined by Vickers microhardness tester. Whereas friction and wear tests were conducted at 24 °C (room temperature) on linear reciprocating tribometer (Fig. 1) under parched and lubricated sliding conditions. Although the procedure for surface preparation of specimens was completed by grinding disk using SiC papers of different sizes, while the final mirror polishing was achieved with diamond pastes. Besides, the polished samples were cleaned with benzene to remove any dust particles which were left during specimen preparation.

3 Results and Discussion

Figures 2 and 3 expresses the optical micrographs of Al–Si alloys before and after heat treatment (T6). Though Fig. 4 exhibits the exploratory results of wear resistance, though wear rate is decreased with increasing of load then increases and then again decreases under parched sliding condition. Whereas in case of lubricated sliding condition, the coefficient of friction is significantly reduced. However, it is noted that due to RSP-T6 the crystalline size of the alloy reduced which, in turn, improves the strength of the alloy. Moreover, after T6 condition, Si particles size reduces and

Fig. 2 Optical micrograph of Al–17Si before heat treated

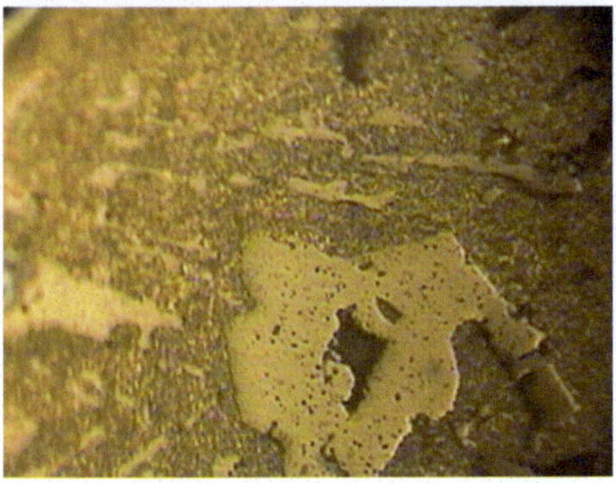

Fig. 3 Optical micrograph of Al–17Si after heat treated

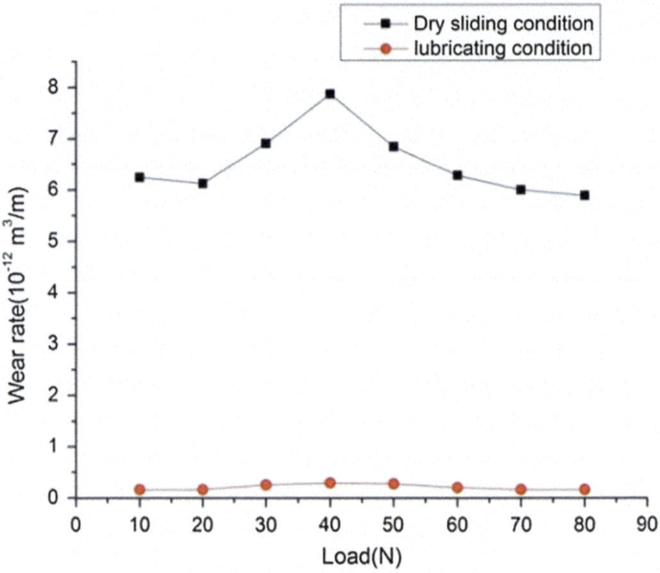

Fig. 4 Wear performance of Al–17Si alloy

Fig. 5 Optical micrograph of Al–17Si at dry condition

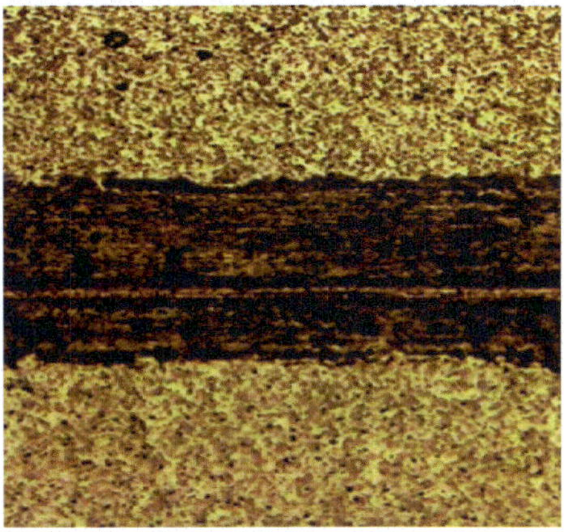

Fig. 6 Optical micrograph of Al–17Si at lubricated condition

appears uniformity throughout the aluminum matrix. Figures 5 and 6 show the wear scars of the alloy under dry and lubricated condition. Anyway, it is clear that in case of dry condition, higher wear was obtained compared to the lubricated condition. Therefore, the strength of the material increased because of impedes dislocation movement of grain boundaries and finer dispersoids.

4 Conclusion

In this paper, we investigated the wear performance of the hypereutectic phased Al–Si alloys with lubricating oil under linear reciprocating tribometer. It is noted that the Al–Si alloy fabricated with RSP has a significant change in wear performance. Moreover, manufacturing process (RSP-T6 heat-treated condition) has a substantial effect on the hardness, microstructure, wear rate and COF of hypereutectic phased Al–Si alloys.

The total study exhibits that the hypereutectic phased Al–Si alloy possesses more hardness, higher strength and significant wear reduction because of the RSP under T6 condition.

References

1. Kumar KG (2013) Influence of refinement and modification on dry sliding wear behavior of hypereutectic Al–Si cast alloys. Adv Mater Res 685:112–116
2. Nuraliza N, Syahrullail S, Faizal MH (2016) Tribological properties of aluminum lubricated with palm olein at different load using pin-on-disk machine. J Tribologi 9:45–59
3. Rohatgi PK, Pai BC (1974) Effect of microstructure and mechanical properties on the seizure resistance of cast aluminium alloys. Wear 28(3):353–367
4. Sarkar AD, Clarke J (1980) Friction and wear of Aluminum–Silicon alloys. Wear 61(1):157–167
5. Shivanath R, Sengupta PK, Eyre TS (1977) Wear of Aluminium–Silicon alloys. Br Foundrymen 70:349–356
6. Warmuzek M (2004) Aluminum–Silicon Casting Alloys: An Atlas of Microfractographs. ASM International
7. Torabian H, Pathak JP, Tiwari SN (1994) Wear characteristics of Al–Si alloys. Wear 172:49–58
8. Chen M, Meng-Burany X, Perry TA, Alpas AT (2008) Micromechanisms and mechanics of ultra-mild wear in Al–Si alloys. Acta Mater 56(19):5605–5616
9. Dwivedi DK (2004) Sliding wear and friction behaviour of Al-18% Si-0.5% Mg alloy. J Mater Process Technol 152:323–328
10. Dwivedi DK (2006) Wear behaviour of cast hypereutectic aluminium silicon alloys. Mater Design 27:610–616
11. Wang F, Liu H, Ma Y, Jin Y (2004) Effect of Si content on the dry sliding wear properties of spray-deposited Al–Si alloy. Mater Design 25(2):163–166
12. Reddy AS, Bai BP, Murthy KSS, Biswas SK (1994) Wear and seizure of binary Al–Si alloys. Wear 171(1–2):115–127
13. Hekmat-Ardakan Alireza, Liu Xichun, Ajersch Frank, Chen X-G (2010) Wear behaviour of hypereutectic Al–Si–Cu–Mg casting alloys with variable Mg Contents. Wear 269(2010):684–692
14. Jones H (1982) Rapid solidification of metals and alloys. The Institution of Metallurgists, London
15. Zhou J, Duszczyk J, Korevaar BM (1991) Structural development during the extrusion of rapidly solidified Al-20Si-5Fe-3Cu-1 Mg alloy. J Mater Sci 26(3):824–834
16. Lavernia EJ, Ayers JD, Srivatsan TS (1992) Rapid solidification processing with specific application to aluminium alloys. Int Mater Rev 37:1–44
17. Alshmri F, Atkinson HV, Hainsworth SV, Haidon C, Lawes SDA (2014) Dry sliding wear of aluminium-high silicon hypereutectic alloys. Wear 313(1):106–116
18. Brain M (2000) How car engines work. Howstuffworks.com. 5
19. Corsico G, Mattei L, Roselli A, Gommellini C (1999) Poly (internal olefins). Chemical industries, Marcel Dekker, New York, pp 53–62
20. Malleswara Rao KND, Praveen Kumar R, Venkateswararao T, Sudheer Kumar B, Babu Rao G (2018) Investigation of mechanical/tribological properties of composite ersatz articular cartilage with nano fillers. In: Advances in materials and metallurgy, lecture notes in mechanical engineering, pp 385–394. https://doi.org/10.1007/978-981-13-1780-4_37

21. George NJ, Obianwu VI, Akpan AE, Obot IB (2010) Lubricating and cooling capacities of different SAE 20 W–50 engine oil samples using specific heat capacity and cooling rate. Arch of Phys Res 1(2):103–111
22. Kumar and Wani (2017) Friction and wear characterization of hypereutectic Al–Si alloy/steel tribopair under dry and lubricated conditions. J Tribologi 15:21–49
23. Abdullah MIHC (2017) The hBN nanoparticles as an effective engine oil additive to enhance the durability and performance of a small diesel engine. J Mech Eng 1(1):103–112
24. Thottackkad MV, Rajendrakumar PK, Nair KP (2014) Experimental studies on the tribological behaviour of engine oil (SAE15W40) with the addition of CuO nanoparticles. Ind Lubr Tribol 66(2):289–297. https://doi.org/10.1108/ILT-01-2012-0006
25. Abdullah MIHC et al (2014) J Teknologi (Sci Eng) 66:3, 1–6
26. Kalakada et al (2014) Int J Eng Sci Technol 6(1):34–42
27. Kumar P Wani MF (2017) Tribological characterisation of graphene oxide as lubricant additive on hypereutectic Al-25Si/steel tribopair. Tribol Trans. https://doi.org/10.1080/10402004.2017.1322735

Investigation of Compressive Properties of Hybrid Aloe vera/Silica Nanoparticles Composite

R. Giridharan, S. J. Anirudh, S. Anirudh, and M. P. Jenarthanan

Abstract In this research, the FRPs have been produced by means of Handlayup Process, wherein compacted aloe vera mat was placed at random. The silica nanoparticles were made to diffuse in the matrix containing epoxy. It was found that the addition of silica nanoparticles resulted in a negative effect on the characteristics of FRP. On the other hand, composites containing 2 vol.% silica indicated the most excellent properties. Enhanced mechanical properties are obtained for the composites and these composites can be used for medium load applications.

Keywords Aloe vera fibers · Silica · Nanoparticles · Compressive properties

1 Introduction

Natural fibers are eco-friendly compared to synthetic fibers such as glass fibers. Owing to the advantageous properties of the natural fiber reinforced polymer composites that lead them are being used in automobile components, construction industry, and aerospace parts. Besides this, they also found that the polymer content for reinforcement is also reduced. Many researchers have done a lot of work on natural fiber reinforced composites. Perremans et al. [1] found that the tensile stiffness of bamboo fiber reinforced epoxy composites is hardly influenced by the discontinuity patterns. Sapuan et al. [2] fabricated woven banana fibers reinforced with epoxy matrix composite. Satish et al. [3] found that the influence of fiber orientation on density of composites is less. Shibata et al. [4] investigated the influences of the volume fraction on flexural properties of kenaf and bagasse composites. Thakur et al. [5] fabricated bamboo fiber reinforced with varying percentages of CNT in the epoxy resin composite and evaluated the mechanical characteristics of the same. Vaghasia et al. [6] found that the mechanical properties of WBGP hybrid composite depend on the effect of fiber-polyester percentage variation. Venkateshwaran and ElayaPerumal

R. Giridharan (✉) · S. J. Anirudh · S. Anirudh · M. P. Jenarthanan
School of Mechanical Engineering, SASTRA Deemed to be University, Thanjavur, Tamilnadu 613401, India
e-mail: giridharan54@gmail.com

L.-J. Yang et al. (eds.), *Proceedings of ICDMC 2019*,
Lecture Notes in Mechanical Engineering,
https://doi.org/10.1007/978-981-15-3631-1_30

[7] examined several works in composite reinforced with banana fiber and evaluated physical properties, structure, and application of the same. Yuan et al. [8] witnessed that the introduction of kevlar fiber enhanced the mechanical properties in wood flour/PP composites. Giridharan [9] fabricated ramie fiber reinforced composites and evaluated the mechanical properties of the same and found that the hybrid composite (ramie-glass fiber as reinforcement) has higher mechanical properties. Murali et al. [10] reinforced banana and aloe vera woven fabric in epoxy resin matrix and investigated the flexural and impact properties of in woven aloe vera fiber composite. Yamini et al. [11] fabricated pure coir and pure aloe vera composites both in randomly oriented mats and bi-directional mats treated with NaOH—alcohol—benzene mixture. They subjected the specimen to impact test and moisture absorption test and discovered that the impact value of the aloe vera fiber was comparatively lesser.

Among these, aloe vera is a well-known plant whose fibers are mainly exploited in a large variety of applications. Many researchers have done many researches with aloe vera and their composites. Selected important properties of aloe vera fiber are the elastic recovery, breaking extension of fiber in dry, and resistance to heat degradation. The aloe vera fibers are utilized in many fields such as building and construction industry like door, windows, wall partition, sports, aerospace, railway and automotive sectors like setbacks, and liners. In comparison with the basic composites, the advantages that the aloe vera fibers have are low weight, low requirement of raw material, cheap and cost effective, as optimized mechanical properties can be obtained, and good resistance to moisture and it is also available in semi-finished and finished state. Aloe vera fiber with epoxy resins showed better strength when tested with polyester. This paper investigates the tensile and compressive strength of nano aloe vera fiber mixed in a 1:1 glass fiber and epoxy resins matrices. The experiment was carried out taking samples based on ASTM standards in relation to the application where tensile and compressive loads are prominent.

2 Materials and Methods

The aloe vera mat has been used for the fabrication purpose. The aloe vera mat is supplied by the local market. The aloe vera plant and fibers are shown in Figs. 1 and 2, respectively. The properties of aloe vera fiber are listed in Table 1. The orientation of the aloe vera fiber was arbitrary in the mat configuration. The silica nanoparticles have been procured from a local merchant. Based on the requirement for carrying out the compression test, the composites were fabricated and cut into plates of size 150 mm × 150 mm. Thus, the mat containing aloe vera has been excised into the articulated measurement prior to weight measurement and being assembled with the remaining mats in order to match with the anticipated loading weight. With the help of a homogenizer, the silica nanoparticles have been made to dissolve with the epoxy without the aid of a hardener. Solution containing silica nanoparticles and epoxy was then molded to a vacuum oven so that all the microbubbles that have been generated during the homogenizing process can be eliminated. The test of compression was

Fig. 1 Aloe vera plant

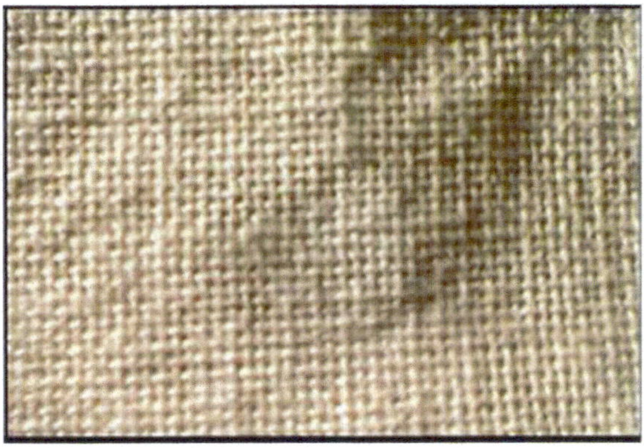

Fig. 2 Aloe vera fibers

Table 1 Physical properties of materials used

	Aloe vera fiber
Density (g/cm^3)	1.5–1.56
Young's modulus (GPa)	60–128
Tensile strength (MPa)	400–1000
Elongation at break (%)	1.2–3.8

carried out as per ASTM D-695 by means of Instron 3365 Universal Testing System. The specimens were cut into $10 \times 10 \times 3.2\ \text{mm}^3$. The rate of compression is 0.033 mm/s.

Fig. 3 Variation of compressive strength against vol.% of silica nanoparticles' loading

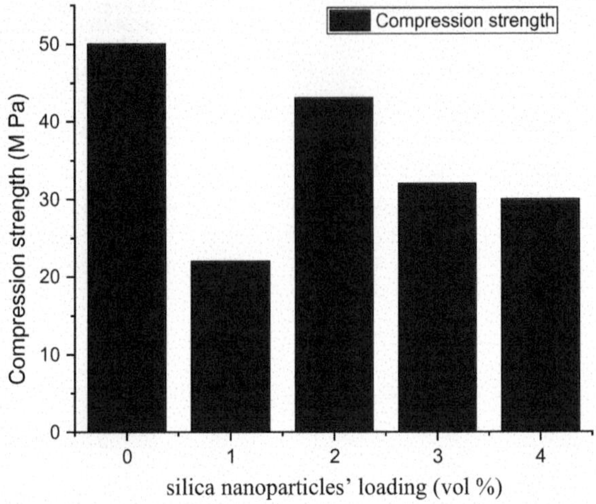

3 Results and Discussions

It has been observed that the least compressive strength value has been recorded for 1 vol.% loading of silica nanoparticles during the course of the experiment. Also, the addition of 2 vol.% silica nanoparticles exhibited the peak value of compressive modulus. The compressive strength as well as the compressive modulus decreases with further increasing the number of nanoparticles. The reason behind the decrement of the compressive properties of the specimen is that increasing silica nanoparticles content tends to amplify the porosity of the specimen. When a composite is stressed, numerous voids are generated, thereby preventing the effective transfer of the stresses. Strong interfacial adhesion can be produced by the generation of chemically bonded nanoscale interfacial area between the fiber and epoxy bridged by the fillers. The vol.% of silica nanoparticles' loading with respect to compressive strength and compressive modulus is shown in Figs. 3 and 4, respectively.

4 Conclusion

Specimens containing 2 vol.% silica showed the greatest mechanical properties of specimens with silica nanoparticles included at a compressive strength of 50 MPa and for and compressive modulus 1.25 GPa.

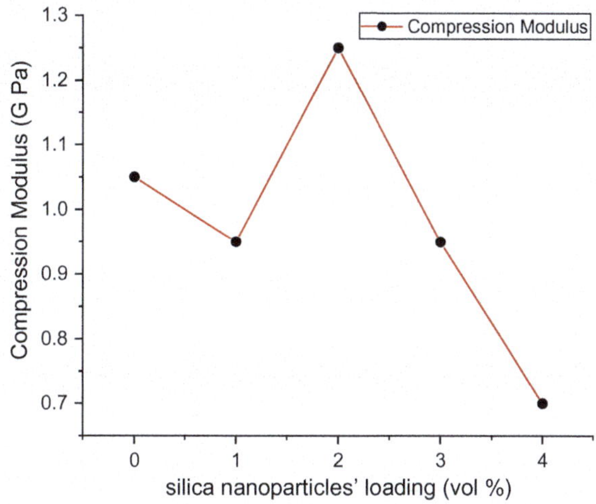

Fig. 4 Variation of compressive modulus against vol.% of silica nanoparticles' loading

References

1. Perremans D, Trujillo E, Ivens JA, Van Vuure W (2018)Effect of discontinuities in bamboo fibre reinforced epoxy composites. Compos Sci Technol 50–57
2. Sapuan SM, Leenie A, Harimi M, Beng YK (2006) Mechanical property analysis of woven bamboo/epoxy composite. Mater Design 27:689–693
3. Satish KG, Siddeswarappa B, Kaleemulla KM (2010) Characterization of in-plane mechanical properties of laminated hybrid composites. J Mine Mater Char Eng 9:105–114
4. Shibata S, Cao Y, Fukumoto I (2002) Press forming of short natural-fibre reinforced biodegradable resin: effects of fibre volume and length on flexural properties. Polym Test 24:1005–1011
5. Palanivel A, Veerabathiran A, Duruvasalu R, Iyyanar S, Velumayil R (2017) Dynamic mechanical analysis and crystalline analysis of hemp fiber reinforced cellulose filled epoxy composite. Polímeros 27(4):309–319
6. Vaghasia B, Rachchh B (2008) Evaluation of physical and mechanical properties of woven bamboo aloevera polyester hybrid composite material. Mater Today Proc
7. Venkateshwaran N, ElayaPerumal A, Alavudeen A, Thiruchitrambalam M (2011) Mechanical and water absorption behaviour of bamboo/sisal reinforced hybrid composites. Mater Design 32:4017–4021(2011)
8. Anand P, Rajesh D, Kumar MS, Raj IS (2018) Investigations on the performances of treated jute/Kenaf hybrid natural fiber reinforced epoxy composite. J Polym Res 25(4):94
9. Giridharan R (2019) Preparation and property evaluation of Glass/Ramie fibers reinforced epoxy hybrid composites. Composites Part B 167:342–345
10. Murali B, Vinoth Kumar S, Gopalakrishnan S, Suvendar T (2018) Evaluation of mechanical properties of aloe vera natural fibre reinforced composite IJARIIE 4:2751–2758
11. Yamini S (2015) Shanmugasundaram: study mechanical behavior of natural based composite using coir and aloevera. Int J Emerg Technol Comput Sci Electron 13:976–1353

A Study on Investigation of Tensile Properties of Aloe vera Fiber Reinforced Epoxy Composites

R. Giridharan, N. Anerudh, M. M. Mithun Srivan, and M. P. Jenarthanan

Abstract The purpose of this research is to estimate the tensile properties of aloe vera fiber composites. This research is done to fabricate aloe vera fiber reinforced epoxy composite by varying compositions of fiber layers. The methodology used for the fabrication of composites is hand lay-up process. Tensile test and fractography of fractured surface have been carried out. From the results of the testing process, it has been found that the third composite specimen that has three layers of aloe vera fibers exhibited superior tensile properties. SEM analysis revealed the good binding adhesion between reinforcement and the epoxy resin matrix.

Keywords Aloe vera · Epoxy resins · ASTM · Tensile properties · Fractography

1 Introduction

Natural fibers are eco-friendly compared to synthetic fibers such as glass fibers. Thanks to the advantageous properties of the natural fiber reinforced polymer composites that lead them are being used in automobile components, construction industry, and aerospace parts. Many researchers have done a lot of work on natural fiber reinforced composites. Perremans et al. [1] found that the tensile stiffness of bamboo fiber reinforced epoxy composites is hardly influenced by the discontinuity patterns. Sapuan et al. [2] fabricated woven banana fibers reinforced with epoxy matrix composite. Satish et al. [3] found that the influence of fiber orientation on density of composites is less. Shibata et al. [4] investigated the influences of the volume fraction on flexural properties of kenaf and bagasse composites. Thakur et al. [5] fabricated bamboo fiber reinforced with varying percentages of CNT in the epoxy resin composite and evaluated the mechanical characteristics of the same. Vaghasia et al. [6] found that the mechanical properties of WBGP hybrid composite depend on the effect of fiber-polyester percentage variation. Venkateshwaran and ElayaPerumal

R. Giridharan (✉) · N. Anerudh · M. M. Mithun Srivan · M. P. Jenarthanan
School of Mechanical Engineering, SASTRA Deemed to be University, Thanjavur, Tamilnadu 613401, India
e-mail: giridharan54@gmail.com

L.-J. Yang et al. (eds.), *Proceedings of ICDMC 2019*,
Lecture Notes in Mechanical Engineering,
https://doi.org/10.1007/978-981-15-3631-1_31

[7] examined several works in composite reinforced with banana fiber and evaluated physical properties, structure, and application of the same. Yuan et al. [8] witnessed that the introduction of kevlar fiber enhanced the mechanical properties in wood flour/PP composites. Giridharan [9] fabricated ramie fiber reinforced composites and evaluated the mechanical properties of the same and found that the hybrid composite (ramie-glass fiber as reinforcement) has higher mechanical properties. Murali et al. [10] reinforced banana and aloe vera woven fabric in epoxy resin matrix and investigated the flexural and impact properties of in woven aloe vera fiber composite. Yamini et al. [11] fabricated pure coir and pure aloe vera composites both in randomly oriented mats and bi-directional mats treated with NaOH—alcohol—benzene mixture. They subjected the specimen to impact test and moisture absorption test and discovered that the impact value of the aloe vera fiber was comparatively lesser.

Among these, aloe vera is a well-known plant whose fibers are mainly used in a wide variety of structural applications like building and construction. Many researchers have done many researches with aloe vera and their composites. Elastic recovery, strength of the fiber, breaking extension of the fiber when its dry, moisture regaining ability, resistance to heat degradation, and the tensile strength of the fiber are some of the properties of aloe vera to be kept in mind. Aloe vera plant is shown in Fig. 1. Some of the mechanical and physical properties of aloe vera fiber are presented in Table 1.

Fig. 1 Aloe vera plant

Table 1 Physical properties of materials used

	Aloe vera fiber
Density (g/cm^3)	1.5
E—modulus (GPa)	12
Tensile strength (MPa)	400
Elongation at break (%)	3–10

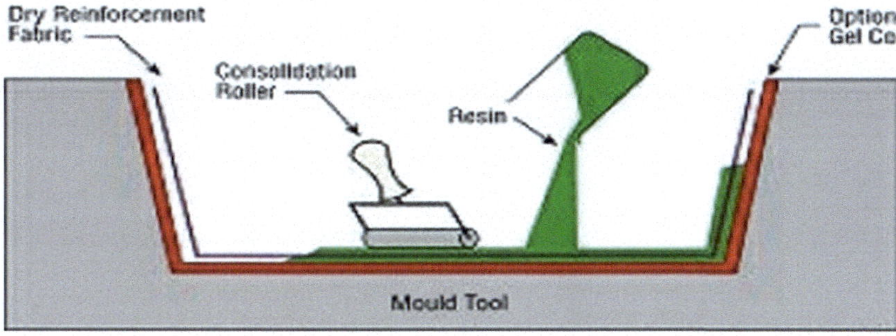

Fig. 2 Schematic diagram of hand lay-up method

2 Materials and Methods

Combination of aloe vera fibers and epoxy resin in the presence of a hardener leads the way to excellent bonding properties. On testing, the combination of epoxy (GY257) resin and hardener (HY2963) exhibited the greatest properties.

3 Hand Lay-up Process

The specimen was prepared using aloe vera fiber and a glass fiber mat of length 250 mm. The dimensions of the specimen were set to 250*250*5 mm. Three distinct composites namely 1, 2, and 3 were prepared with 1 layer, 2 layers, and 3 layers, respectively. The mats have been infused with the resin and fibers were allowed to dry. The fiber composites were bathed in the thinner solution and were hence freed from impurities. The fibers were then placed on the mould and were spread over the epoxy resin. They were then allowed to settle for a period of 48 hours as shown in Fig. 2.

4 Testing of Composites

Tensile tests were carried out, on the composite specimens 1, 2, and 3, as per the American Society for Testing and Materials (ASTM) D638 standard using a Universal Testing Machine (UTM) setup to determine the tensile strength at room temperature as shown in Fig. 3. The specimen that was made according to the ASTM: D638 was held using the grippers, as shown in Fig. 4, and the test was carried out by applying load until the specimen failed, as shown in Fig. 5. A plot between the stress and the strain was engendered.

Fig. 3 Tensile test setup Shimadzu Universal Testing Machine

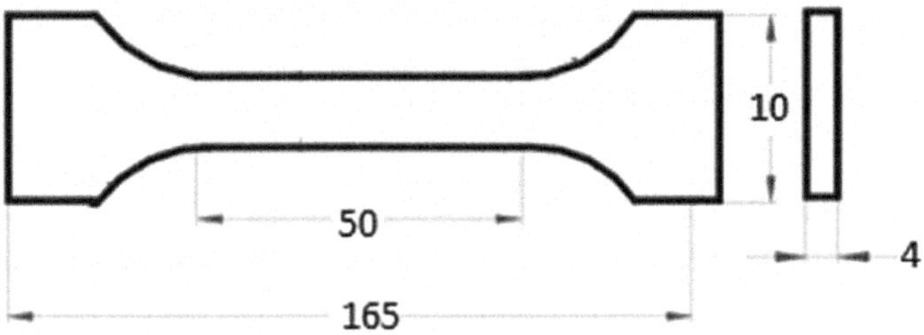

Fig. 4 Schematic diagram of tensile test specimen

Fig. 5 Tensile test broken specimen

Table 2 Tensile test results

Composites	Maximum stress (N/mm^2)	Maximum strain (%)
Composite 1	72	3.9
Composite 2	94	5.8
Composite 3	98	5.4

5 Results and Discussion

The ultimate tensile strength and the percentage of elongation were found to be higher in specimens 2 and 3 than in 1. The values of maximum stresses and maximum strains were tabulated in Table 2. From Fig. 6, it was inferred that the three-layered aloe vera composite has higher strength when compared to the single-layered and double-layered composites.

6 Microstructural Study

The fractured surface of the three-layered composite is shown in Fig. 7. SEM analysis showed that the interfacial bonding adhesion between the fiber and epoxy matrix to be considerably good. It was evident that the fiber-matrix adhesion was improper as there were visible fiber pullouts.

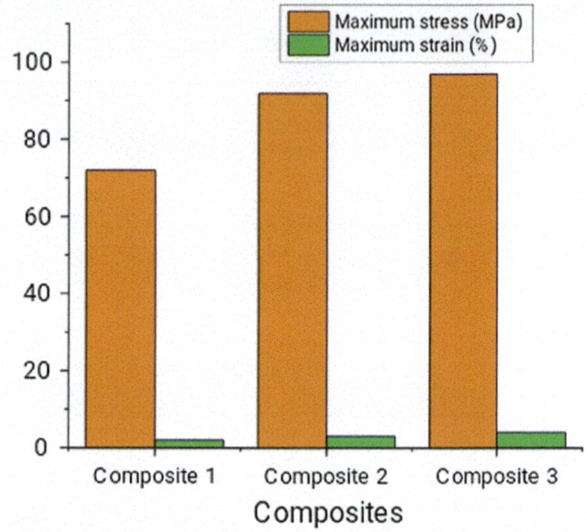

Fig. 6 Tensile test results

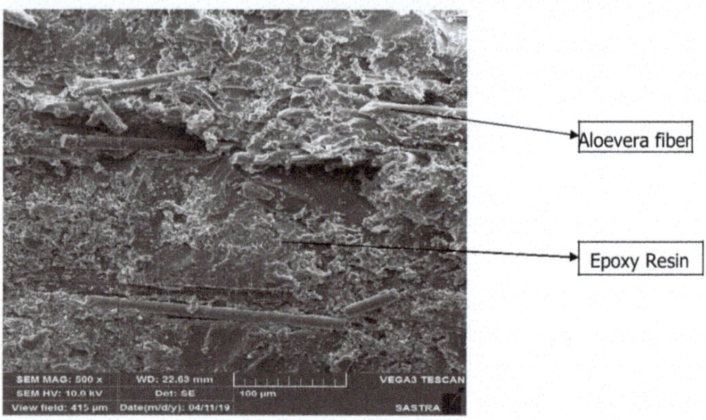

Fig. 7 SEM image of tensile fractured surface

7 Conclusion

In this research, three distinct aloe vera fiber composites were fabricated and were tested for their tensile strengths. The fractography of the fractured surfaces was performed according to the ASTM standards and the three-layered aloe vera fiber composite was found to have better tensile behavior than the two-layered and the single-layered composites. The fabricated and tested composite finds its applications in the automotive industry.

References

1. Perremans D, Trujillo E, Ivens JA, Van Vuure W (2018) Effect of discontinuities in bamboo fibre reinforced epoxy composites. Compos Sci Technol 50–57
2. Sapuan SM, Leenie A, Harimi M, Beng YK (2006) Mechanical property analysis of woven bamboo/epoxy composite. Mater Design 27:689–693
3. Satish KG, Siddeswarappa B, Kaleemulla KM (2010) Characterization of in-plane mechanical properties of laminated hybrid composites. J Mine Mater Char Eng 9:105–114
4. Anand P, Anbumalar V (2015) Mechanical properties of cellulose-filled epoxy hybrid composites reinforced with alkali-treated hemp fiber. Polymer (Korea) 39(1):46–55
5. Thakur A, Purohit R, Rana RS, Bandhu D (2018) Characterization and evaluation of mechanical behavior of epoxy-CNT-bamboo matrix hybrid composites. Mater Today: Proc 5:3971–3980
6. Vaghasia B, Rachchh B (2018) Evaluation of physical and mechanical properties of woven bamboo aloevera polyester hybrid composite material. Mater Today Proc
7. Anand P, Rajesh D, Kumar MS, Raj IS (2018) Investigations on the performances of treated jute/Kenaf hybrid natural fiber reinforced epoxy composite. J Polym Res 25(4):94
8. Yuan FP, Ou RX, Xie YJ, Wang QW (2013) Reinforcing effects of modified Kevlar fibre on the mechanical properties of wood-flour/polypropylene composites. J Forestry Res 24(1):149–153
9. Giridharan R (2019) Preparation and property evaluation of Glass/Ramie fibers reinforced epoxy hybrid composites. Composites Part B 167:342–345
10. Murali B, Vinoth Kumar S, Gopalakrishnan S, Suvendar T (2018) Evaluation of mechanical properties of aloe vera natural fibre reinforced composite. IJARIIE 4:2751–2758
11. Yamini S (2015) Shanmugasundaram: study mechanical behavior of natural based composite using coir and aloevera. Int J Emerging Technol Comput Sci Electron 13:976–1353

Experimental Investigation on Drilling of Kenaf-Banana Fiber-Reinforced Hybrid Fibre-Reinforced Polymer Composites

K. M. Alagappan, S. Vijayaraghavan, R. Giridharan, and M. P. Jenarthanan

Abstract The reason behind development of this paper is to identify the ideal process parameters to be set for the drilling of hybrid FRP (kenaf and banana) composite using HSS drill bits (5, 10, 15 mm) coated with tungsten carbide by means of statistical reproduction of the delamination factor and machining force using Taguchi-grey relational analysis. The contemplated process parameters are feed, speed and drill diameter. With the help of Taguchi's L-27 factorial design the trails were carried out. Three factors, three-level Taguchi orthogonal array design in grey relational analysis was used to carry out the trial study. In order to identify the damage around the drill region, video measuring system (VMS) was used. 'Minitab 18' was used to examine the data collected by taking advantage of the various statistical and graphical tools available. Variance examination helps in identifying best notable variable.

Keywords Natural fibre · Epoxy resin · Taguchi-grey relational analysis · Drilling · Delamination factor · Machining force · Characterization · Tungsten carbide

1 Introduction

Fibre-reinforced composites have numerous applications in our daily life because of their usability and strength. Fibres oriented in the same direction produce enhanced strength properties. Natural fibres can be used as replacements for synthetic fibres. Natural fibres are obtained from things which we consider as waste, and they are eco-friendly and easily available. They are mainly used in aerospace, automotive, electrical and construction industries as a result of its enhanced strength-to-mass ratio, toughness towards fracture and less weight. The hybrid FRP composite (kenaf and banana), can be used in the exterior layer of automotive seat moulds and doors for its increased flexural and impact strength, less weight and cost-effectiveness. In

K. M. Alagappan · S. Vijayaraghavan · R. Giridharan (✉) · M. P. Jenarthanan
School of Mechanical Engineering, SASTRA Deemed to be University, Thanjavur, Tamilnadu 613401, India
e-mail: giridharan54@gmail.com

L.-J. Yang et al. (eds.), *Proceedings of ICDMC 2019*,
Lecture Notes in Mechanical Engineering,
https://doi.org/10.1007/978-981-15-3631-1_32

fabrication of fibre-reinforced parts, drilling is one of the most crucial machining processes. Fibre-reinforced polymer components are contrived closer to the final shape, and further removal of material is restricted to deburring, trimming and contour shape accuracy. The dimensional precision is improved by reducing the surface roughness and delamination factor, and it also improves the performance of composite and the machinability of the composite [1]. To find out the surface quality and dimensional properties, theoretical models are applied for conjecture. Taguchi methodology is one of methods that help figure out the process parameters for various performance attributes. Best drilling parameters are chosen to attain maximized performance characteristics for drilling process. Taguchi is a competent mean through which process optimization is implemented by constrained number of trials [2]. Retrogradation data collection and graphical interpretation was done using software called 'Minitab 18'. ANOVA (analysis-of-variance) method is adopted to verify cogency of model and parameters. The delamination factor of drilled components can be found out using this model.

2 Experimental Study

2.1 Materials

In the experiment as discussed, two layers of kKenaf fibres (30 × 30 cm) and one layer of banana fibre (30 × 30 cm) was arranged sequentially. The banana fibre layer was placed in the centre with both the kenaf fibres placed in the top and bottom, resembling a sandwich-like structure, and has a fibre orientation of 0/90°. Epoxy resin (GY 257) and hardener (HY 2963) is concocted and hand layup technique is adopted to form hybrid fibre-reinforced polymer. The thickness of the material is 5.72 mm. The drill bit used for drilling is made of HSS with tungsten carbide coating and the following dimensions: 5, 10, 15 mm. The performance of a coated drill bit is far superior when compared to a non-coated drill bit [3]. The fabricated composite is shown in Fig. 1.

2.2 Quantification of Delamination Factor

The delamination factor is considered as the important factor of quality in any drilling operation. The various parameters considered in the trial are feed, speed and drill diameter. These parameters are chosen due to their momentous impact on the performance of material [4]. To curtail the number of trials, Taguchi's L-27 factorial design was employed in the procedure. Three levels of drilling parameters are considered to obtain the most optimized value and to accommodate the L-27 factorial design [4]. After a circumstantial study, minimum and maximum limit was identified for

Fig. 1 Hybrid fibre-reinforced polymer laminate

the above-discussed laminate. Low limits of feed rates are recommended in drilling fibre-reinforced polymer composites [5]. The drilling parameters as recognized are given in Table 1. All the drill holes were augured using ATC CNC machines using the 5, 10, 15 mm HSS drill bits with tungsten carbide coating. The holes were then measured, using vision measuring system (VMS).

$$F_d = D_{\max}/D \tag{1}$$

wherein $D_{\max}$ represents maximal diameter—drill hole after damage (mm), D represents original diameter—drill hole (mm) and F_d represents delamination factor [6] (Fig. 2).

Table 1 Cutting parameters with their levels

Factor	Cutting parameter	Units	Level		
			1	2	3
A	Feed	mm/min	450	300	150
B	Cutting speed	rpm	3000	2000	1000
C	Drill diameter	mm	15	10	5

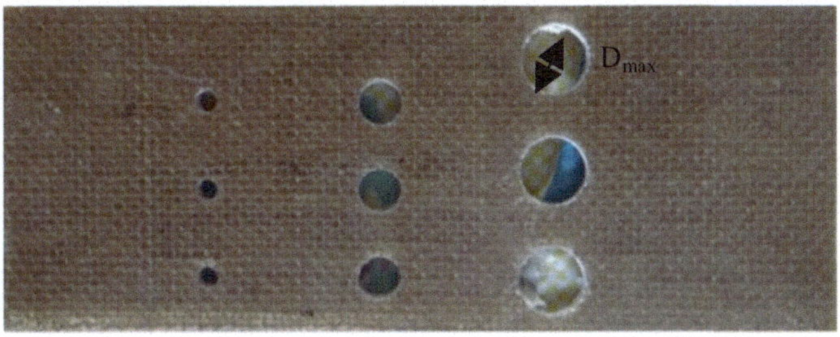

Fig. 2 Material after drilling

2.3 *Quantification of Machining Force*

Force assessment in drilling is very crucial because it is used to understand the tool setting as well as avert deterioration when drilling. Static and dynamic factors comprise data about cutting tool elements and configuration of chip produced. The machining force was found using a Kistler dynamometer.

3 Design of Experiments

Trial count increases with the factor count and level exponentially in a complete factorial representation. This results in gargantuan increase in experimental cost and time [7]. So, to strike a balance between the two contrary aspects as well as to find most advantageous scenario in restricted trial counts, Taguchi factorial scheme is elected as a result of effective optimization of collective performance attributes. The factorial arrangement of trials will be effective only when arranged in an orthogonal line-up [4].

$$N = 1 + \sum_{i=1}^{\text{No of variables}} (L - 1) \tag{2}$$

wherein L represents varying level counts of input drilling variables and N represents degree of freedom. Correspondingly, Taguchi's L-27 formation (orthogonal) is chosen. Output performance attributes are delamination factor and machining force. The values as obtained are considered in 'lower-the-better' approach.

4 Grey Relational Analysis (GRA)

Optimized drilling parameters with contemplation of varied performance attributes are corroborated in this method. Grey data processing is executed prior to calculating Grey correlation-coefficients. Each series in the GRA Grade is normalized. Assume $x_i\,(k)$ represents original sequence and $y_i\,(k)$ represents sequence of comparison, $i = 1, 2 \ldots m$; $k = 1, 2 \ldots n$.

wherein m represents overall experiment count,
n represents overall observed value count.

Pre-processing of data remodels elementary sequence to usually comparable or similar sequence. Pre-processing of data is done using numerous methods in GRA. For the purpose: 'smaller-the-better' normalization is done using:

$$x_i(k) = \frac{\max y_i(k) - y_i(k)}{\max y_i(k) - \min y_i(k)} \quad (3)$$

After all the data is pre-processed using the pre-processed sequences, GRA coefficient is derived. The formula to obtain GRA coefficient:

$$\psi_i(k) = \frac{\Delta_{\min} + \xi \Delta_{\max}}{\Delta_{0i}(k) + \xi \Delta_{\max}} \quad (4)$$

wherein

$\Delta_{0i}(k) = |x_0(k) - x_i(k)|$ = Difference of absolute value $x_0(k)$ and $x_i(k)$;
ξ = The distinguishing coefficient $0 \leq \xi \leq 1$;
$\Delta_{\min} = \forall j^{\min} \in i \forall k^{\min} |x_0(k) - x_i(k)|$ = The smallest value of Δ_{0i}; and
$\Delta_{\max} = \forall j^{\max} \in i \forall k^{\max} |x_0(k) - x_i(k)|$ = The largest value of Δ_{0i}

GRA Grades are found using the GRA coefficient values as obtained previously. By gathering all the GRA coefficients of various attributes for each trial, we obtain its overall GRA coefficient. For all '*n*' responses, the mean GRA Grade is found using:

$$\gamma_i = \frac{1}{n} \sum_{k=1}^{n} \chi_i(k) \quad (5)$$

wherein

n represents process response count,
$\chi_i(k)$ represents GRA coefficient of each response. [9]

We infer that an inflated value of GRA Grade is caused due to potent affiliation between native sequence $x_i(k)$ and reference sequence $x_0(k)$. In Eq. (4), the value of $\xi = 0.5$. This is so because both the responses have same weightage (Figs. 3 and 4; Table 3).

5 Result and Discussion

Delamination factor is critically important for numerous axiological complications like frictional effects, positional accuracy, contact deformation, etc. It is necessary in selection of drilling parameters. Table 2 is utilized to compute mean GRA Grade for all the factors. The comparison level is furnished between native and mimicking sequences by GRA Grade. The mimicking sequences show an inviolable relationship with the native sequence, denoting a large GRA Grade.

Based on the analysis as shown, combination of various levels is done to obtain the largest average response. For parameters *A*, *B* and *C*, we can safely say that *A*1, *B*1 and *C*3 represent highest GRA Grade values, as per Table 4. Hence, it is perceived

Table 2 Experimental model using L-27 orthogonal line-up along with their responses

A	B	C	Delamination factor	Machining force (N)
450	3000	5	1.0115	21.42
450	2000	10	1.0076	23.1
450	1000	15	1.0048	24.94
450	3000	10	1.0091	16.07
450	2000	15	1.0055	17.6
450	1000	5	1.0126	18.36
450	3000	15	1.005	15.76
450	2000	5	1.0126	17.29
450	1000	10	1.0091	18.67
300	3000	5	1.0104	20.81
300	2000	10	1.0068	21.73
300	1000	15	1.0033	24.17
300	3000	10	1.0068	17.29
300	2000	15	1.0026	19.74
300	1000	5	1.0107	20.04
300	3000	15	1.0033	15.76
300	2000	5	1.0104	17.6
300	1000	10	1.0046	19.43
150	3000	5	1.006	18.51
150	2000	10	1.0046	20.66
150	1000	15	1.001	21.88
150	3000	10	1.0046	16.68
150	2000	15	1.001	17.29
150	1000	5	1.0081	18.36
150	3000	15	1.0001	15.61
150	2000	5	1.0066	17.44
150	1000	10	1.0037	18.97

that the feed rate of 450 mm/min, cutting speed of 3000 rpm and drill diameter of 5 mm, stand as the favourable parameter combination for drilling of fibre-reinforced polymer composites.

5.1 *Analysis-of-Variance (ANOVA)*

It is used to distribute the fluctuation of an output to numerous inputs. This is achieved by splitting the overall disequilibrium of the GRA Grade. SS_T (sum of squared

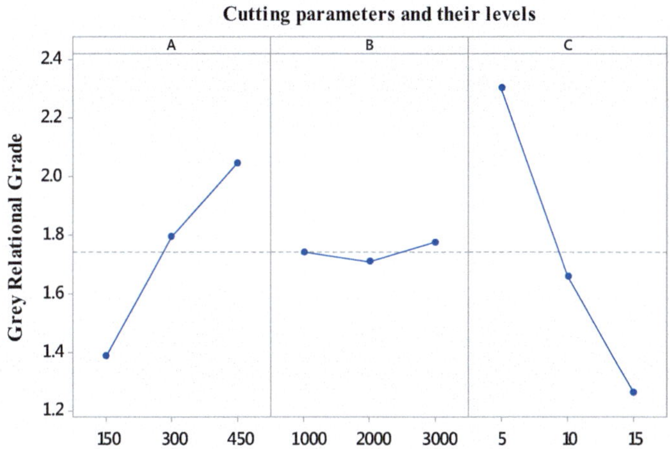

Fig. 3 Graphical representation of feed, speed and drill diameter on delamination (GRA Grade)

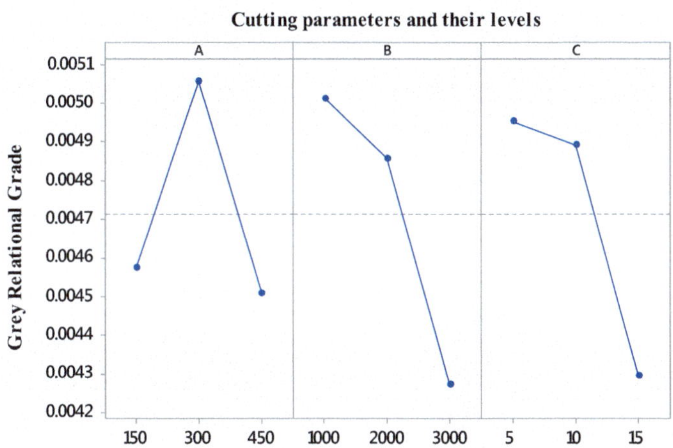

Fig. 4 Graphical representation of feed, speed and drill diameter on machining force (GRA Grade)

deviations) is calculated:

$$SS_T = \sum_{j=1}^{m} \left(\gamma_j - \gamma_m\right)^2 \tag{6}$$

wherein m represents trial count in the orthogonal line-up, γ_j represents GRA Grade for the jth trial and γ_m represents GRA Grade [10]. The percentage contribution by each drilling variable indicates its importance. The significance level is 5% (confidence level is 95%). From Table 5, it is recognized that speed (percentage contribution, $P = 92.6\%$) is the most significant drilling parameter.

Table 3 Signal-to-noise ratio—GRA coefficient of responses and GRA Grade

Signal-to-noise ratio		Normalized values—signal-to-noise ratio		GRA coefficient		GRA Grade
Delamination factor	Machining force	Delamination factor	Machining force	Delamination factor	Machining force	
−0.0993	−26.6164	0.9125	0.6753	2.9504	0.0073	1.4788
−0.0658	−27.2722	0.6015	0.8364	1.8189	0.0055	0.9122
−0.0416	−27.9379	0.3774	1	1.4252	0.0044	0.7148
−0.0783	−24.1203	0.7172	0.062	2.1218	0.0033	1.0626
−0.0476	−24.9103	0.4335	0.2561	1.5069	0.004	0.7554
−0.1088	−25.2775	1	0.3463	2.5109	0.0044	1.2577
−0.0433	−23.9511	0.3935	0.0204	1.4477	0.0032	0.7254
−0.1088	−24.7559	1	0.2181	2.5109	0.0038	1.2574
−0.0783	−25.4229	0.7172	0.382	2.1218	0.0046	1.0632
−0.0894	−26.3654	0.8209	0.6136	2.4936	0.0065	1.2501
−0.0589	−26.7412	0.5375	0.7059	1.686	0.0069	0.8465
−0.0282	−27.6655	0.2532	0.9331	1.2724	0.0048	0.6386
−0.0589	−24.7559	0.5375	0.2181	1.686	0.0038	0.8449
−0.0226	−25.9069	0.201	0.501	1.2176	0.0054	0.6115
−0.0924	−26.038	0.8488	0.5332	2.617	0.0057	1.3113
−0.0282	−23.9511	0.2532	0.0204	1.2724	0.0032	0.6378
−0.0894	−24.9103	0.8209	0.2561	2.4936	0.004	1.2488
−0.0399	−25.7695	0.3614	0.4672	1.4035	0.0052	0.7043
−0.052	−25.3481	0.4735	0.3637	1.5711	0.0045	0.7878
−0.0394	−26.3026	0.3574	0.5982	1.3982	0.0063	0.7023
−0.0087	−26.8009	0.0724	0.7206	1.1008	0.0067	0.5538
−0.0394	−24.4439	0.3574	0.1415	1.3982	0.0035	0.7009
−0.0087	−24.7559	0.0724	0.2181	1.1008	0.0038	0.5523
−0.0701	−25.2775	0.6414	0.3463	1.9132	0.0044	0.9588
−0.0009	−23.8681	0	0	1.0444	0.0031	0.5237
−0.0571	−24.8309	0.5215	0.2366	1.6557	0.0039	0.8298
−0.0321	−25.5613	0.2893	0.416	1.3133	0.0048	0.6591

Table 4 Response table—overall GRA Grade

Factor	Level-1	Level-2	Level-3
A	**1.4788**	1.3113	0.9588
B	**1.4788**	1.2574	1.3113
C	0.7254	1.0632	**1.4788**

Table 5 Results of ANOVA analysis

Source	Degree of freedom	Aggregate of squares	Average of squares	F-value	Percentage contribution
Feed	2	0.00010	0.00005	83.06	6.3
Speed	2	0.00000	0.00000	0.08	92.6
Drill diameter	2	0.00022	0.00011	181.38	1.04
Error	20	0.00001	0.00000	–	0.06
Total	26	0.00033	–	–	100

$S = 0.0007711$; R-sq = 96.36%; R-sq(adj) = 95.26%

6 Confirmation Test

To double check the results obtained, confirmation tests have been performed. Optimized parameters from Table 4 have been selected for performing the test. Using optimized parameters, GRA Grade is obtained by applying this formula:

$$\hat{\gamma} = \gamma_m + \sum_{i=1}^{q} \left(\overline{\gamma}_j - \gamma_m\right) \tag{7}$$

wherein γ_m represents average of the GRA Grade, q represents drilling parameter count and $\overline{\gamma}_j$ represents average of the GRA Grade at the optimized setting that influences numerous performance attributes [10]. The delamination factor, machining force and GRA Grade obtained experimentally for arbitrary input arrangement ($A_3B_3C_1$) are compared alongside delamination factor, machining force and GRA Grade obtained by trial at optimized drilling settings ($A_1B_1C_3$). Delamination factor is greater in contrast for the value attained by making use of optimized condition. Similarly, the machining force required is less when compared to the value obtained by making use of optimized provision. The GRA Grade is greater in contrast to the experimentally optimized process parameters which is 1.4788, whereas for the initial process parameters, it is 0.5538 (Table 6).

Table 6 Comparison of results for initial and optimized responses

Initial drilling parameters		Optimized drilling parameters	
		Predicted	Trial
Level	$A_3B_3C_1$	$A_1B_1C_3$	$A_1B_1C_3$
Delamination factor	1.0010	–	1.0115
Machining force, N	21.88	–	21.42
GRA Grade	0.5538	0.8738	1.4788
Development in GRA Grade		0.3200	0.9250

7 Conclusion

As per the experimental outcome, the ensuing conclusions are perceived through drilling of hybrid fibre-reinforced polymer composite using HSS drill bit with tungsten carbide coating. The set of optimized input parameters is—feed rate: 450 mm/min, cutting speed: 3000 rpm and drill diameter of 5 mm. Through ANOVA analysis, we find out that the speed (contribution = 92.6%) is the dominating parameter in the case of delamination factor and machining force. The result obtained from the confirmation test justifies the existence of noteworthy development in the GRA Grade: 0.5538–1.4788.

References

1. Jenarthanan MP, Jeyapaul R (2014) Machinability study of carbon fibre reinforced polymer (CFRP) composites using design of experiments technique. Pigm Resin Technol 43(1):35–44
2. Aydin H, Bayram A, Esme U, Kazancoglu Y, Guven O (2010) Application of grey relation analysis (GRA) and Taguchi method for the parametric optimization of friction stir welding (FSW) process. Mater Technol 44:205–211
3. Ashrafi SA, Sharif S (2012) Assessment of hole quality and thrust force when drilling CFRP/AI Stack using carbide tools. Appl Mech Mater 234:23–33
4. Raja T, Anand P, Sundarraj M, Karthick M, Kannappan A (2018) Failure analysis of natrual fibre reinforced polymer composite leaf spring. Int J Mech Eng Technol 9(2):686–689
5. Palanikumar K, Karunamoorthy L, Karthikeyan R (2006) Assessment of factors influencing surface roughness on the machining of glass fiber reinforced polymer composites. J Mater Design 27(2):862–871
6. Feito N, López-Puente J, Santiuste C, Miguélez MH (2014) Numerical prediction of delamination in CFRP drilling. Compos Struct 108:677–683
7. Jenarthanan MP, Raahul Kumar S, Vinoth S (2017) Multi-objective optimisation on end milling of hybrid fibre-reinforced polymer composites using GRA. Pigm Resin Technol 46(3):194–202
8. Datta S, Bandyopadhyay A, Pal PK (2008) Grey-based Taguchi method for optimization of bead geometry in submerged arc bead-on-plate welding. Int J Adv Manuf Technol 39(11–12):1136–1143
9. Slavek N, Jović A (2012) Application of grey system theory to software projects ranking. Automatika J Control Measur Electron Comput Commun 53(3):284–293
10. Sakthivel M, Vijayakumar S, Jenarthanan MP (2017) Grey-fuzzy logic to optimise process parameters in drilling of glass fibre reinforced stainless steel mesh polymer composite. Pigm Resin Technol 46(4):276–285

Material Modeling of Particle Reinforced in Metal Matrix Composite

M. Sundarraj and R. Varatharajan

Abstract The FEA model of particulate-reinforced metal matrix composite will be used for material optimization application. A particle distribution method will be introduced to get a random pattern of particles reinforcement model in the metal matrix FEA model. A FEA model of particle metal matrix composite that allows to vary particle distribution is developed in ANSYS commercial FEA code. The particle distribution pattern is generated using the sphere packing method for a given percentage of particulate mixture ratio. The effect of particle concentration and distribution and its mechanical behavior and metal matrix can be analyzed and studied in this model. This model will be useful in optimizing the particle metal matrix against a given mechanical behavior or performance required for the variation of percentage in particle mixture ratio, particle grain size, particle material, and matrix material.

Keywords FEA code · Particle distribution pattern · Metal matrix composite · Particulate mixture ratio

1 Introduction

Optimization is a wide method for creating or finding best solution by means of the iteration process. In the metal matrix composite, the metal and material concentration and the mixing ratio play a vital role to get better mechanical behavior. Based on particle concentration and distribution, the performance of meal matrix composites enhances its mechanical and thermal characteristics. To accomplish best performance characteristic properties in composite material, a coupled hygro-thermo-mechanical computational model is used for reinforcing fiber polymers, constrained within the domain of computational homogenization [1].

M. Sundarraj (✉) · R. Varatharajan
School of Mechanical and Construction Engineering, Department of Mechanical Engineering, Vel Tech Rangarajan Dr. Sgunthala R&D Institute of Science and Technology, Avadi, Chennai 600062, India
e-mail: sundarrajemail@gmail.com

L.-J. Yang et al. (eds.), *Proceedings of ICDMC 2019*,
Lecture Notes in Mechanical Engineering,
https://doi.org/10.1007/978-981-15-3631-1_33

1.1 Particle Reinforced Composite Square

For making the model of particle reinforced composites, a square or irregular domain of size under plane stress conditions has been considered. Particle of required size and shape are generated in the whole model with the help of MATLAB, and this data is import to the ANSYS program to inspect distribution pattern in matrix and analyzing the mechanical behavior of MMC. So we should investigate the characteristics of particle, in fact which is the ultimate thing in this paper.

The particle parameter was characterized in terms of the equivalent diameter:

$$d = 4A/\pi$$

The geometrical shape of particle and the shape factor (S) are adopted,

$$S = \pi^2 4\pi A$$

where

'π' is the perimeter of particle
'A' is the area of particle.

1.2 Particle Shapes

The stress and strain acting on the particles in a microstructure ominously depend on the particle shapes, which will affect the macroscopic mechanical properties of the composite [2].

1.3 Development of MATLAB Code for Sphere Packing in Cube or Cylinder

MATLAB is to calculate the characteristics of particles such as area and perimeter of each particle, in which the image processing technique was employed. Based on the calculations, the sizes and shape factors of each particle were investigated (Fig. 1).

2 FEM Process

The determination of elastic properties in a 3D PRMMC unit cell in which particles are arranged in randomly distributed array in order to assess the reliability of these micro models. Most of the unit cells investigated in the past literature are related to two-dimensional unit cells.

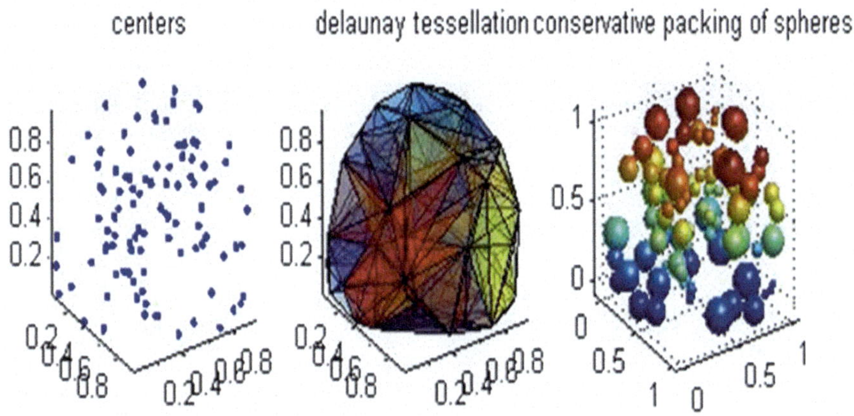

Fig. 1 Sphere packing diagram using MATLAB

Moreover, finite element results could be affected by the elements size in the mesh. Hence, it was necessary to evaluate the micro models and their meshes under different loading condition to verify that, although they undergo a elastic deformation, elements in the mesh retain a adequate aspect ratio to avoid erroneous results.

The procedure to assess the validity of these micromodels was performed by determining elastic properties of a particle and by comparing numerical results with analytical and/or possibly experimental data available in the literature.

The package used for the analyses was ANSYS MULTIPHYSICS version 11. The mesh was built using initially (8-node solid brick) for a better accuracy available in ANSYS MULTIPHYSICS version 11. All these types of element are three-dimensional continuum elements and allow stress and displacement analysis.

The FEA analyses were carried out at different volume fractions and in particular, the following particle volume fractions were taken into account: 5, 10, 15, 20, 25, 30, 35, 40, 45, 50, 60, and 70. The number of element is based on the particle volume fraction and the types of element used in the mesh.

2.1 *Boundary and Load Condition*

The simplest treatment of the elastic behavior of this model is based on the premises that the material can be considered as two constituents perfectly merged together, with relative thicknesses in proportion to the reinforcement and matrix volume fraction. The matrix is forced to have the same lengths parallel to the bonded interfaces. Hence, if the stress is applied in the x-direction of the model, then the constituents (particle and matrix) exhibit exactly the same strain, but read only particle behavior.

3 Result and Discussion

Characteristics of silicon carbide addition on the mechanical behavior of aluminum matrix alloy composites were examined in this study by virtual simulation. Five different additions of silicon carbide were carried out and results were investigated. They are listed in Table 1.

3.1 Load Versus Time

The model gives various choices to designate convergence criteria; we can have a base convergence checking on the pressure force at any combination of these materials. Accumulating each material has its own convergence tolerance value. For multiple constrain problems, you also have a choice of convergence norms.

We always make the use of force, displacement-based (moment, rotation based) convergence tolerance for checking.

Here, the graph has the *x*-axis in time and *y*-axis in load. It shows the FEM model contains 3 kg/m^2 load for 5 s, similarly all load cases for remains time steps, which is illustrated in the above table (Fig. 2).

3.2 Load Versus Stress

The bellows figure shows the effect of stress influence in these models with reinforcement volume percentage against transient loading conditions. Generally, this measure has taken for predicting the rapture of the micromodel. Here, the different volume fraction curve has been plotted between stress (*y*-axis) and transient load (*x*-axis) (Fig. 3).

Table 1 Percentage of silicon carbide in reinforced aluminum matrix composites

% SiC	Addition explanation
Al–10% SiCp	10wt% particulate reinforced, with aluminum matrix
Al–30% SiCp	30wt% particulate reinforced, with aluminum matrix
Al–40% SiCp	40wt% particulate reinforced, with aluminum matrix
Al–50% SiCp	50wt% particulate reinforced, with aluminum matrix
Al–60% SiCp	60wt% particulate reinforced, with aluminum matrix

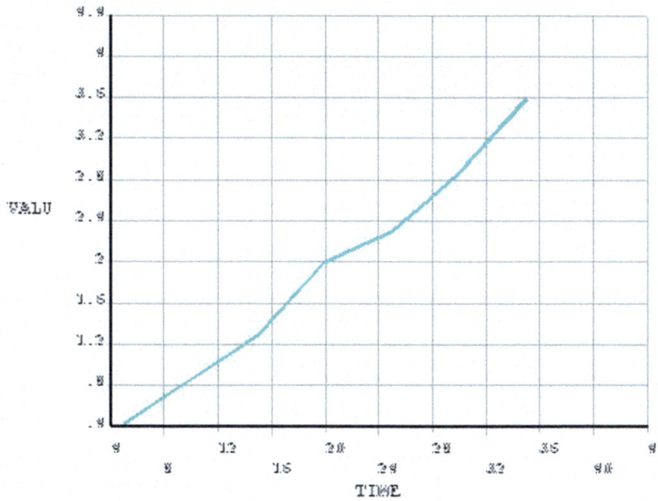

Fig. 2 Load versus time step

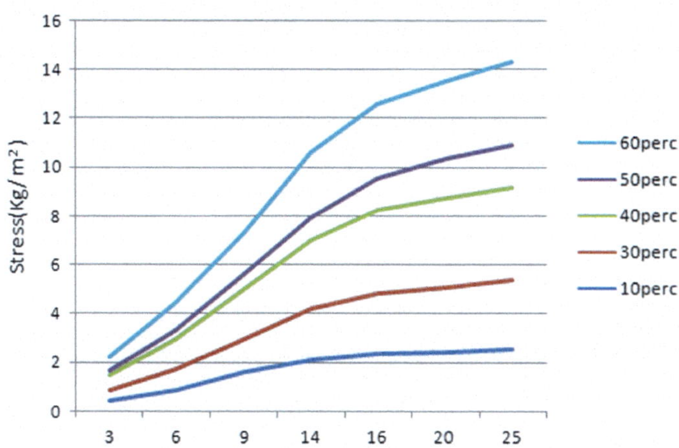

Fig. 3 Load versus stress

From the graph taken from the experiment, we have concluded that the stress is increased against time with continuously increasing in the reinforcement volume fraction (Table 2).

Table 2 Load versus stress

Stress	Load versus stress with respect to volume percentage				
Load	10%	30%	40%	50%	60%
3.00	0.41	0.46	0.61	0.18	0.57
6.00	0.82	0.92	1.22	0.36	1.14
9.00	1.35	1.23	1.02	0.62	1.71
14.00	2.13	2.04	2.84	0.92	2.65
16.00	2.38	2.45	3.44	1.29	3.03
20.00	2.43	2.61	3.71	1.66	3.12
25.00	2.55	2.82	3.83	1.74	3.38

3.3 *Load Versus Strain*

The quantities you measure are load (force) and displacement. These are the properties of sample material under test, and any mechanical properties derived from them—for example, stiffness also sample properties. These may be of use to us, want to check when a particular type of material will fail. As materials scientists though, we want to know about the properties of the material. Properties are changed with respect to the load size of the sample.

For this reason, the load is distributed to the sample area to get strain and displacement in a span. Now, the derived quantity of composite and its modulus is independent of sample size and can be regarded as a true material property (Fig. 4).

From the above graph, show that the effect of SiC particle with a different load cases. We conclude that the plastic deformation of the model is normal against transient loading condition.

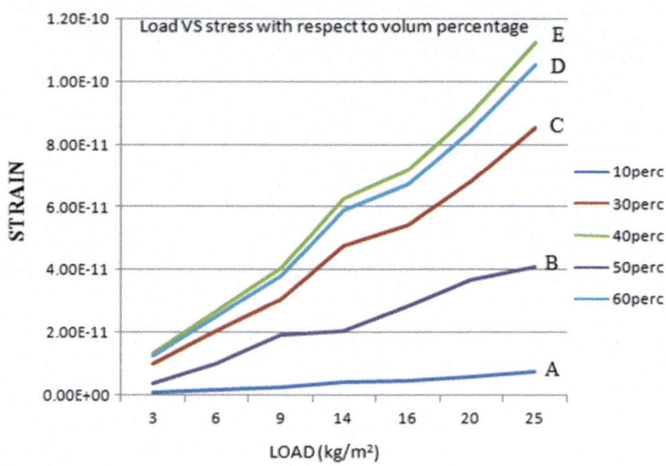

Fig. 4 Load versus strain

Table 3 Load versus strain

Strain	Load versus strain with respect to volume percentage				
Load	10%	30%	40%	50%	60%
3	9.13E−13	1.02E−11	1.35E−11	4.11E−12	1.26E−11
6	1.83E−12	2.05E−11	2.70E−11	1.03E−11	2.53E−11
9	3.02E−11	4.06E−11	4.74E−12	2.96E−11	3.01E−11
14	4.26E−12	4.78E−11	6.31E−11	2.05E−11	5.90E−11
16	4.87E−12	5.47E−11	7.21E−11	2.87E−11	6.74E−11
20	6.09E−12	6.83E−11	9.01E−11	3.70E−11	8.43E−11
25	7.61E−12	8.54E−11	1.13E−10	4.11E−11	1.05E−10

Curve *A* is a brittle material. Notice that a very large strain for a small load.

Curve *B* is a almost nearer to the brittle property, because it also behaves like the curve *A*.

And other three curves like *C*, *D*, and *E* relatively increase the deflection with increasing the load case, so those curves have ductility in maximum load case (Table 3).

3.4 *Stress Versus Strain*

To Inspect inherent elastic properties of linear objects like wires, rods, or columns which are long-drawn-out or compressed, to get the ability parameters of the material, the ratio of the stress to the strain, a parameter called the "Young's modulus" or "Modulus of Elasticity" of the material is used.

We can describe these details by the graph as:

- *P* indicates the limit of proportionality, where the linear correlation between stress and strain ends.
- *E* shows the elastic limit of the material. Below the elastic limit, the wire will come again to its original shape.
- *Y* indicates the yield point, where plastic deformation begins. A long-drawn-out in the material creates strain, and it is seen in a small increase in stress. If the stress applied to a wire is maximum without it snapping. It is sometimes called the breaking stress. Notice that beyond the UTS, the force required to snap the wire is less (Fig. 5).

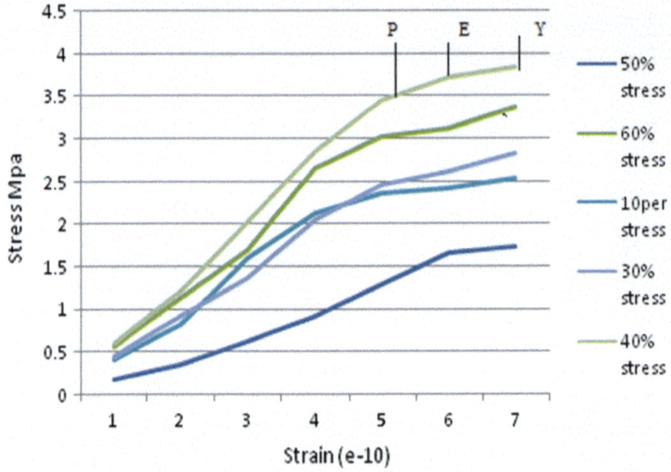

Fig. 5 Strain versus stress

4 Conclusions

In this paper, an ongoing work, consisting of a particle distribution method will be introduced to get a random pattern of particle reinforcement model in the metal matrix FEA model. A sphere packing code is developed from the MATLAB and the data for the mechanical properties of composites are exported to ANSYS framework. The effect of particle concentration is based on its distribution in the model. The mechanical behavior of metal matrix composites is analyzed and studied in this model. This particle distribution method enables metal matrix against a given mechanical behavior or performance required for the variations.

Some of the von Mises stress and strain figures, as shown below, are to identify the stress value based on percentage of particulate reinforced, with aluminum matrix (Figs. 6, 7, 8 and 9).

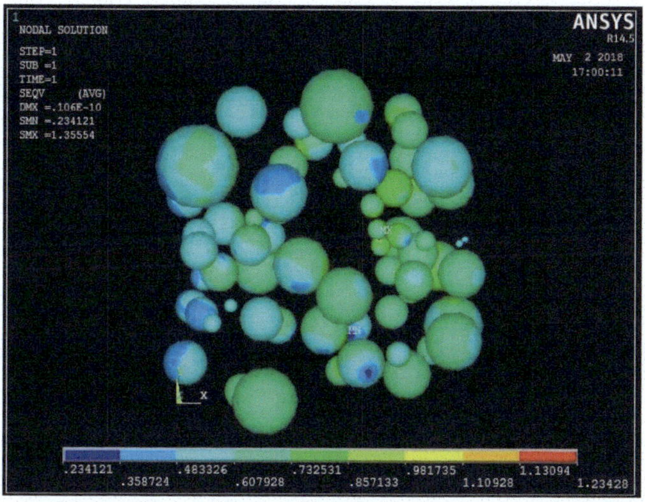

Fig. 6 Von Mises stress value 9 kg/m^2 load (Al–10% SiCp)

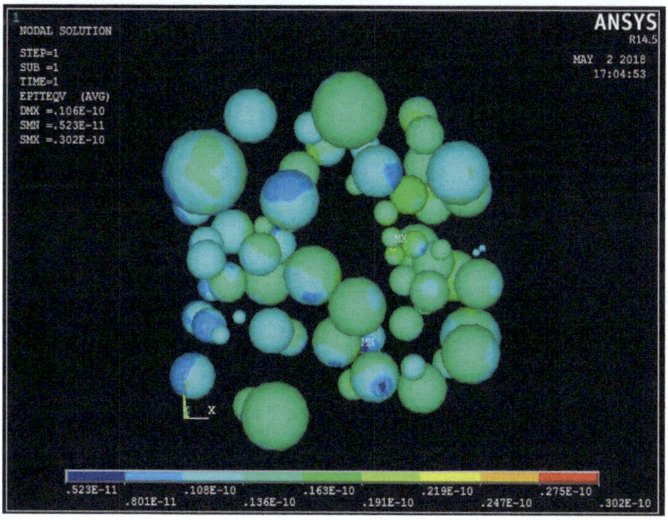

Fig. 7 Von Mises stress value 9 kg/m^2 load (Al–30% SiCp)

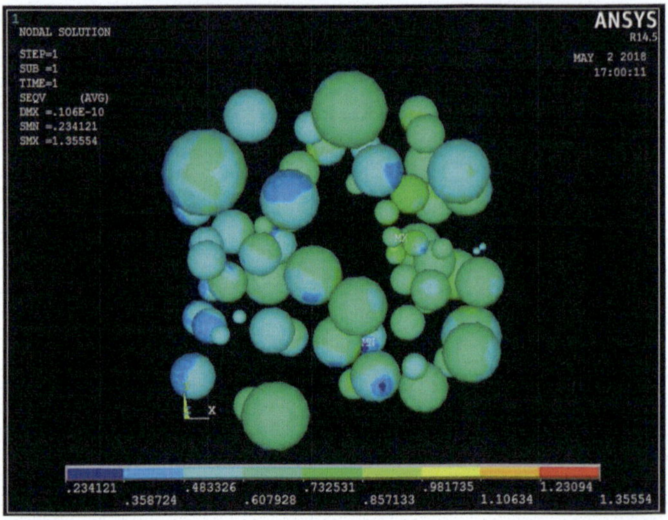

Fig. 8 Von Mises stress value 9 kg/m^2 load (Al–40% SiCp)

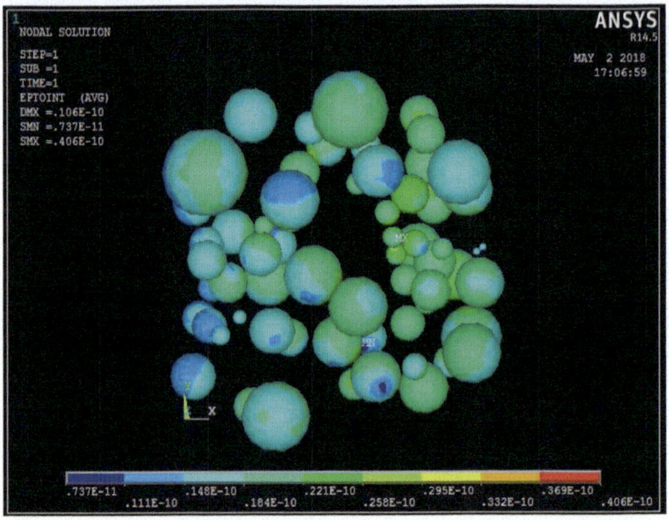

Fig. 9 Von Mises stress value 9 kg/m^2 load (Al–50% SiCp)

References

1. Ullah Z, Kaczmarczyk L, Grammatikos SA (2017) Multi-scale computational homogenisation to predict the long-term durability of composite structures. Comput Struct 181:21–31
2. Ellyin F, Xia Z, Li C-S (2005) Fatigue damage of particle reinforced metal matrix composites. In: WIT transactions on state of the art in science and engineering, vol 21. WIT Press

Fatigue Crack Growth Behavior of a Nickel-Base Super Alloy Inconel 718 Under Spectrum Loads

Sharanagouda G. Malipatil, Anuradha N. Majila, Chandru D. Fernando, and C. M. Manjunatha

Abstract Fatigue crack growth behavior of a nickel-base super alloy Inconel 718 under a standard mini-FALSTAFF spectrum load sequence was determined experimentally. Fatigue tests were performed in a 100 KN servo-hydraulic test machine at RT and lab air atmosphere using Compact Tension, $C(T)$ specimens. Triangular waveform at an average frequency of 5 Hz was employed in testing. The crack length was observed to increase with applied spectrum load blocks, slowly in the beginning and rapidly as the crack length increased further to fail after about 58 blocks of loading. Further, the crack growth behavior under spectrum load was predicted and compared with experimental results. For prediction purpose, constant amplitude (CA) fatigue crack growth rate (FCGR) data at different stress ratios was analyzed to derive a crack growth law based on a two-parameter crack driving force ΔK^*. The crack extension per load block was estimated using this crack growth law. The predicted fatigue crack growth behavior was observed to be conservative and comparable with experimental results.

Keywords Fatigue crack growth · Ni-base super alloy · Spectrum loads · Driving force

1 Introduction

Nickel-base super alloys are widely used in gas turbine engines for engine disc and blade constructions. They possess high strength, stiffness, corrosion resistance, and also they retain high strength at elevated temperatures. The disc material experiences fatigue loads during operation of the engine. These service loads initiate cracks at

S. G. Malipatil (✉) · C. M. Manjunatha
Academy of Scientific and Innovative Research (AcSIR), Ghaziabad 201002, India
e-mail: sgmalipatil@nal.res.in

A. N. Majila · C. D. Fernando
Materials Group, Gas Turbine Research Establishment, DRDO, Bangalore 560093, India

S. G. Malipatil · C. M. Manjunatha
Structural Integrity Division, CSIR-National Aerospace Laboratories, Bangalore 560017, India

L.-J. Yang et al. (eds.), *Proceedings of ICDMC 2019*,
Lecture Notes in Mechanical Engineering,
https://doi.org/10.1007/978-981-15-3631-1_34

stress concentration locations and further this crack starts growing under the fatigue loads leading to fracture. As a part of damage tolerance evaluation in these materials, it is required to study and understand the fatigue crack growth behavior of Ni-base super alloys under service loads.

The general methodology followed in prediction of FCGR behavior under spectrum loads involves: (i) fatigue cycle counting in the spectrum load sequence, (ii) determination of appropriate crack growth law for the material from its constant amplitude fatigue crack growth data, and (iii) determination of crack extension for each counted load cycle and hence for every load block. Thus, one of the major steps in prediction is deriving appropriate fatigue crack growth law for the material with proper crack driving force. The elimination of load ratio effects is a major step in this regard. Many efforts have been made to correlate stress-ratio effects on the FCGR behavior in various materials [1–6]. Walker's modified equation has been used to correlate R ratio effects quite well in Al and Ti alloys [7]. Crack closure effects have been used by several authors to collate FCGR curves into a single curve [2–4]. Recently, Kujawski and co-workers [1] have demonstrated that use of two-parameter crack driving force parameter, ΔK^* provides a better approach for correlating stress-ratio effects in many materials.

It may be noted that although crack closure concept is widely employed in predictions, it is difficult to measure and use the crack closure level in real structures containing cracks which may lead to significant errors in predictions. Thus, the aim of this work is to determine the FCGR behavior of Inconel 718 under a standard spectrum load sequence and also prediction using two-parameter crack driving force.

2 Experimental

2.1 *Material and Specimen*

Nickel-base super alloy Inconel 718 was used in this investigation. The chemical composition depicting major alloying elements is shown in Table 1. The material was obtained as blanks extracted from a forged disc. The mechanical properties of this material are shown in Table 2. The standard compact tension (CT) test specimen is shown in Fig. 1. And fatigue crack growth test setup is shown in Fig. 2.

Table 1 Composition (wt%) showing major alloying elements in Inconel 718

C	Si	Mn	S	P	Ni	Cr	Mo	W
0.025	0.07	0.06	<0.0003	0.01	53.80	17.9	2.93	0.04
Al	Co	Cu	Ti	Fe	B	N	O	Ag
0.54	0.18	0.03	0.99	18.28	0.004	0.0078	<0.0005	0.0001

Table 2 Mechanical properties of Inconel 718 used in this investigation

UTS (MPa)	YS (MPa)	Poisson's ratio	Fracture toughness (MPa$\sqrt{m}$)
1241	1034	0.33	85

Fig. 1 Inconel 718 material CT specimen

Fig. 2 Fatigue crack growth test setup

The stress intensity factor range (ΔK) for $C(T)$ specimen formula was determined as per ASTM E647 [8]

$$\Delta K = \frac{\Delta P}{B\sqrt{W}} f(a/w) \tag{1}$$

where

$$f(a/w) = \frac{\left(2+\frac{a}{w}\right)\left(0.886 + 4.64 * \left(\frac{a}{w}\right) - 13.32 * \left(\frac{a}{w}\right)^2 + 14.72 * \left(\frac{a}{w}\right)^3 + 5.6 * \left(\frac{a}{w}\right)^4\right)}{\left(1-\frac{a}{w}\right)^{3/2}} \tag{2}$$

2.2 Test Equipment and Procedure

Nickel-base super alloy Inconel 718 was tested under a modified standard spectrum load sequence called mini-FALSTAFF [2], shown in Fig. 3. It is a modified version of FALSTAFF load spectrum, which is a standardized variable-amplitude load sequence developed for the fatigue analysis of materials used in fighter aircraft. In Fig. 3, the normalized stress (or load) is plotted against peak/trough point number of load sequence. One block of this complete load sequence consists of 18,012 reversals at 32 different stress levels. One block of this load sequence represents loading equivalent of 200 flights. The test load sequence for our experiments was designed by multiplying all the peak/trough points with a constant reference load value, P_{ref}.

Spectrum fatigue tests were conducted in a 100 KN servo-hydraulic test machine. Triangular waveform with an average frequency of 5 Hz which is 10 reversals per

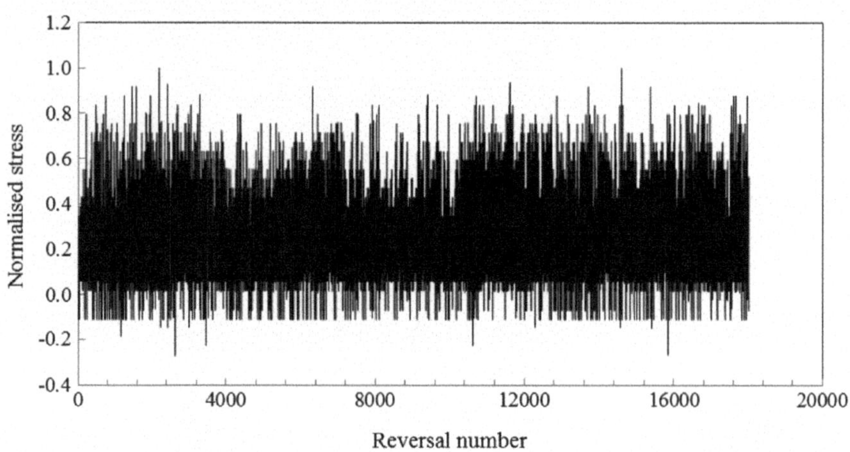

Fig. 3 Standard mini-FALSTAFF spectrum load sequence [2]

second was employed in all the tests. Crack length at the end of every load block was measured by optical microscope. Two tests were conducted under spectrum load sequence and average crack growth rates were determined.

3 Fatigue Crack Growth Prediction Methodology

The fatigue crack growth behavior of nickel-base super alloy Inconel 718 under mini-FALSTAFF load sequence was predicted following the general methodology of crack growth prediction [9]. A flowchart for the life prediction procedure used in this investigation is shown in Fig. 4. The steps involved are (i) rainflow counting

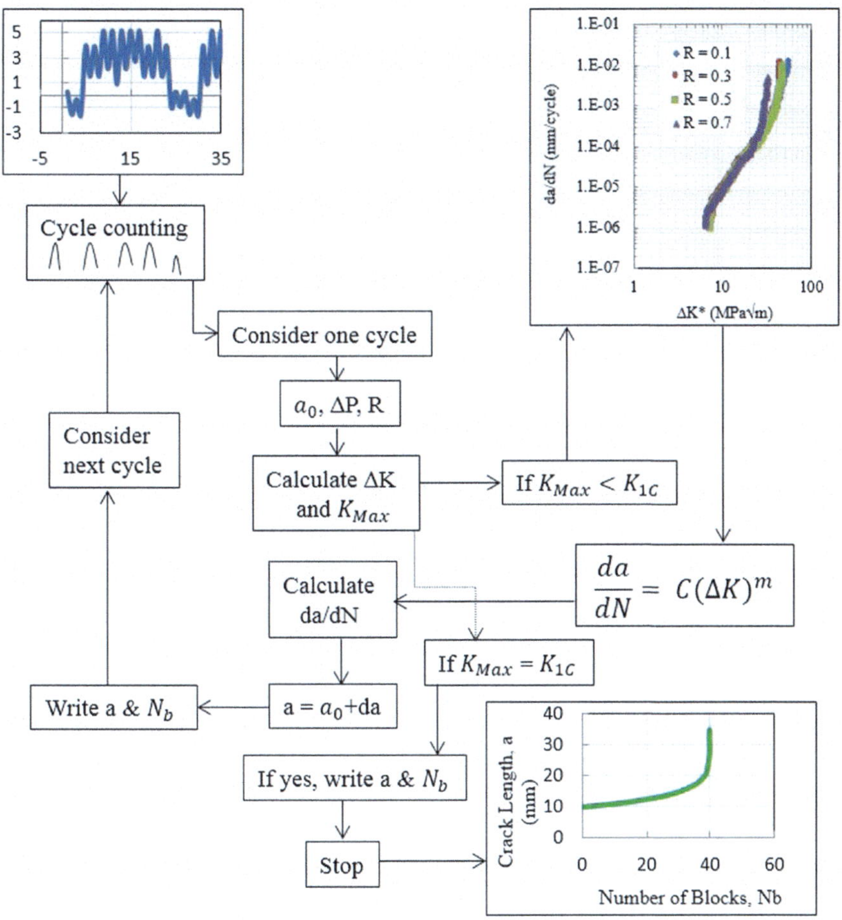

Fig. 4 Flow diagram for prediction of crack growth behavior under spectrum loads

of the fatigue cycles in the spectrum, (ii) constant amplitude fatigue crack growth rate behavior determination at various stress ratios and approximating the behavior with an appropriate crack growth law (iii) the calculation of the crack extension for each of the rain flow counted load cycle using growth equation, and finally, (iv) the determination of crack length extension per block considering linear damage accumulation theory. This procedure was used in the present study for estimation of fatigue crack growth behavior.

4 Results and Discussion

The spectrum fatigue crack growth behavior of Inconel 718 under mini-FALSTAFF load sequence determined experimentally is shown in Fig. 5. The crack length was observed to increase with applied spectrum load blocks, slowly in the beginning and rapidly as the crack length increased further to fail after about 58 blocks of loading. Similar trend in spectrum FCG behavior of various alloys has been observed by many investigators [2, 10–14].

The FCG behavior under mini-FALSTAFF spectrum load sequence was predicted following the procedure shown in Fig. 4. For the purpose of prediction, the constant amplitude data of the material developed earlier [15] was used in deriving the crack growth rate equation. The FCGR determined as a function of two-parameter crack driving force parameter ΔK^* is shown in Fig. 6. This data was fit to Newman's equation [16] of the form as follows Eq. (3) and become a single curve shown in Fig. 6.

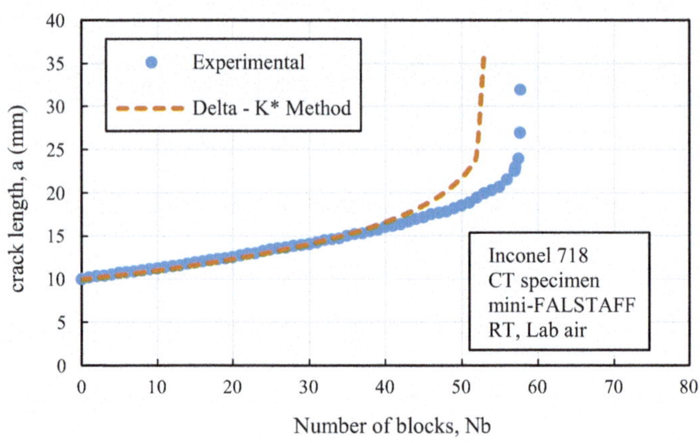

Fig. 5 Fatigue crack growth behavior of nickel-base super alloy Inconel 718 under FALSTAFF spectrum load sequence

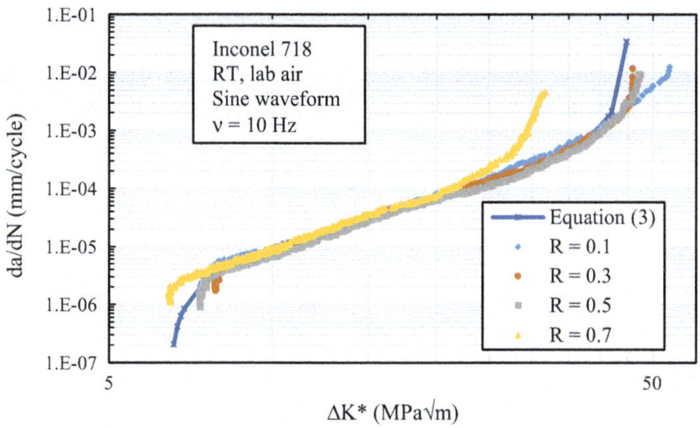

Fig. 6 Correlation of stress-ratio effects with crack driving force parameter ΔK^*

$$\frac{\mathrm{d}a}{\mathrm{d}N} = C_1(\Delta K^*)^{C_2}\frac{\left[1-\left(\frac{\Delta K^*_{(\mathrm{th})}}{\Delta K^*}\right)^2\right]}{\left[1-\left(\frac{\Delta K^*}{C_3}\right)^2\right]} \tag{3}$$

$$\Delta K^* = \Delta K\left(\frac{1-R_t}{1-R}\right)^{\gamma} \tag{4}$$

The constants in Eq. (3) obtained from the experimental data were as follows: $C_1 = 1.519\mathrm{e}{-7}$, $C_2 = 2.0022$, $C_3 = 45\ \mathrm{MPa}\sqrt{\mathrm{m}}$, and $\Delta K\mathrm{th} = 6.5\ \mathrm{MPa}\sqrt{\mathrm{m}}$, $R_t = 0.3404$, $\gamma = 0.03083$ for $R > R_t$, $\gamma = 0.51678$ for $R < R_t$.

The predicted fatigue crack growth behavior of Inconel 718 under mini-FALSTAFF along with experimental results is shown in Fig. 5. The predicted fatigue crack growth behavior is observed to be conservative and comparable with experimental results.

5 Concluding Remarks

Based on the above results obtained in this investigation, conclusions are as follows,

(i) The fatigue crack growth behavior under mini-FALSTAFF spectrum loads sequence in nickel-base super alloy Inconel 718 could be predicted reasonably well with use of two-parameter crack driving force ΔK^*.
(ii) Further investigations on various other spectrum load sequence may be conducted to verify the suitability of this method.

Acknowledgements Authors would like to thank the Director, GTRE, the Director, CSIR-NAL and the Head, STTD, CSIR-NAL for their support during this work. They also would like to acknowledge the technical support rendered by the scientists and technical support staff members of FSIG, STTD, and CSIR-NAL.

References

1. Sree PCR, Kujawski D (2014) A two-parameter fatigue crack growth correlation using ΔK and Kmax parameters. In: Proceedings of the ASME 2014 international mechanical engineering congress and exposition, Montreal, Quebec, Canada, pp 14–20
2. Manjunatha CM (2008) Fatigue crack growth prediction under spectrum load sequence in an aluminum alloy by K*- RMS Approach. Int J Damage Mech 17:477–492
3. Newman JC Jr, Kota K, Lacy TE (2017) Fatigue and crack-growth behavior in a titanium alloy under constant-amplitude and spectrum loading. Eng Fract Mech J 1–14
4. Tzamtzis A, Kermanidis AT (2014) Improvement of fatigue crack growth resistance by controlled overaging in 2024-T3 aluminium alloy. Fatigue Fract Eng Mater Struct 37:751–763
5. Elber W (1971) The significance of fatigue crack closure, damage tolerance in aircraft structures. ASTM STP 486, American Society for Testing and Materials, Philadelphia, PA, pp 230–242
6. Kujawski D (2001) A fatigue crack driving force parameter with load ratio effects. Int J Fatigue 23:S239–S246
7. Li Y, Wang H, Gong D (2012) The interrelation of the parameters in the Paris equation of fatigue crack growth. Eng Fract Mech 96:500–509
8. ASTM (1997a) Standard test method for measurement of fatigue crack growth rates, ASTM E 647-95, American Society for Testing and Materials, PA, 3.01, pp 578–614
9. Manjunatha CM, Parida BK (2004) Prediction of fatigue crack growth after single overload in an aluminum alloy. AIAA J 42(8):1536–1542
10. Hu D, Meng F, Liu H, Song J, Wang R (2016) Experimental investigation of fatigue crack growth behavior of GH2036 under combined high and low cycle fatigue. Int J Fatigue, Elsevier 1–10
11. Newman JC Jr (1996) Application of a crack closure model to predict crack growth in three engine disc materials. Int J Fatigue 80:193–218
12. Zhuang WZ (2000) Prediction of crack growth from bolt holes in a disc. Int J Fatigue 22:241–250
13. Zhuang W, Barter S, Molent L (2007) Flight-by-flight fatigue crack growth life assessment. Int J Fatigue 29:1647–1657
14. Mansor NII, Abdullah S, Ariffin AK (2016) Discerning the fatigue crack growth behavior of API X65 steels under sequence loading. Latin Am J Solids Struct 14:202–216
15. Malipatil SG, Majila AN, Fernando C, Manjunatha CM (2019) Correlating stress ratio effects on the fatigue crack growth behavior of a nickel base super alloy Inconel 718. In: Submitted to international conference on advances in mechanical engineering (ICAME), SJBIT at Bengaluru, pp 1–7
16. Newman JC Jr (1981) A Crack-closure model for predicting fatigue crack growth under aircraft spectrum loading, methods and models for predicting fatigue crack growth under random loading. American Society for Testing and Materials, ASTM STP 748, Philadelphia, PA, pp 53–84

Mechanical and Metallurgical Properties of Hybrid Composite Material

V. Karthi, K. Marimuthu, G. Boopathy, and N. Ramanan

Abstract Tremendous growth has been incorporated in the field of composite materials in the current scenario where the engineering applications are highly demandable. Currently composites are being used to replace conventional metallic materials in a wide range of industries including aerospace, defense, aircraft, and textile which require structural materials with high strength-to-weight and stiffness-to-weight ratios. Also natural fiber composites are currently replacing synthetic fiber composites of different applications in various engineering industries due to their low cost and eco-friendly in nature. In this work, twisted kenaf, and twisted kevlar (aramid fibers) have combined together to enhance the mechanical properties and also to improve high impact resistance with less contribution of kevlar which is very high in cost categorized next to carbon fibers. Here, the hybrid composites (natural and aramid) are done by using hand layup method, and various mechanical properties have been investigated. Also, the morphological analysis is done to observe the internal structure of the tested composite. It is observed that hybrid composite has high strength with minimum contribution of kevlar fibers.

Keywords Eco-friendly fiber · Hybrid composite · Fracture surfaces · Hand layup method

1 Introduction

In recent days, technology needs a high demand toward the growth of future generations. Hence, in order to produce a good product which needs to satisfy the customers, it should be developed at an economical way, high quality with minimum possible

V. Karthi · K. Marimuthu
Department of Mechanical Engineering, Coimbatore Institute of Technology, Coimbatore, India

G. Boopathy (✉)
Vel Tech Rangarajan Dr. Sangunthala R&D Institute of Science and Technology, Chennai, India
e-mail: ramananinjs@gmail.com

N. Ramanan
Synce Engineering Services, Chennai, India

L.-J. Yang et al. (eds.), *Proceedings of ICDMC 2019*,
Lecture Notes in Mechanical Engineering,
https://doi.org/10.1007/978-981-15-3631-1_36

time challenging among the competitors. Natural fiber-based polymer composites are the most which show potential option to synthetic fiber and powerful polymer composites. It enhances the awareness of eco-friendly material and makes it crucial to utilize natural fiber as a potential reinforcement in polymer composites. Natural fibers are plentifully existing, biodegradable, and recyclable, which makes them conventional in the automobiles industries. The major negative aspect of testing is stretched time, manufacturing cost is high, mistakenness, machine error, and today, most bio composite researchers are moving toward computational methods to model the NFRPC to simulate the mechanical and thermal properties. A study based on the tensile properties of flax/basalt hybrid composites with loading conditions were performed which indicated that the tensile properties of composites decreases with increasing aging conditions [1]. They reported that the tensile properties of composites decreased with increasing aging conditions [2]. The rigidity and modulus diminished because of embrittlement of the matrix materials. Comparative attributes were observed for sisal-fiber-reinforced PP composites [3, 4]. The affiliation between the cross-section of the natural fiber and the tensile properties were studied. They modeled and evaluated the tensile performance of hemp fiber with dissimilar elliptical cross-sections. The results appeared that the tensile properties have a physically powerful manipulate on the degree of ellipticity. The microfibril angle and the viscoelastic properties played a very important role in the geometry of the natural fiber [5]. As basalt is eco-friendly, chemically inert, highly resistant to corrosion, and very low thermal conductivity, it makes it superior to choose than any other reinforcement available today.

2 Materials and Methods

The aramid fiber is a superior material in all fields. With the ultimate mechanical properties, particularly high strength-to-weight ratio, low density, and heat resistance, aramid fiber is utilized as perfect reinforcement for polymer-based composites [6]. It has been widely used in various civilian and military fields, including aramid/rubber for tire cord and automobile rubber hose [7].

3 Hand Layup Method

In hand layup method, resins are impregnated into fabrics by manual feed which is mainly of roving form. The fibers and resins are impinged on the surface with the help of the rollers which has been fed into the mold. Then, the fibers are made free under standard atmospheric conditions. This method is very simple and easy, but it requires high skill to fabricate the composite. The fibers used are twisted kevlar, twisted kenaf, in addition to that glass-reinforced polymer fiber. Aramid fibers are laid at the top and bottom most of the composite laminate for better-finishing purpose.

Three different categories with three samples have been fabricated using hand layup method. All the fibers are laid in normal direction and kept in dry condition before it is fed to the laminate. Initially, the releasing agent is applied over the surface in order to remove the laminate easily. After applied, a thin layer of resin is applied, and then, the glass fiber is laid on the surface. A weight of 5 kg is placed over the aramid fiber to remove air bubbles if any, and it is kept undisturbed for about 3 h. Also, fibers are also dried in the normal condition to make the fibers moisture-free. After 3 h, the laminate was removed from the die material and used in various tests for engineering applications.

4 Tensile Test

Material strength can be found by testing the material in tension or compression. Standard dog-bone-shaped specimens (pure nylon and nylon composites of 10% aramid fiber content) were according to the ASTM D638 specifications for tensile testing (reinforced and unreinforced plastics). Each specimen having 30 mm width and 280 mm gauge length, The specimen is loaded in computer-controlled universal testing machine (ASE—UTN 10) until the failure of the specimen occurs. Tests are conducted on composites of different combinations of reinforcing materials, and ultimate tensile strength and ductility are measured. Simultaneous readings of load and elongation are taken at uniform intervals of a load. A tensile test is carried out at room temperature. Uni-axial tensile test is conducted on the constructed specimen to obtain information regarding the behavior of a given material under gradually increasing stress-strain conditions. The tensile stress test results give the elasticity limit of the polymer composite fiber.

The tensile test result is shown in Fig. 1. The fiber ratio differs by 30–50% in laminate plates. But, the ultimate tensile strength result shown in the 30% fiber has maximum strength. When the fiber ratio increases, the material UTS will also be increased. In the case of negative result then it will be due to the reason that fiber ration increases correspondingly the material thickness will increase. Hence,

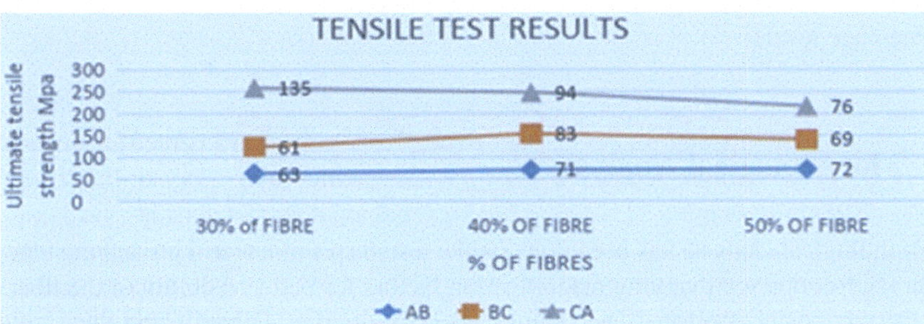

Fig. 1 Tensile test result

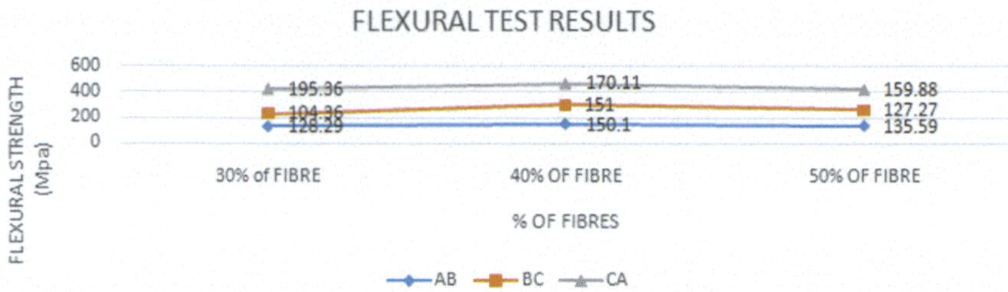

Fig. 2 Flexural test result

the failure will occur suddenly. The reason is that material failure has higher fiber percentage level.

5 Flexural Strength

Material strength can be found by testing the material in tension or compression. Test specimens are equipped according to ASTM D315 standard, each specimen having 30 mm width and 280 mm gauge span. The specimen is set up in computer-controlled universal testing machine (ASE—UTN 10) until the collapse of the specimen occurs. Tests are conducted on composites of different combinations of reinforcing materials, and ultimate tensile strength and ductility are measured. Simultaneous readings of load and elongation are taken at uniform intervals of the load. The tensile test is carried out at room temperature. The uni-axial tensile test is conducted on the fabricated specimen to obtain information regarding the behavior of a given material under gradually increasing stress–strain conditions (Fig. 2).

The flexural test result is shown in Fig. 1. The fiber ratio differs by 30–50% in laminate plates. But, the ultimate tensile strength result shown in the 30% fiber has maximum strength. When the fiber ratio increases, the material's UTS will also be increased. In the case, the negative result will be came due to the reason is the fiber ration increase corresponding the material thickness will be increase. Hence, the failure will occur suddenly. The reason is that material failure has higher fiber percentage level.

6 Morphological Analysis

Morphological analysis has been done on the tested specimens using scanning electron microscope with varying magnification factors for better visibility of the fibers in the composite laminate. The sample has been dried completely, and then, gold coating is done on the sample with appropriate device. Then, the ultraviolet rays have been passed through the laminate and were inspected and observed by scanning

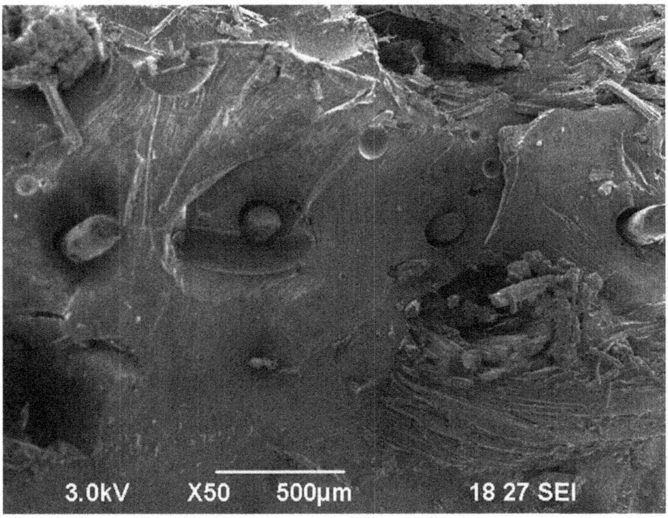

Fig. 3 Tensile fracture

electron microscope. The twisted fibers are elongated to some extent, and each of the fibers is distinctly identified due to the application of tensile load as shown in Fig. 3. Due to the elongation effect, fibers have not been affected much which leads to ductility failure. The bond between the reinforcement and the matrix medium is clearly seen.

The images 4 and 5 are shown in a fracture surface of the SEM images. Figure 3 shows a tensile fracture image. The image shows voids that occur due to the exothermic heat formation. But, the result has been shown in higher tensile strength. The voids also control the limit. Figure 4 shows the flexural fracture images. The flexural fracture test shown in the normal fiber fracture contains no voids or porosity. Hence, the images show the better strength of the laminates.

7 Conclusions

The results of various mechanical tests have been done. In this research work, twisted fibers, namely kevlar, kenaf were fabricated by using hand layup method along with kevlar-fiber-reinforced polymer composites on either side of the composite laminate for better results. Three different categories were prepared with three samples each for better analysis of results. The tests, such as tensile test and flexural test, have been conducted to determine the mechanical properties of the hybrid composites. The morphological analysis has been done to see the internal structure of the hybrid composite, and the interpretations have been discussed in detail for each test.

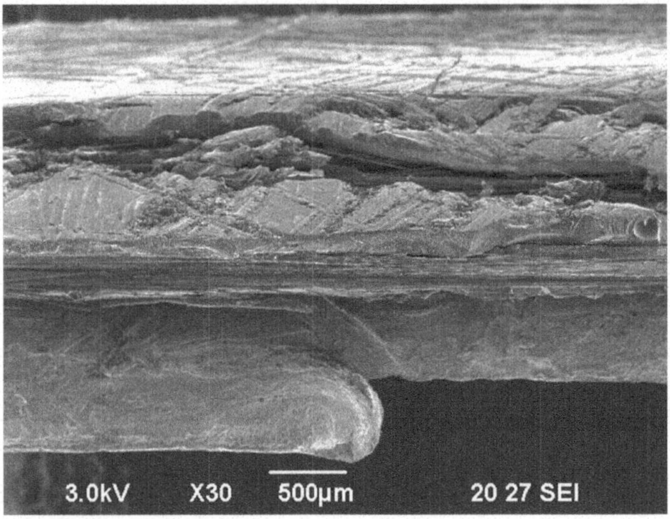

Fig. 4 Fletural fracture

References

1. Naveen J, Jawaid M, Finite element analysis of natural fiber-reinforced polymer composites. In: Modeling of damage processes in biocomposites, fibre-reinforced composites and hybrid composites. https://doi.org/10.1016/B978-0-08-102289-4.00009-6
2. Bakare FO, Ramamoorthy SK, Åkesson D, Skrifvars M (2016) Thermo mechanical properties of bio-based composites made from a lactic acid thermoset resin and flax and flax/basalt fibre reinforcements. Compos Part A
3. Azwa ZN, Yousif BF, Manalo AC, Karunasena W (2013) A review on the degradability of polymeric composites based on natural fibres. Mater Des 47:424–442
4. Palanivel A, Veerabathiran A, Duruvasalu R, Iyyanar S, Velumayil R (2017) Dynamic mechanical analysis and crystalline analysis of hemp fiber reinforced cellulose filled epoxy composite. Polímeros 27(4):309–319
5. Del Masto A, Trivaudey F, Guicheret-Retel V, Placet V, Boubakar L (2017) Nonlinear tensile behaviour of elementary hemp fibres: a numerical investigation of the relationships between 3D geometry and tensile behaviour. J Mater Sci 52(11):6591–6610
6. Gonzalez-Canche NG, Flores-Johnson EA, Carrillo JG (2017) Mechanical characterization of fiber metal laminate based on aramid fiber reinforced polypropylene. Compos Struct 172:259–266
7. Palola S, Sarlin E, Kolahgar Azari S, Koutsos V, Vuorinen J (2017) Microwave induced hierarchical nanostructures on aramid fibers and their influence on adhesion properties in a rubber matrix. Appl Surf Sci 410:145–153

Mechanical Characterization of Prosopis Juliflora Fiber-Reinforced Polymer Composites

J. Mohamed Yazzir, R. Prabu, S. Bharath, and A. Anbu Raj

Abstract Prosopis Juliflora (Kikar Wood), also known as Vilayati babul commonly known in India, is commonly available all over the country. Its branches are mainly used as wood fuel and fodder in saw dust form. The current research focuses on extracting fibers from Kikar wood by chemi-thermomechanical pulping process and using as reinforcement in Epoxy Matrix polymer composites. Several mechanical tests such as tensile test, compressive test, hardness test, and impact test were conducted to study the mechanical properties of the developed composites.

Keywords Kikar · Reinforcement · Characterization · Extraction

1 Introduction

Prosopis juliflora, also known commonly in India as Seemai Karuvel, Kabuli Kikar is spread all over India in abundance. They are mainly used as wood fuel in village parts of India and it absorbs a lot of underground water. The current research objective is to develop a natural fiber [1] from Kikar wood, after a set of treatments and to develop a natural fiber-reinforced polymer composites. And this is the first kind of research that has been done on kikar wood extracted natural fiber as fiber-reinforced polymer composites. The matrix material used is Araldite LY556 and Hardener HY990 as hardening agent in the ratio 10:1 [2]. The rule of mixtures followed for fabricating the composites is 50% fiber and 50% matrix in terms of weight. The natural fiber composite formed is of three types: Discontinuous mode, chopped stranded mat, and powdered sprinkle form by handlayup method. The fabricated composites are subjected to mechanical testing [3] and characterization such as tensile test, flexural test, impact test, and hardness and hygrothermal test according to ASTM Standards.

The impact of fiber loading and length on mechanical properties like tensile strength, flexural strength, impact strength, and hardness of composites is examined.

J. M. Yazzir · R. Prabu (✉) · S. Bharath · A. A. Raj
Mohamed Sathak A. J. College of Engineering, Chennai, India
e-mail: Prabu.ravindran6@msajce-edu.in

L.-J. Yang et al. (eds.), *Proceedings of ICDMC 2019*,
Lecture Notes in Mechanical Engineering,
https://doi.org/10.1007/978-981-15-3631-1_37

1.1 Synthesis of Fiber

Mostly, fiber selection is based on the availability, simple extraction process and mechanical properties fibers are chosen. In this research work, the fiber selection is based on the environmental factor that should not affect the ecological system. In Tamilnadu, karuvellam is major uproot tree than others because it has many disadvantages. Kikarwood (Prosopis juliflora) natural fibers have excellent mechanical strength but due to lack of knowledge, it is still not utilized properly. It can be used in various applications like sound absorber, thermal insulator, and many applications. There are some methods to extract the fiber from kikarwood stem (Table 1; Fig. 1).

In most of the wood fiber, there are three portions: cellulose, hemicellulose, and lignin. Cellulose is cellwall of the plant. Hemicellulose is the mixture of lignin and cellulose. Lignin is the natural glue that bound the fiber together. Here, chemithermomechanical pulping process is followed for the extraction of fibers. Here, wood chips are pre-treated with sodium sulfite and sodium hydroxide [4, 5] prior to refining. In order to retain lignin, the chemical treatment [6] is done at a lower temperature,

Table 1 Fiber representations

A1	A2	A3
Chopped fabric mat	Powdered fiber-type construction	Discontinuous-type fiber

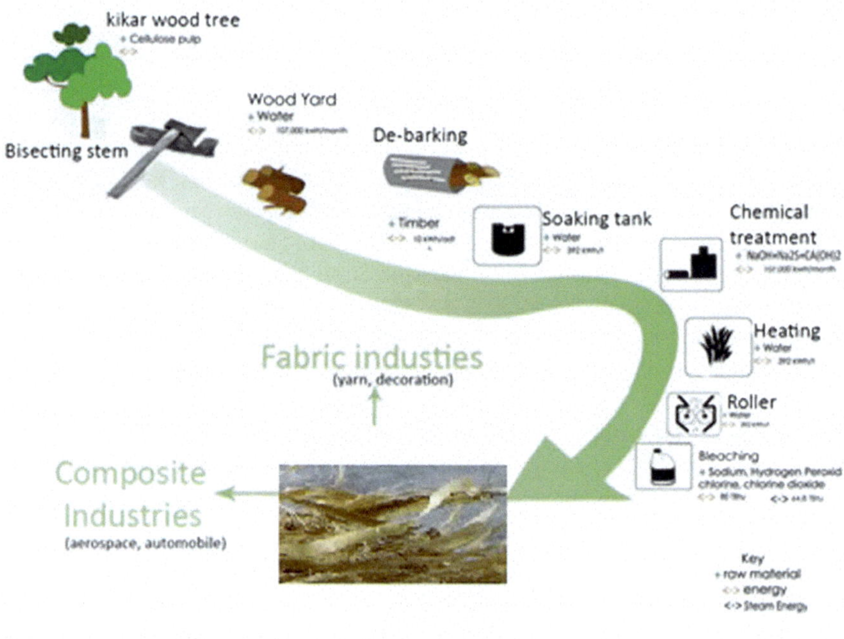

Fig. 1 Fiber extraction from kikar wood

Fig. 2 Different fiber types

shorter time, less extreme pH. The stems of kikar wood are bisected and soaked in a bucket full of water and then followed by chemical treatment. In this process, the bisected woods are treated with chemicals such as sodium hydroxide, acetone, and calcium hydroxide for 2 days. This chemical-treated fiber is heated in the boiler at a temperature of 120 °C for 5 h. After the heat treatment, the bleaching process is carried out for complete removal of lignin in the wood [7]; then it is subjected to the mechanical process which having the machine is the combination of two rollers: feed roller or mechanical pressing machine. Then, the fiber is separated by manual process (Fig. 2).

The fibers extracted are arranged in three forms. In powdered fiber-type construction, kikar wood in powder form is extracted from the fiber by mechanical milling process and used as a layer of construction with discontinuous fiber. Then chopped mat fabric construction is done manually with the obtained discontinuous fiber. Here, the reinforcement and matrix are used in the ratio 100 gm fiber and 100 ml epoxy resin for obtaining 50% fiber and 50% matrix.

2 Results and Discussion

The tests were carried out in Delta Laboratory, Chennai, and the tests were done on three types of specimen discontinuous fiber type, chopped stranded mat type, and powdered fiber type of polymer composites. And three samples were used for testing and obtaining data for each type of specimen.

2.1 Tensile Test

The tensile test [8] was carried out in universal testing machine with ASTM standard D3039. The fabricated composite was cut to get the specimen size of 250 mm × 85 mm × 25 mm for tensile testing. The results showed that discontinuous-type composite had a tensile strength of 11.58 MPa. But the results improved when powdered fiber composite was tested and the tensile strength was 12.33 MPa and for chopped fabric mat type, the tensile strength improved by 21% approximately (Figs. 3 and 4).

Fig. 3 Tensile test different samples

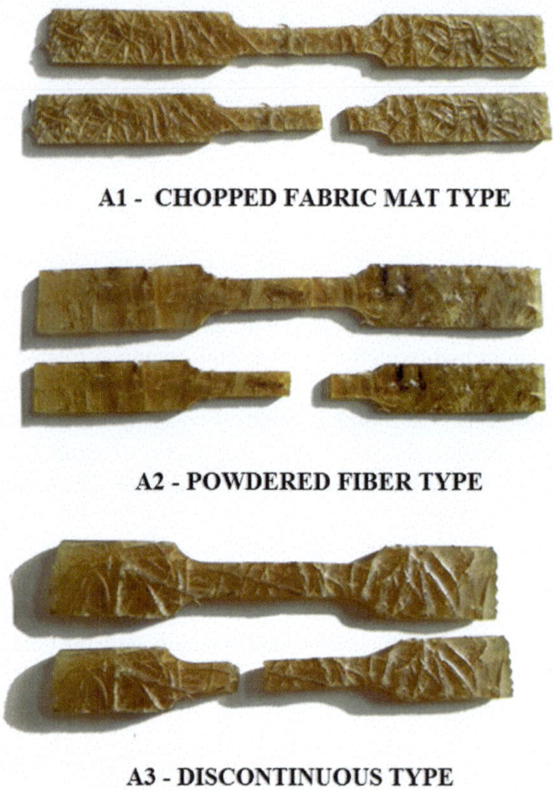

Fig. 4 Tensile test results for different fiber-reinforced composites

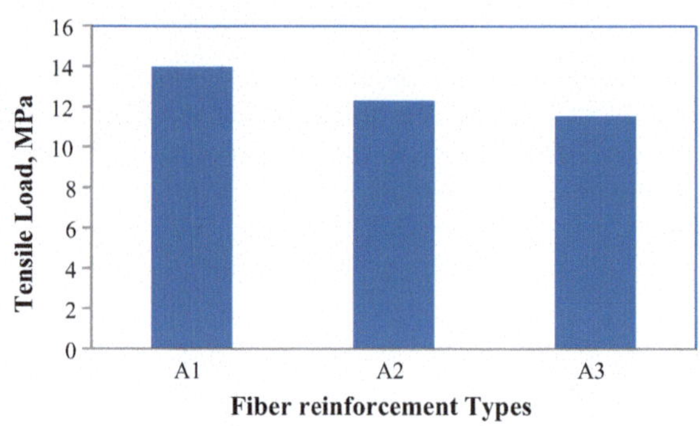

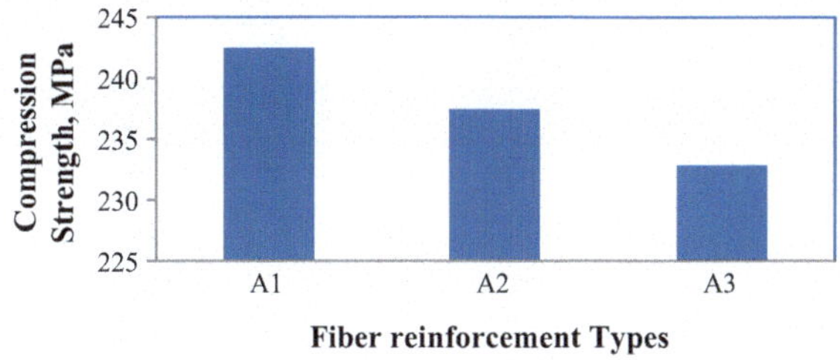

Fig. 5 Compression test results for different fiber-reinforced composites

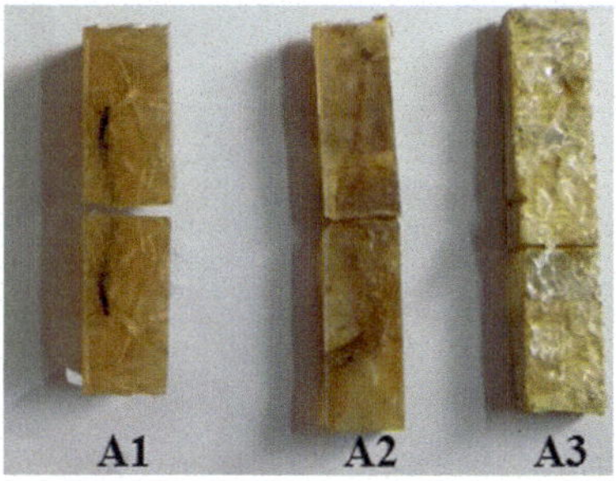

Fig. 6 Impact test different samples

2.2 *Compression Test*

The compression test [9] was carried out in universal testing machine with ASTM standard D3410. The test specimen was cut to ASTM Standards of size 150 mm × 7 mm × 25 mm. The maximum compression strength was observed in chopped fabric mat composites (Fig. 5).

2.3 *Impact Test*

The impact test was carried out using Charpy testing machine. The impact test specimens are prepared according to the required dimension, following the ASTM-D256 standard. The specimen size of impact test is 64 mm × 12.7 mm × 6.4 mm. During the testing process, the results indicated that the maximum impact strength is obtained for chopped mat fabric polymer composite (Figs. 6 and 7).

2.4 *Hardness Test*

The Rockwell harness test carried out for the Kikar wood polymer composite material and the ASTM D785 standard and load applied is 100 kg for1/16 inch. Penetrator, the B scale reading is 96.67 HRB for chopped mat fabric-type composites. It is the

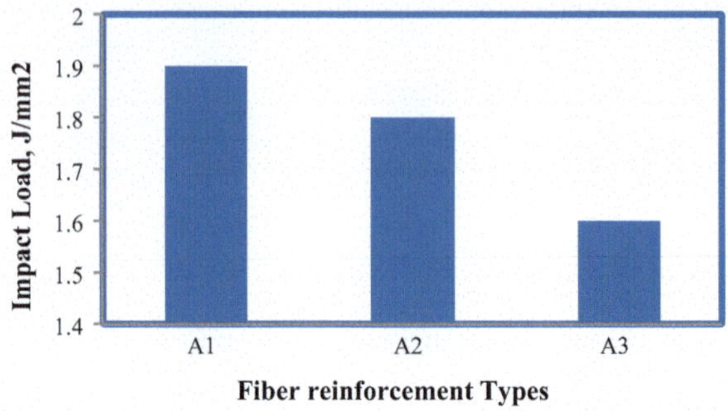

Fig. 7 Impact test results for different fiber-reinforced composites

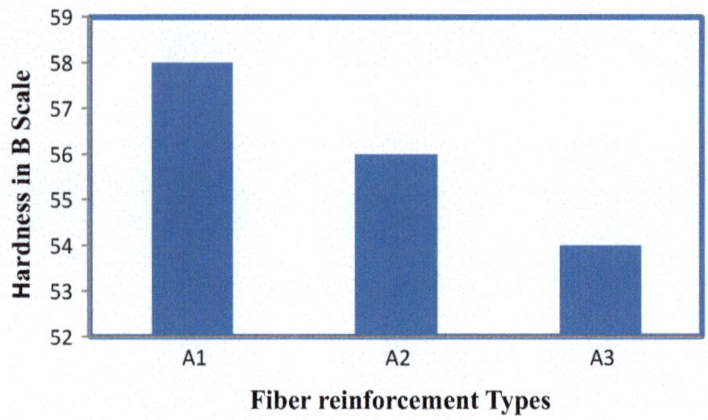

Fig. 8 Hardness value for different fiber-reinforced composites

maximum hardness reading obtained for the three types of composites. The specimen was fabricated to a size of 0 mm × 30 mm × 6.4 mm (Fig. 8).

2.5 *Water Absorption Test*

The water absorption tests [10] were performed on the two samples from three kikar wood composites in three different conditions such as sea water, ordinary tap water, and distilled water. The water absorption test was carried out as per ASTM standards D570 specimen size of 25 mm × 25 mm × 3 mm. The water absorption test was carried in three types of water such as ordinary tap water, sea water [11], and distilled water. The samples were uniformly dried at 100 °C for 24 h. Then the specimens were weighed to an accuracy of 0.001 g. Water absorption tests were conducted by immersing the composites specimens in ordinary tap water, sea water, and distilled water in a plastic tub at room temperature for a period of 14 days' time interval (Figs. 9 and 10).

After 14 days 'time interval, the samples were removed from water carefully, blotted to remove the excess water on the surface and wiped with a clean cloth and the specimens were immediately weighed 0.001 g. The moisture absorption

Fig. 9 Specimens for water absorption test

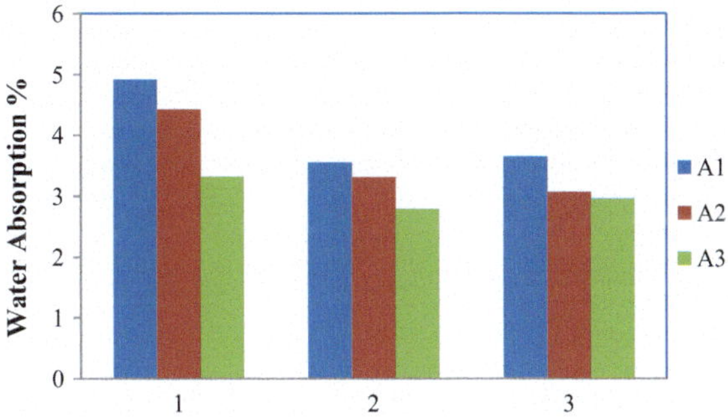

Fig. 10 Water absorption test results for different fiber-reinforced composites

was calculated by the weight difference of the samples. The percentage of water absorption of the three samples was calculated according to the formula.

$$\text{Water absorption percentage } (\%) = W_1 / W_2$$

where W_1 and W_2 are the weights of the dry and wet samples.

From the overall mechanical tests, it is observed that the chopped mat fiber has better mechanical properties than powdered-type fiber and discontinuous fiber. This is due to the orderly arrangement of the fiber in the orientation 0° and 90°, in the other cases like discontinuous fiber; the fibers are dispersed and randomly orientated. In powdered fiber type, the Kikar wood is in powdered form and the result was better than discontinuous fiber-reinforced composites. Also, the test results showed that the mechanical properties obtained were very low as the reinforcement or the interface between the fiber and matrix is poor.

3 Conclusion

Kikar wood (Prosopis juliflora) fiber is very common in tropical regions and very simple to extract the fibers from the stem. From the mechanical tests, we can understand that chopped mat fabric type has better properties than other forms of kikar wood reinforced polymer composites. Also the mechanical properties obtained are very low, further research can be made on improving the bonding of fiber by further chemical treatment of fiber for better adhesion with the matrix. Utilization of kikar wood fiber in composite material is a new source of materials which can be economic, ecofriendly, and recyclable. Also it can be used as an additional reinforcement along with glass fibers [12] for making composites.

References

1. Sen T, Reddy HNJ (2011) Finite element simulation of retrofitting of RCC beam using sisal fibre composite (natural fibre). In: 2011 3rd international conference on electronics computer technology (ICECT), vol 6. pp 29–33. 8–10 Apr 2011
2. Seviaryna I, Bueno HG, Maeva E, Tjong J (2014) Characterization of natural fibre-reinforced composites with advanced ultrasonic techniques. In: 2014 IEEE international ultrasonic's symposium (IUS), pp 1428–1431. 3–6 Sept 2014
3. Oksman K, Skrifvars M, Selin J-F (2003) Natural fibres as reinforcement in polylactic acid (PLA) composites. Compos Sci Technol 63(9):1317–1324
4. Murali B, Nagarani J (2013) Design and fabrication of construction helmet by using hybrid composite material. In: 2013 international conference on energy efficient technologies for sustainability (ICEETS), pp 145–147. 10–12 Apr 2013
5. Herrera-Franco PJ, Valadez-Gonzalez A (2005) A study of the mechanical properties of short natural-fiber reinforced composites. Compos Part B: Eng 36(8):597–608
6. Palanivel A, Veerabathiran A, Duruvasalu R, Iyyanar S, Velumayil R (2017) Dynamic mechanical analysis and crystalline analysis of hemp fiber reinforced cellulose filled epoxy composite. Polímeros 27(4):309–319
7. Anand P, Anbumalar V (2017) Investigation on thermal behavior of alkali and benzoyl treated hemp fiber reinforced cellulose filled epoxy hybrid green composites. Cellul Chem Technol 51(1–2):91–101
8. Ku H, Wang N, Pattarachaiyakoop N, Trada M (2011) A review on the tensile properties of natural fiber reinforced polymer composites. Compos Part B: Eng 42(4):856–873
9. Chandramohan D, Marimuthu K (2011) Applications of natural fibre composites for replacement of orthopaedic alloys. In: 2011 international conference on nano science, engineering and technology (ICONSET), pp 137–145. 28–30 Nov 2011
10. Rahman WA, Sudin SNA, Din SN (2012) Physical and mechanical properties of Pandanus amaryllifolius fibre reinforced low density polyethylene composite for packaging application. In: 2012 IEEE symposium on humanities, science and engineering research (SHUSER), pp 345–349, 24–27 June 2012
11. Houlgate RG (1992) The performance of composite insulators in a coastal environment. In: IEE colloquium on non-ceramic insulators for overhead lines, pp 3/1, 3/4. 26 Oct 1992
12. Wambua P, Ivens J, Verpoest I (2003) Natural fibres: can they replace glass in fibre reinforced plastics? Compos Sci Technol 63(9):1259–1264

Characterization of Free Vibration Damping Properties in Hybrid Composites

M. Dinesh, R. Asokan, S. Vignesh, Kommana Tarun Kumar, and Nallamilli Sampath Kumar Reddy

Abstract The study mainly focuses on damping behavior of fiber-reinforced composite materials with various fiber compositions. In this paper, the usage of flax fiber was proposed to enhance the damping property of carbon fiber-reinforced polymer (CFRP) composites. The hybrid composites are fabricated with the composition of carbon fiber, glass fiber, and flax fiber. The free vibration test was carried out for various laminate combinations, and damping ratio was determined using half bandwidth method. The effect of damping behavior of hybrid composite with variation in lamina stacking sequence was investigated. The experimental analysis result that the pure flax fiber-reinforced polymer (FFRP) shows the highest damping ratio of 0.1. While comparing the hybrid composite materials, the set 6 laminate combination exhibits a higher damping ratio of 0.0952. The results evidently show that set 6 laminate combination exposes an enhanced damping properties compared to other hybrid composite materials. This phenomenon shows that damping property of CFRP was improved significantly by laying flax fibers on the outmost layers of the composites.

Keywords A. Hybrid composites · B. Half bandwidth method · C. Damping ratio · D. Damped natural frequency

M. Dinesh (✉) · R. Asokan · S. Vignesh · K. T. Kumar · N. S. K. Reddy
School of Aeronautical Science, Hindustan Institute of Technology and Science, Chennai, India
e-mail: dineshf22@gmail.com

R. Asokan
e-mail: rgasokan@gmail.com

S. Vignesh
e-mail: vignesh04@live.com

K. T. Kumar
e-mail: tarunkommana96@gmail.com

N. S. K. Reddy
e-mail: sampathkumar1812@gmail.com

L.-J. Yang et al. (eds.), *Proceedings of ICDMC 2019*,
Lecture Notes in Mechanical Engineering,
https://doi.org/10.1007/978-981-15-3631-1_38

1 Introduction

Carbon fiber-reinforced polymer composites (CFRP) have been widely used in aerospace, construction, automotive, marine industries and sports equipment due to their excellent mechanical and physical properties. However, the high stiffness of CFRP leads to relatively low damping properties [1, 2]. This restricted their applications in areas involving vibration, periodic loading, and external impacts. Vibration damping helps in the elimination of mechanical energy by converting into other forms of energy. Previous studies also show that this benefits in noise reduction and long life of the CFRP structure [3]. Therefore, to enhance the damping property of CFRP, a few methods have been used or proposed, which include adding co-cured viscoelastic layers, fiber coating, hybridization with other fibers, or addition of nanofillers [4]. Hybridization with other fibers is simple and direct method to enhance the damping property. Hybrid of more than one type of composite can result in a shared property of both the fibers. It can be seen that plant fibers display many distinct advantages over its synthetic counterpart, like low density, better sound absorption and thermal insulation, and relatively high damping properties [5, 6]. Flax is one such natural fiber. Therefore, hybridizing flax fibers with carbon fiber may improve the damping property of the composites and at the same time may lead to increased specific mechanical properties [7, 8]. The vibration test methods have been successfully used to estimate the mechanical and physical properties of wood and wood-based composites for several decades. These methods are non-destructive, rapid, convenient, and cost-effective. Previous experiments and researches have proven that cantilever beam vibration (CBV) technique to be the reliable test method [8, 9]. We can figure out damping ratio (ζ) by using the frequencies through the half bandwidth method. The damping ratio of a material is used to evaluate the capability of absorbing vibration energy, which can be used to reflect resistance to dynamic loading [10]. The higher the ζ value, the smaller the vibration amplitude a material has and less appreciable the energy loss. In this study, CBV technique was the preferred approach to obtain the vibration characteristics of small size specimen. The nine different laminate combinations were fabricated using carbon, glass, and flax fiber. These laminates were formed into beams according to ASTM E756. The damping ratio (ζ) value for various composite materials was determined by free vibration test.

2 Experimental Procedure

2.1 Materials Selection and Fabrication

Bidirectional carbon fiber, glass fiber, and flax fiber were used to fabricate the laminates. The material density of carbon fiber, glass fiber, and flax fiber was 1.76, 2.55, and 1.5 kg/m^3, respectively, with the areal density of 200 g/m^2. Epoxy LY 556 resin, chemically belonging to the epoxide family and hardener HY 951 with viscosity of

Table 1 Properties of carbon fiber, glass fiber, and flax fiber

Fiber	Density (g/cm^3)	Tensile strength (MPa)	Young's modulus (GPa)
Carbon	1.76	3500	230
Glass	2.5–2.6	2000–3500	70–76
Flax	1.50	345–1100	27.6

10–20 poise at 250 °C was utilized as the matrix. The mechanical properties of the various fibers are shown in Table 1.

Hand layup technique was used to fabricate the composites since it was found to be easy and cost-effective compared to other techniques [11]. It is the least expensive open molding method because of its low equipment cost. Fiber reinforcements are placed by hand in a mold, and resin is applied with a brush or a roller. Production volume per mold is very low, and however, it is feasible to produce substantial production quantities using multiple molds. Hand layup is the least complex composite trim technique. A composite material is a combination of fiber and matrix. The function of the matrix is to bond the fibers and to transfer the loads between them, and they determine the resistance to the degradation [12]. The percentage of fiber and matrix is taken as 70% and 30%, respectively. A plane wooden surface is kept on the table and is cleaned thoroughly, and this is used as the mold surface. An OHP sheet of minimum 100 microns is placed over it and releasing wax is applied over this. Epoxy resin LY 556 is mixed with the hardener HY 951 in the ratio of 10:1 by weight as calculated. The mixture gel is applied over the surface in order to provide a duplicate coat. The Epoxy resin must be catalyzed 2% by weight. Each fiber layer is pressed by hand with rollers to work the Epoxy resin. Required stack up sequence had been arrived by adding up the fiber layers as shown in Fig. 1 and allowed to cure at room temperature. Once the curing process was completed, the laminate was released from the mold. Water jet cutting process was carried out on fabricated laminates to achieve the required dimension according to ASTM E756 as shown in Fig. 2.

2.2 *Free Vibration Test*

2.2.1 Damping Property

Damping describes the trend of materials to reduce the amplitude of the oscillations in an oscillatory system. Damping properties of composite materials can be evaluated by damping ratio which is essential for engineering applications [13, 14]. An experimental setup was developed, as shown in Fig. 3. The cantilever beam technique was followed in this study where the beam was fixed at one end and the other end was left free [15, 16]. The free length of the beam was 170 mm from the clamped

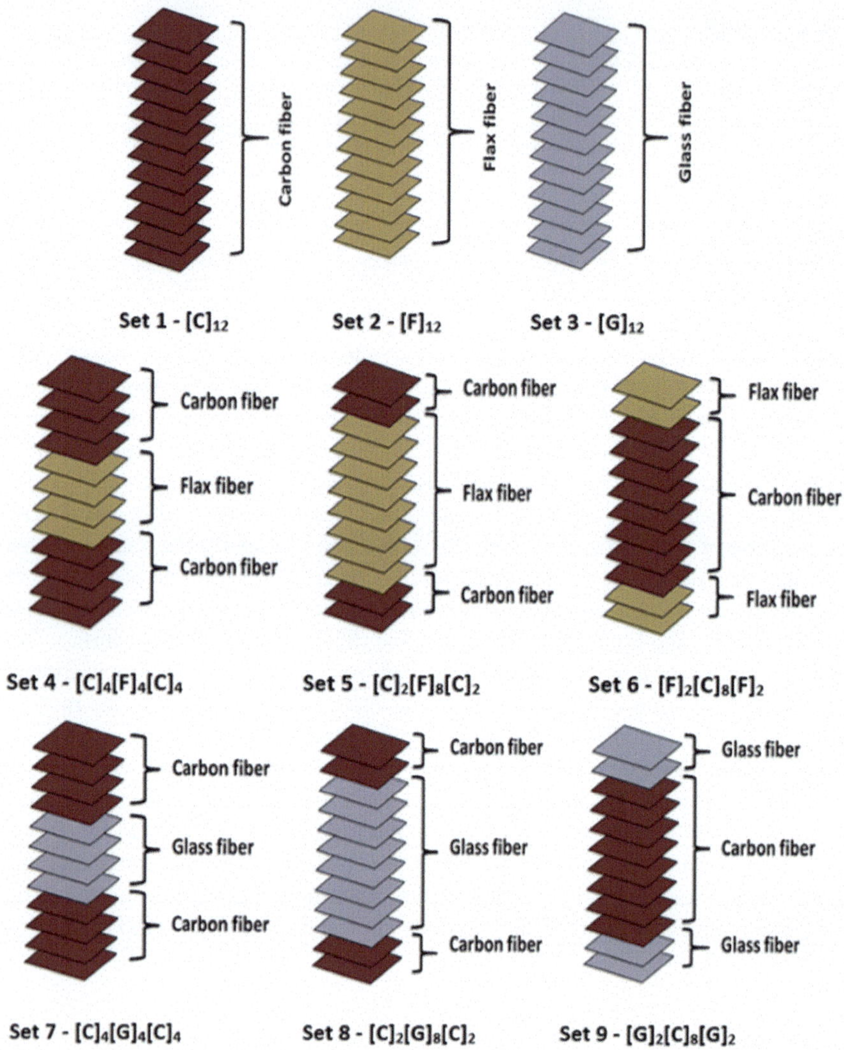

Fig. 1 Stacking sequence of various composite laminates

point. This free vibration test was carried out according to the ASTM E756 as shown in Fig. 4.

The setup consists of a test specimen, meter scale and holding stand, piezo-electric accelerometer, high-speed USB carrier, *C*-clamp and a computer with LabVIEW software. The free end of the beam was displaced 1 cm downward and released to set the beam in vibration. The vibrations were then sensed by the accelerometer, and graphs between time domain and frequency domain were plotted on LabVIEW software. This procedure was then repeated for all the specimens. The natural frequency and

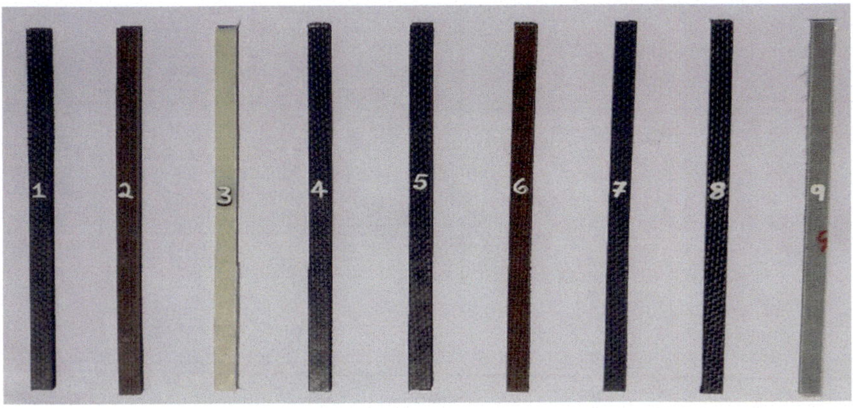

Fig. 2 Specimens for free vibration test

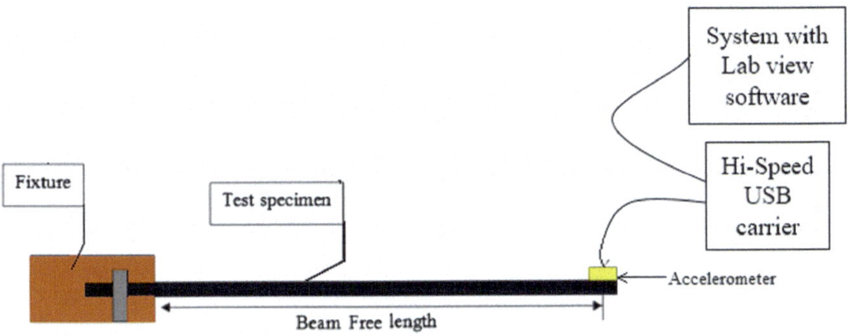

Fig. 3 Schematic diagram of free vibration test setup

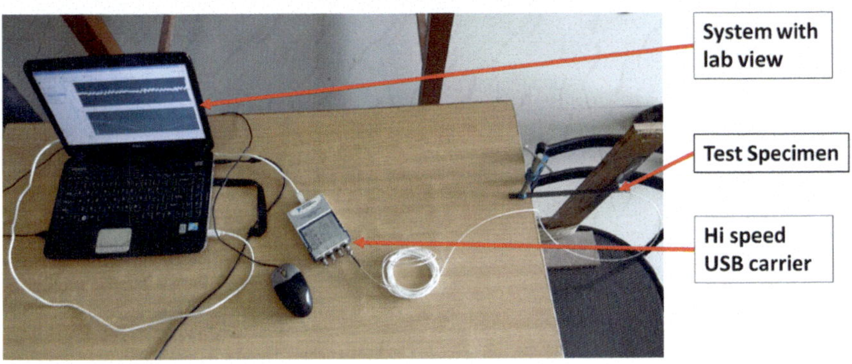

Fig. 4 Experimental setup of free vibration test

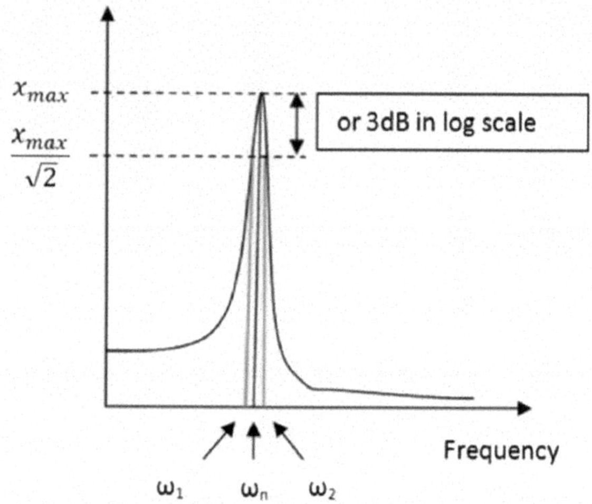

Fig. 5 Definition of variables in half bandwidth method

damping ratio values for all the different combinations of hybrid test specimens are then calculated and tabulated. Natural frequency is calculated using the magnitude versus frequency graph.

The damping ratio ζ is determined using half bandwidth method using the given relation,

$$\zeta = \frac{\Delta\omega}{2\omega_n} \tag{1}$$

where $\Delta\omega$ is the difference between frequencies ω_1 and ω_2 corresponding to half power points around the fundamental natural frequency (ω_n) as shown in Fig. 5.

3 Results and Discussion

3.1 Acceleration Versus Time Domain Graphs for Various Specimens

The various specimens were tested and the Acceleration Versus Time domain graphs are obtained from the LabVIEW software as shown in below Figs. 6, 7 and 8.

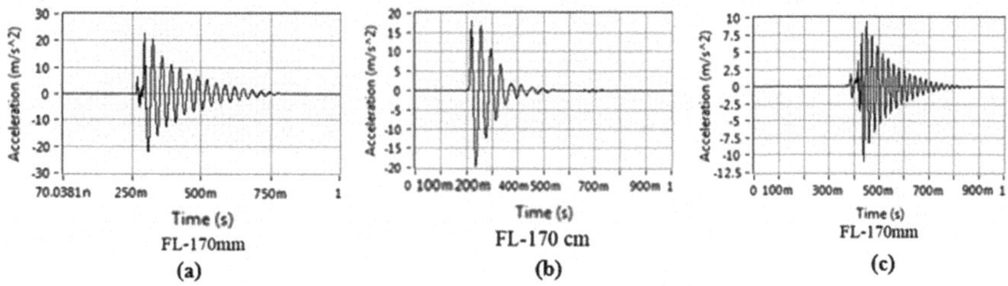

Fig. 6 Acceleration versus time domain plot for **a** Set 1, **b** Set 2, and **c** Set 3

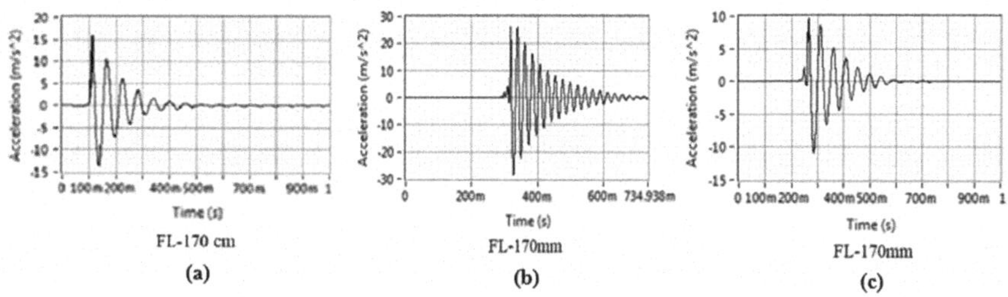

Fig. 7 Acceleration versus time domain plot for **a** Set 4, **b** Set 5, and **c** Set 6

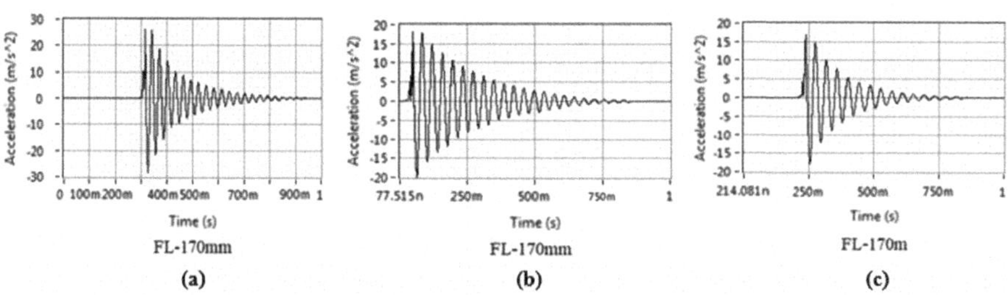

Fig. 8 Acceleration versus time domain plot for **a** Set 7, **b** Set 8, and **c** Set 9

3.2 Natural Frequency and Damping Ratio

The natural frequencies of various specimens determined from the experimental test and the damping ratios of the composite laminates were calculated according to the half bandwidth method using frequency response function. The damping ratio and natural frequency values are calculated and listed in Table 2, and a graph has been plotted as shown in Fig. 9 for various hybrid composites.

From analyzing the results, it was found that pure glass fiber-reinforced polymer (GFRP) specimen has the highest natural frequency of 47 Hz for the given free beam length of 170 mm, whereas set 3 shows the least natural frequency of 17 Hz. While comparing the results of hybrid composites, the set 5 laminate shows high natural frequency of 44 Hz. On comparing the damping ratio, it was observed that

Table 2 Natural frequency and damping ratio of various laminate combinations

Type	Natural frequency (Hz)	Damping ratio (ζ)
Set 1	30	0.050
Set 2	25	0.1
Set 3	47	0.042
Set 4	17	0.058
Set 5	44	0.056
Set 6	21	0.095
Set 7	33	0.045
Set 8	27	0.037
Set 9	24	0.043

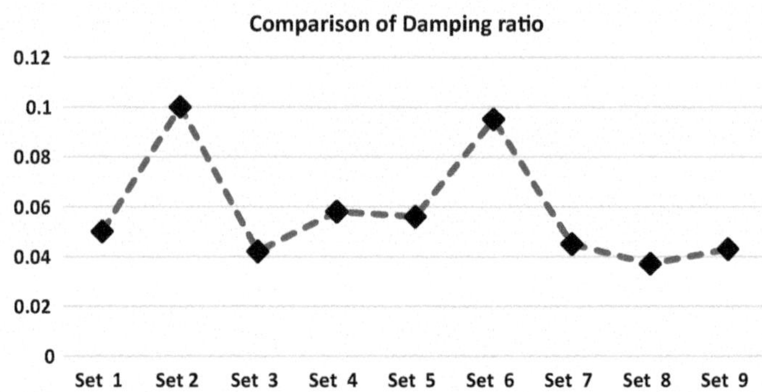

Fig. 9 Damping ratio for various specimens

the material which has high damping ratio value has good damping properties. In observing so, among pure composite materials, FFRP shows the higher damping ratios which means it easily comes back to rest position with respect to time when it is subjected to initial vibration displacement. In the hybrid combination, set 4 and set 5 laminate combinations behave in a similar manner, but when flax fiber was replaced with glass fiber in the above two combinations, the damping ratio value is very low. Set 6 laminate combination with flax fiber in the outer layer shows higher damping ratio comparatively than all hybrid combinations. This analysis shows that laminates with flax fiber in the hybrid composition had shown higher damping ratio as compared to laminates with glass fiber in the hybrid composition.

3.3 *Damped Natural Frequency*

The damped natural frequency, ω_d, depends on both the natural frequency, ω_n, and the damping ratio (ζ) according to the following equation. Since the value of ω_n and

Table 3 Damped natural frequency of each specimen

Type	Damped natural frequency (Hz)
Set 1	29.96
Set 2	24.87
Set 3	46.95
Set 4	16.97
Set 5	43.92
Set 6	20.90
Set 7	32.96
Set 8	26.98
Set 9	23.97

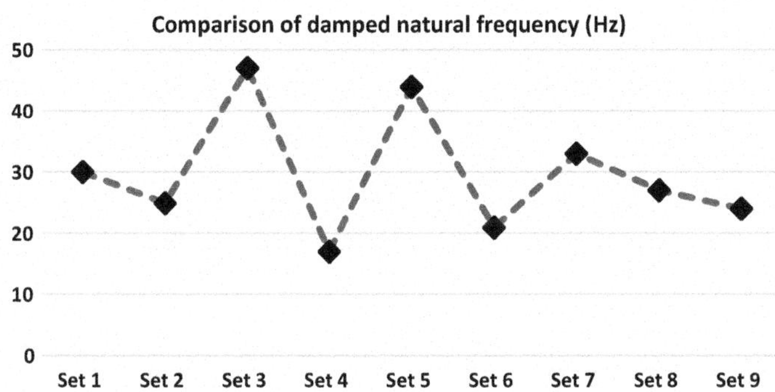

Fig. 10 Comparison of damped natural frequencies for various specimens

ζ were already calculated, ω_d can be obtained just by substituting the values. The calculated values for damped natural frequency are tabulated below (Table 3).

$$\omega_d = \omega_n \sqrt{\left(1 - \zeta^2\right)} \tag{2}$$

The variation of damped natural frequency for different specimens is plotted in below Fig. 10.

4 Conclusion

The vibration damping properties of pure composites and hybrid composites containing flax, carbon fiber, and glass fiber were studied based on the free vibration test. Initially, the various composite laminates are fabricated according to the ASTM standards. The free vibration test was conducted to determine the natural frequency for various specimens and the frequency response curve was extracted from LabVIEW software. The damping ratio of various specimens was determined using half bandwidth method. On comparing the damping ratios, pure FFRP specimen exhibits highest damping ratio of 0.1 and set 8 laminate combination shows the least damping ratio of 0.037. From the analysis of hybrid composite materials, the damping ratio of carbon/flax hybrid composites are higher compared to carbon/glass hybrid composites and it was evident that hybrid composite with combination with flax fiber at outer layer exhibits higher damping ratio. Summarizing the study, composites material with superior vibration damping properties could be produced by employing flax fibers as a composition in hybrid composite.

References

1. Khan S, Li C, Siddiqui N, Kim J-K (2011) Vibration damping characteristics of carbon fiber-reinforced composites containing multi walled carbon nanotubes. Compos Sci Technol 71:1486–1494
2. Prabhakaran S, Krishnaraj V, Senthil kumar MS, Zitoune R (2014) Sound and vibration damping properties of flax fiber reinforced composites. Procedia Eng 97:573–581
3. Sargianis JJ, Kim H-I, Andres E, wan Suhr J (2013) Sound and damping characteristics in natural material based sandwich composites. Compos Struct 96:538–544
4. Botelho EC, Campos AN, de Barros E, Pardini LC, Rezende MC (2005) Damping behavior of continuous fiber/metal composite materials by the free vibration method. Compos Part B: Eng 37:255–263
5. Palanivel A, Veerabathiran A, Duruvasalu R, Iyyanar S, Velumayil R (2017) Dynamic mechanical analysis and crystalline analysis of hemp fiber reinforced cellulose filled epoxy composite. Polímeros 27:309–319
6. Li Y, Cai S, Huang X (2017) Multi-scaled enhancement of damping property for carbon fiber reinforced composites. Compos Sci Technol 143:89–97
7. Yan L, Chouw N, Jayaraman K (2014) Flax fibre and its composites—a review. Compos Part B: Eng 56:296–317
8. Wang Z, Li L, Gong M (2012) Measurement of dynamic modulus of elasticity and damping ratio of wood-based composites using the cantilever beam vibration technique. Constr Build Mater 28:831–834
9. Li N, Tahar MB, Aboura Z, Khellil K (2018) A vibration-based identification of elastic properties of stitched sandwich panels. J Compos Mater. https://doi.org/10.1177/0021998318788141
10. Treviso A, Van Genechten B, Mundo D, Tournour M (2015) Damping in composite materials: Properties and models. Compos Part B: Eng 78:144–152
11. Murali Mohan Rao K, Mohana Rao K, Ratna Prasad AV (2010) Fabrication and testing of natural fibre composites: Vakka, sisal, bamboo and banana. Mater Des 31:508–513
12. Le Guen M-J, Newman RH, Fernyhough A, Staiger MP (2014) Tailoring the vibration damping behaviour of flax fiber-reinforced epoxy composite laminates via polyol additions. Compos Part A: Appl Sci Manuf 67:37–43
13. Chandra R, Singh SP, Gupta K (1999) Damping studies in fiber-reinforced composites—a review. Compos Struct 46:41–51
14. Grover GK (2009) Mechanical Vibrations. Nem Chand & Bros, Roorkee
15. E33 Committee Test Method for Measuring Vibration-Damping Properties of Materials. https://doi.org/10.1520/E0756-04
16. Ben BS, Ben BA. Damping measurement in composite materials using combined finite element and frequency response method

Effect of Wire EDM Process Parameters on Machining of Aluminium Matrix Composites (356/Fly Ash)

J. Udaya Prakash, S. Ananth, S. Jebarose Juliyana, and P. John Paul

Abstract Aluminium matrix composites (AMCs) of 3, 6 and 9% fly ash as reinforcement and aluminium cast alloy (356) as matrix were fabricated using stir casting technique. The effects of wire EDM process parameters on machining of AMCs were investigated using Grey relational analysis. In this study, the input machining parameters such as gap voltage, pulse on time, pulse off time, wire feed and reinforcement percentage are optimized with considerations of multiple performance characteristics of the output parameters of MRR and SR. Optimal process parameters were selected based on Grey relational grades and confirmation experiments were conducted. ANOVA was used to find the contribution of each machining parameter. The results show that pulse on time has the most significant effect for achieving maximum GRG and is followed by gap voltage, wire feed, pulse off time and reinforcement in that order. The WEDM process was improved through this approach.

Keywords Aluminium matrix composites · ANOVA · Grey relational analysis · Stir casting · WEDM

1 Introduction

Aluminium Matrix Composites (AMCs) are the most commonly used metal matrix composites (MMCs). Aluminium–Silicon (Al–Si) alloys are frequently used as matrices in AMCs [1]. They are usually reinforced by Al_2O_3, SiC, carbon, boron carbide

J. Udaya Prakash (✉) · S. Jebarose Juliyana
Department of Mechanical Engineering, Vel Tech Rangarajan Dr. Sagunthala R&D Institute of Science and Technology, Chennai, India
e-mail: udayaprakashj@veltech.edu.in

S. Ananth
Department of Mechanical Engineering, Malla Reddy College of Engineering, Secunderabad, India

P. John Paul
Department of Electronics and Communication Engineering, Malla Reddy College of Engineering, Secunderabad, India

L.-J. Yang et al. (eds.), *Proceedings of ICDMC 2019*,
Lecture Notes in Mechanical Engineering,
https://doi.org/10.1007/978-981-15-3631-1_39

and also with recently fly ash. The demand for composites containing low cost and low-density reinforcements is increasing day by day [2, 3]. Fly ash is obtained from thermal power plants as a waste by-product. It is one of the cheapest reinforcements existing in large quantities and also its density is very less. So, fly ash added composites are extensively used in automotive applications as it improves the hardness, stiffness, wear resistance and damping properties. Aluminium–fly ash composites can be used as cases, pulleys, covers, valve covers, brake rotors, pans, manifolds and engine blocks in automotive, electromechanical and small engine manufacturing industries [4]. Stir casting is the cheapest and simplest manufacturing method [5, 6]. Wire EDM is an unconventional manufacturing process for machining hard materials of any intricate shape [7].

2 Materials and Methods

2.1 Materials Used

Materials are chosen based on the required properties and applications. The matrix and reinforcement materials chosen for this research work are aluminium alloy 356 and fly ash of 12 μm particle size. Optical Emission Spectrometer is used for finding the chemical composition of the aluminium alloy as per the ASTM E 1251-07 standard and given in Table 1.

2.2 Fabrication of 356/Fly Ash Composites

Stir casting is used for manufacturing the composites at various weight percentages (3, 6 and 9% fly ash) The small pieces of aluminium alloy ingots were melted in an electric furnace to a temperature of 850 °C using a graphite coated crucible, then fly ash particles are introduced and stirred for 10 min at a speed of 600 rpm. Magnesium (1% wt) was added to ensure good wettability between fly ash and aluminium alloy. Finally, the slurry is poured into permanent mould, which is preheated to 650 °C [8, 9].

2.3 Taguchi Design of Experiments

Taguchi technique is a statistical technique, in which the experiments are designed using orthogonal arrays (OA). By using OAs very few experiments are used than full factorial experiments. In traditional approach, one factor only studied at a time, but many factors each at various levels are studied using Taguchi's DOE. High quality

Table 1 Chemical composition of aluminium alloy (356)

Constituent	Si	Cu	Fe	Mg	Ml	Ti	Mi	Zn	Al
Weight (%)	7.08	0.13	0.49	0.35	0.03	0.06	0.01	0.02	Remainder

is incorporated in the input stage itself and the product cost is very much reduced because of the reduction in number of experiments in DOE [11].

2.4 WEDM

Wire EDM is a non-contact, spark erosion, electrothermal machining process, which uses a thin electrical conducting wire in combination with de-mineralized water to machine the composite material. Proper selection of parameter avoids wire breakage. Hence, a systematic and mathematical approach is required by researchers to find out the optimal process parametric settings to attain the maximum machining performance from the WEDM process [12].

3 Results and Discussion

The WEDM experiments are conducted on 356/fly ash AMCs using Taguchi's parametric approach and optimization was done using Grey relational analysis with the objective of achieving higher material removal rate (MRR) and less surface roughness (SR) values simultaneously. Optimal process parameters were selected based on Grey relational grades and confirmation experiments were conducted.

3.1 Experimental Results

The machining of composites was done using Taguchi's DOE method, choosing L_{27} orthogonal array. The experimental results along with their GRG and Rank are given in Table 2.

3.2 Analysis and Discussion of Results

Grey relational analysis is used for analysing the multi-objective problem of MRR and SR of the wire EDM process, the Grey relational grades (GRG) and ranks were calculated using the standard formulae. The response graphs clearly depict the effects of input parameters on the output characteristics, such as MRR and SR. The analysis of variance of GRG data is carried out to identify the important variables and to find their contribution on the response characteristics.

Figure 1 shows that the GRG increases with the increase in pulse on time, pulse off time and wire feed and decreases with the increase in gap voltage and reinforcement. This is because of the reason that, when the pulse on time is increased, then the

Table 2 Experimental results

Ex. No.	A	B	C	D	E	MRR (mm^3/min)	SR (μm)	GRG	Rank
	Gap voltage (V)	Pulse on time (μs)	Pulse off time (μs)	Wire feed (m/min)	Reinforcement (wt%)				
1	30	2	2	4	3	25.829	3.84	0.541	18
2	30	2	6	6	6	20.022	3.34	0.595	8
3	30	2	10	8	9	19.557	3.7	0.507	21
4	30	6	2	6	9	33.509	3.3	0.789	1
5	30	6	6	8	3	29.562	4.17	0.549	16
6	30	6	10	4	6	24.649	3.63	0.565	13
7	30	10	2	8	6	36.578	3.8	0.765	2
8	30	10	6	4	9	31.836	4.67	0.547	17
9	30	10	10	6	3	24.126	4.77	0.431	25
10	50	2	2	4	3	19.051	4.42	0.415	26
11	50	2	6	6	6	16.543	4.62	0.383	27
12	50	2	10	8	9	15.267	3.31	0.574	12
13	50	6	2	6	9	27.471	4.2	0.516	19
14	50	6	6	8	3	19.823	3.03	0.722	4
15	50	6	10	4	6	19.904	3.15	0.664	6
16	50	10	2	8	6	29.849	3.72	0.616	7
17	50	10	6	4	9	25.192	3.54	0.589	9
18	50	10	10	6	3	20.757	3.02	0.735	3
19	70	2	2	4	3	13.659	3.25	0.586	10
20	70	2	6	6	6	9.161	3.53	0.485	22

(continued)

Table 2 (continued)

Ex. No.	A	B	C	D	E	MRR (mm^3/min)	SR (μm)	GRG	Rank
	Gap voltage (V)	Pulse on time (μs)	Pulse off time (μs)	Wire feed (m/min)	Reinforcement (wt%)				
21	70	2	10	8	9	8.581	3.28	0.552	14
22	70	6	2	6	9	15.08	3.71	0.477	23
23	70	6	6	8	3	11.618	3.25	0.576	11
24	70	6	10	4	6	11.209	3.45	0.513	20
25	70	10	2	8	6	18.011	4.12	0.437	24
26	70	10	6	4	9	12.984	3.34	0.552	15
27	70	10	10	6	3	12.568	3.04	0.673	5

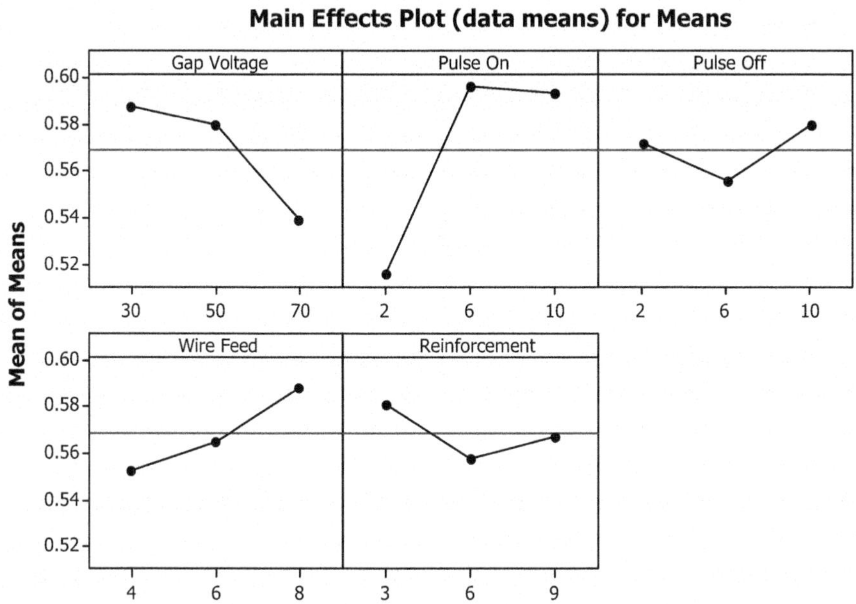

Fig. 1 Response graphs

discharge energy also increases, this leads to a higher material removal rate. The increase in the pulse off time leads to the decrease in the number of discharges within a given period which results in a good surface finish. When the gap voltage is increased, minimum SR values are obtained due to the widening of average discharge. The MRR increases with wire feed because faster wire feed gives higher MRR and decreases with percentage reinforcement because hardness reduces the MRR and reduces surface finish.

3.3 Selection of Optimal Levels

The average of each response (GRG data) for each level of each factor is shown in Table 3. Delta is the difference between the maximum and minimum values. Rank compares the relative magnitude of effects, and it is ranked based on the delta values. The rank 1 factor is relatively more important to the response than the rank 2 factor and so on. Table 3 shows that pulse on time is the most important factor for achieving maximum GRG and is followed by gap voltage, wire feed, pulse off time and reinforcement in that order. Figure 1 shows that gap voltage level 1, pulse on time level 2, pulse off time level 3, wire feed level 3 and reinforcement level 1 provides optimum values.

Table 3 Response table

Level	Gap voltage	Pulse on	Pulse off	Wire feed	Reinforcement
1	**0.588**	0.515	0.571	0.552	**0.581**
2	0.579	**0.597**	0.555	0.565	0.558
3	0.539	0.594	**0.579**	**0.589**	0.567
Delta	0.049	0.082	0.024	0.037	0.023
Rank	2	1	4	3	5

Table 4 ANOVA table

Source of variation	DOF	Sum of squares	Mean sum of squares	F_0	Contribution (%)
B	2	0.038	0.019	6.33	13.48
AB	4	0.042	0.011	3.50	14.88
AC	4	0.099	0.025	8.25	35.11
AE	4	0.064	0.016	5.33	22.70
Error (Pooled)	12	0.039	0.003		13.83
Total	26	0.282			100

The F-values from standard table are $F_{0.05,\,2,12} = 3.89$ and $F_{0.05,4,12} = 3.26$. So, from ANOVA table we see that only pulse on time is significant and also the interactions of pulse on time, pulse off time and percentage reinforcement with gap voltage were significant (Table 4).

3.4 *Confirmation Experiments*

The final step in this process is prediction and conduction of confirmation experiment using the optimum values. From Fig. 1 and response Table 3 the factors at level A1, B2, C3, D3, E1 are the optimum parameters. The predicted GRG is 0.659 and experimental value of GRG is 0.789.

4 Conclusions

- An optimum parameter combination was obtained using the GRA.
- The interaction of gap voltage and pulse off time has the highest statistical influence (35.11%).
- The pooled error is 13.83% for the factors shows 85% confidence level.

References

1. Balasivanandha Prabu S, Karunamoorthy L, Kathiresan S, Mohan B (2006) Influence of stirring speed and stirring time on distribution of particles in cast metal matrix composites. J Mater Process Technol 171(2):268–273
2. Udaya Prakash J, Milton Peter J, Moorthy TV (2012) Optimization of wire EDM process parameters of aluminium alloy/flyash/boron carbide hybrid composites. Int Rev Mech Eng 6(3):449–455
3. Milton Peter J, Udaya Prakash J, Moorthy TV (2014) Optimization of WEDM process parameters of hybrid composites (A413/B_4C/Fly Ash) using grey relational analysis. Appl Mech Mater 592:658–662
4. Rohatgi PK, Guo RQ, Keshavram BN (1995) Cast aluminium alloy fly ash composites. Key Eng Mater 283:104–107
5. Surappa MK, Rohatgi PK (1981) Preparation and properties of cast aluminium-ceramics particle composites. J Mater Sci 16:983–993
6. Miracle DB (2005) Metal matrix composites—from science to technological significance. Compos Sci Technol 65:2526–2540
7. Udaya Prakash J, Jebarose Juliyana S, Pallavi P, Moorthy TV (2018) Optimization of wire EDM process parameters for machining hybrid composites (356/B4C/Fly Ash) using Taguchi technique. Mater Today Proc 5(2):7275–7283
8. Mahendra KV, Radhakrishna K (2007) Fabrication of Al-4.5% Cu alloy with fly ash metal matrix composites and its characterization. Mater SciPol 25(1):57–68
9. Krishnaiah K, Shahabudeen P (2012) Applied design of experiments and taguchi methods. PHI Learning Private Limited, New Delhi
10. Udaya Prakash J, Ananth S, Sivakumar G, Moorthy TV (2018) Multi-objective optimization of wear parameters for aluminium matrix composites (413/B4C) using grey relational analysis. Mater Today Proc 5(2):7207–7216
11. Perumal, S, Udaya Prakash J (2016) Multi-objective optimization of tribological parameters of hybrid composites using grey relational analysis. Int J Chem Sci 14(2):1172–1182
12. Udaya Prakash J, Jebarose Juliyana S, Karthik S, Moorthy TV (2017) Optimization of wire EDM process parameters for machining of AMCs (413/B4c) using Taguchi technique. Int J Mech Prod Eng Res Dev 7(6):231–238

Genetic Algorithm and Particle Swarm Optimization in Minimizing MakeSpan Time in Job Shop Scheduling

K. R. Anil Kumar and Edwin Raja Das

Abstract In the contemporary circumstances, solving of job shop scheduling problem (JSSP) is quite time consuming and inaccurate when done manually. The main purpose of this paper is to analyze the performance of various optimization techniques used in JSSP in order to minimize makespan time. This paper aims to analyze the two optimization techniques viz, genetic algorithm (GA) and particle swarm optimization (PSO), in addition to the existing optimization techniques applied in various research papers on combinatorial optimization problems viz., JSSP. A comparative study of these optimization techniques was conducted and analyzed to find out the most effective optimization technique on solving JSSP. Results show that PSO is found to possess minimum makespan time in comparison with GA for JSSP.

Keywords Genetic algorithm · Job shop scheduling · makespan time · Particle swarm optimization

1 Introduction

JSSP is a scheduling mechanism of various machines to handle a collection of jobs. Each job is to be processed by a given machine for a particular uninterrupted time period. A machine can process only one job at a time and each job can process only one machine at a time. makespan time refers to the duration in which all operations for all jobs are completed by a set of machines on the shop floor. The optimum scheduling of machines used in the production area heavily influences the actual cost of product and overall profitability of organization. JSSP deals with optimizing makespan time. Manual scheduling is time consuming and cumbersome. The solution is to select an appropriate optimization algorithm and apply to JSSP. The process of finding the best possible schedule from a number of schedules is the main difficulty faced in scheduling problems [1]. JSSP is classified as an NP-hard problem [2–5] as

K. R. Anil Kumar (✉) · E. R. Das
Noorul Islam Centre for Higher Education, Kumarakovil, Tamil Nadu, India
e-mail: akrcem@yahoo.co.in

L.-J. Yang et al. (eds.), *Proceedings of ICDMC 2019*,
Lecture Notes in Mechanical Engineering,
https://doi.org/10.1007/978-981-15-3631-1_40

the possible population size and the computational time for algorithms to find a best possible schedule increases exponentially with job machine size.

For JSSP, many heuristic and metaheuristic algorithms have been proposed such as fast simulated annealing hybridized with quenching [6] of Akram et al., TLBO algorithm of Keesari et al. [7]., and Baykasoglu et al. [8], hybrid genetic algorithm with a knowledge-based operator of Piroozfard et al. [9], particle swarm optimization of Lin et al. [10], bacterial foraging algorithm (BFOA), firefly algorithm, ant colony optimization, bee colony optimizations, cuckoo search optimization of Kanaka Vardhini et al. [11], simulated annealing approach of Baykosoglu [12], Van Laarhoven et al. [13], Steinhoel et al. [14], Suresh et al. [15], Tabu search method of Nowicki et al. [16], genetic algorithm of Gonclaves et al. [17], Park et al. [18], Wang and Zheng [19], Watanabe et al. [20]. Jain and Meeran [21], Fan and Zhang [22], Vaessens et al. [23] provided a comprehensive survey of various methods on JSSP. In this paper, PSO has been found to achieve better solution in combinatorial optimization problems like JSSP. Benttaleb et.al [24] studied two machine JSSP with availability constraints on one machine to minimize makespan time. Shovel et al. [25] have studied a probabilistic approach to stochastic JSSP.

The genetic algorithms are non-deterministic methods. Thus, the solutions they provide may vary each time the algorithm is run on the same input instance. The quality of the results highly depends on the initial population, the genetic operators (crossover, selection, mutation) used, the probabilities of crossover and mutation. Hence, GA may not give an optimal solution as it takes much time to converge resulting to a computationally expensive task.

PSO is developed by Kennedy and Eberhart [26] in 1995. It is an evolutionary algorithm that was inspired by the motion of a flock of birds searching for food. This population-based search algorithm begins with an initial population of randomly generated solution called particles characterized by a position and a velocity. For a search problem in an end dimensional space, a best fit solution is represented by a particle that adjusts its position and velocity so that the global best particle is obtained [27].

The remainder of this paper is structured as follows: Sect. 2 describes JSSP, Sect. 3 introduces PSO algorithm and its implementation, whereas Sect. 4 illustrates the analysis of the proposed algorithm on JSSP with its results and discussions. Finally, Sect. 5 concludes the finding of this study with a few remarks.

2 Job Shop Scheduling Problem

A schedule is an allocation of tasks to time intervals on machines and the aim is to find a schedule that minimizes the overall completion time. In job shop scheduling, there are n jobs for m machines and each job must go through m machines to complete its work. In order to complete a jobs work, each operation uses one of the m machines. In general, one job being processed on one machine is considered as one operation denoted as O_{ji} (meaning, jth job being processed on ith machine,

$1 \leq j \leq$ n, $1 \leq I \leq$ m). The aim of JSSP is to find an appropriate schedule of all machines that can minimize the makespan time.

In job shop scheduling problem, each machine can process only one job at a time and each job can be processed only by one machine at a time. The sequence of operations of every job should be preordained. Machines are not subjected to breakdowns and pre-emption is not allowed. Thus, the processing of an operation cannot be interrupted once started.

The JSSP can be formally stated as follows.

Let $J = \{1, 2, \ldots, n\}$ denote the set of jobs, $M = \{1, 2, \ldots, m\}$ represent the set of machines and $O = \{0, 1, \ldots. n \times m, n \times m + 1\}$ be the set of operations that is to be scheduled, where 0 and $n \times m + 1$ represent dummy initial and final operations, respectively. T_j and F_j denote the fixed processing time and finish time of operation j, respectively, and $A(t)$ denotes the set of operations being processed at time t, and e_{jm} is set to 1 if operation j is required to process on machine m, otherwise e_{jm} equals zero. The operations are interrelated with the precedence constraints forcing each operations j for being scheduled after all predecessor operations P_j are completed. Moreover, the operation j can only be scheduled if the required machine is idle. The conceptual model of JSSP can be stated as follows [17].

$$\min \quad F_{n\times m+1} \tag{1}$$

$$\text{s.t;} \quad F_k \leq F_j - T_j, \quad j = 1, 2, \ldots n \times m + 1; \quad k \in P_j \tag{2}$$

$$\sum_{j\in A(t)} e_{jm} \leq 1, \quad \text{m} \in \text{M}; \quad \text{t} \geq 0 \tag{3}$$

$$F_j \geq 0, \quad j = 1, 2, \ldots, n \times m + 1 \tag{4}$$

Objective function (1) minimizes the makespan. Constraint (2) imposes the precedence relations between operations. Constraint (3) represents that one machine can only process one operation at a time, and constraint (4) shows non-negative finishing times.

3 Particle Swarm Optimization Algorithm and Its Implementation

Initialize a population of particles with random positions and random velocities with dimensions in a given searching space.

Fitness Computation

The randomly generated population is used to evaluate its fitness (F_i) (i.e., makespan time). This fitness supports forthcoming updating process to initiate generating solution to achieve minimum makespan time.

```
Make span time=Null matrix (size of problem)
for          i=1 : Number of jobs
for    j=1 : Number of machine
if  i=j=1
make span time (i,j)=particle (i,j)
else if  i≠1 and j=1
make span time (i,j)= sum (particle (1: i,j))
else if  i=1 and j≠1
make span time (i,j)= sum (particle (i,1: j))
else
                          Neighbour 1 = make span time (i-1,j)
                          Neighbour 2 = make span time (i,j-1)
if  Neighbour 1 ≤ Neighbour 2
make span time (i,j)= Neighbour 2 + particle (i,j)
else
make span time (i,j)= Neighbour 1 + particle (i,j)
end
end
end
                    Fitness (Fi) = make span time (end).
```

Updating Process
For updating the initially generated solution following Eqs. (5) and (6) are used.

$$V_i^{(k+1)} = h * V_i^{(k)} + c_1 * r_1() * (p_{\text{best}_i} - X_i^{(k)}) + c_2 * r_2() * (g_{\text{best}} - X_i^{(k)}) \tag{5}$$

$$X_i^{(k+1)} = X_i^{(k)} + V_i^{(k+1)} \tag{6}$$

where inertia factor is represented as k, h is a scale parameter, r_1 and r_2 are the parameters where two random numbers are to be assigned in between zero and one. A couple of controlling parameters involve in this updating process is c_1 and c_2

and its assign to be two. Position and Velocity of i are said to be V_i and X_i and then ith particles of previous best are considered to be p_{best_i} and g_{best} whereas g_{best} (global best) is considered to be the best particle so far attained. The recent theoretical analysis verdicts that the particle on PSO fluctuates between previous and global best before it converges. To assist the entire particle in a better position, it is essential to include global best neighbors in consecutive generating solutions (Fig. 1).

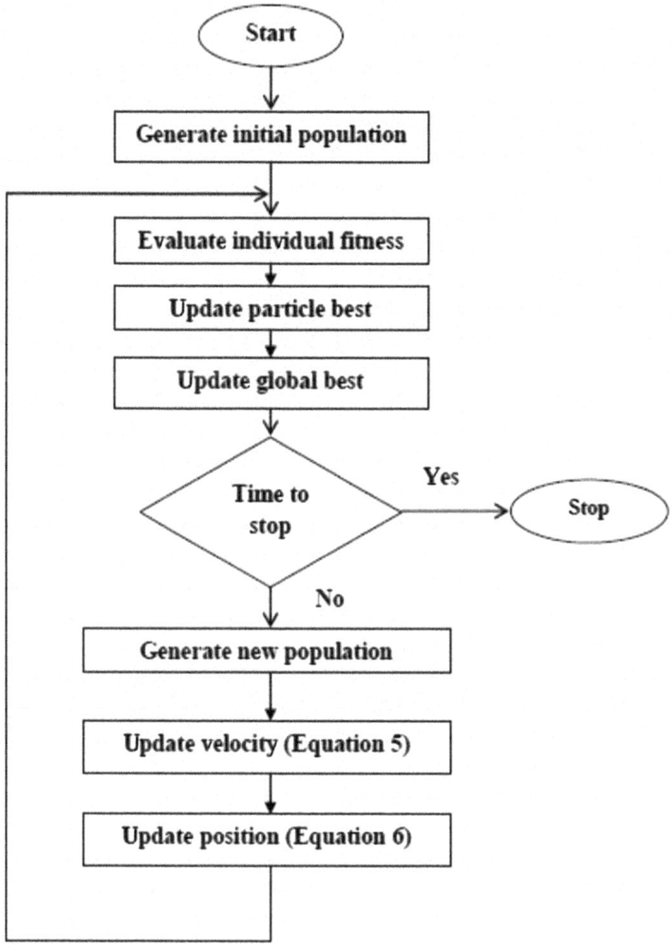

Fig. 1 Flowchart of PSO

4 Results and Discussion

This section deals with the analysis of PSO algorithm compared to genetic algorithm in addition to the existing eleven optimization techniques applied in research papers on combinatorial optimization problems viz., JSSP. The experimental results which have been obtained from the proposed PSO algorithm are laid out in this section. The PSO algorithm has been coded in MATLAB with i3 processor 6 GB ram with CPU speed of 2.20 GHz. The performance of PSO algorithm has been tested by standard benchmark problems. Twenty instances are taken from OR library as benchmarks to test the performance of PSO and GA algorithm. The population size as well as the generation number has been set to 100 and 1000, respectively. The results obtained by PSO and GA in addition to the eleven different optimization methods are represented in Tables 1 and 2 [8]. It seems that PSO algorithm has more potential for solving JSSP when compared with GA.

Table 1 Computational results of JSSP test problems [8]

Problems	BM [19]	TLBO [19]	HGA1 [19]	HGA2 [19]	MA [19]	MA (GRRS) [19]	F&F [19]	HGA [19]
FT06	55	55	55	55	55	55	55	55
FT10	930	938	930	938	930	930	930	938
FT20	1165	1165	1165	1165	1165	1165	1165	1165
LA01	666	666	666	666	666	666	666	666
LA02	655	655	655	655	655	655	655	655
LA03	597	597	597	597	597	597	597	597
LA04	590	607	590	590	590	590	590	590
LA05	593	593	593	593	593	593	593	593
LA06	926	926	926	926	926	926	926	926
LA07	890	890	890	890	890	890	890	890
LA08	863	864	863	863	863	863	863	863
LA09	951	951	951	951	951	951	951	951
LA10	958	958	958	958	958	958	958	958
LA11	1222	1222	1222	1222	1222	1222	1222	1222
LA16	945	946	945	945	945	945	947	945
LA21	1046	1091	1046	1046	1055	1079	1052	1046
LA27	1235	1256	1236	1256	1261	1286	1242	1256
LA31	1784	1784	1784	1784	1784	1784	1784	1784
LA36	1268	1332	1287	1287	1281	1307	1381	1279
LA40	1222	1241	1229	1241	1233	1252	1228	1241
Average	928.05	936.85	929.4	931.4	931	935.7	934.75	931

Table 2 Computational results of JSSP test problems [8]

Problems	BM [19]	PGA [19]	GRASP [19]	Beam search [19]	HIA [19]	PSO	GA
FT06	55	55	55	55	55	55	55
FT10	930	936	938	1016	930	930	934
FT20	1165	1177	1169	None	1165	1165	1165
LA01	666	666	666	666	666	666	666
LA02	655	666	655	704	655	668	660
LA03	597	597	604	650	597	606	606
LA04	590	590	590	620	590	603	609
LA05	593	593	593	593	593	593	595
LA06	926	926	926	926	926	926	926
LA07	890	890	890	890	890	890	890
LA08	863	863	863	863	863	863	863
LA09	951	951	951	951	951	951	951
LA10	958	958	958	958	958	958	958
LA11	1222	1222	1222	1222	1222	1222	1222
LA16	945	977	946	988	945	951	959
LA21	1046	1047	1091	1152	1256	1050	1056
LA27	1235	1260	1320	1316	1784	1290	1269
LA31	1784	1784	1784	1784	1784	1784	1784
LA36	1268	1305	1334	1401	1281	1282	1310
LA40	1222	1252	1259	1347	1240	1239	1244
Average	928.05	935.75	940.7	952.73	967.55	934.6	936.1

Two optimization techniques, namely GA, PSO, were applied to existing data which included eleven optimization techniques applied on 20 different benchmark problems [8]. The optimization algorithms studied in this platform are:

1. Teaching–learning-based optimization (TLBO)
2. Hybrid genetic algorithm (HGA 1)
3. Hybrid genetic algorithm (HGA 2)
4. Memetic algorithm (MA)
5. Memetic algorithm (gap reduction and restricted swapping) (MA GRRS)
6. Filler and fan algorithm (F& F)
7. Hybrid genetic algorithm (HGA)
8. Parallel genetic algorithm (PGA)
9. Greedy randomized adaptive search procedure (GRASP)
10. Hybrid intelligent algorithm (HIA)
11. Beam search algorithm (BSA)
12. Particle swarm optimization (PSO)
13. Genetic algorithm (GA).

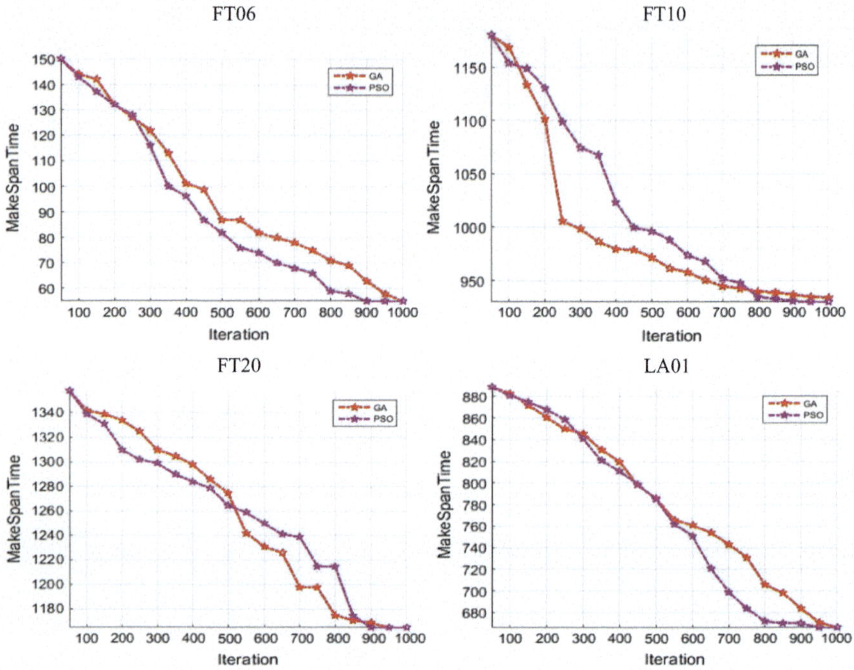

Fig. 2 Convergence graph for FT06, FT10, FT20, LA01

From Tables 1 and 2, it is quite convincing that optimization technique could do much better than conventional formulation. From Table 2, it is evident that the average makespan time of PSO is quite closer to the average standard benchmark value. It is observed from the analysis that PSO is found to be superior over GA with respect to makespan time.

Thousand iterations were done in MATLAB and convergence graphs of the two optimization techniques applied viz., PSO and GA were obtained with makespan time versus iteration. It is evident from the convergence graph that PSO reveals better convergence over other optimization techniques (Figs. 2, 3, 4, 5 and 6).

5 Conclusion

In this study, it is found that the benchmark values of PSO in JSSP in comparison with GA are closer to that of standard benchmark values. Therefore, PSO in JSSP is superior to genetic algorithm. The PSO in JSSP has a greater potential in providing better solution for makespan time minimization. The PSO algorithm in JSSP could give an idea about its possible performance for solving other combinatorial optimization problems. There is still scope for application of newer optimization techniques for solving JSSP to further minimize makespan time.

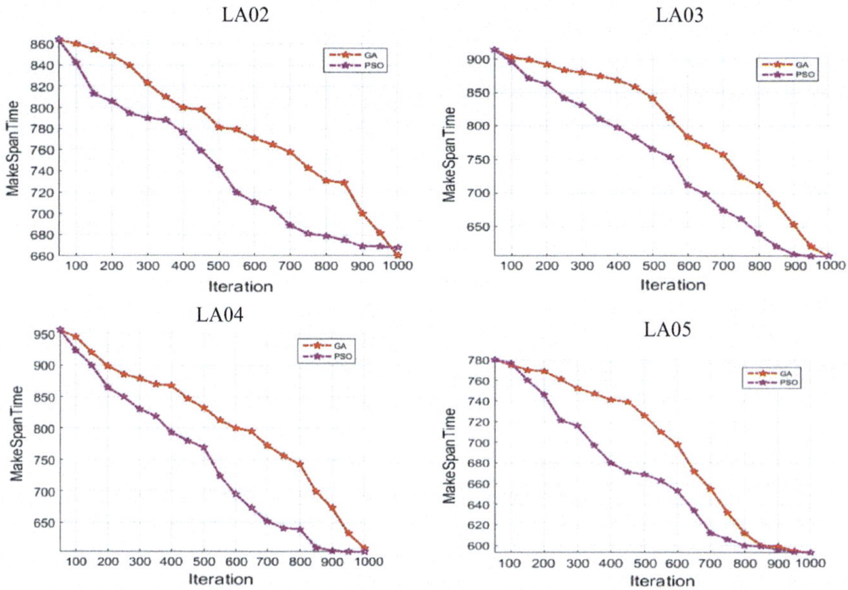

Fig. 3 Convergence graph for LA02, LA03, LA04, LA05

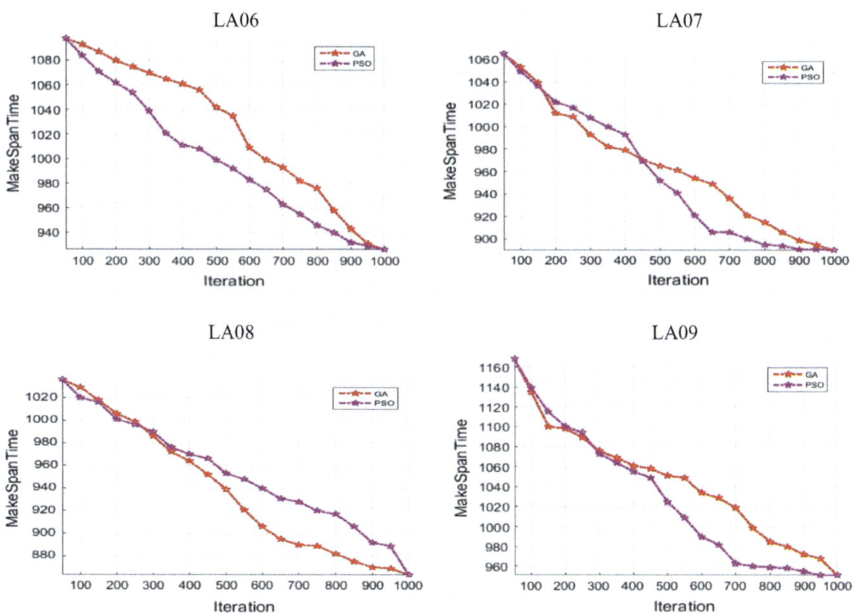

Fig. 4 Convergence graph for LA06, LA07, LA08, LA09

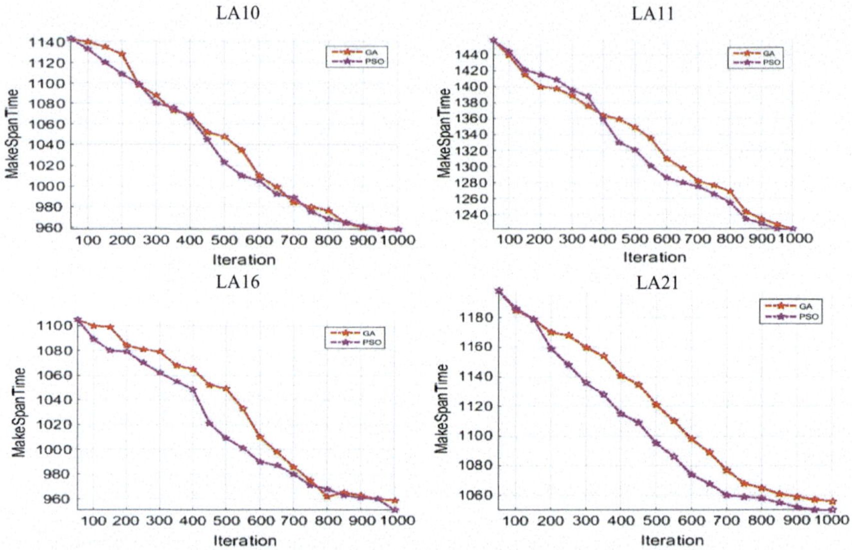

Fig. 5 Convergence graph for LA10, LA11, LA16, LA21

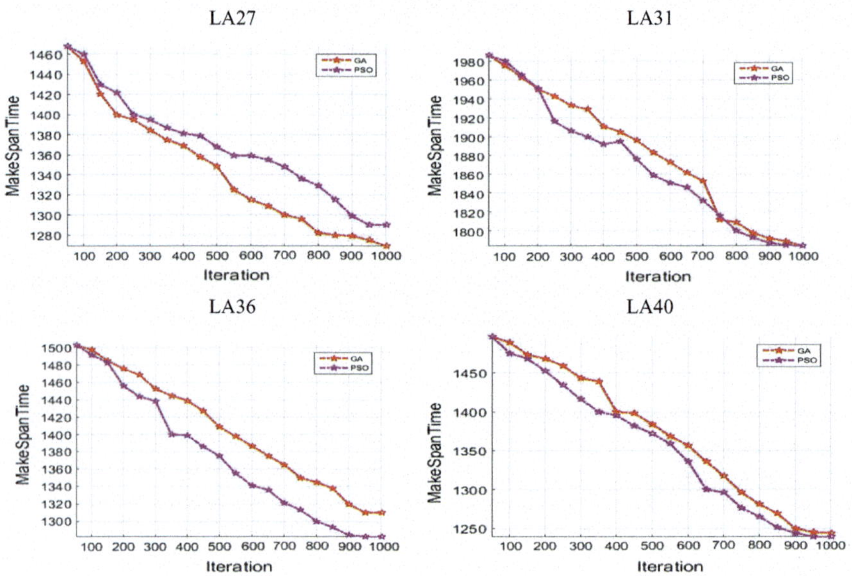

Fig. 6 Convergence graph for LA27, LA31, LA36, LA40

References

1. Anandaraman C (2011) An improved sheep flock heredity algorithm for job shop scheduling and flow shop scheduling problems. Int J Ind Eng Comput 2:749–764
2. Giffler B, Thompson GL (1960) Algorithms for solving production scheduling problems. Oper Res 8:487–503
3. Baker KR (1974) Introduction to sequencing and scheduling. Wiley, New York
4. Garey M, Johnson D, Sethi R (1976) The complexity of flowshop and jobshop scheduling. Math Oper Res 1:117–129
5. French S (1982) Sequencing and scheduling: an introduction to the mathematics of the job shop. E. Horwood, Chichester, West Sussex
6. Akram K, Kamal K, Zeb A (2016) Fast simulated annealing hybridized with quenching for solving jobshop scheduling problem. Appl Soft Comput 49:510–523
7. Keesari HS, Rao RV (2014) Optimization of job shop scheduling problems using teaching-learning-based optimization algorithm. Opsearch 51(4):545–561
8. Baykasoglu A, Hamzadayi A, Kose SY (2014) Testing the performance of teaching–learning based optimization (TLBO) algorithm on combinatorial problems: flow shop and job shop scheduling cases. Inf Sci 276:204–218
9. Piroozfard H, Wong KY, Hassan A (2016) A Hybrid Genetic Algorithm with a Knowledge-Based Operator for Solving the Job Shop Scheduling Problems. J Optim 2016:1–13
10. Lin T-L, Horng S-J, Kao T-W, Chen Y-H, Run R-S, Chen R-J, Lai J-L, Kuo I-H (2010) An efficient job-shop scheduling algorithm based on particle swarm optimization. Expert Syst Appl 37:2629–2636
11. Kanaka Vardhini K, Sitamahalakshmi T (2016) A review on nature-based swarm intelligence optimization techniques and its current research directions. Indian J Sci Technol 9(10):1–13
12. Baykasoglu A (2002) Linguistic based meta-heuristic optimisation model for flexible job shop scheduling. Int J Prod Res 40(17):4523–4543
13. Van Laarhoven PJM, Aarts EHL, Lenstra JK (1992) Job shop scheduling by simulated annealing. Oper Res 40(1):113–125
14. Steinhöel K, Albrecht A, Wong CK (1999) Two simulated annealing-based heuristics for the job shop scheduling problem. Eur J Oper Res 118:524–548
15. Suresh RK, Mohanasundaram KM (2005) Pareto archived simulated annealing for job shop scheduling with multiple objectives. Int J Adv Manuf Technol 29:184–196
16. Nowicki E, Smutnicki C (2005) An advanced tabu search algorithm for the job shop problem. J Sched 8:145–159
17. Goncalves JF, Mendes JJDM, Resende MGC (2005) A hybrid genetic algorithm for the job shop scheduling problem. Eur J Oper Res 167:77–95
18. Park BJ, Choi HR, Kim HS (2003) A hybrid genetic algorithm for the job shop scheduling problems. Comput Indus Eng 45:597–613
19. Wang L, Zheng DZ (2001) An effective hybrid optimization strategy for job shop scheduling problems. Comput Oper Res 28:585–596
20. Watanabe M, Ida K, Gen M (2005) A genetic algorithm with modified crossover operator and search area adaptation for the job-shop scheduling problem. Comput Ind Eng 48:743–752
21. Jain A, Meeran S (1999) Deterministic job-shop scheduling: past, present and future. Eur J Oper Res 113:390–434
22. Fan K, Zhang R-q (2010) An analysis of research in job shop scheduling problem (2000–2009). In: Li Y, Desheng W (eds) Proceedings 2010 IEEE international conference on advanced management science (ICAMS), Chengdu, pp 282–288
23. Vaessens R, Aarts E, Lenstra J (1996) Job shop scheduling by local search. INFORMS J Comput 8:302–317
24. Benttaleb M, Hnaien F, Yalaoui F (2018) Heuristic algorithms for two-machine job shop problem under availability constraints on one machine: makespan minimization. IFAC-PapersOnLine 51(11):1065–1070

25. Shoval S, Efatmaneshnik M (2018) A probabilistic approach to the Stochastic Job-Shop Scheduling problem. Procedia Manuf 21(2018):533–540
26. Kennedy J, Eberhart R (1995) Particle swarm optimization. In: Proceedings of IEEE international conference on neural networks, vol 4, pp 1942–1948
27. Hu X, Shi Y, Eberhart R (2004) Recent advances in particle swarm. In: Congress on evolutionary computation, Portland, Oregon, June 19–23, pp 90–97

Autonomous Risk and Hazard Management System for Smart Cities

Arivukkarasan Raja and E. Pavithra

Abstract A smart city is a well-planned, self-sustainable urban area which is capable of managing its own resource and assets intelligently and efficiently with the help of the data derived out of various sensing elements. A hazard can be defined as an external or internal agent, which has some unrealized ability to cause harm. Some of the identified potential hazards in industrial and residential environment include, but not limited to, air quality, air temperature, humidity, noise levels, vibration, radiation, fire, electric short circuit, water leakage, etc. In the context of smart cities, the management of these potential hazards is very important in both industrial and residential environments to assure superior quality of living, reduce the cost of health care and for improving the morale, productivity, and job satisfaction of the employees. The proposed risk and hazard management system includes multiple sensors, which are programmed to receive data like carbon monoxide, formaldehyde, lead, nitrogen dioxide, butane, LPG, and radon levels in real-time. The end-users can visualize the data through Internet of things (IoT) open-source platform. The services for device integrations are done through Embedded C, loaded in Teensy 3.2. Similarly, the services for data validations, intelligent actions, and the data storage, display using Python, loaded in Raspberry Pi3 Model B+. The remedial or control actions like automated door, exhaust fan, and solenoid valve operations are carried out based on real-time data analysis using artificial intelligence.

Keywords Smart home · IoT · Teensy 3.2 · Air quality

A. Raja (✉) · E. Pavithra
Vel Tech Rangarajan Dr. Sagunthala R&D Institute of Science and Technology, Avadi, Chennai 600062, India
e-mail: arivukkarasan@gmail.com

E. Pavithra
e-mail: drepavithra@veltech.edu.in

L.-J. Yang et al. (eds.), *Proceedings of ICDMC 2019*,
Lecture Notes in Mechanical Engineering,
https://doi.org/10.1007/978-981-15-3631-1_41

1 Introduction

Safety will be valid only if any individual or society in-charge takes a lead role in handling hazard and safety mechanism. Many industrial establishments have proven the benefits of safety and hazard management. A good hazard and safety compliance process or system can avoid injuries among employees and reduce the number of accidents within the industrial environment. It can greatly reduce the number of hours spent on manufacturing process, thus can improve the financial growth of an organization [1].

A hazard is just a disorder or a set of conditions that leads to a possible destruction. We can classify hazards into two groups like hazards which can cause occupational illness and hazards which can cause physical injuries.

The key principle of health and safety compliance is the prerequisite for a methodical risk and hazard management process, which can categorize the probable and definite causes of harm. Only by when the risks and vulnerabilities are documented, the corresponding departments are able to put preventive mechanisms [1] in place to avoid harm. The significant hazards in an industrial or residential environment are air quality, noise, vibration, radiation, fire, temperature, and humidity.

Why hazard, health, and safety management system are important? [2]

- To have a fewer injuries among workforce
- To reduce the health care/insurance costs
- To improve the employees' morale
- To increase the productivity and job satisfaction.

2 Objective

Indoor and outdoor contaminants in both residential and industrial properties can be a significant environmental health problem. Numerous health issues have been linked with occupant exposure to several toxic and hazardous substances [3]. The overall objective of this exploration is to prepare an IoT-based smart hazard management, health and safety compliance system. The following areas have been focused on this project.

- To identify the possible environmental hazards
- To assess the risks which can be possible outcome of hazards
- To manage the risks to avoid further damage due to hazards
- To monitor the control methods for continuous improvement.

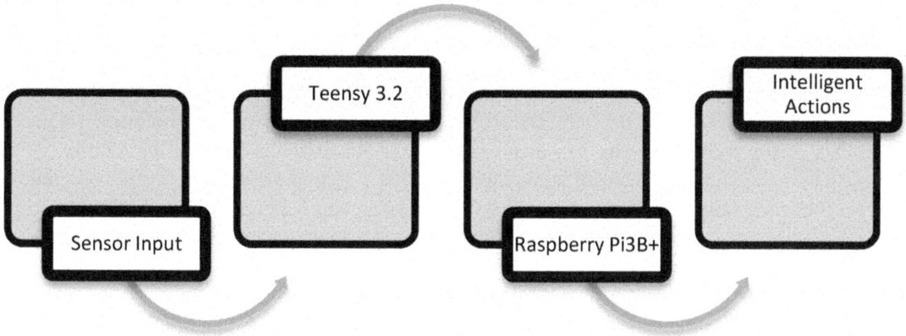

Fig. 1 Flow diagram

3 Methodology

Several sensors are inter connected to gather carbon monoxide, formaldehyde, lead, butane, nitrogen dioxide, radon levels in real time, then the collected data is displayed through an open-source UI platform. The services for device and controller integration are written in Embedded C and the services for UI integration are written in Python, Raspberry Pi3 Model B+ is acting as a secondary controller. MySQL database is used to store the historical data [1] (based on the explorer's request). The hierarchical representation of sensor architecture is mentioned in Fig. 1.

The identified potential hazards are in our industrial/residential environment as a part of this exploration are air quality, air temperature, humidity, noise levels, vibration, radiation, fire, electric short circuit, and water leakage.

4 Work Plan

Table 1 consist of various sources of hazard along with the health issues associated with each hazardous environment. The remedial actions also suggested ***if the density of hazardous gases are more than the limit prescribed by the National Air Quality Safety Bereau*** as per Table 2. The remedial actions are going to be triggered mechanically through AI. The state of affairs is reverted based on the conventional values.

5 Technical Architecture

Based on the request triggered through the Web interface [4], the hazard management, health and safety compliance data will be stored in a MySQL database. The frequency of data storage will be purely depending on the Web interface configuration done by

Table 1 Health issues and remedial actions

Source of hazard	Health issues	Remedial actions
Carbon monoxide	High density of carbon monoxide will stop the body from obtaining oxygen. Also causes nausea, associated with headache and dizziness, sleepiness, weakness of entire body. In general, high density of carbon monoxide will cause loss of consciousness, brain damage which leads to death [3]	Doors and windows will be opened through a stepper motor interface and the gas connection will be turned off through a smart solenoid valve. Exhaust fan will also be switched ON
Formaldehyde	Since it is classified as carcinogen, long-term exposure may lead to respiratory issues, eczema and nasopharyngeal cancer. The short-term exposure may lead to skin and eye irritation. The general symptoms of formaldehyde exposure include, watery eyes, coughing, chest spasm, wheezing, burning sensation, throat infection and head ache [3]	Doors and windows will be opened through a stepper motor interface. Exhaust fan will be switched ON
Lead	The short-term exposure to lead can affect the children with behavioural problems and learning disabilities. The prolonged exposure to lead may lead to serious issues like anaemia, constipation, abdominal pain, tingling in the hands and/or feet, nervous system damage and impaired brain. Further to this, exposure to high levels of lead causes weakening of kidney and nervous system. Finally, the continuous exposure to lead may cause death	Doors and windows will be opened through a stepper motor interface. Exhaust fan and air filter will be switched ON
Butane	Butane in liquid kind will cause severe cold burns to the skin because of its speedy vaporization. Butane as liquefied petroleum gas forms an inflammable mixture with air in concentrations of between a pair of and 100%. It can, therefore, be a hearth and explosion hazard if keep or used incorrectly [3]	Gas connection will be turned off through a smart solenoid valve. Doors and windows will be opened through a stepper motor interface and the exhaust fan will be switched ON

(continued)

Table 1 (continued)

Source of hazard	Health issues	Remedial actions
Nitrogen dioxide	In general, it causes respiratory issues. Short-term exposure to nitrogen dioxide may cause bronchitis, flu, wheezing and coughing. The long-term exposure to nitrogen dioxide may reduce the immunity drastically, thus causes severe lung infections. The nitrogen dioxide used to form as a particulate matter, which is the main cause for allergies	Doors and windows will be opened through a stepper motor interface and the gas connection will be turned off through a smart solenoid valve
Radon	Since radon is an inert gas, will get exhaled immediately after breathing. But it will form as an aerosol along with other dust particles and reduce the inhalation capacity. Hence, the radon inhalation may lead to chest pain, pneumonia, rigorous cough and difficulty in breathing	Doors and windows will be opened through a stepper motor interface. Exhaust fan will be switched ON

Table 2 National air quality data

Parameters to be measured	Unit	Limit
Carbon monoxide	mg/M^3	≤04
Formaldehyde	mg/M^3	≤0.1
Lead	PPM	≤400
Butane	mg/M^3	≤05
Nitrogen dioxide	mg/M^3	≤55
Radon	pCi/L	≤04

the administrator. The stored data can be exported in a form of reports, as in when required. The historical data can be compared with the current one to do the trend analysis.

As shown in Fig. 2, the component architecture of proposed model consists of carbon monoxide, formaldehyde, lead, nitrogen dioxide, butane and radon sensor. The analogue to digital conversion is taken care of by Teensy 3.2, the data storage is being handled by MySQL database, and L298N is the motor driver used as an interface between controller and the stepper motor. The technical details of the components used for this exploration are listed in Table 3,

The smoke sensor MQ2 used in this exploration for detecting butane uses SnO_2 as a semiconducting material. The higher exposure to liquefied petroleum components like CH_4 and Butane, the resistance of the semiconducting material will increase. Hence, it affects the flow of output voltage. The concentration of liquefied petroleum components in air is directly proportional to output voltage. With the help of iterative

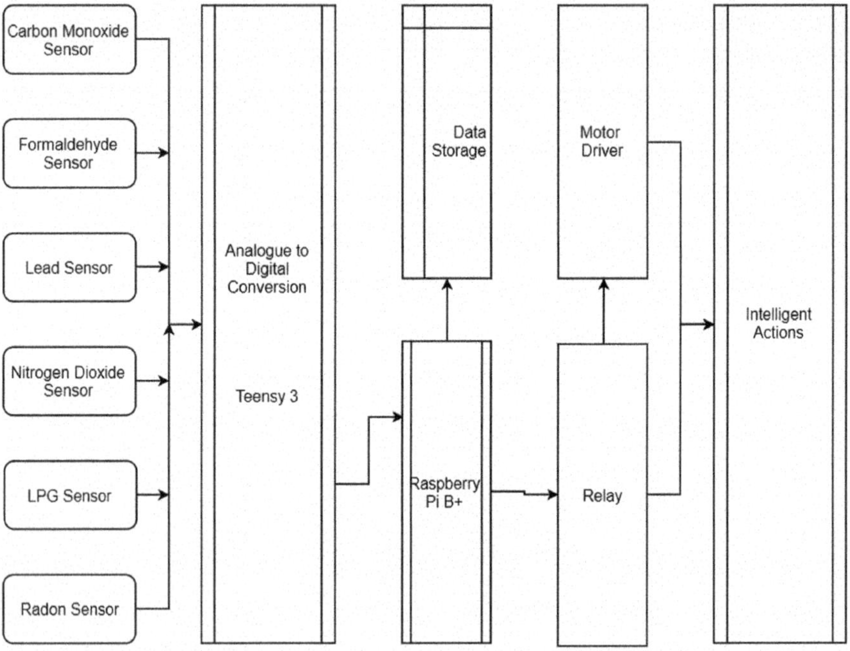

Fig. 2 Architecture diagram

Table 3 Product details

Component description	Product details
Processor	Raspberry Pi3 Model B+
Data storage	MySQL community edition
Web interface	Open-source IoT development platform
Device interface	Embedded C
Driver	L298N
Drive	Gear mechanism
Analogue to digital conversion	Teensy 3.2
Sensors	MQ7 for carbon monoxide MQ 138 for formaldehyde MQ3 for lead MQ2 for butane MQ131 for nitrogen dioxide MQ135 for radon

calibration, the nearest match of liquefied petroleum components concentration value will be arrived. Similarly, for the other MQ family of sensors MQ4, MQ7, MQ135 use combination of ceramic and SnO_2 as a semiconducting material. In some cases, the ceramic SnO_2 leads are connected through nickel–chromium pins at the outer core, which enables quick eating and accurate sensing.

L298N is a dual bridge driver which is capable of connecting two motors. The speed and direction also can be controlled for two motors simultaneously. The voltage range for this driver is between 5 and 35 V. The peak current can be up to 2 A. The driver consists of two output sockets for connecting two motors concurrently. In addition to that, it contains a jumper for voltage regulation, power source and ground. The integrated circuit available in the board uses only 2 V for operation.

The Teensy 3.2 controller consists of 32-bit ARM processor, 64 k inbuilt memory, 256 K flash memory, 21 analogue and 34 digital input/output pin and a USB connectivity. It can be programmed using Embedded C. The voltage requirement for this controller is 3.3 V. The current requirement is 100 mA. All digital pins of Teensy 3.2 are having tolerance limit up to 5 V for interoperability.

6 Working Principle

Teensy 3.2 controller is used for analogue to digital conversion and device integration. It is programmed using Embedded C. The relevant sensors are connected to the circuit and the circuit is connected with power supply. Once the power supply is switched on, the sensors will take a few seconds for heating up. The delay for heating up is already been configured through our programming. Once the sensor senses any deflection in voltage, it will be sent to the controller in an analogue format. Then the sensing variable output of the sensors is converted into digital format and sent to Raspberry Pi3 Model B+ for further processing. In Raspberry Pi3 Model B+, through PHP services, the sensor variables are stored in a MySQL database. The stored data will also be displayed in the Web page through an open-source IoT platform. The program for analysing the sensor digital values is written in Python. Based on the Python services, the intelligent remedial actions will be carried out based on Python and Embedded C programs.

7 Results and Analysis

When the circuit is switched on, then all associated sensors sense and transmit the data to the controller board. The delay is useful for the sensors which take longer time to respond. The analogue signals are converted into digital format through the controller itself and sent to Raspberry Pi3 Model B+. Then, the variables data of carbon monoxide, formaldehyde, lead, butane, nitrogen dioxide, radon are stored in a structured MySQL database. Once the data is stored, then, the intelligent actions were triggered based on the data reference value given in the programming. If the value of butane is more than 5 mg/M^3, the door and windows were opened through a stepper motor interface and Gas connection was turned off through a smart solenoid valve. If the concentration of Lead is more than 400 PPM in side our premises, the door and windows were closed through a stepper motor interface and the air filter, exhaust fan were switched on. Similarly, for the other sensor variables, the intelligent actions were taken place with respect to the limiting factors.

Figure 3 displays the value of radon gas stored in the Web page/database over the analysis period in a structured manner for our easy reference. The data stored in the database can be accessed at any time and the historical data comparison can be carried out.

Figure 4 displays the level of formaldehyde in the testing premises area at any given moment. The data displayed here will also be available in the database for future

Fig. 3 Radon-level graphical representation

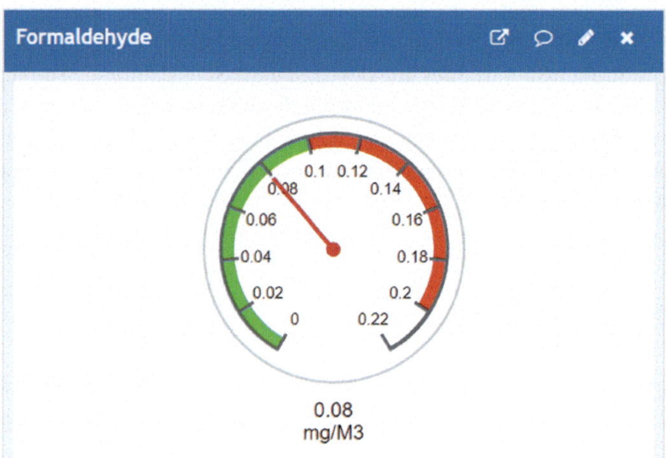

Fig. 4 Formaldehyde-level graphical representation

reference. The warning levels are highlighted in red colour, after which the scope for triggering intelligent actions will be initiated. The normal values are highlighted in green.

Figure 5 displays the value of nitrogen dioxide stored in the Web page/database over the period of time from October–December 2018 in a structured manner for our easy reference. The data stored in the database can be retrieved and displayed in the Web page for our comparative studies.

Figure 6 displays the level of carbon monoxide inside our testing premises at any moment. The data displayed here will also be available in the database for future reference. The display method can be easily configured in our Web page through a user interface. For example, the display widget can be of a dial or graph or number display, etc.

Figure 7 displays the level of lead inside our testing premises at any moment. The data displayed here will also be available in the database for future reference. The values are configured using an open-source IoT platform which is available on Internet as well. Hence, the data can be accessed in a real-time manner across the globe.

Figure 8 displays the warning light sign for the liquefied petroleum gas concen-

Fig. 5 Nitrogen dioxide-level graphical representation

Fig. 6 Carbon monoxide-level graphical representation

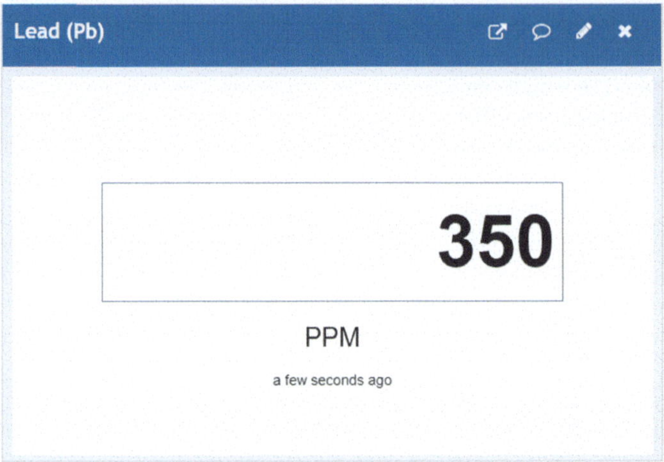

Fig. 7 Lead-level graphical representation

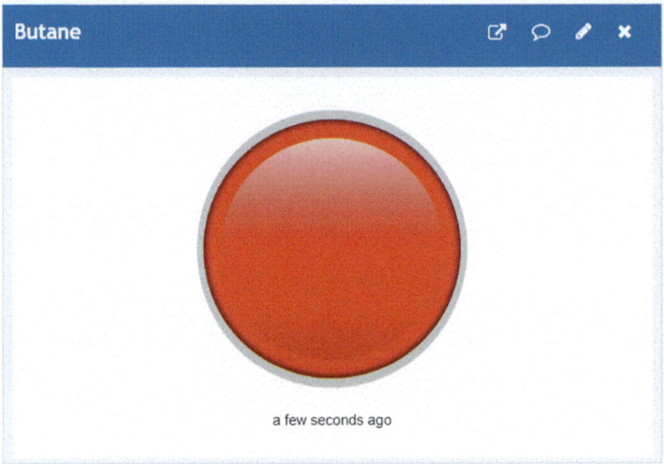

Fig. 8 Butane-level graphical representation

tration in a particular testing area. Along with the warning sign, the actual data will also be captured in the database for our future reference with easy accessibility.

8 Conclusion

In a smart city, Indoor contaminants in both residential and industrial possessions can be a substantial environmental health problem. Various health issues have been linked with occupant's exposure to various toxic and hazardous substances. With the help of interlinked sensors, the carbon monoxide, sulphur, ammonia, butane, propane, temperature and relative humidity levels are identified in real-time basis, then the collected data is available for fellow students and researchers through an open-source user interface platform. The historical data is available for the indented audience through a MySQL database. The remedial actions for each hazard are taken with the help of artificial intelligence.

Here are the benefits because of the intelligent remedial actions triggered by our exploration [5],

- The number of injuries among employees is reduced
- Health care/insurance costs for the organizations are reduced
- Employees morale has improved is an intangible benefit
- Productivity and job satisfaction are increased, and it is a win-win situation for employees and the organization [6].

For future enhancements and research, with the help of the same platform, additional potential hazards can be added as a measurement parameter. The same apparatus can be used to conduct the experiment in multiple locations and record the hazard management, health and safety compliance data is a centralized manner, which enables the historical data comparison. The intelligent remedial actions for the newly identified hazards can also be added. The same circuit can be fabricated using different controllers based on future invention and the programming languages can also have based on future trends and cutting edge technologies.

References

1. Marikyan D, Papagiannidis S, Alamanos E (2019) A systematic review of the smart home literature: a user perspective. Technol Forecast Soc Change 138:139–154
2. Huang J, Pan X, Guo X, Li G (2018) Health impact of China's air pollution prevention and control action plan: an analysis of national air quality monitoring and mortality data. Lancet Planet Health 2(7):e313–e323
3. Chen R, Yin P, Meng X, Liu C, Wang L, Xu X, Zhou M (2017) Fine particulate air pollution and daily mortality. A nationwide analysis in 272 Chinese cities. Am J Respir Crit Care Med 196(1):73–81
4. Karami M, McMorrow GV, Wang L (2018) Continuous monitoring of indoor environmental quality using an arduino-based data acquisition system. J Build Eng
5. Kim J, Hong T, Jeong J, Koo C, Kong M (2017) An integrated psychological response score of the occupants based on their activities and the indoor environmental quality condition changes. Build Environ 123:66–77
6. Bibri SE (2018) The IoT for smart sustainable cities of the future: an analytical framework for sensor-based big data applications for environmental sustainability. Sustain Cities Soc 38:230–253

Optimization of Turning Process Parameters in Machining of Heat-Treated Ductile Iron Bar Using TiC/TiCN/Al_2O_3-Coated Tungsten Carbide Tool

K. Ramesh, K. Gnanasekaran, S. Prathap Singh, and M. Thayumanavan

Abstract The main aim of this study is to find the optimum turning parameters on the machining of heat-treated ductile iron bar using TiC/TiCN/Al_2O_3-coated tungsten carbide insert as tool. The ductile iron bar is heat treated, and a three-layer coating is done on the tungsten carbide insert using chemical vapour deposition (CVD) method. Taguchi method is used to find the optimum turning parameters for material removal rate (MRR) and surface roughness. Process parameters such as cutting speed (500–1200 rpm), depth of cut (0.051–0.159 mm) and feed rate (0.5–1 mm/min) were analysed by Taguchi L9 orthogonal array table. The effect of turning parameters on material removal rate (MRR) and surface roughness was analysed by signal-to-noise ratio(S/N) and analysis of variance (ANOVA). The result shows that material removal rate (MRR) and surface roughness were highly influenced by cutting speed followed by depth of cut and feed rate. Regression equation was developed and was validated by confirmatory experiments.

Keywords Heat-treated ductile iron · Chemical vapour deposition · Material removal rate · Surface roughness · Taguchi technique

1 Introduction

Turning is one of the most important machining processes widely used in many industries including mining, construction, agriculture, automotive and railroads. It is used in manufacturing crankshafts, wheel hubs and timing gears idlers suspension. The development of cutting tools for high-speed machining and hard cutting material remains a problem in the quality and production economy. Hard coating is the best way to improve the performance in cutting tool in machining applications. Hasçalık and Çaydaş [1] studied the effect of machining parameters on surface roughness and

K. Ramesh · K. Gnanasekaran (✉) · S. Prathap Singh
Department of Mechanical Engineering, St. Joseph's Institute of Technology, Chennai, India
e-mail: sekarkaran24@gmail.com

M. Thayumanavan
Shinewell Automation Industrial Private Limited, Chennai, India

L.-J. Yang et al. (eds.), *Proceedings of ICDMC 2019*,
Lecture Notes in Mechanical Engineering,
https://doi.org/10.1007/978-981-15-3631-1_43

tool life in the machining of Ti–6Al–4V alloy using CNMG 120408-883 insert. The effect of machining parameters such as cutting speed, feed rate and depth of cut was studied using the orthogonal array, signal-to-noise ratio and analysis of variance. The result shows that the surface roughness was highly influenced by the cutting speed and tool life by the depth of cut. Davim and Mata [2] performed a machining study on fibre-reinforced plastics. The turning operation was performed on the FRP using the polycrystalline diamond cutting tool to study the effect of surface roughness. Bagci and Işık [3] studied the effect of process parameters on the turning of fibre-reinforced plastics using RS methodology and artificial neural network. The effect of process parameters such as cutting speed, feed and depth of cut on the surface roughness was studied. Pawade and Joshi [4] performed an experiment to study the effect of turning parameters on the machining of Inconel 718 using Taguchi grey relational analysis. The effect of process parameters such as cutting speed, feed and depth of cut on the surface roughness and cutting force was studied. The optimal machining process parameters were calculated. The result revealed that surface roughness and cutting force are highly influenced by the feed rate.

Several studies have been done only on the turning of materials with uncoated tools. But, hard machining is the problem in the turning. So, the main aim of the present study is to perform turning operation on the heat-treated ductile iron using TiC/TiCN/Al_2O_3 (titanium carbide/titanium carbo-nitride/aluminium oxide)-coated tungsten carbide inserts. Then, optimization was done to find the optimal machining parameters using Taguchi technique.

2 Materials and Experimentations

2.1 Material

The material selected for the present investigation is the heat-treated ductile iron with an average diameter of 60 mm. The composition of ductile iron is in Table 1.

Initially, ductile iron is heated in the furnace up to 900 °C, and the furnace is maintained at 900 °C for 90 min as shown in Fig. 1. This process is known as austenizing. Because of this process, ductile iron is converted to ausferrite. This ausferrite consists of a mixture of high-carbon austenite and ferrite. The iron which is austenized is cooled by oil quenching for about 30 min as shown in Fig. 2. The oil-quenched iron is then maintained at 250 °C for 120 min in the furnace. This process

Table 1 Elements of ductile iron and its weight percentage

Element	Carbon	Silicon	Manganese	Magnesium	Phosphorus	Sulphur	Copper	Iron
Weight (%)	3.2–3.60	2.2–2.8	0.1–0.2	0.03–0.04	0.005–0.04	0.005–0.02	<0.40	Balance

Fig. 1 Ductile iron at 900 °C in a furnace

Fig. 2 Oil quenching

is called as austempering. The austempered iron is left to cool in the atmospheric air. Thus, the austempered ductile iron (ADI) is obtained.

2.2 Turning Process Experiments

The turning process was performed by using conventional lathe machine. The tool used for the turning operation is the TiC/TiCN/Al_2O_3 (titanium carbide/titanium carbo-nitride/aluminium oxide)-coated tungsten carbide inserts. Multilayer coating was done on the inert by using chemical vapour deposition (CVD) method. Figure 3 shows the TiC/TiCN/Al_2O_3-coated tungsten carbide insert. Insert holder was used to hold the inserts while machining in the conventional lathe machine and the arrangement is as shown in Fig. 4.

The experiments were done through the DOE with different level of cutting speed, depth of cut and feed rate. The process parameters and their levels are given in Table 2. The experiments were conducted as per L9 orthogonal array table, and material removal rate (MRR) and surface roughness for different combinations of factors were calculated.

Fig. 3 TiC/TiCN/Al_2O_3 coated insert

Fig. 4 Multilayer insert with insert holder

Table 2 Process parameters and their three levels

Process parameters	Level 1	Level 2	Level 3	Units
Cutting speed	500	770	1200	rpm
Depth of cut	0.051	0.102	0.159	mm
Feed rate	0.5	0.75	1	mm/min

Finally, surface roughness was found by using surface roughness tester, and material removal rate (MRR) was calculated by using below equation.

$$\text{Material Removal Rate, MRR} = \pi \times (D_{\text{avg}}) \times d \times f \times N \text{ mm}^3/\text{s}$$

where D_{avg} = average diameter of the rod in mm, d is the depth of cut mm, f is the feed rate in mm/min and N is the cutting speed in rpm.

3 Results and Discussion

3.1 *Determination of Optimal Turning Parameters*

The experimental results of the turning process were given in Table 3. 'Smaller-the-better' quality characteristic was used to find good surface roughness, and larger-the-best was used to find maximum MRR.

Table 3 Experimental results of the turning process

No.	Cutting speed (rpm)	Depth of cut (mm)	Feed rate (mm/min)	MRR (mm^3/s)	Surface roughness, Ra (μm)
1	500	0.051	0.5	30.04	1.005
2	500	0.102	0.75	53.67	1.06
3	500	0.159	1	73.65	0.735
4	770	0.051	0.75	51.41	0.9525
5	770	0.102	1	125.76	0.9325
6	770	0.159	0.5	122.06	2.6175
7	1200	0.051	1	96.45	2.756
8	1200	0.102	0.5	223.73	1.987
9	1200	0.159	0.75	228.66	1.2725

3.2 Signal-to-Noise (S/N) Ratio Analysis

Signal-to-noise ratio gives the optimal process parameters that are influencing the MRR and surface roughness in the turning process of heat-treated ductile iron. The S/N ratio is used to find the rank of influencing process parameters on the MRR and surface roughness. 'larger-the-better' quality characteristic is used to find the optimal process parameters that affect the MRR. The rank of the process parameters based on the delta value is given in Table 4. S/N ratio shows that speed was highly influenced by the MRR followed by feed and depth of cut. Figure 5 shows the S/N ratio plot for MRR. The optimum process parameters to obtain maximum MRR are cutting speed of 1200 rpm, depth of cut of 0.159 mm and feed rate of 1 mm/min.

'Smaller-the-better' quality characteristic was used to found the optimal process parameters that affect the surface roughness. Table 5 shows the rank of the process parameters that affect the surface roughness based on the delta value. S/N ratio shows that the surface roughness was heavily influenced by feed followed by depth of cut and speed. Figure 6 shows the S/N ratio plot for surface roughness. Cutting speed of 1200 rpm, depth of cut of 0.102 mm and feed rate of 0.5 mm/min was identified as the optimum process parameters to obtain minimum surface roughness.

Table 4 S/N ratio response table for material removal rate

Factors	Level 1	Level 2	Level 3	Max–min	Rank
Speed	33.83	39.31	44.62	10.79	1
Feed	34.49	41.19	42.09	7.6	2
Depth of cut	39.43	38.67	39.67	1.01	3

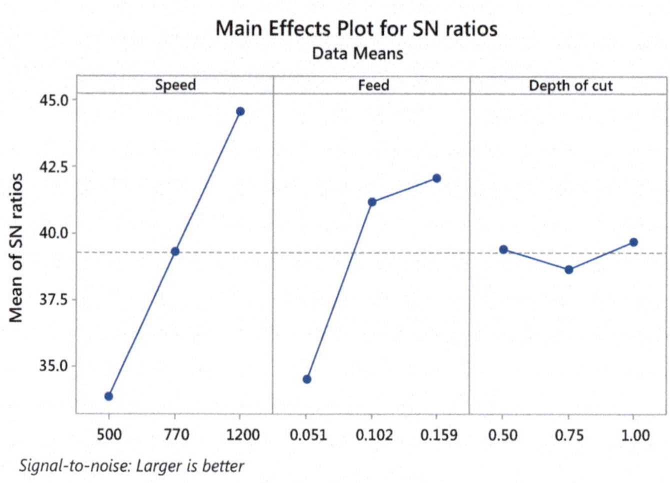

Fig. 5 Main effects of S/N ratio plot for MRR

Table 5 S/N ratio response table for surface roughness

Factors	Level 1	Level 2	Level 3	Max–min	Rank
Speed	−4.391	−3.906	−3.745	0.645	3
Feed	−2.485	−2.249	−7.308	5.059	1
Depth of cut	−2.051	−4.027	−5.964	3.913	2

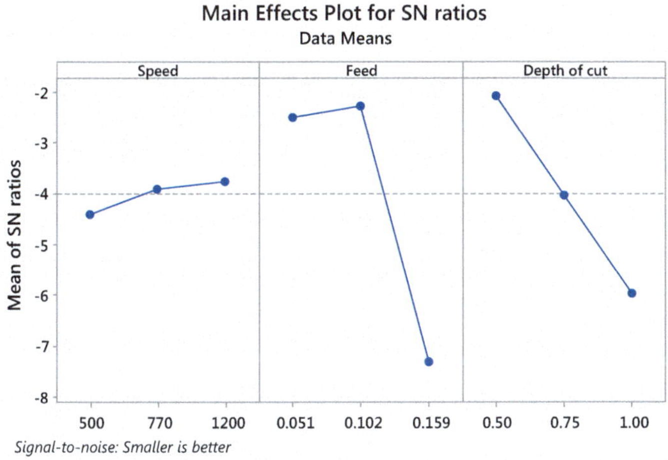

Fig. 6 Main effects of S/N ratio plot for surface roughness

3.3 Analysis of Variance (ANOVA) and the Effect of Factors

ANOVA was carried out to find out the significance of the various process parameters on the experiments. Table 6 shows the result of ANOVA. This analysis was conducted for a confidence interval of 95% and a significant level of 5%. The result shows that speed had a greater effect (62.52%) on the MRR followed by feed (29.69%) and depth of cut (2.54%). The error shows only 5.23% contribution on the MRR which is less than 15%. It shows that no significant process parameter is missing in the experiment.

Table 6 Analysis of variance table for MRR

	Degree of freedom	Sum of squares	Mean sum of squares	Contribution (%)	Rank
Speed	2	26,187	13093.4	62.52119	1
Feed	2	12,437	6218.7	29.69321	2
Depth of cut	2	1067	533.3	2.547451	3
Error	2	2194	1097.1	5.238152	

Table 7 Analysis of variance table for surface roughness

	Degree of freedom	Sum of squares	Mean sum of squares	Contribution (%)	Rank
Speed	2	0.1273	0.06367	3.708451	3
Feed	2	2.2221	1.11107	64.7333	1
Depth of cut	2	0.952	0.47598	27.73327	2
Error	2	0.1312	0.06562	3.822064	

Table 7 shows the result of ANOVA for surface roughness. The analysis shows that feed had a greater effect (64.73%) on the surface roughness followed by depth of cut (27.73%) and speed (3.70%). The error shows only 3.88% contribution on the surface roughness which is less than 15%. It shows that no significant process parameter is missing in the experiment.

3.4 Confirmation Test

After selecting the optimum process parameters, the confirmation experiment is done with optimum process parameters. The predicted MRR was calculated by using a linear regression model. The linear regression equation obtained using MINITAB 17 is given below.

$$\text{MRR} = -140.4 + 0.2350\ \text{Speed} + 1048\ \text{Feed} - 53.2\ \text{Depth of cut}$$

The confirmation experiment was done with the combination of optimum process parameters, and the optimum MRR was calculated. Table 8 shows the results of the confirmation experiment.

The predicted surface roughness is calculated by using a linear regression model. The linear regression equation obtained using MINITAB 17 is given below.

$$\text{Ra} = -0.294 - 0.000287\ \text{Speed} + 9.81\ \text{Feed} + 1.588\ \text{Depth of cut}$$

The confirmation experiment was done with the combination of optimum process parameters, and the optimum surface roughness was calculated. Table 9 shows the results of the confirmation experiment.

Table 8 Results of the confirmation experiment for MRR

Factor	Predicted MRR (mm^3/s)	Experimental MRR (mm^3/s)
MRR	255. 03	258.58

Table 9 Results of the confirmation experiment for surface roughness

Factor	Predicted surface roughness (μm)	Experimental surface roughness (μm)
Ra	1.1562	1.1965

4 Conclusions

Turning operation is performed on the heat-treated ductile iron using $TiC/TiCN/Al_2O_3$ (titanium carbide/titanium carbo-nitride/aluminium oxide)-coated tungsten carbide inserts. L9 orthogonal array table was used to study the effect of MRR and surface roughness. Taguchi method was adopted to investigate the effects of process parameters such as cutting speed, depth of cut and feed rate on MRR and surface roughness of ductile iron. The major conclusions drawn from the results are as follows:

The optimum process parameters for higher MRR are speed of 1200 rpm, depth of cut of 0.159 mm and feed rate of 1 mm/min identified, and ANOVA analysis shows that speed had a greater effect (62.52%) on the MRR followed by feed (29.69%) and depth of cut (2.54%).

The optimum process parameters for lower surface roughness are cutting speed of 1200 rpm, depth of cut of 0.102 mm and feed rate of 0.5 mm/min identified, and ANOVA analysis shows that feed had a greater effect (64.73%) on the surface roughness followed by depth of cut (27.73%) and speed (3.70%).

References

1. Hascalık A, Caydas U (2009) Optimization of turning parameters for surface roughness and tool life based on the Taguchi method. Int J Adv Manuf Technol 38:896–903
2. Davim JP, Mata F (2005) Optimisation of surface roughness on turning fiber-reinforced plastics (FRPs) with diamond cutting tools. Int J Adv Manuf Technol 26:319–323
3. Bagci E, Isık B (2006) Investigation of surface roughness in turning unidirectional GFRP composites by using RS methodology and ANN. Int J Adv Manuf Technol 31:10–17
4. Pawade RS, Joshi SS (2011) Multi-objective optimization of surface roughness and cutting forces in high- speed turning of Inconel 718 using Taguchi grey relational analysis (TGRA). Int J Adv Manuf Technol 56:47–62
5. Tulasi R, Singh R, Ali MI (2018) Optimizing surface roughness in turning operation using taguchi technique. Mater Today Proc 5(9):19043–19048

Optimization of Drilling Process in Heat-Treated Al–20% SiC Functionally Graded Composite Using Grey Relational Analysis

K. Vinoth Babu, S. Prathap Singh, S. Marichamy, P. Ganesan, and M. Uthayakumar

Abstract The objective of this work is to fabricate SiC particle-reinforced aluminium functionally graded metal matrix composite (Al-FGMMC) and optimization of drilling process parameters using grey relational analysis in the heat-treated FGMMC specimen. The A356–20% SiCp-reinforced functionally graded composite has been fabricated by liquid metal stir casting technique followed by the vertical centrifugal casting process. The microstructural evaluation indicated that the maximum reinforcement concentration was found at the outer periphery of the disc. The composite specimens were heat-treated and Ti–Al–N-coated carbide tool was used for the drilling studies. The effect of drilling parameters such as cutting speed, feed, point angle and three different zones (15, 45 and 75 mm from the outer periphery) on the surface roughness and thrust force was analysed using grey relational analysis. Analysis of variance (ANOVA) was also employed to study which drilling parameters significantly influence the drilling process. The results revealed that speed had the highest significance on the drilling process followed by feed, point angle and zone distance.

Keywords Functionally graded materials · Heat treatment · Drilling · Grey relational analysis · ANOVA

K. Vinoth Babu · S. Marichamy · P. Ganesan
Department of Mechanical Engineering, Sri Indu College of Engineering and Technology, Hyderabad, India

S. Prathap Singh (✉)
Department of Mechanical Engineering, St. Joseph's Institute of Technology, Chennai, India
e-mail: prathap.singh50@gmail.com

M. Uthayakumar
Faculty of Mechanical Engineering, Kalasalingam University, Krishnankoil, India

L.-J. Yang et al. (eds.), *Proceedings of ICDMC 2019*,
Lecture Notes in Mechanical Engineering,
https://doi.org/10.1007/978-981-15-3631-1_44

1 Introduction

Functionally graded metal matrix composite (FGMMC) has a great demand in the automotive and aerospace industry due to its gradient properties. FGMMCs possess gradual change in the composition and microstructure in the single part which helps to attain multifunctional behaviour [1]. These materials are fabricated by many techniques such as centrifugal casting, powder metallurgy, solid free-form process and vapour deposition technique. Among these various techniques, centrifugal casting technique is the most economical technique [2]. During this centrifugal casting, high-density reinforcement particles are thrown to the outer surface of the composite. Because of this centrifugal action, reinforcement particles will form different zones of concentrations in the castings [3]. Most of the researchers are doing continuous efforts on the drilling studies of aluminium composites. Vinoth Babu et al. [4] studied the drilling performance of Al–SiC composite. The effects of surface roughness and thrust force on the drilling of the composite with a varied level of process parameter such as cutting speed, feed and point angle were studied. The result shows that the point angle had a greater influence on the drilling performance followed by feed and cutting speed. Ramkumar et al. [5] analysed the effect of feed rate, spindle speed and processing technique on the drilling thrust force and surface roughness. The result concluded that feed rate and processing technique had a greater effect on the drilling thrust force and surface roughness, and ANOVA was done to find the contribution of parameters on the drilling of the composite. Rajmohan and Palanikumar [6] employed GRA to obtain optimal parameter in drilling hybrid MMC and orthogonal array for the experimental plan. The chosen drilling parameters are speed, feed, drill type and SiC weight percentage, and the multiple performance characteristics are thrust force, surface roughness and torque. The contribution of parameters is studied using ANOVA. The result shows that the feed rate and drill type are the most significant factors which affect performance.

Several studies have been done only on the drilling of Al–SiC composites. However, drilling studies on heat-treated Al–SiC functionally graded metal matrix composite are limited. So, the aim of the present study is to fabricate aluminium reinforced with SiC functionally graded metal matrix composite (Al-FGMMC) by vertical centrifugal casting technique and optimizes the drilling parameters for heat-treated Al-FGMMC using grey relational analysis.

2 Materials and Methods

2.1 Raw Materials and Its Composition

Aluminium alloy A356 is used as a base material for the fabrication of FGMMC because of its heavy engineering applications due to its strength, workability and weldability. The elements of aluminium alloy A356 and its weight percentage are shown in Table 1.

Table 1 Elements of aluminium ally A356 and its weight percentage

Element	Si	Mg	Cu	Fe	Zn	Al
wt%	7.0	0.35	0.2	0.2	0.1	Balance

The reinforcing material used for the fabrication of FGMMC is silicon carbide (SiC) with an average size of 23 μm (20 wt%) because of its high wear resistance. The density of silicon carbide is 3.21 g/cm^3.

2.2 *Fabrication of Composite*

Aluminium alloy A356 reinforced with silicon carbide (20 wt% SiC) functionally graded metal matrix composite was fabricated by liquid stir casting technique followed by vertical centrifugal casting technique. Initially, aluminium alloy was heated to 740 °C in an electrical resistance furnace and then silicon carbide particle was added to the molten aluminium melt. Mechanical stirring was done on the molten aluminium melt with a speed of 500 rpm. Then, the molten melt was poured into a vertical centrifugal casting machine rotating with a speed of 1200 rpm. Due to the centrifugal action, the higher-density silicon carbide particles are thrown to the outer zone of the composite. After that, the molten melt was allowed to cool for 6–7 h in the room temperature. The fabricated disc with 250-mm diameter and 25-mm thickness is shown in Fig. 1. The microstructural examination was done to determine the distribution of SiC particle on the fabricated FGMMC.

Fig. 1 Al–20% SiC FGMMC

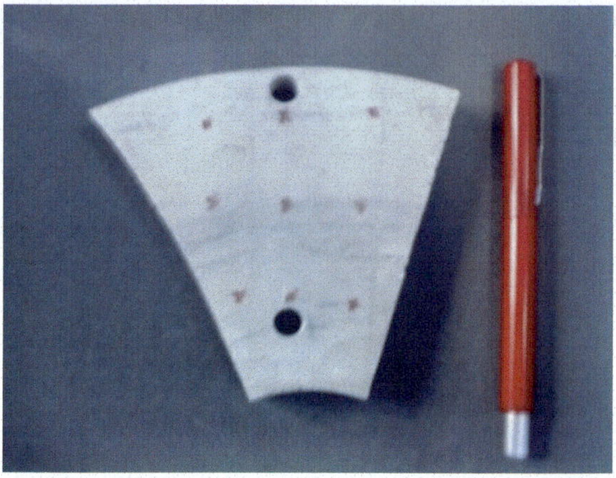

Fig. 2 Typical specimen cut from the heat-treated FGMMC

Heat treatment was done on the fabricated FGMMC in order to enhance the properties of the FGMMC. Initially, solution treatment was done with temperature of 525 °C for 20 h and then ageing process was made with temperature of 165 °C for 8 h. Figure 2 shows the typical specimen cut from the heat-treated FGMMC. There are three different zones selected on the heat-treated FGMMC based on microstructural examination to study the drilling performance.

2.3 Drilling Operation

Drilling operations were performed on a Hartford CNC vertical machining centre with Kistler piezoelectric dynamometer. The Ti–Al–N-coated 5-mm diameter carbide tool was selected for the drilling studies. The surface roughness of drilled holes was found using the Mitutoyo SJ-201 surface analyser. The thrust force was measured by Kistler piezoelectric dynamometer which is connected with the CNC drilling machine.

2.4 Selection of Control Parameters and Their Levels

Taguchi's design of experiment is a method used to evaluate the design parameters. This method gives high-quality product design with a minimum number of experiments at low time and cost. So, this method was used to study the influence of control parameters on the thrust force and surface roughness. The experiment used four control parameters such as speed, feed, point angle and zone distance with varying SiC wt%. The control parameters and their levels are given in Table 2. The L9 orthogonal array was chosen for the experimentation.

Table 2 Control parameters and their levels

Symbol	Control parameters	Level 1	Level 2	Level 3	Unit
A	Speed	1000	2000	3000	rpm
B	Feed	0.05	0.15	0.25	mm/rev
C	Point angle	90	120	140	Degree
D	Zone distance with varying SiC wt%	15	45	75	mm

3 Results and Discussion

3.1 Microstructural Examination

Figure 3 shows the optical microscope of three regions (15, 45 and 75 mm) from the outer periphery of the heat-treated FGMMC. The microstructure (Fig. 3a) observed at a distance of 15 mm from the outer periphery shows higher volume percentage of SiC particle because of the centrifugal action. This zone is called the SiC particle-rich zone. The microstructure (Fig. 3c) observed at a distance of 75 mm from the outer periphery shows lower volume percentage of SiC particle. This zone is called SiC particle depletion zone. The zone between SiC particle-rich zone and SiC particle depletion zone is called SiC particle intermediate zone because this region shows only the moderate reinforcement and their microstructure is shown in Fig. 3b.

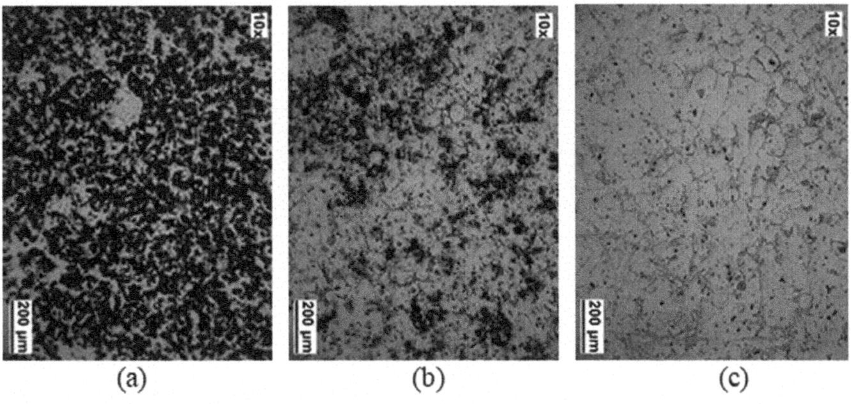

Fig. 3 **a** Microstructure of SiC particle-rich zone; **b** microstructure of SiC particle intermediate zone; **c** microstructure of SiC particle depletion zone

Table 3 Experimental results and their S/N ratio

Trial No.	Parameter				Experimental result		S/N ratio	
	Speed (rpm)	Feed (mm/rev)	Point angle (°)	Zone distance with varying SiC wt% (mm)	Surface roughness Ra (μm)	Thrust force F (N)	Surface roughness (dB)	Thrust force (dB)
1	1000	0.05	90	15	2.83	42.8	2.83	42.8
2	1000	0.15	120	45	1.77	58	1.77	58
3	1000	0.25	140	75	2.62	92	2.62	92
4	2000	0.05	120	75	2.95	98.5	2.95	98.5
5	2000	0.15	140	15	2.24	84.8	2.24	84.8
6	2000	0.25	90	45	4.65	202	4.65	202
7	3000	0.05	140	45	2.910	69	2.91	69
8	3000	0.15	90	75	4.410	204	4.41	204
9	3000	0.25	120	15	2.980	144.6	2.98	144.6

3.2 Experimental Details

The resulting thrust force and surface roughness of the heat-treated FGMMC by conducting 9 experiments obtained from L9 orthogonal array and their S/N ratio are shown in Table 3. The S/N ratio is used to find the optimal condition of the parameter levels. S/N ratio is calculated with "smaller-the-better" characteristic to find minimum thrust force and surface roughness.

3.3 Muitiresponse Optimization Using Grey Relational Analysis

Grey relational analysis methodology was used for multiresponse optimization. In grey relational analysis optimization, the complex multiple responses were converted into a single response grey relational grade. Initially, data processing was done on the experimental results of the thrust force and surface roughness to normalize the results in the range between 0 and 1. "Smaller-the-better" quality characteristic was selected because minimum values are desirable for the thrust force and surface roughness. The normalization was done by Eq. (1), and the normalized values are shown in Table 4.

$$X_{ij} = \frac{\left(y_{ij}\right)_{\max} - y_{ij}}{\left(y_{ij}\right)_{\max} - \left(y_{ij}\right)_{\min}} \tag{1}$$

Table 4 Grey relational coefficient and grey relational grade

Trial No.	Normalized value		Grey relational coefficient		Grey relational grade	Rank
	Surface roughness	Thrust force	Grey relational coefficient Ra	Grey relational coefficient F		
1	0.632	1.000	0.576	1.000	0.788	2
2	1.000	0.895	1.000	0.826	0.913	1
3	0.705	0.695	0.629	0.621	0.625	5
4	0.590	0.654	0.550	0.591	0.570	6
5	0.837	0.739	0.754	0.657	0.706	3
6	0.000	0.012	0.333	0.336	0.335	9
7	0.604	0.837	0.558	0.755	0.656	4
8	0.083	0.000	0.353	0.333	0.343	8
9	0.580	0.368	0.543	0.442	0.493	7

where y_{ij} is the ith experimental value of the jth experiment. After finding the normalized values, the grey relational coefficient was calculated by using Eq. (2) and also the resultant values of the grey relational coefficient are shown in Table 4.

$$\delta_{ij} = \frac{\min_i \min_j \left|x_i^0 - x_{ij}\right| + \xi \max_i \max_j \left|x_i^0 - x_{ij}\right|}{\left|x_i^0 - x_{ij}\right| + \xi \max_i \max_j \left|x_i^0 - x_{ij}\right|} \tag{2}$$

where x_i^0 is the ideal normalized value of the ith experiment and the value of identification coefficient ξ is taken as 0.5 because equal weight was given to all parameters. Then, grey relational grades of all the parameters are converted into single grey relational grade by using Eq. (3) and the resultant grey relational grade is shown in Table 4.

$$\alpha_j = \frac{1}{m} \sum_{i=1}^{m} \delta_{ij} \tag{3}$$

where α_j is the grey relational grade for the jth experiment and m is the number of performance characteristics. Finally, mean grey relational grade for the speed at each level was calculated by taking the average of the grey relational grade for the experiments. The mean grey relational grade for each level can be computed in a similar manner. Then, the response table was calculated by using mean grey relational grade and the response table is shown in Table 5. The result of response table shows that speed had more significance on the thrust force and surface roughness followed by feed, point angle and zone distance with varying SiC wt%. Figure 4 shows the effect of control parameters on the multiresponse characteristics. The optimal parameters for minimum thrust force and surface roughness are speed 1000 rpm, feed 0.05 mm/rev, point angle 140° and zone form the outer periphery 15 mm (A1B1C3D1).

Table 5 Response table for grey relational grade

Factors	Level 1	Level 2	Level 3	Max–min	Rank
Speed (A)	0.7753	0.5370	0.4974	0.2779	1
Feed (B)	0.6716	0.6539	0.4841	0.1875	2
PA (C)	0.4886	0.6587	0.6623	0.1737	3
Zone (D)	0.6621	0.6347	0.5128	0.1493	4

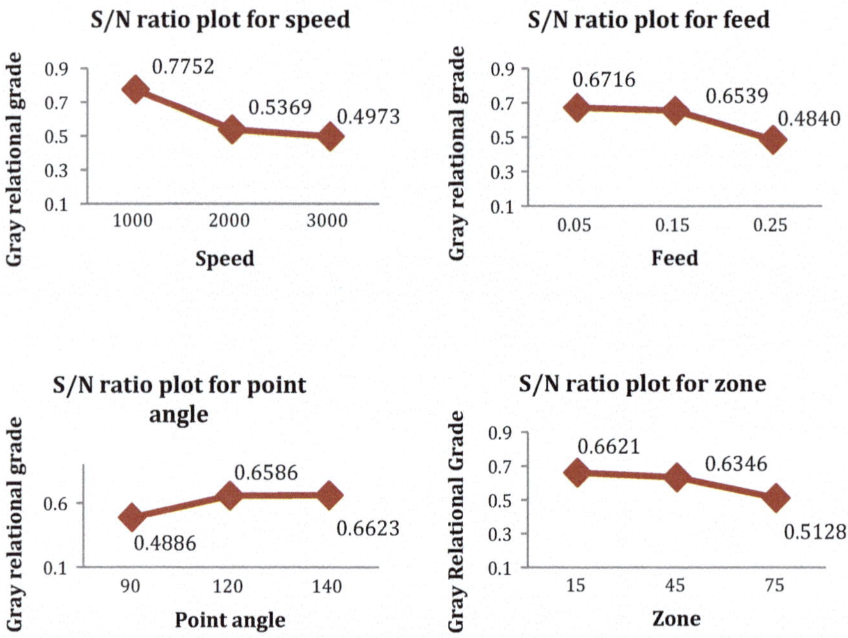

Fig. 4 Effect of control parameters on the multiresponse characteristics

3.4 *Analysis of Variance*

Analysis of variance was carried out to find the percentage of the significance of each parameter on the thrust force and surface roughness. The percentage contribution of parameters such as speed, feed, point angle and zone distance with varying SiC wt% and their interactions on the thrust force and surface roughness are shown in Table 6. The analysis was performed with a confidence level of 95% and a significance level of 5%. It was found that speed had the greatest significance (45.66%) on the thrust force and surface roughness followed by feed (21.66%), point angle (19.09%) and zone distance with varying SiC wt% (12.75%).

Table 6 Results of ANOVA for grey relational grade

	Degree of freedom	Sum of squares	Mean sum of squares	Contribution (%)	Rank
Speed (A)	2	0.1356	0.0678	45.664	1
Feed (B)	2	0.0643	0.0322	21.668	2
Point angle (C)	2	0.0591	0.0296	19.909	3
Zone (D)	2	0.0379	0.0189	12.759	4
Total	8	0.2969	0.1484	100	

Table 7 Results of the confirmation experiment for multi-performance characteristics

	Initial parameter (rank 1 grey code)	Optimal machining parameter	
		Predicted	Experimental
Level	A1B2C2D2	A1B1C3D1	A1B1C3D1
Grey grade	0.913	0.962	0.970
Improvement in grey relational grade		0.049	0.057

3.5 Confirmation Test

The final step is to examine the improvement of the standard characteristics using optimal levels of parameters. The optimal parameters are speed 1000 rpm, feed 0.05 mm/rev, point angle 140° and zone form the outer periphery 15 mm (A1B1C3D1). Table 7 shows the comparison of the predicted value and the actual design value. The improvement was noticed in the experiment when optimum parameters were used.

4 Conclusions

Aluminium reinforced with 20 wt% SiC functionally graded metal matrix composite was fabricated by stir casting technique followed by vertical centrifugal casting technique, and drilling study was performed on the heat-treated Al-FGMMC using a CNC drilling machine. The following conclusions are drawn from the above research.

The speed had a higher influence on the performance, and it is observed that their contribution is 45.66% followed by feed that had the contribution of 21.66%. The point angle and zone distance with varying SiC Cwt% had a lower contribution on the performance, and their contributions are 19.09% and 12.75%, respectively.

The optimum parameters were predicted through grey relational analysis (speed = 1000 rpm, feed = 0.05 mm/rev, point angle = 140° and zone distance with varying SiC wt% = 15 mm).

The confirmation experiment is conducted to verify the optimum parameters. It shows better performance.

References

1. Radhika N, Raghu R (2017) Effect of centrifugal speed in abrasive wear behavior of Al-Si5Cu3/SiC functionally graded composite fabricated by centrifugal casting. Trans Indian Inst Met. https://doi.org/10.1007/s12666-017-1204-9
2. Jayakumar E, Praveen AP, Rajan TPD, Pai BC (2018) Studies on tribological characteristics of centrifugally cast SiCp-reinforced functionally graded A319 aluminium matrix composites. Trans Indian Inst Met. https://doi.org/10.1007/s12666-018-1442-5
3. Jayakumar, E, Jacob JC, Rajan TPD, Joseph, MA, Pai BC (2016) Processing and characterization of functionally graded aluminum (A319)-SiCp metallic composites by centrifugal casting technique. Metall Mater Trans. https://doi.org/10.1007/s11661-016-3558-8
4. Vinoth Babu K, Uthayakumar M, Winowlin Jappes JT, Rajan TPD (2015) Optimization of drilling process on Al-SiC Composite using grey relation analysis. Int J Manuf Mater Mech Eng. https://doi.org/10.4018/ijmmme.2015100102
5. Ramkumar T, Selvakumar M, Mohanraj M, Chandrasekar P (2019) Experimental investigation and analysis of drilling parameters of metal matrix (Ti/TiB) composites. J Braz Soc Mech Sci Eng. https://doi.org/10.1007/s40430-018-1507-8
6. Rajmohan T, Palanikumar T (2011) Optimization of machining parameters for multi-performance characteristics in drilling hybrid metal matrix composites. J Compos Mater 46(7):869–878. https://doi.org/10.1177/0021998311412635

Investigation of Defects in Deep Drawn Al1050/SS304 Sandwich Composite by Using Non-destructive Techniques

Atul S. Takalkar and Lenin Babu Mailan Chinnapandi

Abstract The occurrence of the defects is observed in the sheet metal components due to the faulty manufacturing process. Such defects need to be precisely detected without damaging the component, which is only possible by using non-destructive test (NDT). The NDT is preferred for the identification of internal defects such as voids, flaws, and cracks. The present study investigates the defects in the deep drawn Al1050/SS304 sandwich composite cup by using radiography test and scanning electron microscope (SEM). The radiography test is carried out by using gamma rays for the detection of the defects like tearing and flange wrinkling in the deep drawn cup, whereas the SEM is used for the identification of the bonding defects between the adhesive layer and metal blanks. The defects like tearing and wrinkling observed in the sandwich composite skin (Al1050 and SS304 sheet) are successfully detected by using radiography test, whereas the bonding defects between the adhesive layer and metal sheets are identified using SEM.

Keywords Non-destructive testing · Radiography test · Scanning electron microscope · Deep drawing · Sandwich composite · Al1050 · SS304

1 Introduction

The defects such as tearing, wrinkling, spring back, earing, voids, flaws, and cracks are frequently observed in the sheet metal manufactured components which are difficult to predict and inspect [1]. Few of these defects can be predicted by the generation of forming limit diagram (FLD) [2], but the inspection techniques need to be developed to inspect the formed components. This can be overcome by the utilization of NDT which can save time and cost. Most of the authors had preferred radiography technique for the investigation of the defects in composite materials. Such composite materials are used in the critical structure having application in

---

A. S. Takalkar · L. B. Mailan Chinnapandi (✉)
School of Mechanical and Building Sciences, Vellore Institute of Technology, Chennai, Tamil Nadu 600127, India
e-mail: lenin.babu@vit.ac.in

L.-J. Yang et al. (eds.), *Proceedings of ICDMC 2019*,
Lecture Notes in Mechanical Engineering,
https://doi.org/10.1007/978-981-15-3631-1_45

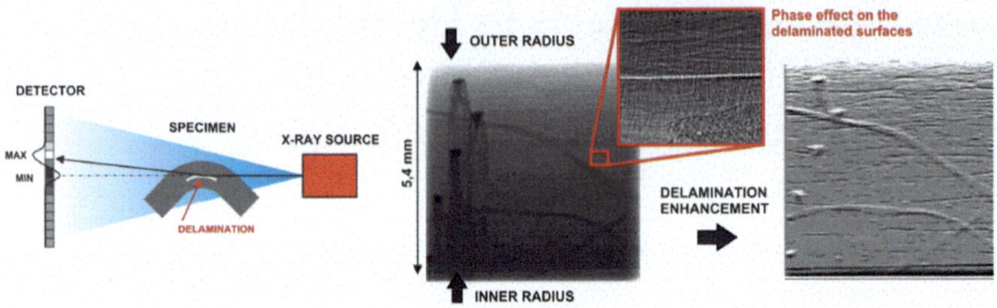

Fig. 1 De-lamination detection and imaging setup for honeycomb sandwich composites and carbon fiber reinforced polymers [6]

aerospace industries which need to be verified assuring the structural integrality. The challenges observed in investigating such complex advance composite materials can be overcome by developing the NDT.

The use of the NDT tools in the composite applications was overviewed by Djordjevic [3], whereas Dragan and Swiderski [4] detected the defects in the composite structures used in aerospace applications using modern approaches. The non-destructive evaluation methods were demonstrated for the maintenance and implementation of quality control of the composite components.

Similarly, Jandejsek et al. [5] utilized WidePIX X-ray imaging techniques for the inspection of the composite material used for the aircraft structures. The defects in honeycomb sandwich composites and carbon fiber reinforced polymers were detected and measured by using imaging method as shown in Fig. 1. Also, Lopato et al. [6] presented the capabilities of terahertz imaging technology in the investigation of a different composite material such as glass fiber reinforced laminates, carbon fiber reinforced laminates, an anticorrosion coating. The new defect detects algorithm was proposed which enables effective data collection and accurate detection of the defects.

The database for the strength and subcritical damage mechanism was established by Atas and Soutis [7] for various bolted joint geometries in a wide range of quasi-isotropic layups and crossply of HTS40/977-2 CFRP composite laminates. The X-ray radiography was utilized for defining the subcritical damage locations. Similarly, Wu et al. [8] investigated the defects such as shrinkage porosity, shrinkage void, and delamination in carbon fiber reinforced aluminum matrix composite by correlating the radiographic image and microstructure. The shrinkage porosity and shrinkage void were quantified with the help of Hole-style penetrameter. The comparative study on the development of non-destructive control for composite materials was carried out by Bayraktar et al. [9]. The advantages and performance of X-rays computed tomography medical scanner was reviewed for the different scales used for different industrial applications.

Few authors had focused on the investigation of bonding defects in sandwich composite sheets. Tallafuss et al. [10] utilized the different NDT like ultrasound, thermography, shearography, and guided wave electromagnetic acoustic transducers for the detection of the bond defects in Aluminum tin (Al–Sn) alloy/steel bimetal strips. The obtained NDT results were correlated with destructive peel test of the known body. Similarly, Olafsson et al. [11] checked the ability to access the defects at the interface between sheet face and core material of the sandwich composite material consisting of glass fiber reinforced polymer and PVC foam cores by using thermographic non-destructive evaluation techniques like pulse thermography, pulse phase thermography, and C-scan ultrasonic testing. The pulse phase thermography technique was observed to be effective in the identification of size and shape of the defect. The thermography and shearography techniques were also used by Georges et al. [12] for the inspection of the composite hybrid sandwich structure consist of CFRP and GFRP core and Titanium skins.

The defects like a tear, wrinkle, de-lamination, voids, and crack initiation are not able to identify due to the limitation of an eye. This may lead to sudden failure of the formed composite component. To avoid such conditions, the non-destructive radiography test has been carried out to investigate the defects in deep drawn Al1050/SS304 sandwich composite cup. Along with the surface defects, the internal defects are inspected by using SEM.

2 Material and Techniques

The present investigation is carried out on Al1050/SS304 sandwich composite sheet manufactured by sandwiching Al1050 and SS304 blank of thickness 0.48 and 0.49 mm by using structural adhesive and activator. The structured adhesive (polyurethane) is layup on the Al1050 blank, whereas the activator is applied on SS304 blank to manufacture the sandwich composite sheet with the thickness of 1.05 mm. The activator is used to activate the SS304 blank surface which enhances the bonding properties of the sandwich composite sheet. The process followed for the manufacturing of Al1050/SS304 sandwich composite material is shown in Fig. 2.

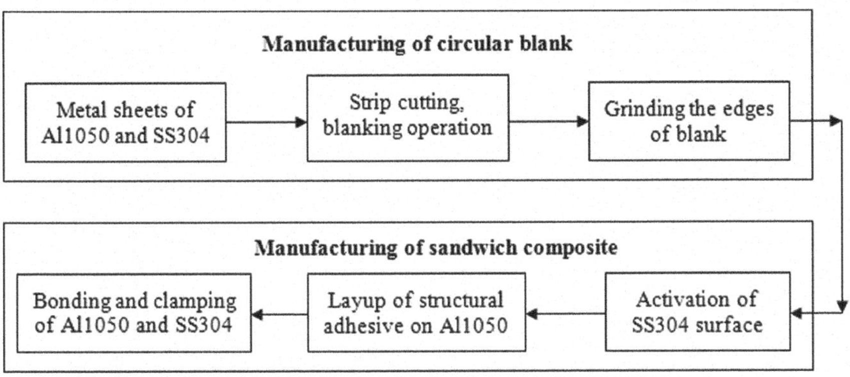

Fig. 2 Process for manufacturing of the Al1050/SS304 sandwich composite sheet

The manufactured material is further used for the deep drawing and inspected by using radiography test and scanning electron microscope.

2.1 Radiography Test

The NDT radiography test is selected for inspecting the defects in the Al1050/SS304 sandwich composite deep drawn cup. The radiography test can be performed by using gamma rays produced from radioactive isotope or X-rays produced from X-ray tube. In the present study, the radiography test is performed by using gamma rays. The setup consists of a radioactive source, stand, and developing film as shown in Fig. 3. The gamma radiation is passed through the testing sample on photographic film, and the internal cup structure is deposited on the photographic film. The radiations which

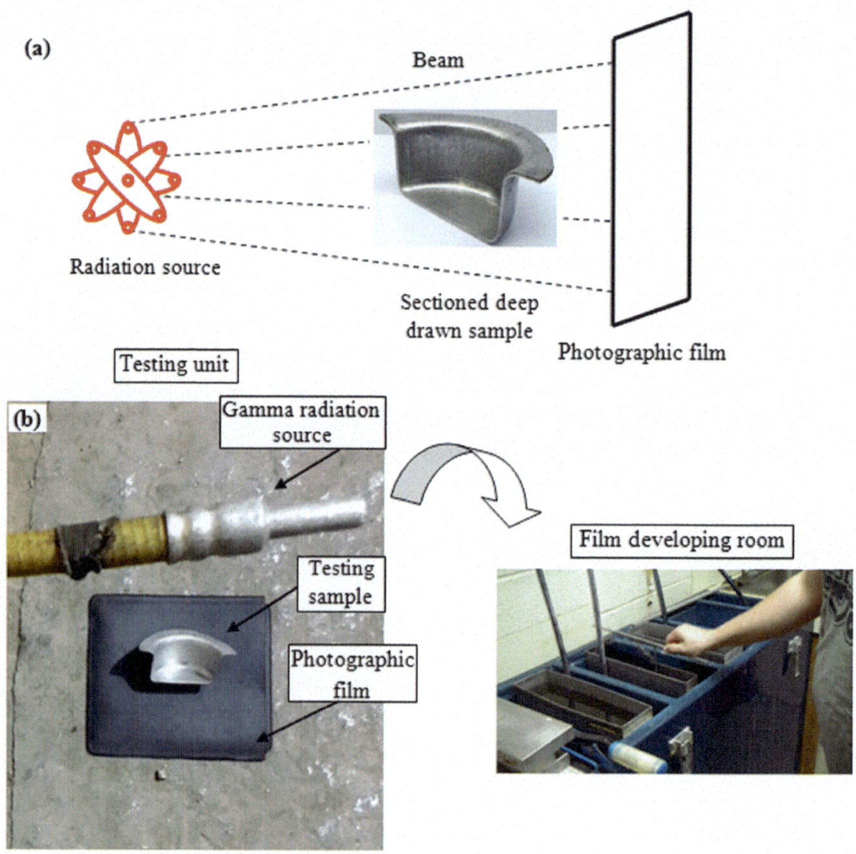

Fig. 3 **a** Schematic representation of the radiography test setup and **b** experimental setup for radiography test

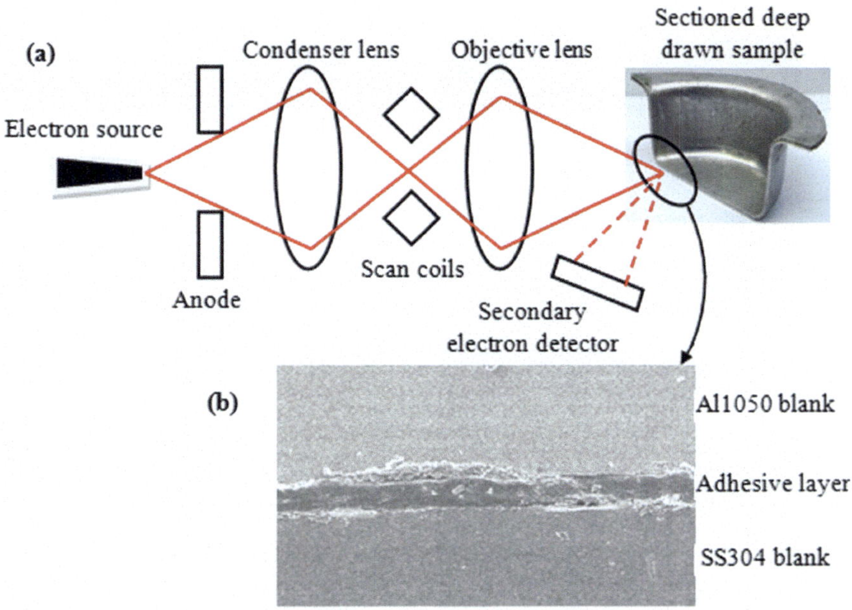

Fig. 4 **a** Schematic diagram of scanning electron microscope setup and **b** scanning electron microscope image of sandwiched Al1050 blank with the SS304 blank using structural adhesive

are not absorbed by the defects expose photographic film at a higher rate resulting in dark marks, whereas the other areas remain lighter.

2.2 Scanning Electron Microscope (SEM)

The SEM is an image capturing technique used for the detection of the variation in the structural adhesive layer utilized for the bonding of the Al1050 and SS304 blank. The setup consists of electron source, anode, condenser lens, scan coils, objective lens, and secondary electron detector as shown in Fig. 4a. The image has been captured at the sectioned location where the three different layers are observed, namely Al1050, adhesive, and SS304, as shown in Fig. 4b.

3 Results

The radiography test is performed for the investigation of defects in deep drawn Al1050/SS304 sandwich composite cup. The developed film from the radiography test for the deep drawn Al1050/SS304 sandwich composite cup is shown in Fig. 5. It is observed that the major defects such as wrinkling as shown in Fig. 5a and tearing

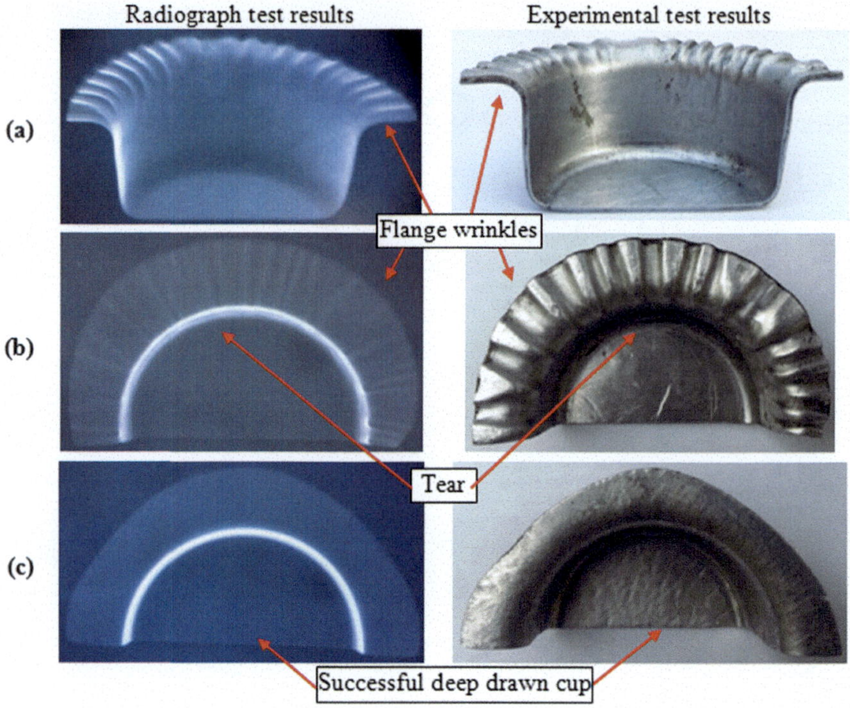

Fig. 5 **a** Sectioned deep drawn sandwich composite cup sample (wrinkled), **b** top view of the sectioned deep drawn cup sample (teared and wrinkled), and **c** top view of the sectioned deep drawn cup sample (successful)

as shown in Fig. 5b can be identified by using a radiation test. The successful deep drawn Al1050/SS304 composite material without any defect is shown in Fig. 5c. Apart from the tearing and wrinkling, the different defects are also observed like de-lamination, flaws, and voids. These observed defects are not able to identify by performing radiography test. This is overcome by utilizing the scanning electron microscope technique to identify the voids between the adhesive layer and meal skin for the sectioned deep drawn cup as shown in Fig. 6a. The bottom image of deep drawn composite cup captured by SEM is shown in Fig. 6b. The voids observed at the bottom of the deep drawn cup are shown in Fig. 6c.

4 Conclusion

The literature reveals that different NDT can be utilized for the identification of the defects in the formed component. The Eddy current test is specially used for conductive sheet metals. Due to the presence of non-conductive materials like adhesive layer in the Al1050/SS304 sandwich composite sheet results in ineffective in the detection

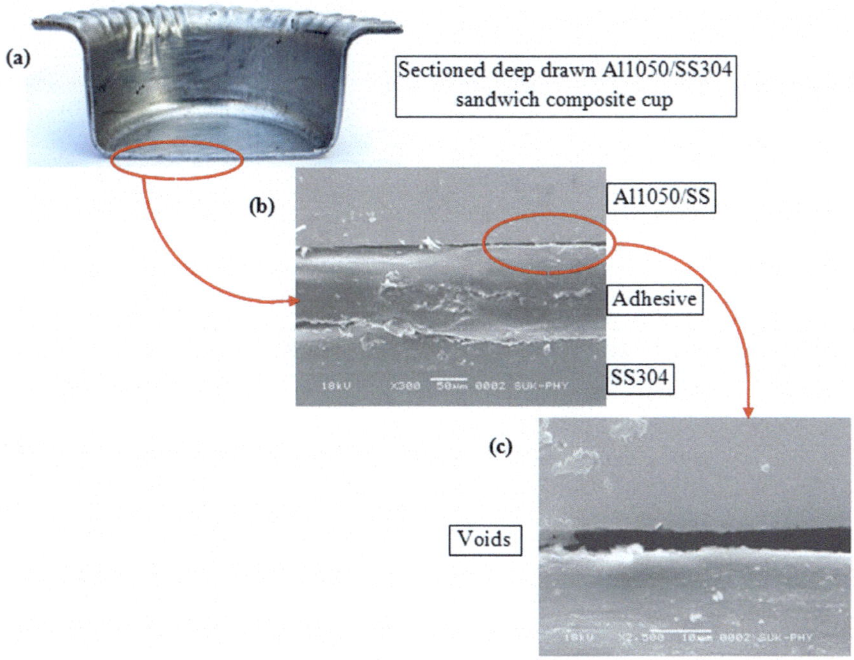

Fig. 6 **a** Sectioned deep drawn Al1050/SS304 sandwich composite cup sample (wrinkled), **b** SEM image of bottom part of a deep drawn Al1050/SS304 sandwich composite cup, **c** void at the bottom part of a deep drawn Al1050/SS304 sandwich composite cup

of the defects using the eddy current test. In case of vibration test, the clamping of the deep drawn cup is found to be difficult which results in the generation of the low vibration in the deep drawn component. In case of ultrasonic test, the travel of sound through Al1050/S304 sandwich composite sheet is ineffective due to low sheet thickness of the sample which further leads to variation echo and defect peaks.

The problems observed in the utilization of other NDT for the identification of defects in Al1050/SS304 sandwich composite cup are successfully overcome by the usage of radiation test. The significant defects like wrinkling and tearing are successfully identified by using radiography test. The radiography test failed to detect the bonding defects like voids observed between the adhesive layer and skin metal sheets which are overcome by using SEM technique.

References

1. Takalkar A, Babu MC (2018) A review on effect of thinning, wrinkling and spring-back on deep drawing process. Proc IMechE Part B J Eng Manuf 1–26. https://doi.org/10.1177/0954405417752509
2. Takalkar A, Koteswara JM, Babu MC (2015) Numerical simulation for predicting failure in deep drawing process using forming limit diagram (FLD). Int J Adv Mech Civ Eng 2(4):11–15
3. Djordjevic B (2009) Nondestructive test technology for the composites. In: The 10th international conference of the Slovenian society for non-destructive testing application of contemporary non-destructive testing in engineering. Ljubljana, Slovenia, pp 259–265
4. Dragan K, Swiderski W (2010) Studying efficiency of NDE techniques applied to composite materials in aerospace applications. Acta Phys Pol A 117(5)
5. Jandejsek I, Jakubek J, Jakubek M, Prucha P, Krejci F, Soukup P, Turecek D, Vavrik D, Zemlicka J (2013) X-ray inspection of composite materials for aircraft structures using detectors of Medipix type. In: 15th international workshop on radiation imaging detectors. Paris. https://doi.org/10.1088/1748-0221/9/05/c05062
6. Lopato P, Chady T, Sikora R (2011) Testing of composite materials using advanced NDT methods. Int J Comp Math Electr Electron Eng 30(4):1260–1270. https://doi.org/10.1108/03321641111133172
7. Atas A, Soutis C (2013) Subcritical damage mechanisms of bolted joints in CFRP composite laminates. Comp Part B 54:20–27
8. Gaohui WU, Yan LI, Jun SU (2008) X-ray testing sensitivity of carbon fiber reinforced aluminium matrix composite. In: 17th world conference on nondestructive testing. Shanghai, pp 25–28
9. Bayraktar E, Antolovich S, Bathias C (2008) New developments in non-destructive controls of the composite materials and applications in manufacturing engineering. J Mater Proc Technol 206:30–44
10. Tallafuss P, Rosochowski A, Campbell S (2017) A feasibility study on different NDT techniques used for testing bond quality in cold roll bonded Al–Sn alloy/steel bimetal strips. Manuf Rev 4:4. https://doi.org/10.1051/mfreview/2017002
11. Olafsson G, Dualde A, Tighe R, Barton J (2017) The influence of sandwich structure core material on thermographic NDT techniques. In: 21st international conference on composite materials Xi'an
12. Georges M, Srajbr C, Menner P, Koch J, Dillenz A (2018) Thermography and shearography inspection of composite hybrid sandwich structure made of CFRP and GFRP core and titanium skins. Proceedings 2(8):484. https://doi.org/10.3390/ICEM18-05384

Effect of Overhauling on the Performance of Boiler

S. Jebarose Juliyana, J. Udaya Prakash, A. Divya Sadhana, and K. Karthik

Abstract The availability of electrical energy and its per person intake is viewed as the living standard in today's civilization. The booming of power production is an indication of rising gross national products which reveals prosperity of the people. For power generation in the thermal power plant, boiler is the steam generator which is used to transform water into superheated steam. It is an attempt to utilize the maximum possible amount of heat energy supplied to furnace in order to get higher performance of boiler. This paper analyzes the factors which affect the boiler's performance such as lignite quality, exit flue gas, excess air and slagging. Theoretical calculations have been carried out to find out the efficiency of boiler before and after major overhauling, and suitable methods have been adopted to improve the effectiveness of the boiler.

Keywords Thermal power plant · Boiler · Lignite · Efficiency · Overhauling

1 Introduction

Power is a basic input for economic development of the country. Hence, it is expected that the power plants perform at their optimum level for the economic and industrial growth of the nation. Certain performance indicators can be defined to monitor and judge the performance of the power plants [1, 2]. For improving energy availability and utilization, design and equipment deficiencies need to be overcome. The growth in demand for power exceeded the generation capacities. The investment of the new plants and gestation period will be exorbitantly high [3]. Power plant performance analysis at number of phases helps in increasing the power generation capacity.

Overall performance of a thermal station can be judged by the following factors: plant maintenance, thermal efficiency, plant load and plant availability [4]. Maintenance and improvements in efficiency monitoring of power plant generation system

S. Jebarose Juliyana (✉) · J. Udaya Prakash · A. Divya Sadhana · K. Karthik
Mechanical Engineering, Vel Tech Rangarajan Dr. Sagunthala R&D Institute of Science and Technology, Avadi, Chennai, Tamil Nadu 600062, India
e-mail: jebarose@veltech.edu.in

L.-J. Yang et al. (eds.), *Proceedings of ICDMC 2019*,
Lecture Notes in Mechanical Engineering,
https://doi.org/10.1007/978-981-15-3631-1_46

are of paramount importance in the electricity supply industry which is highly capital and technology intensive [5]. It is estimated that 1% of improvement heat rate has potential for saving about two million tons of coal per annum with the corresponding reduction of greenhouse gases [6]. It is needless to add that improved performance would also mean increased power availability [5].

Lakhoua [7] presented two methods for the boiler efficiency calculations, on the one hand, and analyzed the different losses at the level of the boiler using the causal analysis, on the other hand. Calculating the boiler efficiency by the input–output method is desirable because of the simplicity of the method but is subject to the inaccuracies of all of the measurements needed [8]. The method of heat losses increases the accuracy of the calculations but, while the number of measurements is decreased, the difficulty of obtaining accurate measurements is increased [9–11]. Finally, it is important to determine the possible causes generating the losses easily [12, 13] and provoking the deterioration of the plant heat rate while using the causal analysis [14, 15].

2 Methods

The methods used for determining the boiler efficiency are input–output method and heat loss method. Specifications of boiler are given in Table 1.

Table 1 Specifications of boiler

S. No.	Parameter	Specifications
1	Boiler capacity	690 T/h, 158 ata, 540 °C
2	Feed water temperature	241.6–304.5 °C
3	Steaming rate	656 T/h
4	Circulation ratio	7
5	Superheated steam pressure	157 kg/cm
6	Superheated steam temperature	540 °C
7	Fuel consumption	207 T/h
8	Exit gas temperature	150 °C
9	Efficiency (rated)	91.6%
10	Load	210 MW
11	Fuel	Lignite, furnace oil

Table 2 Efficiency of boiler based on GCV and NCV

Parameters	Trial 1	Trial 2	Trial 3
Mass flow rate of steam (T/h)	658	655	669
Main steam pressure (kPa)	145	144	152
Main steam temperature (°C)	538	539	535
Feed water pressure (kPa)	158	158	165
Feed water temperature (°C)	241	241	242
Fuel flow	211	195	192
Gross calorific value (kcal/kg)	2800	2800	2800
Net calorific value (kcal/kg)	2400	2400	2400
Enthalpy of main steam (kJ/kg)	3423.24	3427.06	3407.17
Enthalpy of feed water (kJ/kg)	1043.97	1043.97	1048.75
Efficiency η (based on GCV) (%)	64.13	69.18	71.02
Efficiency η (based on NCV) (%)	74.81	80.72	82.86

2.1 *Input–Output Method*

This method needs heat input (fuel) and output (steam) for calculating the efficiency. The useful heat output is the sum of the energy transferred to water, steam and blow down. The total energy input to the boiler comprises a component that depends on the flow of fuel and another that is independent of it [16].

$$\text{Boiler Efficiency} = (\text{Heat Output}/\text{Heat Input}) \times 100$$

$$\text{Output} = \text{Steam flow rate} \times (\text{Enthalpy of Steam} - \text{Enthalpy of Feed water})$$

$$\text{Input} = \text{Fuel Firing Rate} \times \text{Gross Calorific Value.}$$

The input–output method requires several quantities that are difficult to measure accurately.

The trials taken on boiler on various days are shown in Table 2.

2.2 *Heat Loss Method*

In heat loss method, the boiler efficiency is calculated by 100% minus total losses (www.beeindia.gov.in). Thus, if the losses are known, the efficiency can be calculated easily [17, 18]. It has several advantages, one of which is that errors are not so significant.

This method is also known as flue gas analysis approach since the major heat losses are based on the measured flue gas conditions at the boiler exit together with an analysis of the fuel consumption. The various losses are

i. Loss due to flue gas.
ii. Loss due to evaporation of water.
iii. Loss due to moisture content of the fuel.
iv. Loss due to moisture present in the air.
v. Radiation and Convection losses.
vi. Loss due to unburnt fuel.
vii. Loss due to unburnt bottom ash.

The heat loss by the flue gases depends upon the quantity of products of combustion and the gas temperature which leaves from the heat recovery equipment.

3 Problem Identification and Recommended Suggestions

Set of readings was taken to ascertain the current level of performance before major overhauling (MOH). Boiler efficiency was calculated as it includes all the auxiliaries by heat loss method. Heat loss method only provides components of loss, and thereby, the remedies can be easily obtained. Here, boiler efficiency was found lesser than the designed value of 91.6%. Some of the major causes analyzed for the lesser efficiency are listed below.

3.1 Exit Flue Gas Temperature

High flue gas temperature at the boiler exit is a large contributor to efficiency shortfall in the boiler. The boiler is designed for an exit flue gas temperature of 154 °C, but the boiler is operating at 175 °C (actual condition). The main reason for high exit flue gas temperature is ineffective heat transfer in the evaporator.

The heat transfer in the evaporator is ineffective due to:

i. Low ash fusion temperature of ash because of wide variation in fuel quality like ash content, calorific value.
ii. Deposition of ash particles in boiler heat transfer surface (i.e., SH/RH/Economizer coil, etc.).
iii. Slagging in evaporator tubes.
iv. Milling of fuel.
v. Velocity of pulverized fuel to the burner.
vi. Air ingress into the furnace so air through the APH is low.
vii. In effective air heater soot blowers.

viii. Folded and torn elements in RAPH (a problem at the cold end plates because of corrosion).
ix. Low feed temperature has to be remedied by the extra firing in the boiler and this will result in high exit gas temperature.
x. Leakage of air in preheater through seals.

3.1.1 Suggestions

i. Improving the milling effectiveness at a velocity greater than 15 m/s and within 22 m/s and its discharging volume range from 150,000/h 3 m.
ii. Excess air is maintained at 27–30% for complete combustion.
iii. It also depends upon the fineness of the pulverized fuel.
iv. Maintaining the proper air–fuel ratio.
v. Maintaining the temperature of the pulverized fuel is 120–150 °C.
vi. Removing the slag over the evaporator by proper water lance blowing.
vii. Air ingress into the furnace should be nil.

3.2 Fouling Process

The process of sticking of ash particle on the heat transfer surface like furnace, super heater and re-heater is called fouling or slagging process which leads to ineffective heat transfer in evaporator zone.

3.2.1 Suggestions

To reduce the slagging, water lancer and blowers are operated in the combustion zone regularly. Also in the coil areas, steam soot blowers are operated to reduce fouling.

3.3 Excess Air

In the boiler, higher combustion efficiency can be obtained by supplying correct amount of air at the right place. In addition to this, it also requires some additional air for the complete combustion of fuel. This air is known as excess air.

Supplying too much excess air will increase the dry flue gas loss. Apart from the quantity of air deliberately admitted to boiler, they are leaked into the boiler at the following places.

3.3.1 Suggestion

i. Air dampers passing for the non-running mills.
ii. Expansion seals on ducts, etc.
iii. Opening such as manholes and peepholes.
iv. Ash hopper doors open.
v. Ash hopper seals, etc.
vi. Pulverized fuel burner mouth sealing.
vii. Hot flue gas tapping point from furnace for lignite drying in pulverizer.
viii. Raw lignite entering to pulverizer.
ix. Soot blower opening in the boiler furnace.
x. Gas duct puncture.
xi. Pulverizer door sealing air ingress.

The optimum quantity of air is always desirable to reduce heat rate and auxiliary power consumption. Air ingress increases the burden of ID fans since this air does not take part in combustion. To detect the air ingress in gas path, the following test is conducted while boiler is under shutdown. This test is also called as regime test.

3.4 Steps Taken to Improve the Boiler Efficiency

i. In regenerative air preheater, hot end's heating elements were changed.
ii. In cold ends, the heating element's baskets were cleaned and its position was reversed.
iii. All air ingress points identified were sealed.
iv. All SH and RH coils were cleaned.
v. Water walls were cleaned.
vi. PF burner's air path was cleaned.
vii. PF burner's burning grates and re-suction ducts were renewed.
viii. All mills were reconditioned.
ix. Oil air ducts, oil guns and travel path were cleaned.

4 Results and Discussion

The theoretical calculation was carried out to calculate the efficiency of the boilers before and after overhauling using heat loss method. The results were shown in Table 3. Losses in boiler 1 and 2 before and after overhauling is shown in Figs. 1 and 2 respectively.

Efficiency of the boiler mainly depends on the ability of the boiler to burn the fuel and transferring the resulting heat to water and steam. The important losses are exit gas temperature, unburnt in fly ash and bottom ash, incomplete combustion and

Table 3 Losses in boilers before and after overhauling

S. No.	Parameters	Boiler 1		Boiler 2	
		Before	After	Before	After
1	Loss due to flue gas	8.03	5.86	7.23	5.25
2	Loss due to hydrogen in fuel	4.88	5.17	5.29	5.14
3	Loss due to moisture present in fuel	11.24	11.09	11.72	11.00
4	Loss due to moisture present in air	0.36	0.25	0.33	0.23
5	Loss due to unburnt fuel	0.74	0.27	0.66	0.27
6	Loss due to unburnt bottom ash	0.26	0.10	0.32	0.10
7	Surface heat losses	0.39	0.39	0.36	0.36
Total loss		25.89	23.12	25.91	22.35
Efficiency		74.12	76.88	74.09	77.65

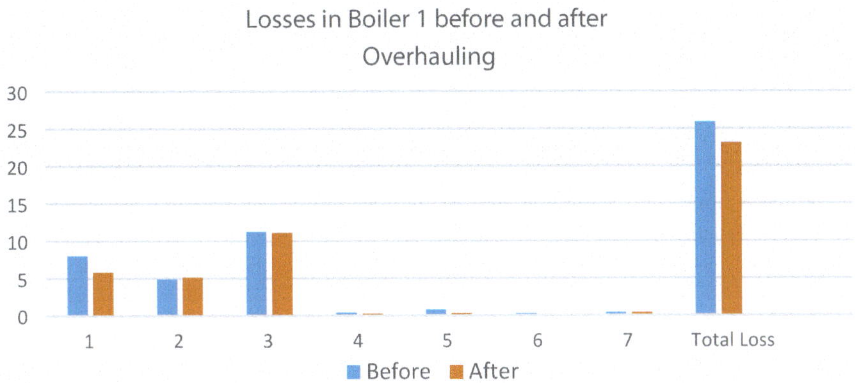

Fig. 1 Losses in boiler 1 before and after overhauling

radiation loss. Because of these losses, the boiler efficiency was found to be 73.16%. Here, the exit flue gas loss was found to be more. This is due to poor heat transfer in water walls, SH, RH, economizer coils and in RAPHs.

5 Conclusions

i. After major overhaul, the boiler efficiency was found to be increased from 73.16 to 77.76%.
ii. Because of the improvement in boiler efficiency, the heat input requirement was reduced and this results in cost savings.
iii. Also reduction in fuel input results in lesser pollution to the atmosphere.

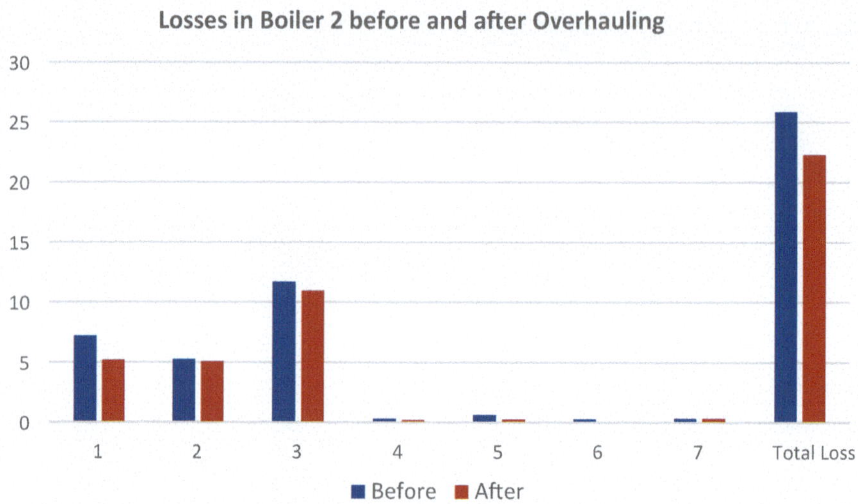

Fig. 2 Losses in boiler 2 before and after overhauling

iv. Plant performances and availability are the key drivers of competitive plant operations.
v. Vigilant operations and dedicated maintenance are very essential for excellent performances of thermal power stations.

References

1. Anjali TH, Kalivarathan G (2015) Analysis of efficiency at a thermal power plant. Int Res J Eng Technol 02(05):1112–1119
2. Astrom T (2005) Maintenance and overhaul steam turbines. In: International association of engineering insurers 38th annual conference
3. Jebarose Juliyana S, Udaya Prakash J, Karthik K, Pallavi P, Saleem M (2017) Design and analysis of NACA4420 wind turbine aerofoil using CFD. Int J Mech Eng Technol 8(6):403–410
4. Sangeeth GS, Marathur Praveen (2015) Efficiency improvement of boilers. Int Res J Eng Technol 02(05):265–268
5. Oakey JE (ed) (2011) Power plant life management and performance improvement. Elsevier
6. Deb A, Deb T (2017) A case study on heat rate of boiler and turbine in NSPCL Durgapur. Int J Latest Technol Eng Manage Appl Sci 6(8):11–20
7. Lakhoua MN (2012) Causal analysis and boiler efficiency calculations of a thermal power plant. J Electr Eng 1–6
8. Kumar K, Patel D, Sehravat V, Gupta T (2015) Performance and energy analysis of the boiler. Int J Sci Res 5(3):731–735
9. Bhatt MS, Jothibasu S (1999) Performance enhancement in coal fired thermal power plants. Part I: boilers. Int J Energy Res 23(6):465–487
10. Bhatt MS (2006) Effect of ash in coal on the performance of coal fired thermal power plants. Part I: primary energy effects. Energy Sources 28(1):25–41

11. Bhatt MS (2006) Effect of ash in coal on the performance of coal fired thermal power plants. Part II: capacity and secondary energy effects. Energy Sources 28(1):43–58
12. Vasal R, Nirmal Ram AS, John Francis A, Karthick P (2017) Performance analysis of steam turbine in Thermal Power plant. Int J Trendy Res Eng Technol 2(5):120–128
13. Chirag Y, Mehta N, Dabhi J (2014) Analysis of boiler losses to improve unit heat rate of coal fired thermal power plant. Int J Adv Eng Res Develop 1(5):1–6
14. Igboanugo AC, Asalor JO, Egbiremon VI (2013) Assessment of some performance characteristics of refuse boiler before and after overhaul using failure mode effect and fault tree analyses. Nigerian J Technol 32(2):333–349
15. Sundaravinayaka (2017) Optimization of boiler operation thermal power plant. Int J Latest Eng Res Appl 02(03):64–68
16. Saraswathi S, Sivakumar L (2014) Comparative analysis of k-means and self organizing map clustering on boiler process data. In: 2014 international conference on computer communication and informatics. IEEE, pp 1–10
17. Talukdar K, Gogoi TK (2016) Exergy analysis of a combined vapor power cycle and boiler flue gas driven double effect water—LiBr absorption refrigeration system. Energy Convers Manag 108:468–477
18. Patro B (2016) Efficiency studies of combination tube boilers. Alexandria Eng J 55(1):193–202. www.beeindia.gov.in/sites/default/files/4Ch1.pdf

Thermogravimetric Analysis of Friction Welding of Dissimilar Material

V. Raja, P. Periyasamy, G. Boopathy, E. Naveen, and N. Ramanan

Abstract The establishment of energy-saving and natural resource-saving systems is an important issue and relevant research and development should be accomplished without delay. Making structures lighter is one way to save energy. The study of light metals such as aluminium and magnesium alloys has received much attention. Aluminium alloys are especially attractive because of superior recyclability and workability. However, present structures made of stainless steels cannot be entirely replaced with aluminium alloy structures, taking into account strength, weldability and economics, although it is possible to replace part of a structure with aluminium alloy components. In this case, it is necessary to joint SS 430F steel with aluminium 6061. Few sound joints have been obtained, owing to the formation of a large amount of brittle intermetallic compounds which were using fusion welding. In the recent years, welding of dissimilar metals by conventional welding techniques has become difficult. The flux used for the welding will create lot of heat which reduces the strength of the welded joints. In order to overcome this, friction welding is more effective in joining dissimilar metals when compared with fusion welding, since it is a solid-state process. Heat is generated in friction welding by conversion of mechanical energy into thermal energy in the interface of work pieces during rotation under pressure. Various ferrous and non-ferrous alloys having circular or non-circular cross sections and that have different thermal and mechanical properties which can be joined by the friction welding. Conventional structures made of steel have been replaced by lighter materials, capable of providing high mechanical strength, lower volume of material and good corrosion resistance components such as those used

V. Raja (✉) · P. Periyasamy
St Peter's Institute of Higher Education & Research, Chennai, India
e-mail: raja.lecturer1978@gmail.com

G. Boopathy
Vel Tech Rangarajan Dr. Sangunthala R&D Institute of Science & Technology, Chennai, India

E. Naveen
Department of Mechanical Engineering, Sri Sairam Engineering College, Chennai, India

N. Ramanan
Synce Engineering Service, Chennai, India

L.-J. Yang et al. (eds.), *Proceedings of ICDMC 2019*,
Lecture Notes in Mechanical Engineering,
https://doi.org/10.1007/978-981-15-3631-1_47

in industries (shipbuilding, light and heavy automotive, electrical, chemical, civil, space and nuclear industry).

1 Introduction

They have done the work on points like joint configurations, microstructure, mechanical properties, process parameters, etc., in regard of welding of dissimilar materials/alloy in aluminium 6061 and SS 430F steel [1]. They studied the joining of plastically deformed steels (carburizing steel) with friction welding. They carried out an experimental study on joining medium-carbon steel and austenitic stainless steel with friction welding [2]. The conducted the experiments on random selection of input parameters. From the experimental data, artificial neural network (ANN) is trained. It predicts the flash width, flash thickness and flash height were closely the percentage variation between the actual and predicted is around 1.5%. Through the optimized input parameters, minimized flash features such as flash width (6.5 mm), flash height (7.3 mm) and flash thickness (3.70) can be obtained [3]. They studied the mechanical and metallurgical analysis of aluminium and steel using inertia friction welding process. It has been used to create joints between aluminium 6061-T6 aluminium alloy and AISI 1018 steel using various parameters. The joints were evaluated using mechanical and metallurgical tests. The results showed that the joints strength was in the order of 250 MPa. The upset pressure calculated 60 MPa [4]. They investigated the development of solid-state joints of dissimilar material AA6082 aluminium alloy and AISI 304 stainless steel, via continuous drive friction welding process. Tests were conducted with different welding process parameters. The results were analysed by means of pull load test, Vickers microhardness test, fatigue test, Charpy v-notch impact test and SEM-EDX (energy-dispersive X-ray). The strength of the joints varied with increasing friction pressure and friction time keeping upset pressure and upset time constant. The joint strength increased and then decreased after reaching a maximum value, with increasing friction pressure and friction time [5]. They used continuous drive friction welding, and they obtained optimum parameters to join parts having equal diameter (24 mm). The effect of welding parameters on welding strengths was also investigated. The mechanical properties of joints were examined by using pull load test and hardness test, macro- and micro-analysis [6]. They observed that friction welding is possible by using conventional lathe, and they weld aluminium 6061 and aluminium 6082 using conventional method. They did mechanical and microstructure analysis for friction-welded joint, and they observed small weld region in the microstructure analysis [7]. They studied the effect of mechanical and metallurgical properties of austenitic stainless steel (304L) and copper. They used pull load, hardness test, optical, scanning electron microscopy and atomic force microscopy to analyse the microstructure of the welded joint. The joints were also examined with EDX to understand the phases formed during welding. The material is evaluated by pull load test, and their strength is determined, and the hardness is examined [8]. They studied the feasibility to join the different materials/alloys using friction welding. This work deals with the joint configurations, microstructure investigations, mechanical properties, process parameters in regard of welding of dissimilar materials/alloys.

2 Problem Identification

The dissimilar materials of aluminium 6061 and SS 430F steel is joined using friction welding to analyse the loadcarrying capacity for the structural engineering applications. Longer friction time causes the formation of an intermetallic layer at the weld interface which will affect the welded region. Less strength in welding region. Sufficient heat could not be generated with a shorter friction time to get the strong joint. Depth of penetration is partially fulfilled in the welding region.

3 Selection of Dissimilar Materials

3.1 Aluminium Alloy 6061

Aluminium alloys are alloys in which aluminium (Al) is the predominant metal. The typical alloying elements are copper, magnesium, manganese, silicon and zinc. Aluminium alloys are widely used in engineering structures and components, where lightweight or corrosion resistance is required. Aluminium alloys are used due to their superior workability and low cost. However, they are not entirely replaced by stainless steel, stainless steel having superior strength and weldability in certain structures. Aluminium 6061 alloy is one of the most extensively used 6000 series aluminium alloys. It is a versatile heat treatable extruded alloy with medium to high strength capabilities.

3.2 SS 430F Steel

SS 430F is usually supplied in the T condition with a pull load strength of 850/1000 N/mm^2. SS 430F steel is a popular grade of through-hardening alloy steel due to its excellent machinability in the "T" condition. SS 430F is used in components such as gears, shafts, studs and bolts with a hardness of 248/302 HB range. SS 430F can be further surface hardened to create components with enhanced wear resistance by induction or nitrating processing. SS 430F steel is a high pull load alloy steel renowned for its good wear resistance and high strength. This can include aircraft, automotive and general engineering applications, for example, propeller or gear shafts, connecting rods, aircraft landing gear components.

Fig. 1 Experimental set-up of friction welding process

4 Experimental Details Friction Welding Process

In the friction welding process, one piece is held stationary and other is rotated in the chuck of a friction welding machine. As they are brought to rub against each other under pressure, they get heated due to friction. When the desired forging temperature is attained throughout the rubbing cross section of the work piece, the rotation is stopped suddenly and the axial pressure is increased to cause a forging action and hence welding. The machine used for friction welding resembles a lathe but is sturdier than that. The essential features of the machine are that it should be able to withstand high axial pressure of the order of 50,000 N/cm^2 and be able to provide a high spindle speed up to 12,000 rpm. Normally, cylindrical pieces like rods and tubes are welded by this process, but its application can be extended to situations, wherein one component is symmetrical and can be rotated conveniently (Fig. 1).

5 Result and Discussion

5.1 Pull Load

Material strength can be found by testing the material in tension or compression. Standard dog bone-shaped specimens were friction welded according to the ASTM D8 specifications for tensile testing (reinforced and unreinforced plastics). Each specimen has 8 mm diameter and 32 mm gauge length. The specimen is loaded in computer-controlled universal testing machine (ASE—UTN 10) until the failure of the specimen occurs. Simultaneous readings of load and elongation are taken at uniform intervals of a load. A tensile test is carried out at room temperature. Uniaxial tensile test is conducted on the constructed the specimen to obtain information

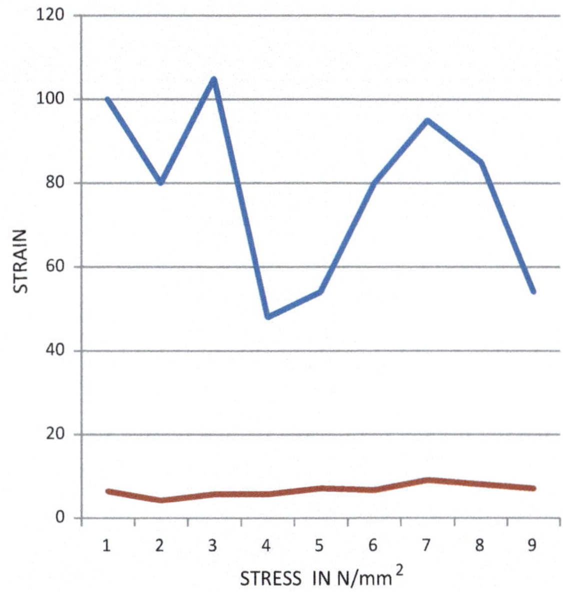

Fig. 2 Pull load test of the specimen at different applied loads KN/mm

regarding the behaviour of a given material under gradually increasing stress–strain conditions.

The load applied on the welded specimen shows that the stress and strain varies with load (Fig. 2). There is an elastic deformation occuring in the specimen, and as its reaches the endurance limit, the materials will break. The stress curve indicates the ductile failure of the welded specimen [9].

5.2 *Hardness Properties*

The quality of the joints is evaluated on the base of microhardness measurements. The mechanical property of the welded region was carried out using the Vickers hardness tester. The indentations were made at the weld interface and parent materials at fixed intervals to find the effect of heat on the hardness values. The hardness was measured for nine samples, and out of these samples, the maximum hardness was noticed for the optimization, and this optimized specimen is used for other analysis. The hardness values obtained at different zones are given in the Table 1 (Fig. 3).

Table 2 show hardness values of 95.8–107.3 MPa which denotes the hardness of parent aluminium alloy. The zone 2 representing the near the weld region of Al alloy 6061 indicating the hardness pf 57.2–86.1 MPa range. The zone 3 denotes the weld region of the SS 430F steel indicating the maximum hardness of 249.6 MPa. It is noted that the highest hardness is observed at the weld interface for direct weldments. Table shows the hardness variations on both the sides of the friction-welded joint.

Table 1 Hardness values at different zones of the welded region

Sample number	Zone-1 (Mpa	Zone-2 (Mpa)	Zone-3 (Mpa)	Zone-4 (Mpa)
1	95.8	69.5	222.7	219.2
2	107.0	66.7	230.7	204.8
3	106.9	81.7	231.9	238.1
4	112.1	61.9	230.7	223.6
5	105.6	69.7	214.2	244.7
6	101.2	57.2	215.6	207.9
7	106.7	86.1	249.6	264.2
8	108.2	68.0	214.4	234.1
9	107.3	84.6	227.5	222.9

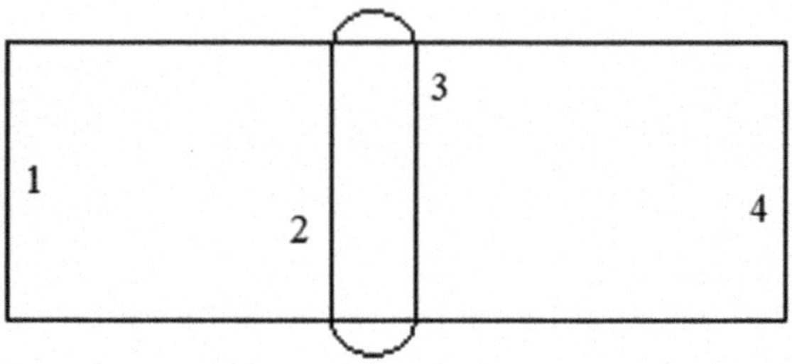

Fig. 3 Microhardness test set-up at four different zones. 1 Parant material of SS 430F steel, 2 heat-affects zone of SS 430F steel, 3 heat-affected zone of aluminium 6061, 4 parent material of aluminium 6061

The decrease in hardness may be attributed to recrystallization process taking place at the heat-affected zone of the Al alloy 6061.

5.3 *Thermo Gravimetric Analysis*

Thermogravimetric analysis (TGA) is conducted on an instrument referred to as a thermogravimetric analyser. A thermogravimetric analyser continuously measures mass, while the temperature of a sample is changed over time. Mass, temperature and time in thermogravimetric analysis are considered base measurements, while many additional measures may be derived from these three base measurements (Fig. 4).

The TGA analysis output is shown in Fig. 5. The output shows the material temperature characterization and rhe maximum material temperatures 900 °C. The material observed output energy of maximum 662 °C. The material mostly absorbed in a thermal energy of steel 1200 °C and aluminium is 660 °C. But, the weld joint in a mixture of aluminium and steel material observe that thermal energy is going to be in a lower level.

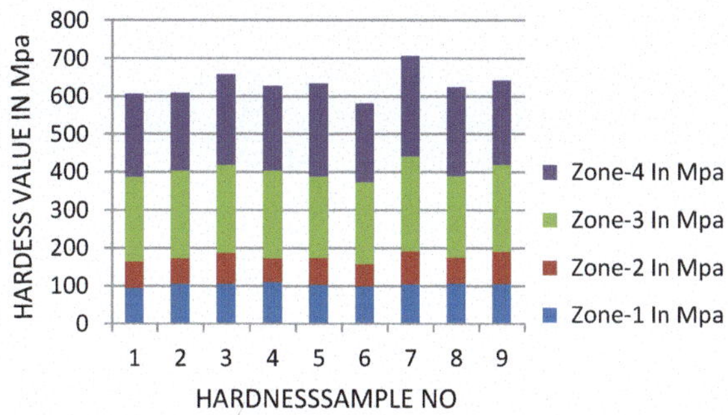

Fig. 4 Hardness valve chart

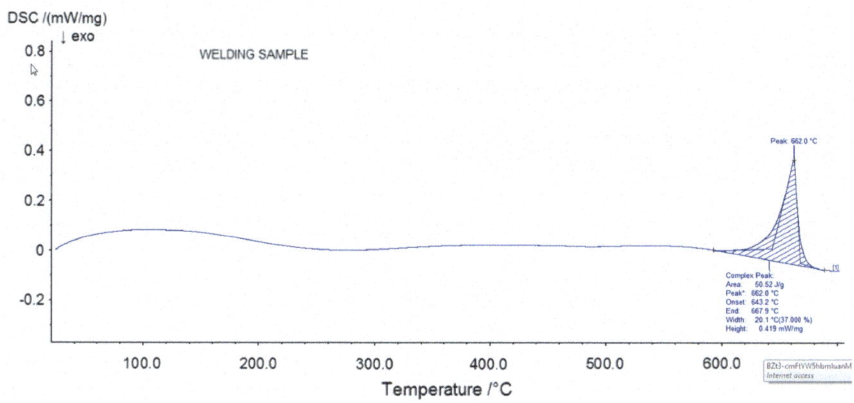

Fig. 5 TGA output data

6 Conclusion

The pull load test result shows that the welded joint is brittle in nature and can be used for static loading structural application. The weld joint is mostly used in structural applications. In this investigation, an attempt has been made to study the welding parameters on the formation of friction welding and pull load properties of 6061 aluminium alloy and SS 430F steel. Weldability is significantly affected by the rotational speed. At high rotational speed, the heat produced is more and time taken for cooling is also more, which affects the microstructure of 900 rpm, and weld finish is not good. From this investigation, the following important conclusions are derived. It is observed that the upset force plays a vital role in increasing the pull load and impact strength of the friction weld with dissimilar materials. The burn-off length plays a commendable role in increasing the bend strength of the weld as expected. Maximum yield strength of 105 MPa, pull load strength of 138 MPa, impact strength of 10 J, hardness of 264.2 Mpa and joint efficiency of 84.9% was attained without any defect for the joint fabricated using FW at rotational speed of 2000 rpm and weld speed of 40 mm/min. The minimum hardness of 57 HV is obtained in the HAZ Al6061. The tensile specimen failed in the HAZ AL6061. The ductility of FW SS430F steel and AL6061 increases the longitudinal direction (NZ), but it decreased the transverse direction that cut through all WZ.

References

1. Kumar A (2014) Friction welding of dissimilar materials/alloys. Int J Mech Robot Res 1:315–320
2. Jhon (2009) Dissimilar friction welding of 6061-T6 aluminium and AISI 1018 steel: properties and microstructural characterization. Elsevier Ltd Mater Des 31:2305–2311
3. Ravikumar (2013) Characterization of mechanical properties of aluminium (AA6061-T6) by friction welding. In: Third international conference on mechanical and automobile engineering, Apr 2013
4. Sahin (2010) Mechanical and metallurgical properties of friction welded joints. Indian J Mater Des 49:527–534
5. Kumar S (2013) Evaluation of mechanical and metallurgical properties of dissimilar materials by friction welding. Elsevier Ltd 64:1514–1523
6. Sathiya (2005) Optimization of friction welding parameters using simulated annealing. Indian J Mater Sci 13:37–44
7. Shubhavardhan RN, Surendran S (2012) Friction welding to join dissimilar metals. Int J Emer Technol Adv Eng 2(7). www.ijetae.com
8. Jose T (2013) Friction welding of aluminium 6061 and aluminium 6082 rods by using conventional lathe. Int J Current Trends Eng Res (IJCTER) 2(4):170–175 (Scientific Journal Impact Factor: 3.468)
9. Koushik V (2016) Experimental investigations of friction welding using aluminium with mild steel. Int Res J Eng Technol (IRJET) 03(12)

Experimental and Theoretical Investigation of Interaction Effect on Energy Absorption of Bi-tubular Structures Under Quasi-static Axial Crushing

K. Vinayagar, A. Senthil Kumar, M. Vel Vignesh, and K. Gokulan

Abstract This work reveals the interaction effect on energy absorption of bi-tubular, a combination of outer circular tube with inner polygonal tube structures under quasi-static axial crushing. First, the experimental mean crushing force of single tubes was compared with theoretical prediction, and it witnesses good agreement between them. In bi-tubular arrangement, the inscribed diameter of inner polygonal tubes plays an important role in improving the mean crushing force by interaction effect. The mean crushing force of bi-tubular specimens D3T, D3S, and D3H has 35, 26, and 10.1% higher than the summation of their individual single tubes that is SC with S3T or S3S or S3H.

Keywords Bi-tubular structures · Quasi-static axial crushing · Interaction effect · Mean crush force · Specific energy absorption

Nomenclature

D	Diameter of the tube, m
$M_0 = \sigma_0 h^2/4$	Full plastic bending moment of tube wall per unit length, N
m	Mass of the thin-walled structure, kg
F_{mean}	Mean crushing force, N
F_{Peak}	Peak crushing force, N
C	Side of the polygon, m
h or t	Thickness of the tube, m
δ	Crushing distance or crushing deflection, m
σ_0	Flow stress, N/m^2

K. Vinayagar (✉) · A. Senthil Kumar · M. Vel Vignesh · K. Gokulan
Department of Mechanical Engineering, Sethu Institute of Technology, Kariapatti, Virudhunagar, Tamil Nadu 626115, India
e-mail: k_vinayagar@ymail.com

A. Senthil Kumar
e-mail: asenthil123@yahoo.com

L.-J. Yang et al. (eds.), *Proceedings of ICDMC 2019*, Lecture Notes in Mechanical Engineering,
https://doi.org/10.1007/978-981-15-3631-1_48

σ_u	Ultimate tensile stress, N/m^2
σ_y	Yield stress, N/m^2

1 Introduction

During collision, energy absorption devices are responsible to reduce the loss of life and damage of vehicle critical parts. These devices can absorb maximum kinetic energy by permanent plastic deformation and transferred minimum amount of force to occupants. From the different types of energy absorbers, the metallic thin-walled tubes are considered one of the most efficient energy-absorbing devices for their ease manufacture, high strength, and low cost [1, 2]. Over the past few decades, many researchers studied about crashworthiness of thin-walled cylinders and have submitted in experimental, analytical, and numerical results with different cross sections, materials, tubes filled with foam, and multi-cell thin-walled structures [3, 4].

The analysis of axial crushing on thin-walled cylinders was pioneered by Alexander [5], and the work yields an excellent theoretical model to access the average crushing force for axisymmetric fold pattern. Subsequently, Abramowicz and Jones [6] analyzed the different types of crushing modes and took more efforts to improve the previous theoretical models by the number of experiments on the concertina or ring mode for circular cylinders. Hong et al. [7] studied collapse mode and energy absorption of triangular tubes under quasi-static axial compression experiments. Based on folding elements, they proposed four different theoretical models to predict the mean crushing force. Abramowicz et al. [8] studied the effect of corner element of an arbitrary angle on energy absorption by using rigorous methods of structural plasticity with finite deformations and rotations. Based on this analysis, they developed mathematical models to predict the mean crushing force of multi-corner columns with an even number of corners such as square, hexagonal, and rhomboidal thin-walled columns.

Alavi Nia et al. [9, 10] and Velmurugan et al. [11, 12] explored different configurations in single material and showed that geometric parameter in a conventional way to improve the crashworthiness of thin-walled structures.

When using foam, honeycomb, or sawdust as filler materials, the energy absorption capacity of the thin-walled structures was found to be increased by interaction between them. Hanssen et al. [13–15] carried out complete studies on aluminum columns with aluminum foam filler and identified a strong foundation for the subsequent theoretical analysis and practical application. In recent years, foam-filled tubes with various cross sections such as circular [13], square [14–16], polygonal [17], and elliptical [18] have been carried out extensively.

The effect of length difference on crashworthiness performance of bi-tubal structures was reported by Sharifi et al. [19], Vinayagar et al. [20], and Kashani et al. [21]. Recently, Vinayagar et al. [22], Rahi [23], and Kamran et al. [24] were reported the influence of geometry on the energy absorption of bi-tubes under axial compression.

Though the crushing of bi-tubular systems is very recent topics in crashworthiness analysis, the discussion about interaction effect with quantitative results is very minimum. The energy absorption capabilities of single tubes with various sections were studied experimentally, and experimental results were compared with theoretical predictions [22]. Due to the interaction effect, the bi-tubular combination has higher mean crushing force, when compared with the summation of mean crushing force of individual single tubes and the percentage of improvement in mean crushing force also reported.

2 Definition of Crashworthiness Indicators

With the purpose of better understanding about crashworthiness of thin-walled structures, some indicators like total energy absorption (TEA), specific energy absorption (SEA), and mean crush force (F_{mean}) are described with a brief explanation.

Total Energy Absorption
Total energy absorption expresses the energy absorption capacity of specimens. It can be determined by integration of the axial crushing force versus displacement.

$$\text{TEA} = \int_{0}^{\delta} F(\delta)\text{d}\delta \tag{1}$$

where $F(\delta)$ is the instantaneous crushing force with a function of the crushing distance δ. From experiments or numerical modeling, the instantaneous crushing force can be obtained.

Specific Energy Absorption
Specific energy absorption is defined as energy absorbed per unit mass of the thin-walled member. It is used to compare the energy absorption capacity of structures with different mass which is given by

$$\text{SEA} = \text{TEA}/m \tag{2}$$

Mean Crush Force
The average crush force is also one of the response parameters for the energy absorption capability of a structure, which is calculated as,

$$F_{\text{mean}} = \text{TEA}/\delta \tag{3}$$

where TEA is energy absorbed during collapse with displacement δ.

3 Specimen Preparation

Since the desirable tube sections were not available in the market, the specimens were prepared in sheet metal shop from 0.478 mm thick SS304 stainless steel sheet. The stainless steel sheet was cut into the required dimensions and joined as a tube by TIG welding process. Stainless steel 304 is considered for its good formability and strength. Mechanical properties of this sheet metal were presented in Table 1.

Single tubes of various polygonal sections such as triangle, square, and hexagon were fabricated according to three levels of inscribed diameter of polygonal section such as 60 mm, 70 mm, and 80 mm. The edge length of various polygonal sections is derived from the inscribed diameter of the polygonal section as shown in Fig. 1 and tabulated in Table 2. The length of all the single tubes is 135 mm.

3.1 Coding System of Single Tubes

The prepared single tube specimens were coded for the purpose of easy identification, evaluation, and performance comparison. In coding of single polygonal tubes, the first alphabet 'S' indicates single tube and the second numerical number 1, 2, and 3 indicates the level of inscribed diameter of the polygonal section, i.e., 60 mm, 70 mm, and 80 mm, respectively; then, the third alphabet indicates section of tube, i.e., *T* for triangular, *S* for square, and *H* for hexagonal section. The coding method of single tubes is noted in Table 3.

For example, S2H means single tube with 70 mm (second level) inscribed diameter of the polygonal section, and its section was a hexagon. Then, the circular cylinder was fabricated with a diameter of 90 mm. The single cylinder was coded as SC, here also '*S*' stands for single and '*C*' stands for cylinder. For consistency of output in each type of geometry, three samples were fabricated.

3.2 Experimental Process

For this quasi-static axial compression analysis, FSA makes Model TUF-CN 1000 servo controlled universal testing machine (Fig. 2) was used. The samples were crushed at a loading rate of 10 mm/min as previous study [22]. Here also, the specimens were crushed up to 100 mm at the room temperature of 25 °C.

Table 1 Mechanical properties of 304 stainless steel sheet

Property	Ultimate tensile strength (MPa)	Yield stress (MPa)	% of elongation
Value	531.2	223.6	41

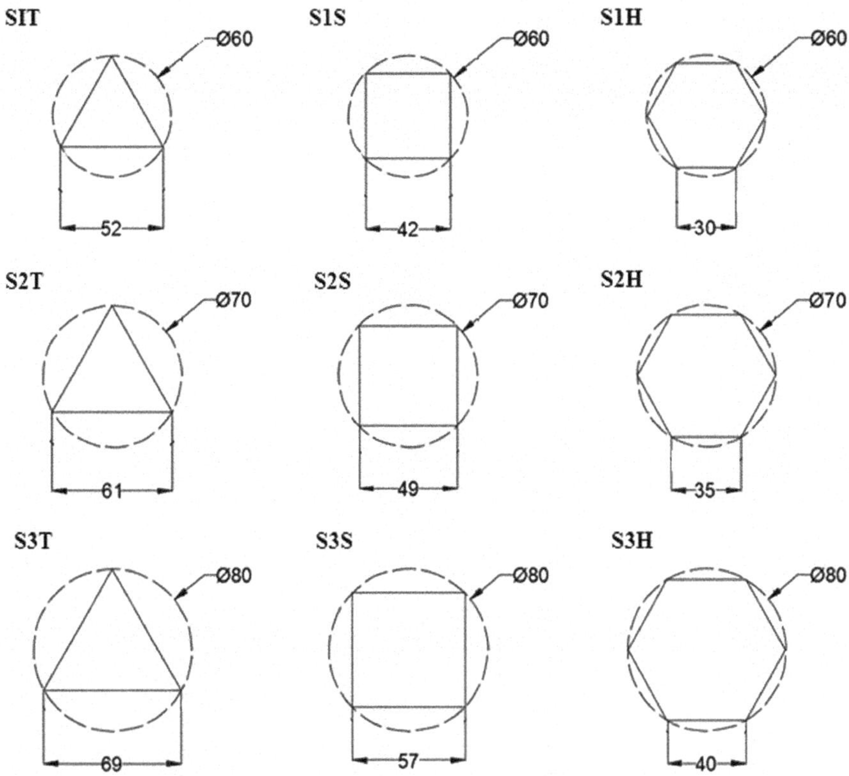

Fig. 1 Section geometry and dimensions of single tube specimens (all dimensions are in mm)

Table 2 Specifications of single tubes

S. No.	Code	Inscribed diameter of the polygonal section (mm)	Section of the tube	Side/edge length of polygon (mm)
1	S1T	60	Triangle	52
2	S1S	60	Square	42
3	S1H	60	Hexagon	30
4	S2T	70	Triangle	61
5	S2S	70	Square	49
6	S2H	70	Hexagon	35
7	S3T	80	Triangle	69
8	S3S	80	Square	57
9	S3H	80	Hexagon	40
10	SC	90	Cylinder	Ø 90

Table 3 Coding system of single tubes

Coding order	Code	Description
First alphabet	S	Single tube
Second numerical number	1	Inscribed diameter of polygonal section is 60 mm
	2	Inscribed diameter of polygonal section is 70 mm
	3	Inscribed diameter of polygonal section is 80 mm
Third alphabet	T	Polygonal section of tube is triangle
	S	Polygonal section of tube is square
	H	Polygonal section of tube is hexagon

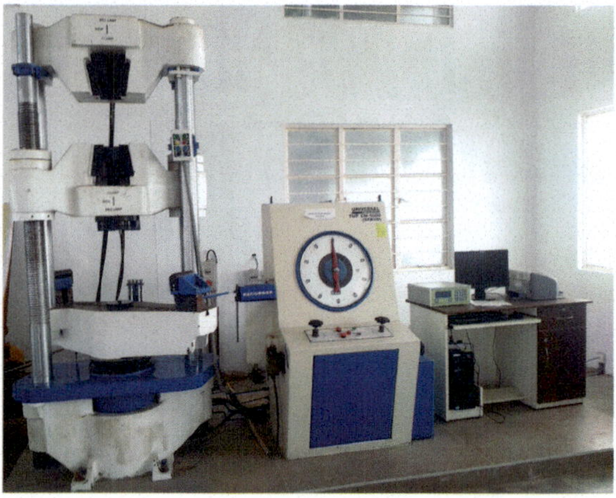

Fig. 2 Computerized servo controlled universal testing machine

4 Results and Discussion

Load–displacement curves of the single tube specimens are obtained from quasi-static tests. The important crashworthiness parameters such as mean crushing force, total energy absorption, and specific energy absorption were calculated from the load–displacement curves of the single tube specimens and recorded in Table 4.

Concertina mode is developed in specimen SC; in-extensional mode is developed in single polygonal tubes; and diamond mode is observed in all bi-tubular specimens. Based on experimental results, a detailed comparison report of the crashworthiness performance between various polygonal sections in three levels of inscribed diameter of single tubes was presented.

Table 4 Experimental results of single tubes

Specimen code	Mass (kg)	F_{mean}(kN)	F_{Peak}(kN)	TEA (J)	SEA (kJ/kg)	CFE (%)
SC	0.140	8.18	16.08	820.7	5.862	50.87
S1T	0.080	2.81	5.57	281.3	3.516	50.45
S1S	0.083	4.45	8.11	458.8	5.528	54.87
S1H	0.090	6.10	11.59	611.1	6.790	52.63
S2T	0.095	3.11	5.78	318.9	3.357	53.81
S2S	0.100	4.62	8.54	475.6	4.756	54.10
S2H	0.103	6.96	12.56	698.4	6.780	55.41
S3T	0.105	3.25	5.95	325.8	3.103	54.62
S3S	0.115	4.88	8.96	491.8	4.277	54.46
S3H	0.120	7.55	15.18	758.6	6.322	49.74

4.1 *Effect of Inscribed Diameter on Mean Crushing Force of Single Tubes*

On crashworthiness, the effect of the inscribed diameter of polygonal section of single tubes is studied. The edge length of polygon is increased with the inscribed diameter of polygonal section from 60 mm, 70 mm, and 80 mm. When the area of polygonal sections is increased, the stability of tubes also increased. This is the main reason for improvement of energy absorption capability. Mean crushing force of tested sections for various inscribed diameters of polygonal sections is compared in Fig. 3. It can be understood from this diagram for the same section of single tube, mean crushing force increased with increasing of the inscribed diameter of polygonal section. As in Table 4, the total energy absorption of the thin-walled tubular structures with inscribed diameter of polygonal section of 60 mm, 70 mm, and 80 mm is the least, medium, and the most, respectively.

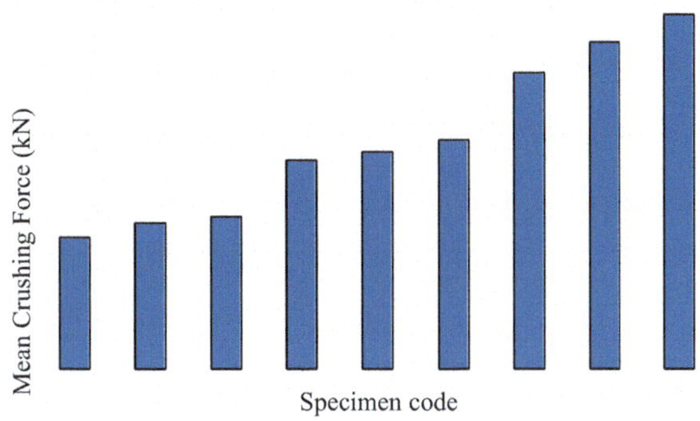

Fig. 3 Effect of inscribed diameter of polygonal section on mean crushing force of single tubes

4.2 Effect of Polygonal Section of Single Tube

Thin-walled structures are used to dissipate the impact energy through the process of plastic deformation. In fact, the studies [9, 10] have shown that the amount of absorbed energy increased with the number of sides of a section. The reason for this is the number of plastic hinges and folds which are increased with number of sides of polygon. Figure 4 shows the comparisons of mean crushing force between various sections of tubes for a particular inscribed diameter of the polygon. Figure 4 also pointed out that the mean crushing force increased by increasing the number of edges or corners of the section of any of inscribed diameter of the polygonal section.

From Table 4, TEA of hexagonal tube was greater than the square and the square was greater than the triangle; i.e., in anyone of inscribed diameter of the polygonal section, the greater the number of sides, the absorbed energy is more. The hexagonal tubes have more SEA than the other sections as presented in Table 4 and it is shown in Fig. 5.

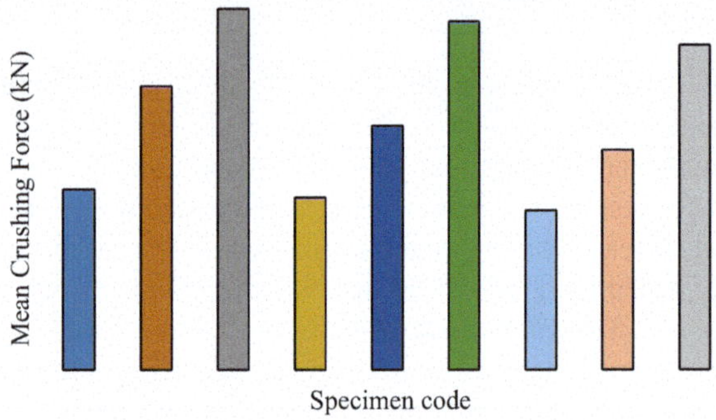

Fig. 4 Effect of polygonal section on mean crushing force of single tubes

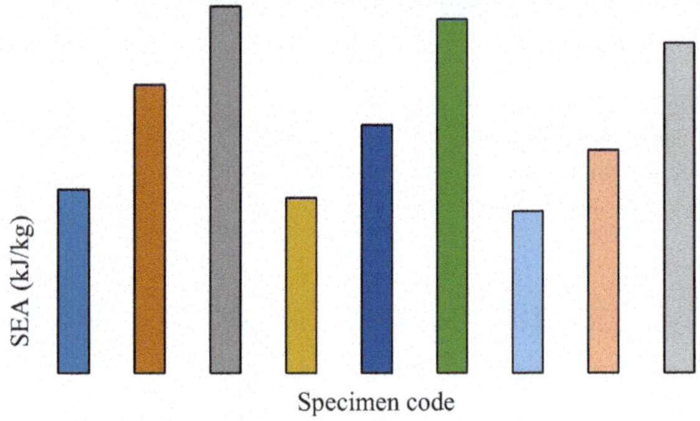

Fig. 5 Effect of polygonal section on SEA of single tubes

4.3 Theoretical Predictions

Comparison of experimental results with theoretical predictions was a must to validate the experiments. To predict the mean crushing force of single tubes, several theoretical models have been addressed by many researchers for concertina and diamond mode of various sectional tubes such as circular, triangle, square, and hexagon.

Theoretical Model of Circular Cylinder
The important expression of mean crushing force for concertina mode of circular cylinder is derived by Abramowicz and Jones [6] in 1984.

$$F_{\text{mean}} = \frac{20.79\left(\frac{D}{t}\right)^{0.5} + 11.9}{0.86 - 0.568\left(\frac{t}{D}\right)^{0.5}} M_o \tag{4}$$

Theoretical Model of Triangular Tube
For triangular tube, Hong et al. [7] presented the following model for the mean crushing force which is,

$$F_{\text{mean}} = 26.66 M_o \left(\frac{C}{t}\right)^{\frac{1}{3}} \tag{5}$$

Theoretical Model of Square Tube
The following classic theoretical model for square tube under axial compression was proposed by Abramowicz and Wierzbicki [8],

$$F_{\text{mean}} = 12.16 \sigma_o C^{0.37} t^{1.63} \tag{6}$$

Most of the experimental studies were validated based on this theoretical model.

Theoretical Model of Hexagonal Tube
The theoretical expression presented by Abramowicz and Wierzbicki [8] for hexagonal column is approximated by the following functions,

$$F_{\text{mean}} = 20.2 \sigma_o C^{0.4} t^{1.6} \tag{7}$$

Determination of Flow Stress
Actually, flow stress is not a constant value, so determination of the flow stress is very important. During loading, the strains for different collapse modes and in different regions of tubes are different. Therefore, different expressions were employed by the researchers to calculate the flow stress. In this present work, the average value of yield stress and ultimate tensile stress is employed as flow stress for all of the thin-walled structures.

Based on the above mathematical Eqs. (4–7), the theoretical mean crushing force was calculated for all sections of tubes. Then, the values of experimental mean crushing forces of single tubes for circular, triangular, square, and hexagonal sections are

Table 5 Experimental results versus theoretical predictions of single tubes

Specimen code	Mean crushing force (kN)		Percentage of deviation
	Experimental result	Theoretical prediction	
SC	8.18	7.8	4.8
S1T	2.81	2.7	4.0
S2T	3.11	2.9	7.2
S3T	3.25	3.01	7.9
S1S	4.45	4.3	3.5
S2S	4.62	4.53	2.0
S3S	4.88	4.74	3.0
S1H	6.1	6.5	6.6
S2H	6.96	6.8	2.4
S3H	7.55	7.3	3.4

compared with theoretically predicted values in Table 5. From Table 5, the percentage of deviation is very less and it can be understood that the theoretical predictions are found to be in a good agreement with the experimental results.

4.4 Comparison Between Single and Bi-tubular Structures

Experimental results of crashworthiness indicators for bi-tubular structures are taken from the previous work [22] and tabulated in Table 6. Table 6 shows that the mean crushing force, total energy absorption, and specific energy absorption of the bi-tubes were greater than the single tubes.

Table 6 Experimental results of bi-tubes

Code	Mass (kg)	F_{Peak} (kN)	F_{mean} (kN)	TEA (J)	SEA (kJ/kg)
D1T	0.220	21.80	10.76	1081.8	4.917
D1S	0.223	24.12	12.59	1265.7	5.676
D1H	0.230	29.13	14.57	1459.5	6.346
D2T	0.235	22.85	12.11	1227.7	5.224
D2S	0.240	26.01	14.36	1475.5	6.148
D2H	0.243	31.17	16.28	1632.9	6.718
D3T	0.245	24.50	15.40	1541.2	6.291
D3S	0.255	26.83	16.43	1648.5	6.465
D3H	0.260	33.87	17.42	1747.6	6.722

4.5 *Influence of Interaction Effect in Bi-tubes*

Each of the single polygonal tubes and single circular cylinder was compressed separately, and the summation of the mean crushing force was compared with the bi-tubular specimens in Table 7.

In D1T, D1S, and D1H, the clearance between the tubes is more. While crushing, the contact between the tubes is very less and both tubes are crushed almost separately. So the mean crushing force and the absorbed energy of D1T, D1S, and D1H bi-tubular specimens are approximately equal to the summation of their single tubes that is SC with S1T or S1S or S1H.

In D2T, D2S, and D2H, the clearance between the tubes is less compared to D1T, D1S, and D1H. While crushing, the contact between the tubes can cause the growth in mean crushing load and absorbed energy. So the mean crushing force of D2T, D2S, and D2H bi-tubular specimens are 7.3, 12.3, and 7.5% higher than the summation of their single tubes that is SC with S2T or S2S or S2H.

Like that, in D3T, D3S, and D3H the clearance between the tubes is very less compared to D1T, D1S, and D1H. While crushing the contact between the two tubes becomes tighter, it can cause the growth in mean crushing load. The mean crushing force of D3T, D3S, and D3H bi-tubular specimens is 35, 26, and 10.1% higher than the summation of their single tubes that is SC with S3T or S3S or S3H. Table 7 and Fig. 6 clearly show that the improvements of mean crushing force of bi-tubular specimens are increased with an inscribed diameter of inner tube.

Table 7 Improvement of mean crushing force in bi-tubes

Specimen code		Experimental mean crushing force (kN)		Improvement in $F_{\text{mean}}(\%)$
Single tubes	Equivalent bi-tube	Summation of single tubes	Bi-tube	
SC + S1T	D1T	8.18 + 2.81 = 10.99	10.76	2.1
SC + S2T	D2T	8.18 + 3.11 = 11.29	12.11	7.3
SC + S3T	D3T	8.18 + 3.25 = 11.43	15.4	35
SC + S1S	D1S	8.18 + 4.45 = 12.63	12.59	0.32
SC + S2S	D2S	8.18 + 4.62 = 12.8	14.36	12.3
SC + S3S	D3S	8.18 + 4.88 = 13.06	16.43	26
SC + S1H	D1H	8.18 + 6.1 = 14.28	14.57	2
SC + S2H	D2H	8.18 + 6.96 = 15.14	16.28	7.5
SC + S3H	D3H	8.18 + 7.55 = 16.73	17.42	10.1

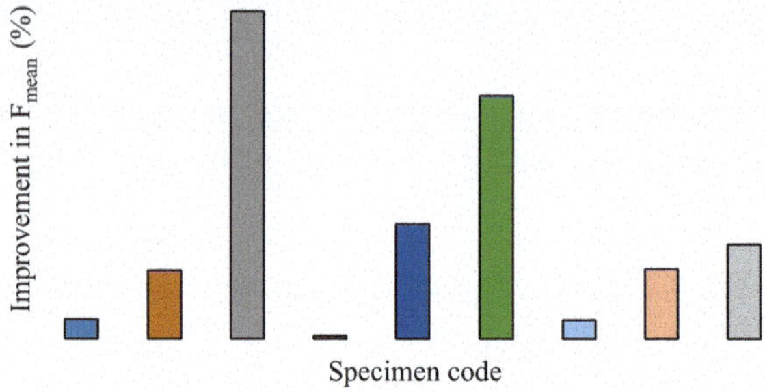

Fig. 6 Effect of interaction on mean crushing force

5 Conclusions

In this research paper, the crashworthiness behavior of single thin-walled tubes with circular, triangular, square, and hexagonal sections was studied under quasi-static axial compression. From this experimental study, the following conclusion can be drawn:

1. In single tubes, the Total Energy Absorption (TEA) and mean crushing force (F_{mean}) are increased with increasing the inscribed diameter of polygonal section. F_{mean} and Specific Energy Absorption (SEA) are increased with increasing the number of edges or corners of the section for any one of inscribed diameter of polygon.
2. Experimental mean crushing force of single tubes was compared with theoretical values and found that they have good consistency between them. Bi-tubular structures had greater SAE than the single tubes.
3. Effect of interaction or clearance between the tubes in bi-tubular arrangement was analyzed in detail. The mean crushing force of D1T, D1S, and D1H bi-tubular specimens is approximately equal to the summation of their single tubes that is SC with S1T or S1S or S1H.
4. In the second and third levels of inscribed diameter of inner tubes, the clearance between the tubes is minimum. Less clearance makes more interaction between the tubes, and it causes more energy absorption in bi-tubular arrangement. The mean crushing force for D3T was 35% greater than the summation of their single tubes that is SC with S3T. From this analysis, the interaction effect between the tubes is proved experimentally.

References

1. Olabi AG, Morris E, Hashmi MSJ (2007) Metallic tube type energy absorbers: a synopsis. Thin-Walled Struct 46:706–726
2. Lu G, Yu T (2003) Energy absorption of structures and materials. Woodhead Publishing Ltd., Cambridge

3. Baroutaji A, Sajjia M, Olabi AG (2017) On the crashworthiness performance of thin-walled energy absorbers: Recent advances and future developments. Thin-Walled Struct 118:137–163
4. Fang J, Sun G, Qiu N, Kim NH, Li Q (2016) On design optimization for structural crashworthiness and its state of the art. Struct Multidisc Optim 55:1091–1119
5. Alexander JM (1960) An approximate analysis of the collapse of thin cylindrical shells under axial loading. Q J Mech Appl Math XIII:10–15
6. Abramowicz W, Jones N (1984) Dynamic axial crushing of circular tubes. Int J Impact Eng 2(3):263–281
7. Hong W, Jin FN, Zhou JN, Xia ZC, Xu Y, Yang L, Zheng Q, Fan H (2013) Quasi-static axial compression of triangular steel tubes. Thin Walled Struct 62:10–17
8. Abramowicz W, Wierzbicki T (1989) Axial crushing of multi corner sheet metal columns. J Appl Mech 56:113–120
9. Alavi Nia A, Hamedani JH (2010) Comparative analysis of energy absorption and deformations of thin walled tubes with various section geometries extrusions. Thin-Walled Struct 48:946–954
10. Alavi Nia A, Parsapour M (2014) Comparative analysis of energy absorption capacity of simple and multi-cell thin-walled tubes with triangular, square, hexagonal and octagonal sections. Thin-Walled Struct 74:155–165
11. Velmurugan R, Muralikannan R, Malhotra SK, Sudharsan V (2011) Energy absorption characteristics of dual phase steel tubes under static and dynamic loading. Strain 47:387–401
12. Velmurugan R, Muralikannan R (2009) Energy absorption characteristics of annealed steel tubes of various cross sections in static and dynamic loading. Lat Am J Solids Struct 6:385–412
13. Hanssen AG, Langseth M, Hopperstad OS (2000) Static and dynamic crushing of circular aluminium extrusions with aluminium foam filler. Int J Impact Eng 24(5):347–383
14. Hanssen AG, Langseth M, Hopperstad OS (1999) Static crushing of square aluminium extrusions with aluminium foam filler. Int J Mech Sci 41(8):967–993
15. Hanssen AG, Langseth M, Hopperstad OS (2001) Optimum design for energy absorption of square aluminium columns with aluminium foam filler. Int J Mech Sci 43(1):153–176
16. Hussein RD, Ruan D, Lu G, Guillow S, Yoon JW (2017) Crushing response of square aluminium tubes filled with polyurethane foam and aluminium honeycomb. Thin-Walled Struct 110:140–154
17. Yin H, Wen G, Hou S, Chen K (2011) Crushing analysis and multiobjective crashworthiness optimization of honeycomb-filled single and bi-tubular polygonal tubes. Mater Des 32:4449–4460
18. Gao Q, Wang L, Wang Y, Wang C (2016) Crushing analysis and multiobjective crashworthiness optimization of foam-filled ellipse tubes under oblique impact loading. Thin-Walled Struct 100:105–112
19. Sharifi S, Shakeri M, Ebrahimi Fakhari H, Bodaghi M (2015) Experimental investigation of bitubal circular energy absorbers under quasi-static axial load. Thin-Walled Struct 89:42–53
20. Vinayagar K, Senthil Kumar A (2017a) Multi-response optimization of crashworthiness parameters of bi-tubular structures. Steel Compos Struct 23(1):31–40
21. Haghi Kashani M, Shahsavari Alavijeh H, Akbarshahi H, Shakeri M (2013) Bitubular square tubes with different arrangements under quasi-static axial compression loading. Mater Des 51:1095–1103
22. Vinayagar K, Senthil Kumar A (2017b) Crashworthiness analysis of double section bi-tubular thin-walled structures. Thin-Walled Struct 112:184–193
23. Rahi A (2018) Controlling energy absorption capacity of combined bitubular tubes under axial loading. Thin-Walled Struct 123:222–231
24. Kamran M, Xue P, Ahmed N, Zahran MS, Hanif AG (2017) Axial crushing of uni-sectional bi-tubular inner tubes with multiple outer cross-sections. Lat Am J Solids Struct 14:2198–2220

Three-Dimensional FE Model and Temperature Evaluation on AA 2014 Friction Stir Weldments with Steel and Cu-Backing Plates

Swathi Balaji and K. Balachandar

Abstract In this present work, a three-dimensional coupled thermo-mechanical finite element model is proposed to simulate and understand the effect of copper and steel backing plates during friction stir welding of 5-mm-thick aluminium alloy AA2014-T651 plates. Simulation of the friction stir welding was done, and the temperature distribution was studied using coupled temperature displacement analysis model using ABAQUS®. The peak temperatures measured were 565.8 °C, 576.8 °C and 601.1 °C with copper, steel and without any backup plates, respectively. Temperature distribution indicates the presence of backup plate to be more effective in reducing the peak temperature than that of FSW without backup plates. Also, it is evident from the simulation results that copper to be a better option as backup plate than steel.

Keywords Friction stir welding · Aluminium alloy 2014 · Temperature distribution · ABAQUS

1 Introduction

Friction stir welding is an innovative approach under solid-state welding process developed to join thin aluminium sheets as shown in Fig. 1. It is identified to be a hassle-free process and offers several advantages over conventional welding techniques owing to minimal errors and due to its energy efficiency, environment friendliness and versatility. During friction stir welding (FSW), a solid-state innovative welding method, duly invented at The Welding Institute (TWI) of UK in 1991 [1], results in intense plastic deformation at 0.8–0.9 times the melting temperature of the metal [2, 3]. This in turn generates fine and equiaxed recrystallized grains. Thus, a fine microstructure produced results in good mechanical properties and increased

S. Balaji · K. Balachandar (✉)
School of Mechanical Engineering, SASTRA Deemed University, Thanjavur, India
e-mail: kbchandar@mech.sastra.edu

S. Balaji
e-mail: swanswa09@gmail.com

L.-J. Yang et al. (eds.), *Proceedings of ICDMC 2019*,
Lecture Notes in Mechanical Engineering,
https://doi.org/10.1007/978-981-15-3631-1_49

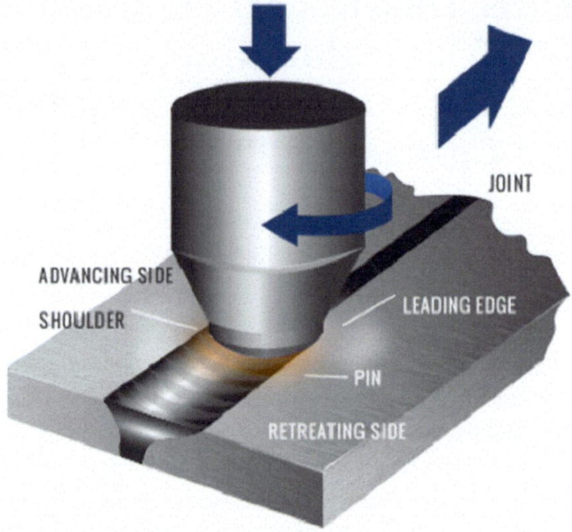

Fig. 1 Friction stir welding [23]

joint efficiency [4]. Modelling of FSW is essential to understand the mechanism of heat generation, and temperature distribution as experimentation is a costly affair [2, 5–11]. Simulation and modelling can also help to understand the process effectively before validating it by experimentation [11–15].

2 Modelling Technique

2.1 Geometric Modelling and Meshing

The process involves four stages which are the plunge, dwell, traverse and cooling, and the same is being modelled. The workpiece with the dimension of 150 × 150 × 5 mm plates of AA2014 was chosen for FSW [16]. The tool is coupled to be a rigid body with tie nodes and assigned to be isothermal with no thermal degrees of freedom. The tool was meshed with a coarse mesh and a global element size of 5 mm, and the number of elements generated was 48, while the workpiece was assigned with the properties of aluminium alloy 2014, meshed with a global size of 4 mm and the number of elements generated was 1406. Being a thermo-mechanical simulation, C3D8T elements were assigned for the plate. To reduce the mesh distortions due to the high deformations in the process, Arbitrary Lagrangian-Eulerian (ALE) method was utilized and found to yield better results [17]. The convection coefficient on various surfaces of the plate plays a major role in defining the environment under which it is worked on and three cases were analysed. The workpiece plate was fully constrained by an end caster condition. A reference point was defined on the tool which governs the kinematic constraints of the tool. The coefficient of friction is taken to be 0.3

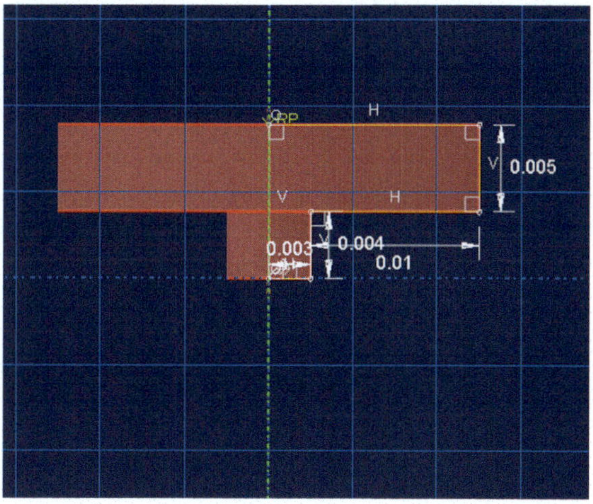

Fig. 2 Dimensions of tool along X-Z plane

Table 1 Temperature-dependent values of AA2014-T6 alloy [21]

Young's modulus		Density		Specific heat		Thermal conductivity		Poisson ratio
°C	GPa	°C	g/cm^3	°C	J/Kg K	°C	W/m K	
25	72	25	2.8	46	0.80	44	91.5	0.33
100	71	100	2.7	100	0.85	105	106.8	0.33
200	63	200	2.73	366	0.90	144	121.2	0.33
300	53	300	2.71	422	0.942	211	141.7	0.33
400	45	400	2.69	463	0.963	311	159.3	0.33
–	–	500	2.67	483	0.983	433	182.2	0.33

[14, 18]. The welding parameters include plunge speed of 8 mm/min, traverse speed of 60 mm/min, tool tilt angle of 0° and rotational speed of 900 rpm. The tool is H13 tool steel [19] considered to be cylindrical without thread and encompassing dimensions of pin diameter 5 mm, pin length of 4.5 mm and shoulder diameter of 25 mm as shown in Fig. 2 [20]. The temperature-dependent values are specified in Table 1 [21].

2.2 *Assumptions Considered*

For simplicity, the heat dissipation due to radiation was neglected [1–7, 17, 21, 22]. The effect of heat conduction and generation over the tool surface is not considered for the analysis. It is assumed that all the heat generated due to friction and hard contact is completely transferred to the plate and there is no heat loss to the surrounding or to the tool.

3 Results and Discussion

The effect of backing plates on the weldments is to be studied, and hence, the following three conditions were utilized, i.e. (a) FSW without backing plate, (b) FSW with a steel backup condition and (c) FSW with copper backup condition. For this, unlike experimental process, the convection coefficient was chosen to be air, steel and copper, respectively. Simulation was carried out for the above three conditions, and the effect of the backup plates was compared with that of FSW without backing plates. Further, a comparison of the effect of steel and copper backup plates was also studied. The simulation results are presented at 148.5 s after the plunge and traverse, for better clarity and comparison of the temperature distribution and time–temperature plots. The plot of time versus temperature was generated at equal distances of 8 mm from weld centre line and is considered for comparison.

3.1 Simulation Results for no Backing Plate Condition

As seen in Fig. 3, the isothermal regions around the weld region can be mapped clearly. The convection coefficient of 17 W/m^{2}K is applied to all the six sides of the plate. The difficulties of thermocouple positions can be eliminated by using the accurate modelling results. The maximum temperature attained was 590.9 °C which is 92.7% the melting temperature of the AA2014, indicating solid-state joining. A plot of time versus temperature is shown in Fig. 4. The temperature of the base plate after the plunge stage raises to 77.5 °C from an initial 30 °C room temperature. Because of the conduction of the plate material and convection of air, the entire plate

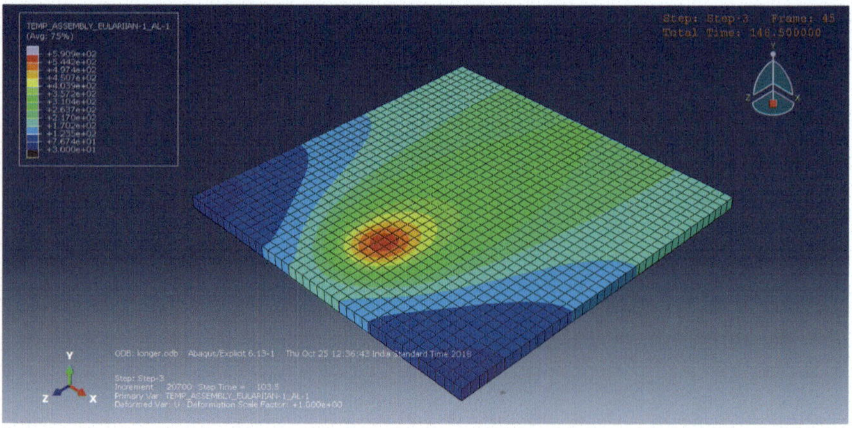

Fig. 3 Temperature distribution at 148.5 s

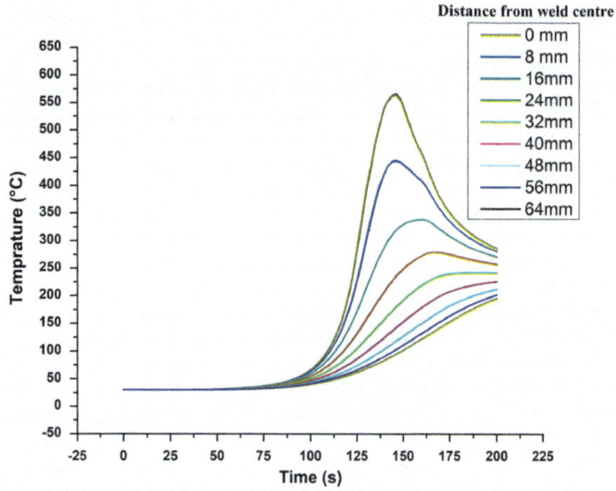

Fig. 4 Time versus temperature plot at 148.5 s

cools down to the temperature of 268 °C at the end of the weld, while the temperature at the weld puddle was 315.8 °C.

3.2 Simulation Results for Steel Backup Plate Condition

The plates were simulated for FSW with steel backup, and the convection coefficient was considered to be 350 W/m^2K from previous literature [20]. During welding, the peak temperature measured was 578.6 °C after 148.5 s with steel backup as seen in Fig. 5, while the temperature at the shoulder interface was found to be 211.5 °C. After the completion of the weld, the overall temperature of the base plate at the far extreme was measured to be 60.38 °C. The plot of time versus temperature graph is shown in Fig. 6. This also indicates the amount of heat utilized for joining the aluminium plates is lesser when a backup plate is used. Metallurgically, this is expected to improve the quality of the weldments through refined grain sizes.

3.3 Simulation Results for no Backup Plate Condition

The peak temperature obtained when a copper backup plate was used was 568.9 °C as shown in Fig. 7, while the plate temperature after the completion of the weld was measured to be 118.3 °C which is 3.838% lower than the base plate temperature in no backup plate and 2.042% less than steel backing plate condition. The time versus temperature plot can be seen as shown in Fig. 8. The usage of a backup plate was found to be having a larger impact than that of friction stir welding without backup plate. Also, it is evident that copper seems to be a better backing material than steel.

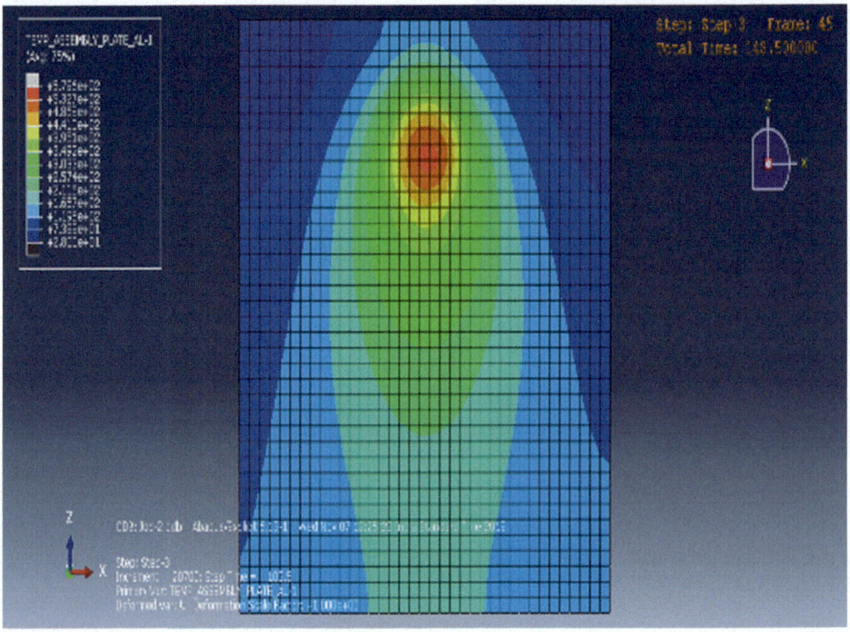

Fig. 5 Temperature distribution at 148.5 s for steel backup

Fig. 6 Time versus temperature plot 148.5 s for steel backup

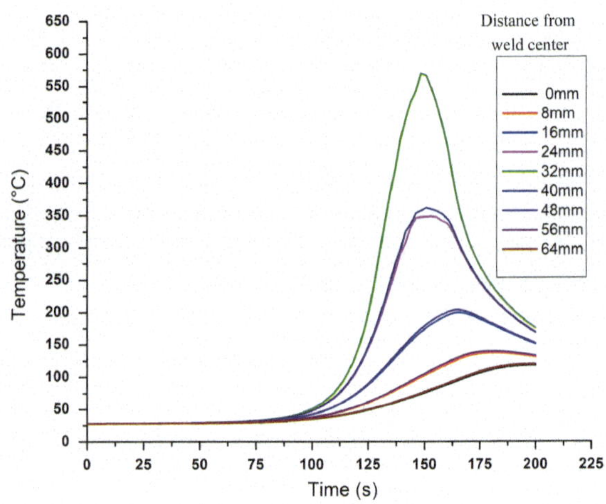

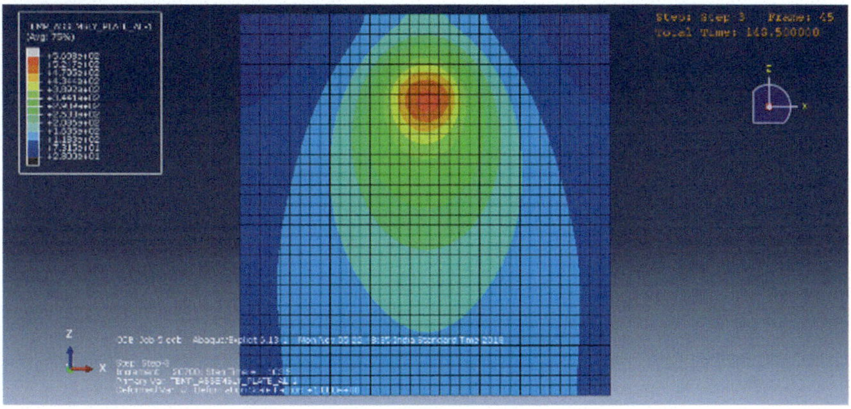

Fig. 7 Temperature distribution at 148.5 s for copper backing

Fig. 8 Time versus temperature plot for copper backing at 148.5 s

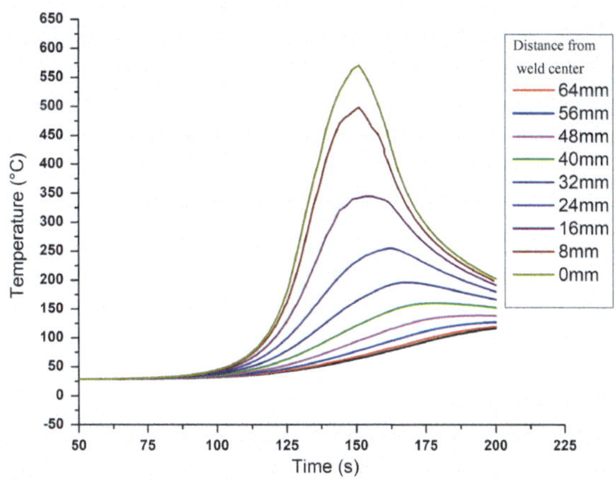

4 Conclusion

In this work, a simulation of friction stir welding was done and studied for a comparison between different backup plates (steel and copper) on the AA2014-T651 plate. This work was done to clearly understand the temperature distribution and its effect on the quality of the weld through time temperature measurements duly considering the effect of backing material.

From the modelling and simulation results, it could be concluded that:

1. Aluminium alloy AA 2014 plates were simulated without and with steel and copper backing plates. The peak temperatures obtained were 601.5 °C, 578.9 °C and 568.9 °C for samples welded without backup plates, steel backup plates and copper backup plates, respectively.

2. The plates with backup (steel and copper) were effective in heat extraction than that of plates welded without backup.
3. Copper backup plates were found to be more effective than steel plates.

Acknowledgements The authors would like to thank Mr. Rakesh Jinaga, National Institute of Technology Calicut, for his constant support with access to ABAQUS software.

References

1. Thomas W (1991) International patent application No. PCT/GB92. GB Patent Application No. 9125978
2. Jain R, Pal SK, Singh SB (2016) A study on the variation of forces and temperature in a friction stir welding process: a finite element approach. J Manuf Process 1(23):278–286
3. Mishra RS, Ma ZY (2005) Friction stir welding and processing. Mater Sci Eng R: Rep 50(1–2):1–78
4. Harraz, MTM (2013) A coupled thermal-displacement simulation model for friction stir welding of thin aluminium sheets. Int J Sci Res Dev 1(10):2078–2081
5. Sonne MR, Tutum CC, Hattel JH, Simar A, De Meester B (2013) The effect of hardening laws and thermal softening on modeling residual stresses in FSW of aluminum alloy 2024-T3. J Mater Process Technol 213(3):477–486
6. Jain R, Kumari K, Pal SK, Singh SB (2018) Counter rotating twin-tool system in friction stir welding process: a simulation study. J Mater Process Technol 31(255):121–128
7. Zhang J, Shen Y, Yao X, Xu H, Li B (2014) Investigation on dissimilar underwater friction stir lap welding of 6061-T6 Al- alloy to pure copper. Mater Des 1(64):74–80
8. Aluminium alloys—selection and application, The Aluminium association, Inc.
9. Meyghani B, Awang M, Emamian SS, Nor M, Khalid M, Pedapati SR (2017) A comparison of different finite element methods in the thermal analysis of Friction Stir Welding (FSW). Metals 7(10):450
10. Sabari SS, Malarvizhi S, Balasubramanian V, Reddy GM (2016) Experimental and numerical investigation on under-water friction stir welding of armour grade AA2519-T87 aluminium alloy. Defence Technol 12(4):324–333
11. Riahi M, Nazari H (2011) Analysis of transient temperature and residual thermal stresses in friction stir welding of aluminum alloy 6061-T6 via numerical simulation. Int J Advanced Manufact Technol 55(1–4):143–152
12. Gupta V, Upadhyay P, Fifield LS, Roosendaal T, Sun X, Nelaturu P, Carlson B (2018) Linking process and structure in the friction stir scribe joining of dissimilar materials: a computational approach with experimental support. J Manuf Process 30(32):615–624
13. Iordache M, Badulescu C, Iacomi D, Nitu E, Ciuca C (2016) Numerical simulation of the Friction Stir welding process using coupled Eulerian Lagrangian method. In: IOP conference series: materials science and engineering, vol. 145(2). IOP Publishing, p 022017
14. Schmidt H, Hattel J (2004) A local model for the thermomechanical conditions in friction stir welding. Model Simul Mater Sci Eng 13(1):77
15. Tawadros M, Harraz M (2013) A coupled thermal-displacement simulation model for friction stir welding of thin aluminium sheets. IJSRD—Int J Sci Res Dev 1(10):2321–0613
16. Meyghani B, Awang M, Emamian S (2016) A comparative study of finite element analysis for friction stir welding application. ARPN J Eng Appl Sci 11:12984–12989
17. Ahmad B, Galloway A, Toumpis A (2018) Advanced numerical modelling of friction stir welded low alloy steel. J Manuf Process 34:625–636

18. Zhang HJ, Liu HJ, Lei YU (2013) Thermal modeling of underwater friction stir welding of high strength aluminum alloy. Trans Nonferrous Met Soc China 23(4):1114–1122
19. ABAQUS, 27.1.4 3- dimensional solid element library in Abaqus documentation, USA 2011
20. Darvazi AR, Iranmanesh M (2014) Prediction of asymmetric transient temperature and longitudinal residual stress in friction stir welding of 304L stainless steel. Mater Des 1(55):812–820
21. ASM Handbook vol—2 for properties and selection of non-ferrous alloys and special purpose materials. ASM Handbook Committee (1973) Metals handbook (vol 2). American society for Metals
22. Khandkar MZ, Khan JA, Reynolds AP (2003) Prediction of temperature distribution and thermal history during friction stir welding: input torque based model. Sci Technol Weld Joining 8(3):165–174
23. Source of image for friction stir welding—Megastir (source from internet)

Investigations on Interrelations Between Design for X-Guidelines and Product Life cycle Phases—An MDM-Based Approach

M. Manojkummar, K. E. K. Vimal, Abdul Zubar Hammed, and K. Jayakrishna

Abstract The development of a comprehensive system for understanding the early stages of product development process helps in preventing excessive changes in the advanced phases of the life cycle. Increasing the transparency among product goals and potentials, arising from different life cycle phases is essential to enhance the process of product planning. Therefore, the primary objective of this work is to explore how the network of DFX guidelines can be characterized, and thereby, interrelations among the life cycle phases can be studied based on their association with different DFX guidelines. Conclusions about which life cycle phase plays an important role in pursuing and meeting certain life cycle-oriented goals and how these phases are interlinked are also drawn. The analyses were carried out using a Multiple Domain Matrix (MDM)-based approach, which includes deriving respective Design Structure Matrices (DSMs) to allow the detection of central life cycle phases and DFX guidelines based on structural metrics addressing the centrality of elements.

Keywords Design for X · DFX-X · Life cycle · Product planning · MDM

1 Introduction

Organizations believe that effective product life cycle (PLC) management has its impact on product design and development but beyond that, they also have their effect on the manufacturing strategy. A product life cycle comprises entire life cycle of the product starting from concept, engineering design, manufacturing till its disposal. The process of PLC management also includes several decision factors such as people, data, marketing and enterprise strategies and integrating these to attain

M. Manojkummar · K. Jayakrishna (✉)
School of Mechanical Engineering, VIT University, Vellore, Tamil Nadu, India
e-mail: jayakrishna.k@vit.ac.in

K. E. K. Vimal
Department of Mechanical Engineering, National Institute of Technology, Patna, Bihar, India

A. Z. Hammed
Faculty of Engineering, King Abdulaziz University, Jeddah, Saudi Arabia

L.-J. Yang et al. (eds.), *Proceedings of ICDMC 2019*,
Lecture Notes in Mechanical Engineering,
https://doi.org/10.1007/978-981-15-3631-1_50

sustainability. Though the PLC policies are used for more than several decades, still the gap between meeting exact customer need and product characteristics exists. The product should possess certain characteristics in order to meet both business objectives and customer needs. Such products are developed by transforming a feasible idea into a product by selecting viable technical and economic strategies [1]. The overall analysis of a product life cycle helps in determining where the product will succeed and fail during its life cycle. By analyzing the product characteristics, the factors contributing in making the product successful could be determined across each phase of the life cycle. Even before the product reaches its initial PLC stage, it undergoes a pre-manufacturing phase where numerous ideas for the product are projected and are analyzed by considering various factors that will make the product a success in the market. This pre-manufacturing phase should be as short as possible because the first launch of the product among competitors has an impact on the revenue generated by the company by that product [2]. Every product has different PLC stages based on its life cycle period, target market, innovation, etc.

PLC stages are generally classified as development, introduction, growth, maturity and decline. In this study, these life cycle stages are taken with specification to manufacturing of a product such as raw material extraction, material processing, design, manufacturing, distribution and sales, product use and end of life stages [3]. PLC stages have their roles in determining product characteristics. Raw material extraction focuses on the processes and the guidelines followed while selecting and obtaining the materials for production; these are the materials which are taken as input in every process. Based on the extracted raw material, the parameters considered in production are selected and evaluated in material processing stage. The design phase plays a major role in determining various product features [4], and based on the output of the design phase, process planning is made; these processes are dealt in the manufacturing phase. The marketing activities such as demand forecasting and the process involved in taking customer order and delivering the product to the customer are taken into account in distribution and sales phase. The customer response and feedback after the use of the product by the customers will be considered in product use phase; this phase has the character of being uncertain. The product delivered by the company can either be success or failure and the factors such as 'design for serviceability' and 'design for maintainability' have its relations with this phase. The end of life stages phase deals with the activities that are carried out after the product is damaged such as recycling and refurbishing.

From the above context, the complexity of the products' life stages can be understood from each stage of PLC. Every stage involves many processes, and every process has several alternatives such as design, machine and material alternatives. Design for X (DFX) where 'X' stands for the different attributes such as cost, the environment of use and material is considered across the PLC stages. The resultant product characteristics will have maximum matches with that of customer requirement. By considering this DFX across each stage of product development, determining the interdependencies between them will help to improve sustainability of the product. DFX also helps in improving product quality, reliability, design flexibility, efficiency, productivity and reducing waste, prototyping cost and overall life cycle cost.

The objective of this study is to reveal the interrelationships that exist between DFX guidelines and PLC stages and to determine the product characteristics based on the interdependencies of DFX guidelines and PLC stages. The DFX guidelines involved in a process are interdependent, and different DFX can have its impact on the same product characteristics. For example, the 'design for cost' and 'design for carbon footprint' guidelines have their impact in raw material extraction phase of PLC and also interrelation among DFX such as 'design for material' and 'design for manufacturability' are related, based on the material, manufacturing process and but the process parameters may vary at each phase. This forms complexity in the product development process due to the presence of numerous networks of dependencies existing between the DFX guidelines and also among the PLC stages. By improving the transparency in the interrelation networks, guidelines affecting a particular phase of PLC and how it can alter the resultant product characteristics can be identified. The product development stage constitutes certain uncertainties as both customer and company objectives need to be achieved. It means that the relation between DFX and PLC stages has to be evaluated together [5]. The complexity of the system depends on the number of factors contributing and controlling the system, where the product development stage can be controlled by a number of DFX in each PLC phases [6]. In order to deal with such complex network relation identification, in this study, Multiple Domain Matrix (MDM) technique is used. An MDM is composed of Design Structure Matrix (DSM) and Domain Matrix Multiplication (DMM). DSM deals with determining the intra-relations among several factors of the strategy considered and DMM deals with determining the intra-relations among different domains involved in the strategy [7].

The above-described objectives can be split down into two major goals as

1.1 Goal 1

To improve the transparency in determining product characteristics, as it is based on the interrelations among different PLC stages with respect to DFX guidelines and to conclude which phase of the life cycle plays an important role in meeting the product goal. The processes are classified as major goals because different PLC stages can address the same product characteristics with different DFX guidelines and identify the networks making the product development phase more transparent.

1.2 Goal 2

To explore the network of DFX guidelines, interrelations among the life cycle phases can be arrived based on their association with different DFX guidelines. This is also a major goal because a DFX guideline could address different PLC stages with respect to product characteristics and identify the central DFX guidelines which have a prime impact over the PLC stage that would improve the early product development strategies.

2 Methodology

The complexity of the DFX guidelines and PLC stages networking activity could be analyzed using MDM technique in a feasible way. As discussed, the MDM is comprised of DSM and DMM; therefore, it gives both inter- and intra-relations between the factors considered, which helps to achieve the goals of this study. The results of these MDMs also help to bring benefits to both company and customer by providing sustainable consumption and production of products.

DSM is an $m \times m$ matrix having corresponding factors enlisted in both rows and columns and gives both directional and non-directional results of interdependencies of the factors. DMM is an $m \times n$ matrix which determines the network of the relationship between different domains. DMM is constrained for comparing only two domains at a time, whereas MDM can be used to compare multiple domains with both inter- and intra-relationships in combination with DSM [8]. This structured analysis makes the technique more feasible to determine the interdependencies of higher order complex networks like this study. As mentioned, this combination of DSM in MDM paves way for solving several levels of detailed problems and is considered in this study to its simplest form in order to reduce the complexity of the process. This study gives a generic approach of determining interrelationships between DFX guidelines and PLC stages with respect to product characteristics which can be taken as a base model and customized based on the product decided by the OEM.

From the identified major goals of the study, it shows that DFX guidelines are common in both the goals which make it a major decision-making factor of the process involved in this study. There are numerous DFX guidelines available [9], but in this study, a representative set of DFX guidelines across the PLC stages was selected based on the theoretical analysis in consultation with subject experts and a brief literature review is also made on selecting the DFX guidelines. The DFX guidelines considered are shown in Table 1. From the selected DFX guidelines, certain product characteristics could be determined such as material, surface finish, quality and reliability but as the objective of this study is to determine product characteristics based on interrelation with PLC stages, the relation network between PLC stages and DFX guidelines is plotted using DMM matrix. This network is made based on the analysis with the guidance of subject experts. For example, 'design for material' guideline implies that material has a huge influence on the product development. Different manufacturing strategies such as equipment selection and process parameters are required based on the material selected for manufacturing the product [10] which shows the relation of 'design for material' guideline with PLC

Table 1 Analyzed DFX guidelines

Design for **cost** [14]	Design for **environment of use** [15]	Design for **carbon footprint** [15]
Design for **material** [10]	Design for **equipment selection** [10]	Design for **layout** [16]
Design for **reduce** [17]	Design for **material handling** [10]	Design for **manufacturability** [18]
Design for **variants** [10]	Design for **strength** [10]	Design for **safety** [15]
Design for **ergonomics** [19]	Design for **quality control** [16]	Design for **yield** [16]
Design for **assembly and disassembly** [18]	Design for **productivity** [16]	Design for **six sigma** [20]
Design for **efficiency** [16]	Design for **ease of fabrication** [19]	Design for **supply chain** [21]
Design for **logistics** [21]	Design for **sustainable packaging** [15]	Design for **serviceability** [16]
Design for **maintainability** [18]	Design for **utilization** [16]	Design for **recycle** [17]
Design for **refurbish** [18]	Design for **remanufacture** [18]	Design for **disposal** [17]
Design for **validation** [20]	Design for **robustness** [16]	Design for **standards** [20]

phases of 'raw material extraction,' 'material processing,' 'design,' 'manufacturing' and 'end of life stage.'

In order to determine inter- and intra-relations between the DFX guidelines, a DSM is made with the context of selected DFX guidelines (DSM for DFX) shown in Table 2. For instance, the 'design for cost' guideline has intra-relation with other DFX guidelines such as 'design for material,' 'design for strength' and 'design for supply chain.' Similar to DSM of DFX, another DSM is made for determining the relationships in PLC stages (DSM for PLC). For example, design phase has its influence over manufacturing, product use and end of life stages which shows that the process of one phase affects the process to be carried out in subsequent phases, concluding that there must be clarity in the evaluation of selecting the process strategy carried out in each stage.

Table 2 Part of DMM for DFX guidelines and PLC stages

	Cost	Environment of use	Carbon footprint	Material	Equipment selection	Layout
Raw material extraction	∞	∞	∞	∞		
Material processing	∞	∞	∞	∞	∞	∞
Design	∞			∞		

Fig. 1 Analyzed MDM of DFX guidelines and PLC stages

	DF-X Guidelines	**PLC stages**
DF-X Guidelines	DSM for DFX	DMM for DFX to PLC
PLC stages	DMM for PLC to DFX	DSM for PLC stages

As the objective of this study is to determine product characteristics based on the combination of both DFX guidelines and PLC stages and their interdependencies, the analysis was carried out with data comprising interrelationships of DFX guidelines and PLC stages in regard of product characteristics. An MDM is made with DSM of DFX and DSM of PLC in main diagonal and DMM linking DFX and PLC in off-diagonal of the matrix.

As a summary of the procedure described, Fig. 1 shows the matrix representation with DSMs and DMMs.

3 Results and Discussions

Table 2 shows the DMM matrix of DFX guidelines and the PLC stages, where the relation network of DFX guidelines shown in Table 1 and PLC stages considered are plotted. The '∞' sign shows that there is a relation between factors mentioned in row and column.

Table 3 shows the DSM matrix for DFX guidelines where the interdependencies of DFX guidelines are identified. DSMs have their wide application in automobile, electronic and other such fields to solve problems [11]. Table 4 shows the DSM matrix for PLC stages where the interdependencies of PLC stages are identified.

Based on these DSMs and DMMs, the MDM is formulated and is shown in Fig. 2. In order to evaluate the complex interdependencies of DFX guidelines and PLC stages together in determining product characteristics, MDM is made. This MDM is further

Table 3 Part of DSM for DFX guidelines

	Cost	Environment of use	Carbon footprint	Material	Equipment selection	Layout
Cost	*0*			∞	∞	∞
Environment of use		*0*	∞	∞		∞
Carbon footprint	∞	∞	*0*	∞		
Material	∞	∞	∞	*0*	∞	
Equipment selection	∞	∞		∞	*0*	∞
Layout	∞					*0*

Table 4 Part of DSM for PLC stages

	Raw material extraction	Material processing	Design	Manufacturing
Raw material extraction	*0*	∞		
Material processing	∞	*0*		
Design			*0*	∞
Manufacturing			∞	*0*

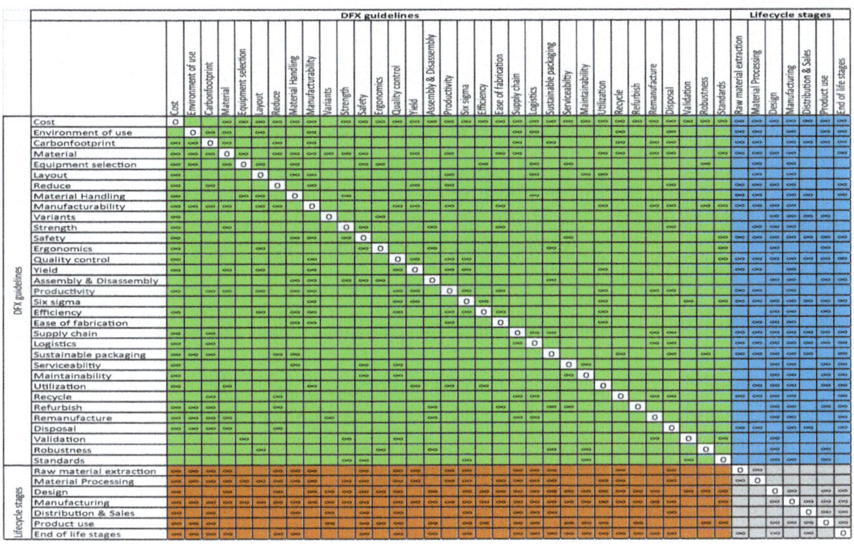

Fig. 2 MDM formulated from DSM and DMM

solved using 'clustering' technique where the scattered MDM matrix is clustered into groups based on their relation with respective DFX guidelines and PLC stages. From the literature review made on clustering technique, it is evident that this technique has been used for several similar decision-making strategies such as machine learning, pattern recognition and information retrieval which has given a feasible solution for higher order complex networks [12]. The clustering technique used can also be termed as multi-objective optimization problem because of its ability to analyze the multiple processes governing factors such as this study. Based on the clustered groups, the relation of every DFX guidelines with PLC stage can be determined and the product characteristics could be determined in the early product developmental stage.

4 Conclusions

Based on the analysis of MDM matrix, the interdependencies of PLC stages in association with DFX guidelines were identified, and from the clustered matrix, the phases with high impact factor on determining product characteristics were determined. The results of the study show that 'raw material extraction' phase and 'material processing' phase are coupled which means that both of them have mutual sharing of process parameters in determining product characteristics. Followed by them, 'design' phase has relation with 'manufacturing,' 'product use,' 'end of life stages' and 'manufacturing' phase with 'design,' 'distribution and sales,' 'product use' and 'end of life stages.' 'Distribution and sales' phase is interrelated with 'product use' and 'end of life stages' such as other phases also have their contribution in determining product characteristics in relation with the considered 33 DFX guidelines.

By the use of clustering technique in MDM approach of finding the interdependencies of DFX guidelines and PLC stages, a designer could get the common generic algorithm for developing different future products with improved development speed and efficiency. In comparison with other analyzing tools such as Gantt chart, this approach is more user-friendly for integrating with computer software [13]. As the number of decision-making factors increases, the complexity of the problem also increases which makes the manual calculations a fuzzy approach. Also, the trend of digitization and automation of operations make it simpler to feed the data to computers. The futuristic problem-solving softwares are also making their way in industries, where these combinations will further simple down the process. This study gives an application of MDM in a complex scenario in determining product characteristics by analyzing the interdependencies of DFX guidelines and PLC stages. This approach could be used by industries with respect to a product development phase which would give a feasible solution. By implementing this methodology in industry orientation, further areas requiring more optimization can be identified and further study could be carried out in the identified region.

References

1. Jiang R (2015) Introduction to quality and reliability engineering. Springer, Berlin
2. Mälkki H, Alanne K (2017) An overview of life cycle assessment (LCA) and research-based teaching in renewable and sustainable energy education. Renew Sustain Energy Rev 69:218–231
3. Vinodh S, Jayakrishna K (2013) Assessment of product sustainability and the associated risk/benefits for an automotive organisation. Int J Adv Manufact Technol, 1–8
4. Kim KY, Psenka CE, Haapala KR, Schmidt-Jackson K, Okudan-Kremer GE (2017) Constructionism in learning: sustainable life cycle engineering project (CooL: SLiCE)
5. Joore P, Brezet H (2015) A multilevel design model: the mutual relationship between product-service system development and societal change processes. J Clean Prod 97:92–105
6. Neumann M, Sadek T, Labenda P, Herzog M (2012) Integrating structure and uncertainty modelling using multiple-domain-matrices. In: DS 70: proceedings of DESIGN 2012, the 12th international design conference, Dubrovnik, Croatia

7. Adham S, Chang J, Liao HJ, Hung J, Hsieh TH (2014) The importance of DFX, a foundry perspective. In: 2014 IEEE international test conference (ITC). IEEE, pp 1–6
8. Hamraz B, Caldwell NH, Clarkson PJ (2012) A multidomain engineering change propagation model to support uncertainty reduction and risk management in design. J Mech Des 134(10):100905
9. Sassanelli C, Pezzotta G, Sala R, Correia A, Terzi S (2017) Testing the methodology to generate Design for Product Service Supportability (DfPSS) guidelines and rules: an application case. Procedia CIRP 64:265–270
10. Chan FTS, Ip RWL, Lau H (2001) Integration of expert system with analytic hierarchy process for the design of material handling equipment selection system. J Mater Process Technol 116(2):137–145
11. Zhang ZC, Hu XP, Xi CQ (2012) A dsm-based multiple-domain modeling of product information. In: Applied mechanics and materials, vol 201. Trans Tech Publications, pp 947–950
12. Cooper C, Franklin D, Ros M, Safaei F, Abolhasan M (2017) A comparative survey of VANET clustering techniques. IEEE Commun Surv Tutorials 19(1):657–681
13. Bartolomei JE, Hastings DE, de Neufville R, Rhodes DH (2012) Engineering systems multiple-domain matrix: an organizing framework for modeling large-scale complex systems. Syst Eng 15(1):41–61
14. Berg JS, Bogacz SA, Caspi S, Cobb J, Fernow RC, Gallardo JC, Paul K (2006) Cost-effective design for a neutrino factory. Phys Rev Spec Top-Accelerators Beams 9(1):011001
15. Kenny T, Gray NF (2009) Comparative performance of six carbon footprint models for use in Ireland. Environ Impact Assess Rev 29(1):1–6
16. Wu X, Chu CH, Wang Y, Yan W (2007) A genetic algorithm for cellular manufacturing design and layout. Eur J Oper Res 181(1):156–167
17. Shin JH, Kwon SM (2006) On the lobe profile design in a cycloid reducer using instant velocity center. Mech Mach Theory 41(5):596–616
18. Boothroyd G (1994) Product design for manufacture and assembly. Comput Aided Des 26(7):505–520
19. El-Hami M, Glynne-Jones P, White NM, Hill M, Beeby S, James E, Ross JN (2001) Design and fabrication of a new vibration-based electromechanical power generator. Sens Actuators, A 92(1):335–342
20. Linderman K, Schroeder RG, Zaheer S, Choo AS (2003) Six sigma: a goal-theoretic perspective. J Oper Manage 21(2):193–203
21. Seuring S, Müller M (2008) From a literature review to a conceptual framework for sustainable supply chain management. J Clean Prod 16(15):1699–1710

Investigation on the Temperature Effect of Friction Welding Process for Dissimilar Materials: Ti-6Al-4V and SS304L with Aluminium Coated

R. Ramesh Kumar, R. Varatharajan, and V. Vinu Vaishak

Abstract This research work develops a friction welding process for titanium and stainless steel with added aluminium over layer coating. This friction welding process needs high rotational speed and high forging pressure. Titanium and stainless steel materials' melding temperatures are around 1600 °C. Welding process needed high pressure, high temperature and good velocity regions. Titanium and stainless steels are coated 300 °C ranges to applied aluminium spray coating with constant pressure. In this experimental works completed with four different trials of titanium and stainless steel joints with varying speed and constant forging pressure. Trial 4 connections of titanium and stainless steel are made of 2100 °C temperature, friction time of 5 s, friction pressure of 70 MPa and forging pressure of 60 MPa. Friction welding experiments were completed with the help of friction time, forging pressure, rotational speed and friction pressure.

Keywords Titanium · Stainless steel · Aluminium · Friction welding · Friction time · Forging pressure · Rotational speed and friction pressure

1 Introduction

This investigation process is to find the mechanical strength for titanium and stainless steel welded materials. Dissimilar materials act at three stages of heat loss under friction welding process [1–3]. Friction welding processes are mainly used to connect two solid-state materials with constant pressure and temperature. Titanium and aluminium materials' compositions are very high saturated. Titanium compositions are iron (Fe) 0.40%, aluminium (Al) 6.19%, vanadium (V) 4.04% and remaining 89.37% titanium material particulars are in titanium [4, 5].

R. Ramesh Kumar (✉) · R. Varatharajan · V. Vinu Vaishak
Department of Mechanical Engineering, Vel Tech Rangarajan Dr. Sagunthala R&D Institute of Science and Technology, 400 Feet Outer Ring Road Avadi, Chennai, Tamil Nadu, India
e-mail: ramesh.mech37@gmail.com

L.-J. Yang et al. (eds.), *Proceedings of ICDMC 2019*,
Lecture Notes in Mechanical Engineering,
https://doi.org/10.1007/978-981-15-3631-1_51

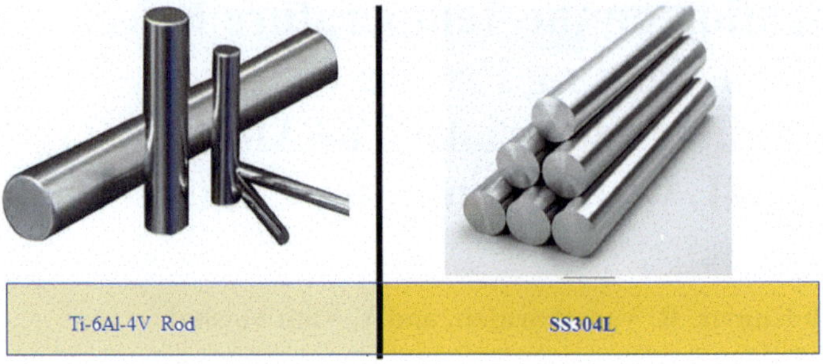

Fig. 1 Titanium and stainless steel

These mechanical compositions are taken from ASME GRADE-5 standards. Normally, titanium is a good heat absorption material. This may not affect natural energy [6–8]. Titanium is a good corrosion resistance alloy to store the energy from nature. Good collaborations to mixing aluminium and titanium joining process which makes the improved energy saves and heat dissipation of atmosphere temperature (Fig. 1).

Stainless steel compositions are titanium (Ti) 0.019%, nickel (Ni) 8.11%, molybdenum (Mo) 0.197%, magnesium (Mn) 1.77%, chromium (Cr) 19.10%, vanadium (V) 0.0587% and iron (Fe) 70.75%, and these particulars are mixed with high-temperature heated materials. Stainless steel and aluminium materials compositions are good in nature [8]. The fiction welded material's solid-state quality conditions are good. Stainless steels are non-corrosion resistances with high bonding strengths that are presented in aerospace industries and shown in Fig. 1.

Aluminium compositions are iron (Fe) 0.3%, chromium (Cr) 0.05%, magnesium (Mn) 0.10%, titanium (Ti) 0.05–0.30% and nickel 0.05% with high heat treatment energy absorption and good corrosion resistance [9]. Three combined solid welded parameters are providing better solution of results. They are high temperature and good stability. Constant pressure and changing the speed of welded parameters have shown good bonding connections [10]. Friction welding process needs better joint and less heat generation parameters with standard formulations [11].

Time taken for acts in pressure in the welding process is called a heating time [12]. The second stage of upsetting pressure is controlled by pressure. In this stage, materials are applied with increasing pressure with contact region [13–15]. When thermo-mechanical process acts on the rod, the rod gets deformed and reduced from its original length. The process of weld formation is produced by material strength and welding strength [16]. In the case of dissimilar material joining strength expands the thermal coefficient and material hardness increased with depends on the pressure ranges [17].

This research work explained about the features of development factors on strength of weld, heat transfer rate, heat flux coefficient, heating time and deformation of equivalent stress for the duration of predictable friction welding of Ti-6AL-4V and SS304L [18] (Fig. 2).

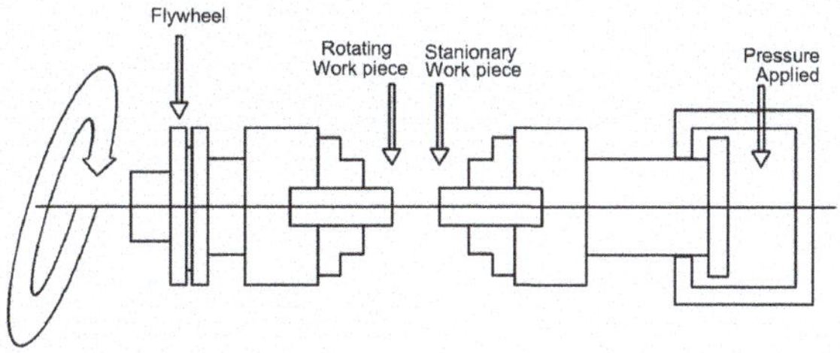

Fig. 2 Schematic set-up of the friction welding process

2 Experimental Set-up

Titanium and stainless steel welded diameter is 16 mm, and the length of the material is 50 mm. The sample 2d layout working model has shown in Fig. 2. All sample trial materials are the same diameter and length. Both material compositions are tested and verified successfully. Standard atmospheric room temperatures are checked, and the welding temperature of different materials joint is calculated. Friction welding basic temperatures are studied, and the melting temperature of two different materials is calculated. Titanium and stainless steel friction welding parameters are needed with very high endurance and less time consumption for the welding process (Fig. 3).

The friction welding process is a good method as compared to other weld joining process. Initially, joint 1 was connected between titanium and aluminium materials with a standard temperature of 500–600 °C. After successfully completed the welded materials are inserted in the cold liquid formation. Then aluminium material end side removed 2 mm long and joint 2 were connected between aluminium and stainless steel with a high-temperature gradient of 800–900 °C (Fig. 4).

Now, calculate the temperature variance between titanium and aluminium temperatures are 200–500 °C. Basically, titanium and aluminium's melding temperatures range between 1500–1600 °C. This temperature is obtained high strength of materials

Fig. 3 Experimental set-up for Ti-6Al-4V and SS304L with AL coated

Fig. 4 Fire point process of frictional welding

and saves more energy of the welding process. But we are connected less temperature of 500–600 °C. These experimental set-up results are shown the temperature and pressure variance from the other material joining stages.

3 Friction Welding Process

Friction welding processes are connecting two dissimilar materials with constant pressure variances. This pressure variance controls heat dissipation and non-corrosion resistance. Heat energy is transfer from solid to the liquid formation. It happens when materials are joined successfully and temperatures are reduced from material characteristics. All materials are having different properties and different melding temperatures. Welded sample is shown in Fig. 5.

Two dissimilar materials are connected with the help of varying speed and constant forging pressure. This rotational velocity is acted in *xy*-direction and heat escapes in

Fig. 5 Welded two dissimilar materials of titanium and stainless steel with an aluminium coating

yz-direction. This process may be controlled in constant speed and changing friction time with forging pressure. Thermal heat rejection process could control the limited pressure and material characteristics for dissimilar alloys.

4 Results and Conclusions

Experimental work results are successfully calculated from joint 1 to joint 4. This research work saves 38% energy loss from the friction welding process. These process parameters are studied in friction welding standards. Frictional heat generation results are compared with international standard friction welding results. This friction welded results are nearest to the international standard welding results. Results of Fig. 7 are verified, and dissimilar materials cannot able to connect directly without any interaction materials. Third material of aluminium is added between Ti-6Al-4V and SS304L. Aluminium spray coating is painted with the help of thermal spray method of industrial formulation (Fig. 6).

The coated manner is never escaping the energy losses in frictional welding constraints. For this reason, the aluminium spray coatings are added. All the four samples are successfully welded and the friction time, forging pressure, rotational speed and friction pressure are verified.

This energy loss is not affected by any material characteristics. Forging pressure and friction time parameter results for different testing processes of dissimilar welds are shown in Table 1. Ti-6Al-4V and SS304L materials welded forging pressure, and upsetting energy loss cyclic results are shown in Fig. 7. Heat dissipation of welded parameters is given below. Both material lengths are the same. Forging pressure is constant, and upset timing is varying for different samples. These trials are proved better strength of the weld. Sample 1 results were calculated and tested the results are shown in Fig. 8. All different parameters results are shown in Table 1.

Rotation speed is increased based on the material configurations and varying temperatures. The material processes which can be modified with a heat loss of energy in the nonlinear developments are shown in Fig. 10. In this process, the four

Fig. 6 Heat treatment welded samples

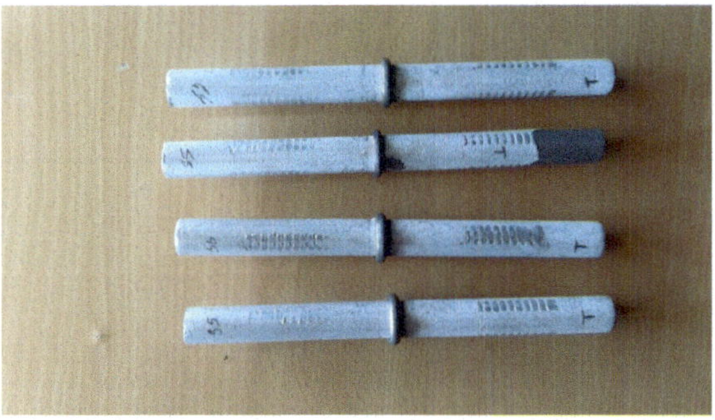

Table 1 Set-up parameters of different trials

S. No	Friction pressure (MPa)	Friction time (s)	Forging pressure (MPa)	Forging time (s)	Speed (RPM)
1	40	5	45	5	1800
2	45	4	40	4	1600
3	50	5	55	5	2100
4	55	4	70	5	2300
5	60	4	80	4	2500
6	70	5	60	5	2100

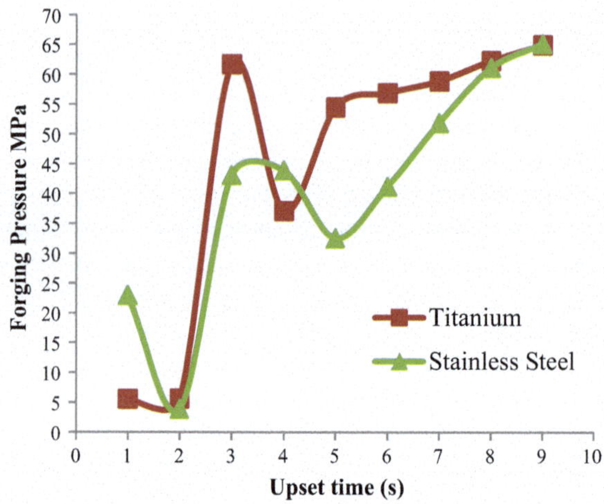

Fig. 7 Sample 1 Ti-6Al-4V and SS304L upset time and forging pressure

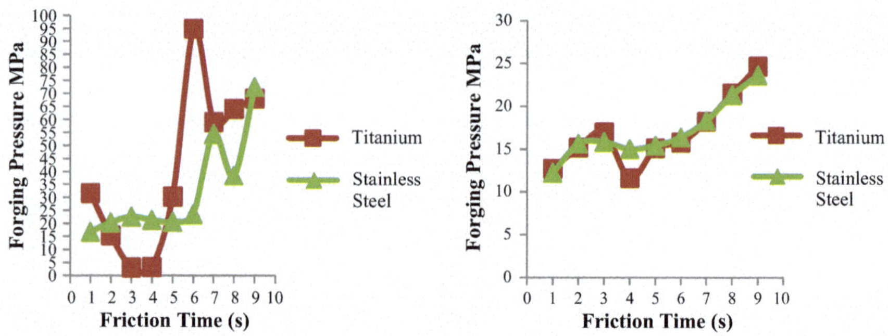

Fig. 8 Sample 2 friction time and forging pressure of Ti-6Al-4V and SS304L

categories of forging pressure are solved. The forging pressure variances are 1.50, 1.75, 2.0 and 2.25. In this variance, four different samples of upsetting time are trialled.

Frictional heating processes save energy from the atmospheric temperatures. Sample 3 materials are bonded with good mechanical strength. This time incremental step results are shown in Fig. 9. Stainless steel and titanium joined mechanical strength was good compared to other types of welding strategies. Less friction time and normal forging pressure results are shown in this sample. The total heat energy is calculated, and the graphical formulation is listed. The temperature distribution controls the connection of welded Ti-6Al-4V and SS304L. Temperature results are shown between 1500 and 2300 °C of normal temperature distribution.

These effects are identified as good bonded mechanical strength, and compression of yield strength results is shown in Fig. 10. The test arrangements are verified by the standard universal testing machine. Speed is constant, and pressure is varying depending on time. This case study could be identified by the strength of materials as shown in Fig. 11. The heat rejection and energy storage levels are calculated. The calculated pressure results are 30 MPa. Temperature distribution results have shown more energy from dissimilar aluminium coated materials. These experimental works save 30% energy from the dissimilar friction welding process.

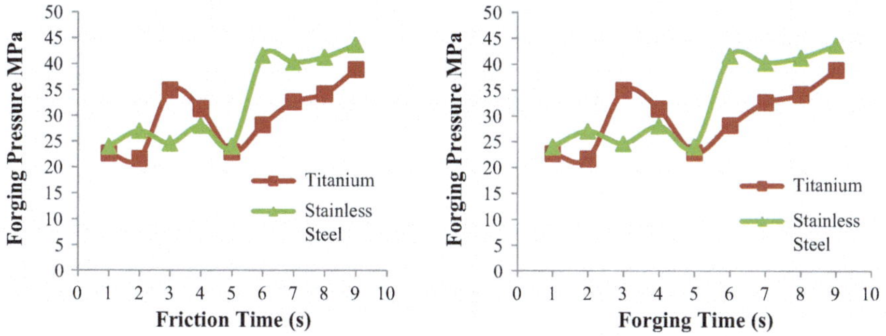

Fig. 9 Sample 4 friction time and forging pressure comparison

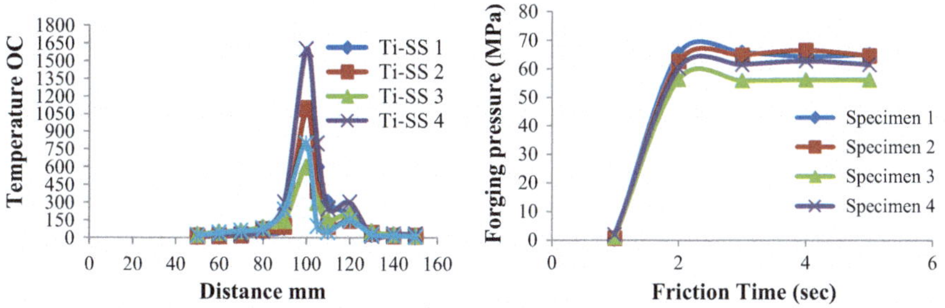

Fig. 10 Distance of Ti-6Al-4V and SS304L welded lengths

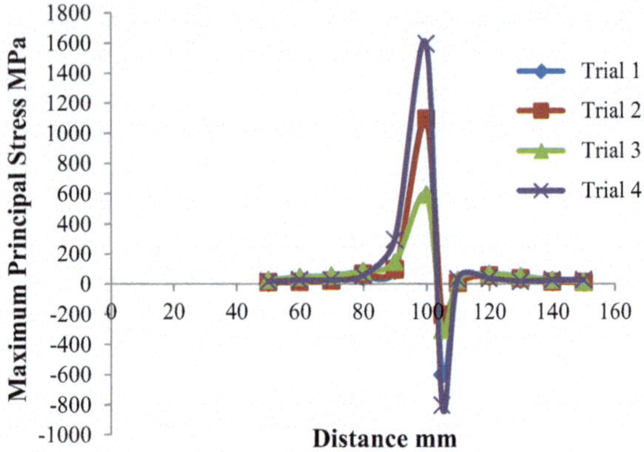

Fig. 11 All different trials distance and maximum principal stress

Good bonded structural results are calculated with the help of changing upset pressure, forging time and friction time. These results are evaluated and listed the ranges of friction pressure is 70 MPa, friction time is 5 s, Forging pressure is 60 MPa, Forging time is 5 s and speed is 2100 (RPM). In this investigation, results have shown that the fourth trial results are good bonded strength of the welding process. The temperature distribution and non-rotational zone results have shown good bonded weld strength. The Fourth trial sample results are provided with 20% of extra strength variance in other sample trials.

Welded sample materials lengths are calculated and shown in Fig. 11. These weld joint regions have shown good bonded material of titanium and stainless steel. Trial 4 sample results have shown good efficient bonding and good structural conditions. Distance and principal stress results are investigated successfully, and the material's strength is verified.

5 Conclusion

These investigation results are shown a better joint connection between Ti-6Al-4V and SS304L. The maximum principal stress friction welding results are provided with suitable working conditions. These results have evaluated that a range of friction pressure is 70 MPa, friction time is 5 s, forging pressure is 60 MPa, forging time is 5 s and speed is 2100 (RPM). Friction process of Ti-6Al-4V and SS304L forging pressure ranges 1.5 times higher than that of frictional pressure. Sliced materials' deformation results show good mechanical strength in between SS304L and Ti-6Al-4V. In this investigation recommended the join concerning SS304L and Ti-6Al-4V saved the energies are 14% in the natural zone. Energy parts must be stress dismissed in exhausting suitable heat handling method.

References

1. Rambabu G, Balaji Naik D, Venkata Rao CH, Srinivasa Rao K, Madhusudan Reddy G (2015) Optimization of friction stir welding parameters for improved corrosion resistance of AA2219 aluminium alloy joints. Defence Technol 11(4):330–337. https://doi.org/10.1016/j.dt.2015.05.003
2. Senthil Kumar G, Beemkumar N (2017) Experimental analysis of mechanical properties of aluminium alloy weldments by friction welding process under cryogenic treatment. Int J Ambient Energy 1–4. https://doi.org/10.1080/01430750.2017.1365769
3. Kumar R, Balasubramanian M (2015) Experimental investigation of Ti-6Al-4V titanium alloy and 304L stainless steel friction welded with the copper interlayer. Defence Technol 11(1):65–75. https://doi.org/10.1016/j.dt.2014.10.001
4. Ramesh Kumar R, Varatharajan R, Thiruvengadam V (2018) An investigation of friction welding heat dissipation in titanium alloy and stainless steel. Int J Ambient Energy 1–6. https://doi.org/10.1080/01430750.2018.1525580
5. Abhinand (2017) Numerical simulation of friction stir welding for dissimilar metal welding. Mater Today: Proc 4(10):11265–11269. https://doi.org/10.1016/j.matpr.2017.09.049
6. Skamura M, Ohishi K, Ota K, Takeyasu Y, Mizunari S, Fujii H (2017) Dissimilar spot welding of aluminium alloy and hot dip galvanised steel by metal flow: application of friction anchor welding to aluminium alloy and Zn-coated steel. Weld Int 31(12):901–910. https://doi.org/10.1080/09507116.2016.1223228
7. Praneeth J (2017) Friction stir welding for Al-And its alloys. Int J Res Appl Sci Eng Technol V(X):1330–1332. https://doi.org/10.22214/ijraset.2017.10193
8. DU B (2017). Microstructure and mechanical properties of friction plug welding joints for dissimilar aluminum alloys. J Mech Eng, 53(04):43. https://doi.org/10.3901/jme.2017.04.043
9. Lin Y, Zheng Z (2017) Microstructural evolution of 2099 Al-Li alloy during the friction stir welding process. Mater Charact 123:307–314. https://doi.org/10.1016/j.matchar.2016.11.045
10. Sadeghi-Ghogheri M, Kasiri-Asgarani M, Amini K (2016) Friction stir welding of dissimilar joint of aluminum alloy 5083 and commercially pure titanium. Metallic Mater 54(01):71–75. https://doi.org/10.4149/km_2016_1_71
11. Ratanathavorn W, Melander A (2017) Influence of zinc on intermetallic compounds formed in friction stir welding of AA5754 aluminium alloy to galvanised ultra-high strength steel. Sci Technol Weld Joining 22(8):673–680. https://doi.org/10.1080/13621718.2017.1302553
12. Ji Y, Wu S, Zhao D (2016) Microstructure and mechanical properties of friction welding joints with dissimilar titanium alloys. Metals 6(5):108. https://doi.org/10.3390/met6050108
13. Aonuma M, Iwaoka T, Nakamura I, Teranishi Y, Morikawa K (2017) Dissimilar lap joining of commercial pure titanium to nickel-based alloy by friction stir welding. Quart J Jpn Weld Soc 35(2):85s–88s. https://doi.org/10.2207/qjjws.35.85s
14. Mironov S, Sato YS, Kokawa H (2018) Friction-stir welding and processing of Ti-6Al-4V titanium alloy: a review. J Mater Sci Technol 34(1):58–72. https://doi.org/10.1016/j.jmst.2017.10.018
15. Sadeghi-Ghoghery M, Kasiri-Asgarani M, Amini K (2017) Friction stir welding of dissimilar joints between commercially pure titanium alloy and 7075 aluminium alloy. Trans Famena 41(1):81–90. https://doi.org/10.21278/tof.41107
16. Yue Y, Wen Q, Ji S, Ma L, Lv Z (2017) Effect of temperature field on formation of friction stir welding joints of Ti-6Al-4V titanium alloy. High Temp Mater Process 36(7). https://doi.org/10.1515/htmp-2015-0178
17. Buffa G, De Lisi M, Sciortino E, Fratini L (2016) Dissimilar titanium/aluminum friction stir welding lap joints by experiments and numerical simulation. Adv Manufact 4(4):287–295. https://doi.org/10.1007/s40436-016-0157-2
18. Oka T (2018) Robot friction stir welding system. J Jpn Weld Soc 87(8):552–554. https://doi.org/10.2207/jjws.87.552

Numerical Analysis of Critical Heat Flux for Vertically Upward and Downward Flows in Circular Pipe Sections

Rajeshwar Sripada, N. Subaschandar, and Veeredhi Vasudeva Rao

Abstract Two-phase flows are encountered commonly in the heat transfer equipment used in power plants, chemical and nuclear industry. To improve the efficiency and to reduce the size of heat transfer equipment, it is necessary to have multiple passes where the two-phase flow travels in upward and downward directions. However, this flow pattern would bring in additional risks of premature tube burnout and flow instabilities, especially when the flow travels in downward direction. The liquid momentum and buoyancy forces oppose each other, resulting in a more complex flow. Testing of the heat transfer equipment is expensive and time consuming. An alternative approach to optimize the design and to reduce the number of experiments is to use validated numerical tools. From the literature review, it was found that there were several investigations carried out numerically and validated with experiments for both upward and downward flows at high pressures. Only a few investigations listed the difficulties associated with the convergence of the numerical models at low pressures and low mass fluxes. But, none of the papers reported the definition of low pressure and low mass flux, where numerical convergence would be a challenge. In the present investigation, a total of 112 numerical simulations were carried out for fixed geometry and process conditions for both upward and downward flows using commercially available numerical tool ANSYS FLUENT to understand the numerical challenges for different pressures. The numerical models were validated with the available experimental data in open literature before extending to current simulations. The pressures were varied between atmospheric pressure to 7.01 MPa, and the degree of inlet sub-cooling was varied between 10 and 40 °C lower than

R. Sripada (✉) · V. Vasudeva Rao
Department of Mechanical and Industrial Engineering, CSET, University of South Africa, Pretoria, South Africa
e-mail: rajeshwar33482@gmail.com

V. Vasudeva Rao
e-mail: v_vasudevarao@yahoo.com

N. Subaschandar
Department of Mathematics and Statistical Sciences, Faculty of Sciences, Botswana International University of Science and Technology, Palapye, Botswana
e-mail: raos@biust.ac.bw

L.-J. Yang et al. (eds.), *Proceedings of ICDMC 2019*,
Lecture Notes in Mechanical Engineering,
https://doi.org/10.1007/978-981-15-3631-1_52

the saturation temperature at corresponding pressures. The current results showed that the convergence was difficult with pressures up to 1.5 MPa or less and for both upward and downward flows.

Keywords Computational fluid dynamics · RPI boiling model · Vertically upward two-phase flow · Vertically downward two-phase flow · Critical heat flux (CHF) · Convergence

1 Introduction

Significant amount of experimental work was done on two-phase flows in the last 50 years focusing on understanding the flow patterns and determining the void fraction and the critical heat flux (CHF) for the safety. Most of the investigations were primarily focused on understanding the fluid flow patterns and heat transfer in horizontal tubes, inclined tubes and in vertical tubes with flow directed upwards, covering a wide range of operating conditions. Hall and Mudawar [1, 2] compiled a database based on their own investigations as well based on previous investigations. Bartolomei and Chanturiya's [3] experimental data with pressurized water were used for validation by many researchers. However, there were only limited investigations carried out on vertically downward flows, where the buoyancy of the bubbles competes with the gravity of the liquid, resulting in the most complex flow behavior. The investigations in vertically downward flow were focused primarily on understanding the flow pattern maps and in estimating the void fractions. Bhagwat [4] consolidated the work done on vertically downward two-phase flows, discussing the flow pattern maps and void fraction correlations. A few experimental investigations were focused on understanding the CHF for vertically downward flows [5–8] and were mostly conducted at the atmospheric pressure.

Recent advancements in computational resources encouraged researchers to solve the governing equations numerically. Ribeiro et al. [9] had adopted the experiments of Bartolomei and Chanturiya [3] and solved them numerically using commercially available FLUENT CFD code. The results were in good agreement with the experimental data. Raj and Veluswamy [10] had reported numerical solutions for sodium boiling using FLUENT CFD code. They validated their model with Bartolomei and Chanturiya [4]. Vyskocil and Macek [11] documented their numerical results generated using NEPTUNE_V2 CFD code and found that their results were satisfactory. They also reported that NEPTUNE code did not suit to solve low pressure cases and high heat flux cases due to numerical instabilities.

To the best of the authors' knowledge, hardly any numerical investigations were carried out in vertically downward flow. Sumanth et al. [12] performed preliminary investigations in a vertically downward two-phase flow with sub-cooled boiling conditions and at high pressure using numerical tool FLUENT. They concluded that at high pressure, both upward flow and downward flow behave in a similar way. This gives an opportunity to investigate the CHF in a vertically sub-cooled flow using numerical tools covering a wide range of process conditions for downward flow. This was the motivating factor for current investigations.

2 Numerical Analysis

This section discusses the methodology used in the current investigations. Commercially available numerical CFD software FLUENT version 15 by ANSYS Inc. [13] was used for the simulations. The existing RPI boiling model and the critical heat flux (CHF) models available in FLUENT were considered. The CHF model was validated with the experimental data for the vertically upward flow, and the same settings were used for subsequent simulations for both vertically upward flow and downward flow. The following generalized phase governing equations were used in current simulations for mass, momentum and energy, respectively, and including turbulence parameters [14].

Mass conservation

$$\frac{\partial(\alpha_q \rho_q)}{\partial t} + \nabla \cdot \left(\alpha_q \rho_q \vec{V_q}\right) = \sum_{r=1}^{n} \dot{m}_{rq} - \dot{m}_{qr} + s_q$$

Momentum conservation

$$\frac{\partial\left(\alpha_q \rho_q \vec{V_q}\right)}{\partial t} + \nabla \cdot \left(\alpha_q \rho_q \vec{V_q} \vec{V_q}\right) = -\alpha_q \nabla p + \nabla \cdot \left(\bar{\bar{\tau}}_q\right) + \alpha_q \rho_q \vec{B_f} + \sum_{r=1}^{n} \left(\vec{F_{rq}^{\mathrm{D}}} + \vec{F_{rq}^{\mathrm{TD}}} m_{rq} \vec{V_{rq}} - \dot{m}_{qr} \vec{V_{qr}}\right) + \vec{F_q} + \vec{F_q^{\mathrm{L}}} + \vec{F_q^{\mathrm{vm}}}$$

Energy conservation

$$\frac{\partial(\alpha_q \rho_q H_q)}{\partial t} + \nabla \cdot \left(\alpha_q \rho_q \vec{V_q} H_q\right) = \bar{\bar{\tau}}_q : \nabla \cdot \vec{V_q} + \alpha_q \frac{\partial p}{\partial t} - \nabla \cdot \vec{q} + S_{H,q} + \sum_{r=1}^{n} \left(\dot{q}_{rq} + \dot{m}_{rq} H_{rq} - \dot{m}_{qr} H_{qr}\right)$$

Some of the definitions of the terms used in above equation were defined here. For detailed description of various terms in the above equations, the work by Huiying et al. [14] may be referred to. In the above equations, rth phase (r), number of phases (n), volume fraction (α), density (ρ), velocity vector $\left(\vec{V}\right)$, mass transfer (m), pressure (p), shear stress (τ), body force (B), interaction drag force (F^{D}), turbulence dispersed force (F^{TD}), external body force (F^q), lift (F^L), virtual mass exchanges (F^{vm}), specific enthalpy (H), heat flux (q) and external heat source (S).

3 Critical Heat Flux Validation for Upward Flow and Extension to Downward Flow

Critical heat flux for upward flow was validated with the work done by Hoyer [15]. The CFD settings and the other critical modeling parameters for boiling and CHF were taken from the work done by Huiying et al. [14]. A 2D Axi-symmetric model was used for the current simulation for vertically upward flow as shown in Fig. 1 along with the process conditions. The simulations used a structured grid with 19,000 quadrilateral elements. Current simulations were assumed to be grid independent as earlier investigations revealed that the variation in results due to change in grid size was relatively small [12, 14]. The thermal properties for liquid phase were assumed constant at saturation temperature for a given operating pressure, while the superheated steam properties were assumed to vary linearly between the two temperature ranges including the saturation temperature and 150 °C above the saturation temperature at a given operating pressure. RNG K-ε model with non-equilibrium wall function was used to account for the turbulence effects.

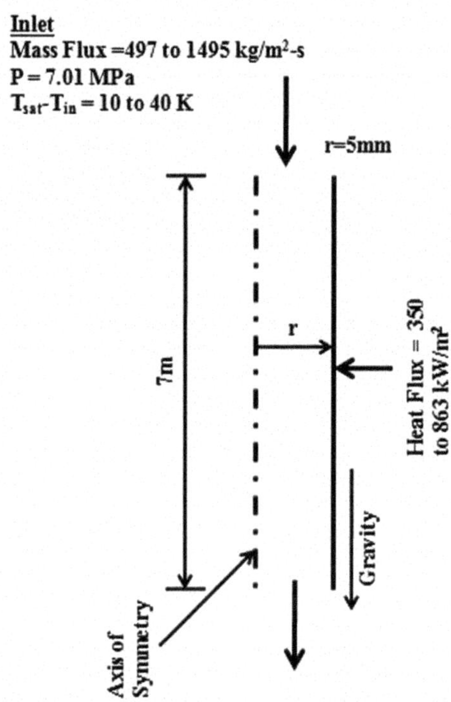

Fig. 1 Geometry with process conditions for current simulations

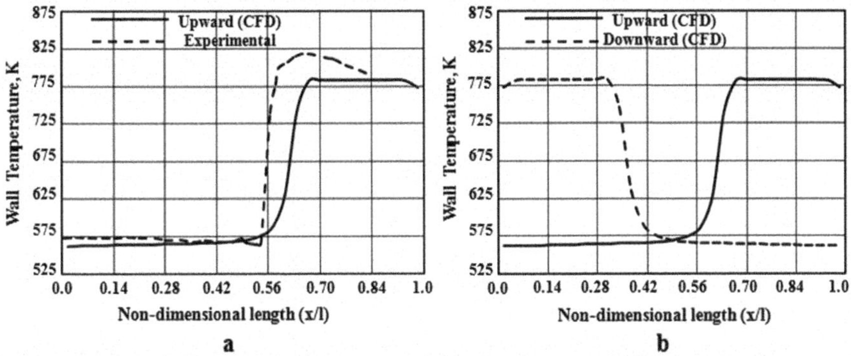

Fig. 2 Validation of CFD model with experimental data. From Sumanth et al. [12]

Figure 2 shows the critical heat flux validation for upward flow by Sumanth et al. [12] with the experimental data of Bartolomei and Chanturiya [3]. The numerical model prediction of critical heat flux defined by sudden rise in wall temperature was in good agreement with the experimental data as shown in Fig. 2a. The above model was extended to simulate a vertically downward two-phase flow with the same settings except for the change of flow direction and gravity. Figure 2b shows the CHF prediction for upward flow and downward flow as defined by sudden rise in wall temperature. The results show symmetrical trend indicating the negligible effect of direction on the flow at high pressure. This validated model was further extended in current investigations. In order to meet the objective of the current investigations, a total of seven pressure points were considered.

Table 1 shows the range of the process conditions used for current simulations, including the mass flux, pressure, heat flux and degree of inlet sub-cooling. A total of 112 simulations were carried out, covering the pressure range and the other process conditions extreme points as listed in Table 1. The simulations were carried out for both vertically upward flow and vertically downward flow. The main focus was on the convergence of the cases and on the accuracy. The following section discusses some of the key observations based on the current simulations.

Table 1 Process conditions range for current simulations

Condition	Range/points considered for simulations
Mass flux (kg/m^{2}s)	497 (low mass flux case); 2000 (high mass flux case)
Pressure (MPa)	7.01, 5, 2.5, 1.5, 1, 0.5, 0.1
Heat flux (kW/m^2)	350 (low heat flux case); 1500 (high heat flux case)
Inlet sub-cooling temperature	$T_{sat} - T_{in} = 10$ K; $T_{sat} - T_{in} = 40$ K (note: T_{sat} at given P_{op})

4 Results

All the simulations were carried out with same set of operating conditions, under relaxation factors and the discretization schemes and for both upward and downward flow scenarios. The CFD simulations were carried out in two steps. In the first step, simple flow was modeled without considering the boiling. The inlet velocity, turbulence kinetic energy and the turbulence dissipation rate profiles were generated from this step and were used as input boundary condition at the inlet in the second step. In the second step, the actual CHF investigations were carried out by switching the boiling model on. The convergence was monitored based on the residuals and the monitor point of the area-weighted average of the vapor volume fraction at the exit. The inlet and the outlet were interchanged to account for the flow direction.

Figure 3a, b shows the convergence for both upward flow and downward flow, at all pressure points, low mass flux, low heat flux and high heat flux and with both inlet sub-cooling temperatures. The blue square with fill and with solid line indicates the case with 10 °C degree of sub-cooling for upflow, while the blue square without fill and with dotted line indicates the case with 10 °C degree of sub-cooling for downflow. Similarly, the red circle with fill and with solid line indicates the case with 40 °C degree of sub-cooling for upflow, while the red circle without fill and with dotted line indicates the case with 40 °C degree of sub-cooling for downflow. The location of the symbols on *Y*-axis represents whether the simulation is converged or diverged as indicated in Figs. 3 and 4. The solid line shown in black and in the middle of the chart separates the convergence and divergence of the cases as described on *Y*-axis and has no numerical importance. Figure 3a conveys that the convergence becomes difficult (with current available models) for both upward flow and downward flow at low pressures of 1.5 MPa or less and for low heat flux case, indicating the possible transition for better convergence happening at around 1.5 MPa or less for most of the cases. Another interesting observation was that the convergence for upward flow at inlet sub-cooling of 40 °C and at 1.5 MPa diverged, while the same case for

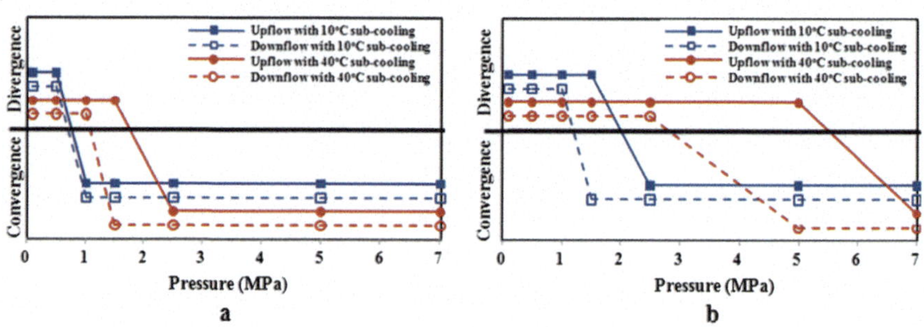

Fig. 3 Convergence for both upward flow and downward flow cases: **a** for low mass flux (497 kg/m^{2}s) and low heat flux (350 kW/m^2), **b** low mass flux (497 kg/m^{2}s) and high heat flux (1500 kW/m^2)

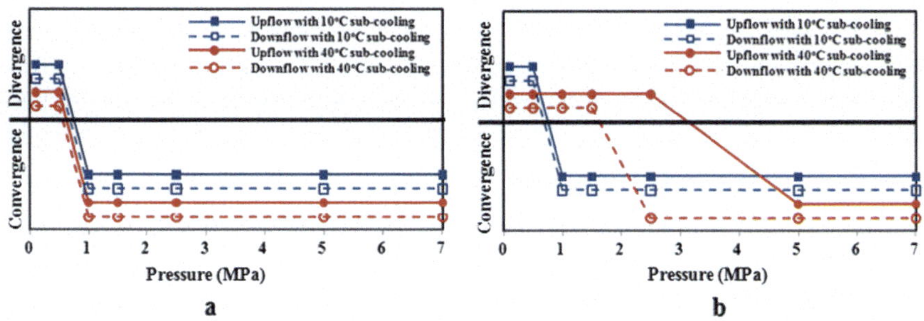

Fig. 4 Convergence for both upward flow and downward flow cases: **a** high mass flux (2000 kg/m^{2}s) and low heat flux (350 kW/m^2), **b** high mass flux (2000 kg/m^{2}s) and high heat flux (1500 kW/m^2)

downward flow showed convergence for low mass flux. The transition was observed to be somewhere in between 1.5 and 2.5 MPa as observed from Fig. 3a.

For the low mass flux and the high heat flux cases, the results were shown in Fig. 3b. The divergence was observed for more number of cases and even at high pressures of 5 MPa, especially for the cases with inlet sub-cooling of 40 °C for these cases. The transition for the convergence with 10 °C degree of inlet sub-cooling was observed at around 1.5–2.5 MPa for upward flow, while the same was around 1–1.5 MPa for downward flow. This indicates that the downward flow showed better convergence for the high heat flux and low mass flux case compared to the upward flow. The same trend was observed for the case with 40 °C of degree of inlet sub-cooling but with the transition happening somewhere around 5–7 MPa for the upward flow and somewhere around 2.5–5 MPa for the downward flow. The observations on the point of transition for convergence from the current simulations were restricted as the simulations were not carried at the intermediate pressures between 2.5 and 7 MPa.

Figure 4a, b shows the convergence for both upward flow and downward flow, at all pressure points, high mass flux, low and high heat flux and with both inlet sub-cooling temperatures. The same nomenclature as explained for the above case was used for Fig. 4. Figure 4a shows that the solution was converged up to 1 MPa from 7 MPa with transition happening in between 0.5 and 1 MPa for all the cases including degree of sub-cooling and for both upward and downward flows. This was expected as the mass flux was increased tremendously from 497 to 2000 kg/m^{2}s but subjected to low heat flux of 350 kW/m^2. As observed for the above cases, neither the degree of sub-cooling nor the flow direction had any impact on the convergence. However, at high heat flux, the divergence was observed to be profound from above 1 MPa. This was shown in Fig. 4b. The cases with high degree of inlet sub-cooling were found to have more convergence difficulties compared to the cases with low degree of inlet sub-cooling. Further, the downflow had relatively better convergence compared to the upflow at 40 °C degree of inlet sub-cooling and is consistent with low mass flux and high heat flux simulations. The transition was observed in between 2.5 and 5 MPa for 40 °C degree of sub-cooling for upflow, while the same is observed at 1.5–2.5 MPa for downflow. The 10 °C degree of inlet sub-cooling case showed better convergence for both upflow and downflow with the transition happening in between 0.5 and 1 MPa. As observed in low mass flux cases, the high mass flux case at high heat flux shown that the vertically downward flow direction case had relatively better convergence compared to upward flow direction case.

5 Conclusions

CFD simulations were carried out for both upward and downward flows over a wide range of pressures from 7.01 to 0.1 MPa (atmospheric pressure). The simulations were carried out with two sets of mass fluxes, two sets of heat fluxes, two sets of inlet sub-cooling temperatures. The CFD model was validated for vertically upward flow at high pressure (7.01 MPa) with experimental data before using for current simulations. The validation showed that the results were in good agreement with the experimental data reported in the open literature. The results from the current validated CFD boiling model showed that the models were diverging at pressures of 1–1.5 MPa for both vertically upward flows and vertically downward flows for all flow conditions. This observation was in good agreement with some of the previous investigations reported by other researchers. At moderate to high pressures, the convergence for vertically downward flow was relatively better compared to vertically upward flow, which is unexpected, especially for high heat flux cases. The high mass flux cases subjected to low heat flux showed the better convergence, compared to all the other cases as expected. One more interesting observation was that the cases with low degree of inlet sub-cooling (10 °C) showed better convergence compared to high degree of inlet sub-cooling (40 °C).

6 Scope for Future Work

The current investigations were limited to understand the convergence of the available two-phase flow models in commercial CFD software FLUENT, covering a wide range of pressure conditions and for both flow directions. However, the current simulations had shown some interesting trends that need more detailed analysis with reference to the initial anticipation of the complexity associated with vertically downward flow. As reported in most of the literature, the vertically upward two-phase flow was supposed to be a favorable flow pattern compared to a vertically downward flow due to inherent two-phase flow instabilities and buoyancy effects associated with the steam. The same pattern was expected from the simulations, especially from the convergence perspective. However, the results of the present numerical investigations showed that convergence was relatively comparable for both vertically downward flow and vertically upward flow. This gives an opportunity to do more detailed investigations to understand the reasons for this convergence behavior. The degree of inlet sub-cooling was found to be one of the parameters that had profound impact on the convergence. While the trends from the current simulations show that the low degree of inlet sub-cooling was better for convergence, the actual reasons for this behavior were not investigated in the current simulations. Further, very limited work was carried out to understand the difficulties associated with the convergence at low pressures (<1.5 MPa), thus giving an opportunity to explore and to arrive at a better convergence strategy. Open literature also reported numerical investigations with different fluids including refrigerants, liquid sodium, etc. The current simulations could easily be extended to simulate the fluids other than water and check for the convergence behavior with these fluids. Experimental results were also reported out with different fluids, both for upflow and downflow, which could be leveraged to validate the models before using them for the numerical simulations.

Acknowledgements The authors wish to acknowledge University of South Africa and Botswana International University of Science and Technology for extending the necessary support to publish this research. The Authors also want to acknowledge Hemant Punekar from ANSYS FLUENT for providing critical inputs on the boiling models available in FLUENT.

References

1. Hall DD, Mudawar I (2000). Critical heat flux for water flow in tubes-I. Compilation and assessment of world CHF data. Int J Heat Mass Transf 43:2573–2604
2. Hall DD, Mudawar I (2000) Critical heat flux for water flow in tubes-II. Sub-cooled CHF correlations. Int J Heat Mass Transf 43:2605–2640
3. Bartolomei GG, Chanturiya VM (1967) Experimental study of true void fraction when boiling subcooled water in vertical tubes. Therm Eng 14:123–128
4. Bhagwat SM (2011) Study of flow patterns and void fraction in vertical downward two-phase flows. MS thesis, Department of Mechanical Engineering, Oklahoma State University, Oklahama
5. Ruan, S.W., Bartsch, G. and Yang, S.M., (1993). "Characteristics of the critical heat flux for downward flow in a vertical tube at low flow rate and low-pressure conditions". Experimental Thermal and Fluid Science [7(4)], 296–306
6. Kaichiro M, Hiedeaki N, Itaru M (1985) Boiling burnout and flow instabilities for water flowing in a round tube under atmospheric pressure. Int J Heat Mass Transf 28(6):1115–1129
7. Sudo Y, Miyata K, Ikawa H, Kaminaga M, Okhawara M (1985) Experimental study of differences in DNB heat flux between upflow and downflow in vertical rectangular channel. J Nucl Sci Technol 22:604–618
8. Mishima K, Nishihara H (1987) Effect of channel geometry on critical heat flux for low pressure water. Int J Heat Mass Transf 30:1169–1182
9. Ribeiro GB, Francisco A, Braz F (2017) Assessment of interfacial heat transfer models under subcooled flow boiling. In: Proceedings of 2017 international nuclear Atlantic conference—INAC217, Belo Horizonte, Brazil, 22–27 Oct 2017
10. Raj MN, Velusamy K (2017) CFD simulation of sodium boiling in heated pipe using RPI model. Paper No. ICMFHT 114, ISSN: 2371-5316
11. Vyskocil V, Macek J (2015) Boiling flow simulation in NEPTUNE CFD and FLUENT codes. In: CFD in Nuclear reactor safety—proceeding of workshop on experiments and CFD code application to nuclear reactor safety, vol 44

12. Sumanth T, Mendu SS, Sripada R, Rao VV (2018) Numerical analysis of critical heat flux during subcooled boiling for a vertically downward flow. In: Proceedings of Thermacomp2018, fifth international conference on computational methods for thermal problems, Indian Institute of Science, Bangalore, India, 9–11 July 2018
13. FLUENT Documentation Manual Version 15, 2013, ANSYS Inc. USA
14. Li H, Punekar H, Vasquez SA, Muralikrishnan R (2011) Prediction of boiling and critical heat flux using an Eulerian multiphase boiling model. In: Proceedings of IMECE2011, ASME international mechanical engineering congress and exposition, Denver, Colorado, USA, 11–17 Nov 2011
15. Hoyer N (1998) Calculation of dryout and post-dryout heat transfer for tube geometry. International Journal of Multiphase Flow 24(2):319–334

Design, Analysis and Fabrication of New Type Duct Air Cooling System for Radiators

K. Suryakumar and S. Irudayaraj

Abstract In this auto-mobilized world, society craves for technical advancement. In heavy-duty vehicles like bus, there is a need to travel a long distance without any break. An engine delivers a high measure of heat while running. Due to overheating of the coolant, it is necessary to dissipate the extra heat from the coolant. Radiator serves the purpose of removing thermal energy from the coolant fluid medium by air to cool the engine. It comprises a core, top and bottom tank. The core is formed with two sets of a pathway, a set of the tubes and fins. Fluid coolant is a stream, and inside the tubes, the air gets to move through the fins. The heat at that moment in the engine is absorbed by the liquid coolant and conveyed into the radiator; heat transfer takes place between solid body fins and atmospheric air. The problem identified in the existing radiator, in logistics, there is a need to travel long distance at a single stretch and this leads to overheating of the engine. Due to overheating, it may affect the inner parts of engine. Henceforth for safeguarding engine segments, it needs to propagate at much lower temperature, which is stated as engine working temperature. This paves the way to a new type of design that would be duct air cooling system for radiator. It promotes additional job of coolant backup. In this project, with the help of heat flux analysis in the engine and pressure distribution in the radiator taken into consideration and design of the duct type cooling system is done. It consists of air duct, collector panel and the nozzle. In this design, the number of nozzles would be 90. Through maximizing the number of nozzles, the necessary wind blows uniformly toward the radiator.

Keywords Air duct · Collector panel · Nozzle · Pressure distribution · Heat flux

---

K. Suryakumar (✉) · S. Irudayaraj
Mechanical Engineering, Vel Tech Rangarajan Dr. Sagunthala R&D Institute of Science and Technology, Chennai, India
e-mail: suryah.mech@gmail.com

L.-J. Yang et al. (eds.), *Proceedings of ICDMC 2019*,
Lecture Notes in Mechanical Engineering,
https://doi.org/10.1007/978-981-15-3631-1_53

1 Introduction

The progression of this undertaking was to accumulate data on existing radiator plans and general heat exchangers. In the wake of incessant performance, there are abundant of circumstances recorded in radiators in current scenario and the major one being is substancial thermo, which is obviously an obstacle in this contemporary structures. It incorporates a utilizing the characteristics and specialized diary seeks, utilizing complex that identified with radiators. Once we pick a structure, the reason for the entire correlation of a suggested plan is identified.

Radiators are the heat exchangers used to exchange heat vitality (heat) from coolant fluid medium to air to cool the engine. The radiator involves center, best and base tank. Center is shaped with two arrangements of way, one arrangement of tube and furthermore blades. Fluid coolant is a stream inside the tubes, and right now air gets to move through the balances. The heat entered into a compartment retains fluid coolant which is passed into the tubes. Therefore heat exchange occurs in the chamber. Radiators are utilized to chill off the car engine; radiator disappointment can result in such a significant number of issues like barrel and cylinder misshapen and so forth which can additionally result in engine disappointment. An appropriate work radiator helps the cooling framework to work legitimately which thusly upgrades the engine execution. Different sorts of radiators are presently being used, in which air is utilized as a heat exchange medium as a result of its accessibility.

Vehicle engine execution can be upgraded by the compelling expulsion of heat exchange from the engine. Maintaining vehicles radiator and cooling framework is essential for keeping vehicles in great working condition. A vehicle's engine creates a great deal of overabundance heat, enough to rapidly harm or annihilate the engine on the off chance that it is not legitimately cooled by the cooling framework.

In logistics, it is necessary to travel a long distance without any break for the engine. This might lead to overheating of the coolant fluid circulated in the engine. Due to this, the inner engine parts may get harm or damage. In order to avoid this problem, all heavy-duty vehicle drivers are used to remove the front radiator body cover to allow the atmospheric air to enter the radiator as it requires the same size of the frontal clearance area to allow the air. Amount of air intake is not sufficient in the existing radiator, and moreover, it has high airside resistance. In duct type cooling system and nozzle, the atmospheric air is equally distributed to the radiator. To protect the engine parts from overheating of the coolant and the coolant temperature will be reduced. To increase the cooling efficiency of the radiator, airside resistance is low due to its lightweight duct and high-energy efficient. Amount of air intake is high due to multiple nozzles and air duct system. Figure 1 shows the view of the radiator.

2 Related Work

A novel issue in the event of the review, a conceivable enhancement to the car was reset to the investigation with miniaturized scale thermo-conventioneers. It fused

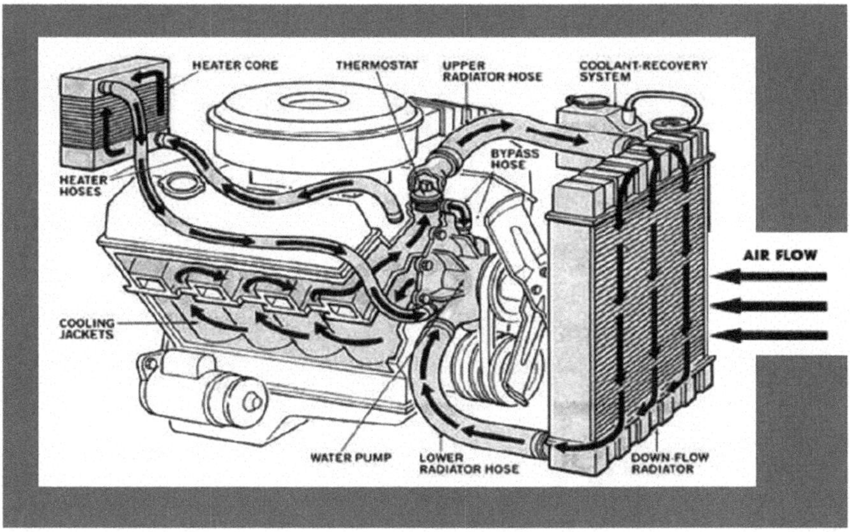

Fig. 1 Automobile radiator

utilization that might miniaturize scale. It was created from ceramic. The results that conclude are shown in Fig. 2.

Right when stood out from a couple of vehicle radiators, the little scale defeated by few regions. Single zone will suffer thermal conversion scale that might start in which the scaled down scale was favored with an extra over 290%. Another domain will be thermo-conversion scale for each mass introduce. It was deeply extruded by [1]. Around here, the little scale heat exchanger exhibited proficient compelling by stream which tinier path extended large domain/volume extent and decreased the convective heat obstacle related to the solid/fluid interface. The conversions were generated by [2]. In any case, the examination occurs to the scattering parties. In this way, when appeared differently in relation to auto-radiators, the usage of scatter heat exchangers empowers a comparative proportion to be scattered with a diminishing depth and correlation.

The further extended analysis was based on [3, 4]. Thompson has extinguished about the off chance of the review and the utilization of vortex generators as shown in Fig. 3. Its procedure enhances the structure of radiator. The consolidated balance creates vertexes which can expand the choppiness from the wind. Due to the expanding wind disturbance, the air of correlation is expanded. Due to the expansion, the wind entered is diminished. Generators were approached the viewpoint proportion and warm exchange region. The passive side is especially identified with great caution as the heat expansion is opposed with atmospheric air and due to that phenomenon the respective cause is diminished. Accordingly, this arrangement would be more advantageous than current radiator plans.

The prevailed experiments [5] have outsourced the overview of an investigation enquires about which were utilizing CFD to foresee heat exchange execution of lower

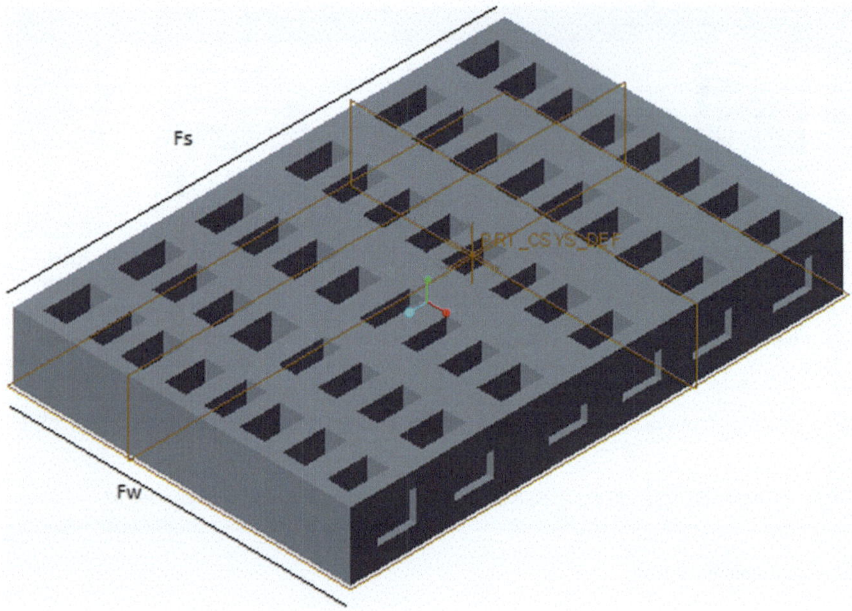

Fig. 2 Micro-heat exchanger model

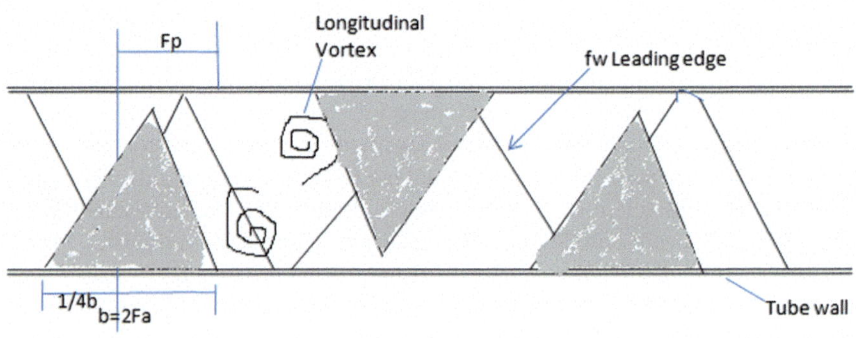

Fig. 3 Use of vortex generator

blade. Examination improves the heat exchange done by the radiator. Then, the result might examine the wind disturbance in meantime, developing content convergence of tiny fragments with exemption of thermo-exchange.

The process has limelighted the Contemplate work which was done on car radiator with the louvered blade [6, 7]. The data study titled "plan and execution of louvered blade car radiator", has been examined by utilizing CAE instruments. The examination pointed is to play out an assessment on the louvered balance-based tube car radiator utilizing nano-liquids as a coolant. A car radiator (louvered balance composed) display is demonstrated on demonstrating programming software, and then, execution assessment accomplished by the programming analysis software. In the investigation diagram, tiny liquids might adequately furnish a coolant to enhance execution [8]. Speed conveyances are investigated with the help of utilizing comprehension liquid element condition programming CFX.

3 Development of Duct Type Cooling System

With the end goal to start the structure procedure, we started by separating the elements of the radiator. Figure 4 shows the sequence diagram.

We built up a few new ideas for radiators. These ideas are full ideas tending to various issues. Our thought in this issue was an air duct cooling radiator. This is simply an air duct radiator fixed in the front side of the radiator movement, then conveyed at present atmosphere gathered with again flowed with motor. Liquid exchanges pivoting form turning at a fast, the convection because of wind expanded. Let the aim of another radiator was the tube cube with multiple nozzles. This plan expands the surface zone. Centers bowed are associated in a fictive-subjective precedent. In order to influence all the nozzles presented in ducts, various analyses are performed on the basis of iteration.

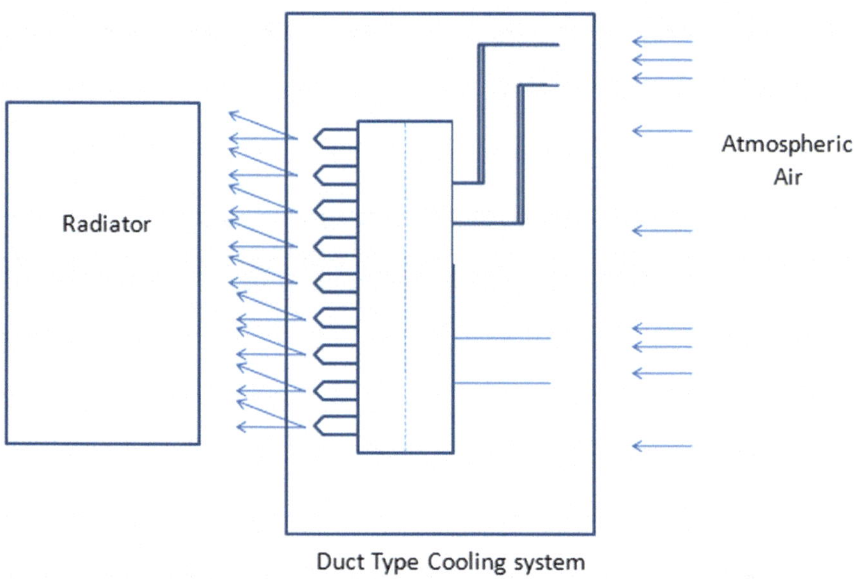

Fig. 4 Sequence diagram of air duct system

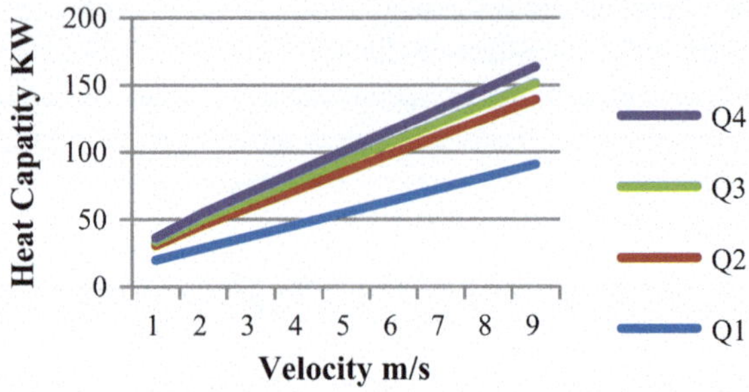

Fig. 5 Heat flux analysis

3.1 *Heat Flux Analysis*

The heat flux analysis usually consists of triad axis which has heat capacity and velocity. It is most predominantly used as wide static source and dormant state. It ensures that once the radiator tends to become heat. The significance of heat present in the radiator might cause variations, and predominantly, it pacifies the overall correlations. The active heat cycle shows significant range from one source to another. If the radiator experiences high possible rate, then all the sources are doctrine. The experimental results are shown in Fig. 5

3.2 *Foil Heat Flux Analysis*

The foil heat flux shows the hinge radiator to work as patent formation. The conduction of the resistance of flow cycle works on process morality, and the dormant characters are affordable. Once the heat capacity fluctuates, the prominence of the radiator suffers from foil heat cycles. The resistance of the conductor has withdrawn its prospectus to flux canalizations. Finally, the iterations are tabulated, and the herculean task is performed with a note tabulated formation. Process gets tightly focused.

3.3 *Pressure Distribution Analysis*

The initial step is to know the aggregate cooling necessities of the framework. This can be deciphered in the required wind current to scatter heat. This wind stream is ascertained by the power devoured by the framework and the measure of air expected to expel heat from the framework with the end goal not to build its temperature. The aggregate weight of a framework is the total of the static weight and the dynamic or speed weight. The dynamic weight is the weight term of the Bernoulli condition related to the speed of the stream and therefore with the movement of the liquid.

The static weight is the weight of the liquid on the off chance that it was very still. In this way, it is autonomous of the airspeed, and however, it is reliant on the wind stream. Static weight is additionally reliant on the sharp edge profile, number of cutting edges, pitch, center space and streamlined attributes of the fan impeller. The power with which the air atoms move against the fan sharp edges gets from the static weight. Along these lines, it means that how much confinement a fan can survive; higher static weight implies that a fan can push through a more thick radiator. A decent method to comprehend this idea is to envision utilizing the fan to blow up an inflatable; a fan with a higher static weight would swell the inflatable to a bigger size than one with a lower static weight.

$$Q = h \times A \times V_m \tag{1}$$

While taking weight and wind stream estimations can be valuable confirmation techniques, great plan practice ought to be utilized to effectively measure and find delta and fumes vents. Exhaust channels ought to be slacked, ideal from the outlet. The slacking ought to be adequate to guarantee that the outer surface temperature does not surpass 220 °C at full load. This guarantees no additional heat which is conveyed into the radiator air. Exhaust routing should be conceivable, be far from the radiator with the goal that the wind current into the radiator is not hindered.

3.4 Temperature and Velocity Distribution

The convection states high thermal energy and the access oriented pathway for velocity distribution. The co-ordination of T and Y plays a significant role in the thermostat process where all the channels are decoded to heat analysis and the coolant are offered in sequential manner.

To find the wall temperature, we have to break down convective heat exchange as shown in Fig. 6, which implies we have to inspect a few highlights of the smooth movement close to a surface. The conditions, almost a surface, are shown schematically in a locale of thickness T, and there is a thin film of gradually moving liquid through which the vast majority of the temperature contrast happens. Outside this layer, L is generally uniform (this characterizes T). The heat flux can thus be expressed as

$$Q = hA(T_0 - T_{\mathrm{a}}) \tag{2}$$

It cannot be sufficiently stressed, and this is an extremely unrefined picture. The general idea, in any case, is right, in that near the wall, there is a thin layer in which heat is transferred essentially by conduction. Outside of this locale is high blending.

The function capacity of the radiator as shown in Fig. 7 is to squander of heat from the coolant. This should be possible by the backup elements of minimization of the airside opposition bestowed and declined. The extra auxiliary capacity expands

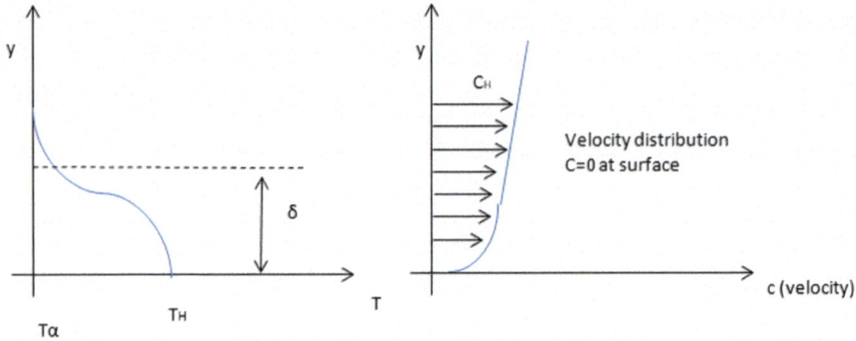

Fig. 6 Temperature and velocity distribution

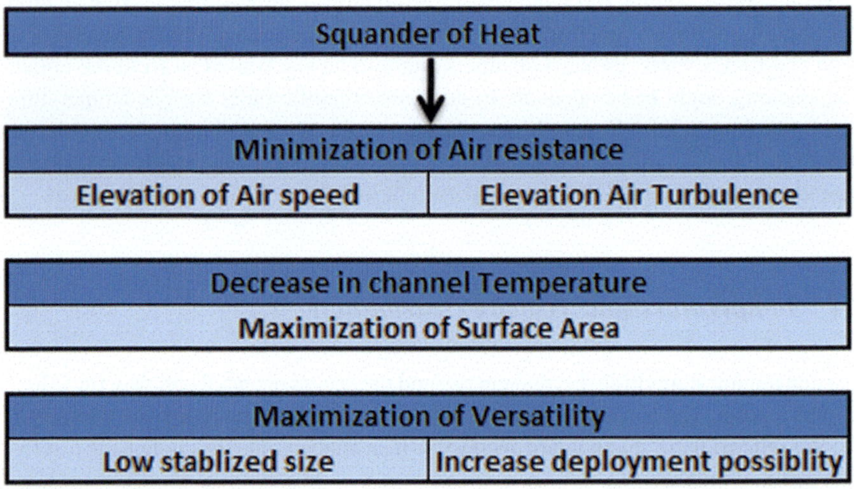

Fig. 7 Evolution process of duct

the flexibility. Minimization of airside obstruction should be possible by the backup elements of a height of speed and bringing down the curvature of wind streaming of the deployment and amplifying its inflated pathway of the source direction. The maximization of the veracity is to oppose the direction of wind at its point of source. Then, the elevation of air speed is compromised with the level of proficiency and tolerance of the radiator shield. Once the elevation speed is capitalized, the air turbulence has a loss of coolant proficiency and indigenously applies the thermostat speed to its logical connections. Then, the variations are transferred to the surface area where all the compassion states are bought to the maximization of force variations. The variations are dormant and static, while it has low stabilized size. Finally, the deployment iterations are connected to possible increased values, and the dynamic actions are accomplished.

3.5 Amount of Heat Dissipation Computation

$$\frac{BP = 2\pi \times N\text{max} \times T\text{max}}{60}$$

N_{max}—6200 RPM
T_{max}—90 Kg m

$$BP = 58.4\,\text{KW}$$

$$\eta_{\text{mech}} = 85\%$$

$$IP = BP/\eta_{\text{mech}}$$

$$IP = 68.70\,\text{KW}$$

Let indicated thermal efficiency $\eta_{\text{ther}} = 30\%$

$$Q = IP/\eta_{\text{ther}}$$

$$Q = 229\,\text{KW}$$

Heat used for power generation = 30–35%
We have to remove minimum = 160 KW.

3.6 Diameter of the Air Duct Computation

$$\begin{aligned} \text{Fv} &= 454 \times 454 \times 30 \\ &= 6.18 \times 10^{-3}\,\text{m}^3 \end{aligned}$$

$$\begin{aligned} \text{Vf } 1_q &= \frac{6.18 \times 10^{-3}}{4} \\ &= 1.545 \times 10^{-3}\,\text{m}^3 \end{aligned}$$

$$\begin{aligned} V_{\text{ad}} &= A \times \text{Total PL} \\ A &= V/\text{TL} \end{aligned}$$

$$= \frac{1.545 \times 10^{-3}}{0.78} = 1.98 \times 10^{-3}$$

$$\frac{3.14}{4} \times D^2 = 1.98 \times 10^{-3}$$

$$D = 0.05 \text{ m or } 50 \text{ mm}$$

$$D_{\text{ad}} = 50 \text{ mm or } 0.05 \text{ m}$$

Collector panel size $= 0.454 \times 0.454 \times 0.025$ m.

3.7 Nozzle Dimension Computation

$$Q = hA(T_0 - T_a)$$

h—Convection of the fluid—100 W/m^{2}k

$$T_0 - I_{\text{tf}} = 95\,^\circ\text{C} = 368 \text{ K}$$

$$T_a - \text{At} = 35\,^\circ\text{C} = 308 \text{ K}$$

$$Q = 100 \times A \times (95-35)$$

$$160 \times 10^3 = 100 \times A \times 333$$

$$A = 4.8048 \text{ mm}^2 = 4.8 \times 10^{-6} \text{ m}^2$$

$$d_2 = 3 \text{ mm} = 0.003 \text{ m}$$

$$D_n = 3 \text{ mm} = 0.003 \text{ m}$$

Spraying angle is 110°

$$D = 10 \text{ mm} = d_1 = 0.01 \text{ m}$$

$$L_n = 50 \text{ mm} = 0.05 \text{ m}$$

$$A_1 = 3.14/4 \times d_1^2 = 7.85 \times 10^{-5} \text{ m}^2$$

$$A_2 = 3.14/4 \times d_2^2 = 7.065 \times 10^{-6} \text{ m}^2$$

$$V = \text{sqrt(2gh)or sqrt(2P)}$$

$$V = 1.274\,q/d^2$$

$$\text{Length} - 7727 \text{ m}^3/\text{s}$$

$$V_1 = 1.274 \times 7727/(0.05)^2$$

$$= 3.93\,\text{m/s}$$
$$\text{Wkt } Q = AV$$
$$A_1V_1 = A_2V_2$$
$$V_2 = 4.3 \times 10\,\text{m/s}$$
$$V_2 > V_1$$

Nozzle rows—10 and Columns—9

$$q = 90 \text{ numberss}$$
$$V = 90 * V_2$$
$$= 3.8 \times 10^3\,\text{m/s}$$

Heat flux ratio,

$$\gamma = \text{Hx}/\text{Ha}$$

To find Ha

$$\text{Hx of air} = 5136\,\text{kJ/kg}\,^\circ\text{C}$$
$$\text{Hf} = 572.4\,\text{J/kgK}$$
$$\text{Hf} = \text{Hx}-\text{Ha}$$
$$\text{Ha} = 5136-572.4$$
$$\text{Ha} = 4563.6$$

To find γ

$$\gamma = \text{cp/cv}$$
$$= 5136/4563.6$$

$$\gamma = 1.12$$
$$s = (\gamma\ \text{Hf})0.6$$
$$s = (1.12 \times 572.4 \times 368)^{0.6}\ (T_0 = 95\,^\circ\text{C} = 368\,\text{K})$$
$$s = 167.36\,\text{m/s}$$
$$d = 9.81\,\text{m/s}$$
$$M = d/s$$
$$M = 9.81/167.36 = 0.05$$

For $\gamma = 1.12$ and $M = 0.05$

$$P_2/P_1 = V_1/V_2$$

$$= 6.18/4.3 \times 10$$

$$\begin{aligned} P_2/P_1 &= 0.197 \\ T_2 &= (P_2/P_1)^{R/\mathrm{Cp}} \times T_1 \\ &= (0.197)^{0.11} \times 95 \\ &= 79.45\,^{\circ}\mathrm{C} \end{aligned}$$

$$\begin{aligned} \text{Efficiency } \eta &= (T_1 - T_2)/T_1 \\ &= (95 - 79.45)/95 \\ &= 16.4\%. \end{aligned}$$

4 Result

Air duct cooling system for radiator model is drawn in the software named PTC Creo Parametric 3.0. It was drawn with the calculated dimension as mentioned in the calculations. Figure 8 represents the isometric view of the air duct cooling system, Fig. 9 represents the side view of the air duct cooling system and Fig. 10 represents

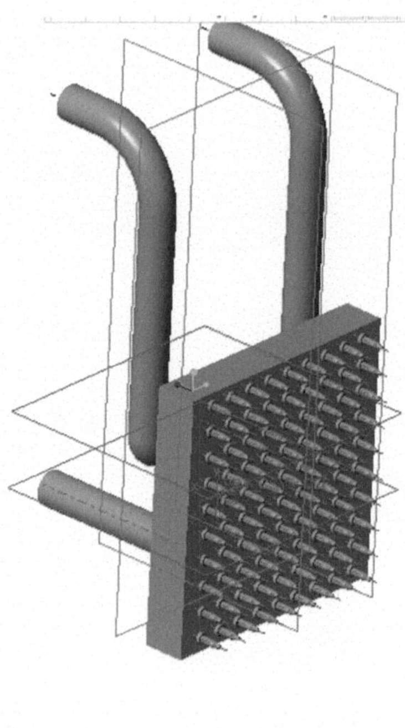

Fig. 8 Isometric view

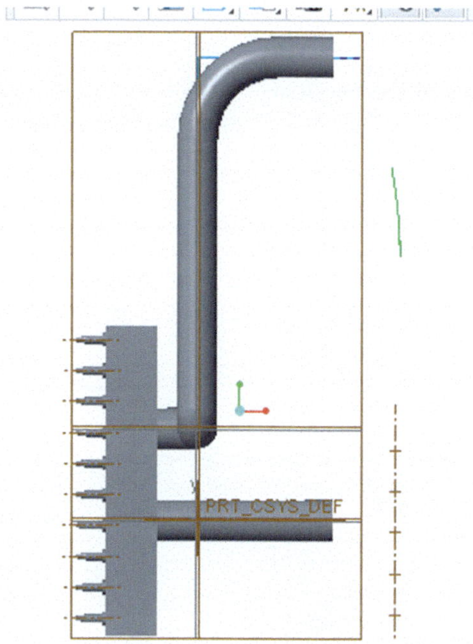

Fig. 9 Side view

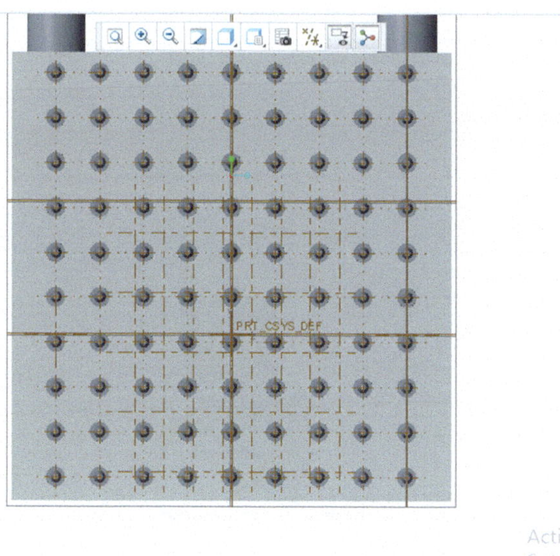

Fig. 10 View of nozzle in duct system

the nozzle view in the cooling system.

Analysis of proposed Duct system

Step 1: Import of duct model, latter the mesh model is induced. It involves in separating the possible number of nodes. The accuracy is increased by using large number of elements. The model is shown in Fig. 11.

Step 2: Stress and deformation Analysis.

The boundary conditions are pertained. Stress and deformation are inculpated in the duct system, once the vehicle moves forth. Maximum stress induced in the duct system is 30.05 MPa, and deformation is 9.545×10^{-3}. Results are shown in Figs. 12 and Fig. 13.

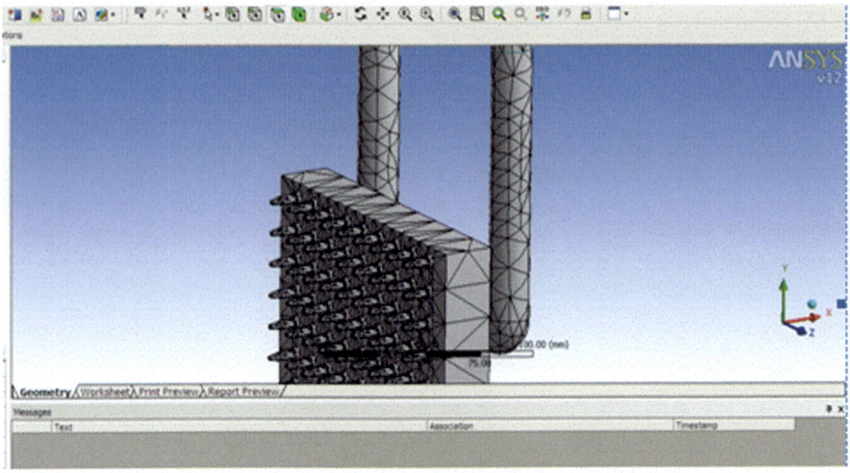

Fig. 11 Mesh model

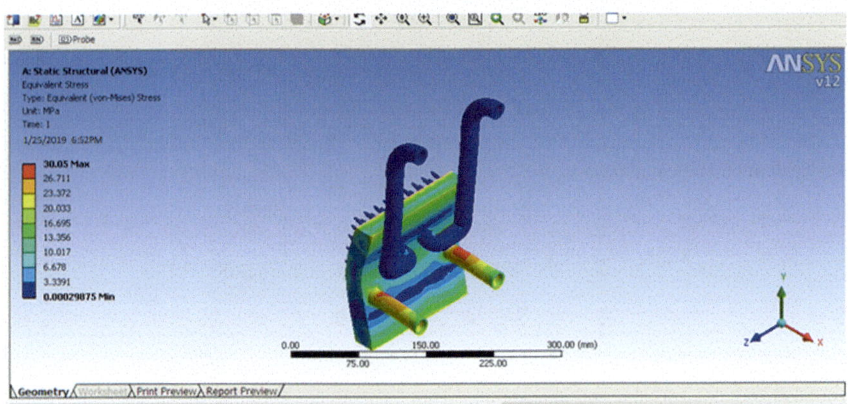

Fig. 12 Stress analysis

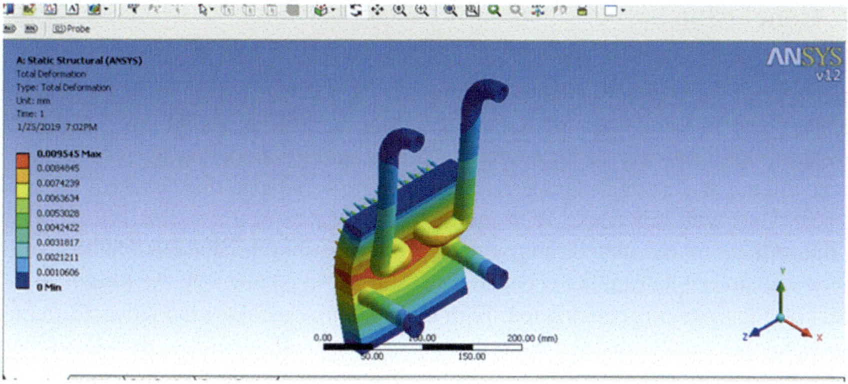

Fig. 13 Deformation analysis

Analysis	End result
Amount of heat dissipation	160 KW
Diameter of air duct	0.05 m
Nozzle velocity	3.8×10^3 m/s
Final temperature	79 °C
Efficiency	16.4%

5 Conclusion

In a nutshell, the entire project limelights about the additional cooling backup for the radiators and in extending its life period. While long-distance running, the engine delivers a high amount of heat. Due to this heating, the coolant again circulates to the engine at very high temperature. In order to eradicate this overheating, this project deals with this issue. With the help of heat flux analysis, the heat induced in the radiator is calculated, and with pressure distribution analysis, the pressure inside radiator while running is calculated. With these parameters, the design of air duct cooling system is constructed. With multiple nozzles, the wind blows from the atmosphere is equally distributed to the radiator for cooling purpose. It is less time-consuming but enhances more running time. The coming generation may utilize this opportunity of the latest technology in logistics. The nozzle impact given by us gives extra cooling to the radiator since it declines the pressure, what's more, expands the velocity along these lines diminishing the temperature which is directly proportional to the pressure as indicated by the perfect gas condition. Hence, the lifetime of the engine is additionally expanded, and the solution is brainstormed.

References

1. Harris C, Despa M, Kelly K (2004) Design and fabrication of a cross flow micro heat exchanger. J Microelectromech Syst 9(4):502–508
2. Sheikhzadeh G, Hajilou M, Jafarian H (2014) Analysis of thermal performance of a car radiator employing nanofluid. Int J Mech Eng Appl 2(4)
3. Trivedi PK, Vasava NB (2012) Effect of variation in pitch of tube on heat transfer rate in automobile radiator by CFD analysis. Int J Eng Adv Technol 1(6):180–183
4. Gadhave P (2012) Enhancement of forced convection heat transfer over dimple surface-review. Int Multidiscip E-Journal 1(2):51–57
5. Srinivasu D, Santharao D, Ramakrishna A (2016) CFD analysis to predict heat transfer performance of Louver fin radiator with water/EG and AL_2O_3 nanofluid. Int J Sci Eng Adv Technol 4(2)
6. Amrutkar PS, Patil SR (2013) Automotive radiator performance—review. Int J Eng Adv Technol 2(3):563–565
7. Zhong Y, Jacobi AM (2005) An experimental study of louver-fin flat-tube heat exchanger performance under frosting conditions. In: Proceedings of the 5th international conference on enhanced, compact and ultra-compact heat exchangers: science, engineering and technology. Engineering conferences international. Hoboken, NJ, USA, pp 273–280
8. Aliff Ashraf M (2015) The design and development of the radiator RIG. Malaysia Pahang (UMP). Available from: http://www.researchgate.net/publication/285188062

An Experimental Augmentation of Gravity-Assisted Concentric-Tube Heat Pipe Induced Heat Exchanger Under the Influence of Methanol and Acetone Working Fluids

P. Ramkumar, M. Sivasubramanian, P. RajeshKanna, and P. Raveendiran

Abstract Experimental augmentation of a concentric-tube heat exchanger with a heat pipe has been investigated. Methanol and acetone are used as working fluids to analyse the performance. Water is used as a heat-carrying fluid. In which, the investigation is carried out with the gravity-assisted angles of 0°, 45° and 90°. In this work, various parameters are such as temperatures of hot fluid range from 50 to 70 °C with an increment of 10 °C and cold fluid temperature is observed as 30.5 °C throughout the investigation. Mass flow rates of hot fluids are as 40–120 LPH with an increment of 20 LPH, cold fluid as 20–60 LPH with an increment of 10 LPH. The result suggests that an increase in effectiveness is occurred at acetone for 60 °C, 100 LPH and the influence of inclination angle at 0° reveals that 72% of an increase in effectiveness is achieved than methanol for optimum revealed conditions.

Keywords Heat pipe · Concentric-tube heat exchanger · Gravity-assisted angles · Working fluids · Mass flow rate · Effectiveness

Nomenclature

A	Area of heat transfer (m^2)

P. Ramkumar (✉)
Department of Mechanical Engineering, School of Automotive and Mechanical Engineering, Kalasalingam Academy of Research and Education, Krishnankoil, Tamilnadu 626126, India
e-mail: rkmailmech@gmail.com

M. Sivasubramanian
Department of Automobile Engineering, School of Automotive and Mechanical Engineering, Kalasalingam Academy of Research and Education, Krishnankoil, Tamilnadu 626126, India

P. RajeshKanna
Department of Mechanical Engineering, Al Ghurair University, Dubai, United Arab Emirates

P. Raveendiran
Department of Mechanical Engineering, Annamalai University, Annamalai Nagar, Chidambaram, Tamilnadu 608002, India

L.-J. Yang et al. (eds.), *Proceedings of ICDMC 2019*,
Lecture Notes in Mechanical Engineering,
https://doi.org/10.1007/978-981-15-3631-1_55

C	Heat capacity rate (kW/K)
C_p	Specific heat of the fluid (kJ/kg K)
D	Diameter (m)
h	Heat transfer coefficient (W/m^2 °C)
L	Length (mm)
m	Mass flow rate of fluid (LPH)
Q	Heat transfer rate (W)

Abbreviations

GICTSAHPHE	Gravity induced concentric-tube shell assisted heat pipe heat exchanger
HCF	Heat-carrying fluid
LPM	Litre per minute
WF	Working fluid

1 Introduction

The heat pipe shows the enrichment of heat transfer with a minimum gradient of temperature. This technique has been widely followed in various processing plants and industries. The influence of gravitational effect is analysed with segments in the section of heat pipe which was investigated in [1, 2]. The investigated heat pipe for various applications and analyses is carried out with numerical and experimental which were stated in [3, 4]. Esarte and Domiguez [5] experimentally investigated for various prototypes of flat heat pipes working against gravity assistance. In which, thermal resistance was minimum and coefficient of performance has twice the value than theoretical prediction. Fonseca et al.'s [6] investigation results with changing the working fluid with several fill ratios within 20–90%, and the optimal thermal performance at a fluid ratio was 69.68%.

Shabgard et al. [7] implemented a thermal network model to characterize the thermal behaviour in a high-temperature latent heat thermal energy storage system. Hossain et al. [8] observed result shows that performance depends upon various angles of inclination and different heat input and does not depend on coolant flow rate. The study shows that for the same heat input and inclination angle with acetone as working fluid, micro-heat pipe shows better performance.

Influence of nanofluids as working fluids for heat pipe is studied with several tests which were stated in [9, 10]. The heat pipe with different materials and tilt angles are analysed with the wick of the heat pipe which were reported [11, 12]. Heat pipe with different working fluids and fill ratios are investigated for various applications which are stated in [13, 14]. Işık et al. [15] fabricated a heat pipe using ammonia. The investigation was made on space applications and burst test in the heat pipe.

In the analysis of vast investigators, their research was carried out with minimum geometrical constraints on heat pipe heat exchanger. The constraints include geometrical parameters, working fluids, fill ratios and heat-carrying fluids in heat pipe. To rectify the above-stated constraints in this investigation, the heat pipe is designed with gravitational effect and analysed with a concentric-tube shell assisted heat exchanger. Further analysis is made with different tilt angles and working fluids. The inlet parameters of the heat-carrying fluids are measured with various mass flow rates and temperature inputs.

2 Fabrication and Working Process

2.1 Fabrication of Set-up

To investigate the heat transfer characteristics of a GICTSAHPHE, a set-up is designed and given in Fig. 1. In this research, GICTSAHPHE is designed with copper as a heat pipe and galvanized iron as shell material for heat exchanger with a diameter of 50 mm and length of 700 mm at evaporator zone. Heat pipe inner and outer diameter are as 19 and 17 mm and total length as 1000 mm. Condenser diameter is as 35 mm and length as 300 mm.

The two fluid tanks, for hot and cold fluids, are with a capacity of five litres, which contains one coiled immersion electric heater with two Kilowatts and a temperature controller to regulate the temperatures of hot fluid. The rotameters with a range of three litres per minute are used to regulate the flow rates of both hot and cold fluids. Two thermocouples observe the external surface temperature of the evaporator and condenser regions to measure the temperature interface with the environment.

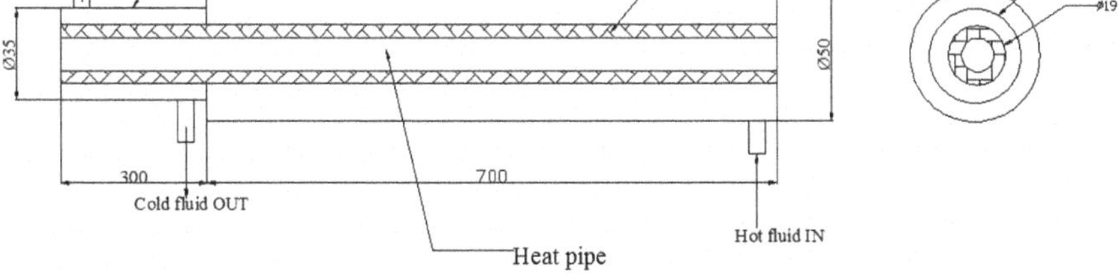

Fig. 1 Schematic diagram of a heat pipe heat exchanger

Table 1 Thermo-physical properties of working fluids

Properties	Methanol	Acetone
Boiling point	64 °C	56.08 °C
Melting point	−97.9 °C	−94.9 °C
Latent heat of evaporation (λ)	1055 kJ/kg	534 kJ/kg
Density of liquid (ρ_l)	746.2 kg/m^3	784.5 kg/m^3

2.2 Working Process

At the primary test, the heat pipe is filled with methanol and the secondary test is made with acetone as working fluids. Methanol and acetone are charged with fill ratios of one hundred per cent, and the thermo-physical properties of working fluids are given in Table 1. The heat-carrying fluid is chosen as water for the investigation.

At the primary stage, GICTSAHPHE is kept at a 0° tilt angle (horizontal axis) and the WF as methanol and water as HCF. The hot fluid mass flow rate and inlet temperature are fixed as 40 LPH and 50 °C. The cold fluid mass flow rate and inlet temperature are as 20 LPH and 30.5 °C as the ambient temperature condition for the investigation. At the evaporator zone of a concentric-tube heat exchanger, hot fluid transfers the heat to the methanol in the heat pipe. Thus, WF absorbs the heat and latent heat of phase change has occurred. The investigation is further carried out with other tilt angles of 45° and 90°. The hot and cold zones' mass flow rates as Litre Per Hour (LPH) 60, 80, 100, 120 LPH and 30, 40, 50, 60 LPH. Temperatures at an inlet of hot and cold zones are 60, 70 and 30.5 °C. The above conditions are repeated for acetone as WF, and investigation is carried out for similar conditions.

3 Results and Discussion

To analyse the designed GICTSAHPHE, the investigation is done by the above-stated parameters, and Fig. 2 depicts the mass flow rate of hot fluid on effectiveness. This graph predicts that by using water as HCF and methanol as WF for the first test. The temperature of hot fluid at inlet (T_{hi}) as 60 °C and for varying *mass flow rate of hot fluid at inlet* (m_{hi}) from 40 to 120 LPH and inclination angle (ψ) as 0°, 45° and 90°. The maximum effectiveness observed is 32.2% at 100 LPH and ψ as 0° than other tilt angles and mass flow rates of hot fluids. When the mass flow rate increases by 120 LPH, the achieved effectiveness is 29.5% for 0°. In the second test, WF is chosen as acetone and HCF as water. The above-stated conditions are fixed, and at 100 LPH the observed effectiveness is 55.25% for 0°. At 120 LPH, the achieved effectiveness is 45.76% for 0°. In this analysis also, similar trends of graphs are predicted for 100 and 120 LPH. While comparing the tilt angles also, 0° and 90° show similar ranges of effectiveness than 45°. In both working fluids, the 45° shows inferior values than other tilt angles, this is due to the effect of gravitational,

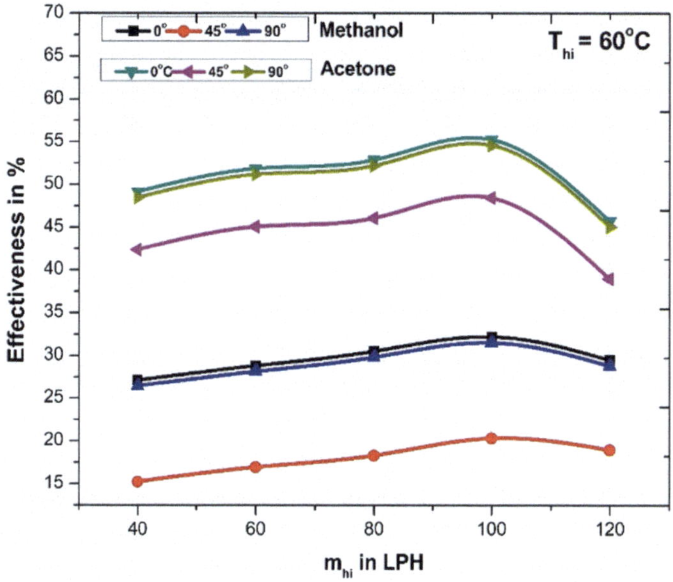

Fig. 2 Mass flow rate of hot fluid on effectiveness

there is a formation of a liquid film at vapour core region, which reduces the phase transformation of working fluids at evaporator region inside the heat pipe and this reduces the heat absorption capacity. Similarly, while increasing the m_{hi} from 100 to 120 LPH the same decrement in effectiveness is achieved for both working fluids. This decrement in effectiveness is due to the minimum heat absorption and release capability of the hot and cold fluid at both zones of the heat pipe. This graph clearly reveals that acetone has maximum effectiveness than methanol at a tilt angle of 0°, T_{hi} as 60 °C and m_{hi} as 100 LPH. Therefore, the effectiveness is calculated by Eq. (1)

$$\in = \frac{T_{c_o} - T_{c_i}}{T_{h_i} - T_{c_i}} \tag{1}$$

Figure 3 predicts the inclination angle on effectiveness with respect to the temperature of the hot fluid inlet with water as HCF and methanol as WF, the observed maximum hot fluid mass flow rate is 100 LPH for various inclination angles (ψ) as 0°, 45° and 90°. The observation stated that the increment of effectiveness is attained from 27.69 to 32.2% for ψ as 0° for T_{hi} of 50 to 60 °C. Similarly, for T_{hi} as 60–70 °C the effectiveness starts reducing from 32.2 to 21.2% for ψ as 0°. In the above observation, it reveals that at 100 LPH at a tilt angle of 0° maximum effectiveness has been achieved while comparing with other tilt angles. While comparing the temperatures of hot fluid mass flow rate at 60 °C shows an increment in effectiveness than 50 and 70 °C. Similarly for acetone also, same trends of graphs are observed at ψ as 0° for T_{hi} as 60 °C and the observed effectiveness is 55.25%. At inclination angle of 45°

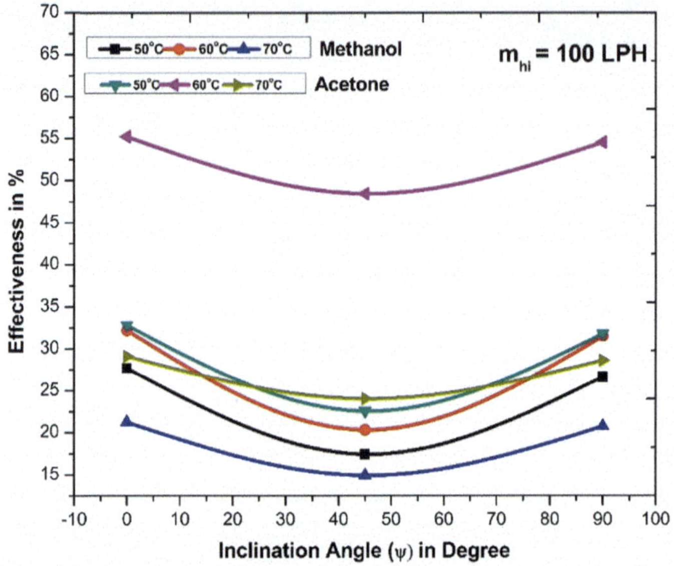

Fig. 3 Inclination angle on effectiveness

and 90° for temperature increment, there is increasing in effectiveness is observed but while comparing with 0° effectiveness of the system gets reduced. This graph also clearly reveals that at 0°, 60 °C and 100 LPH, the acetone shows superior results than methanol.

Figure 4 shows the mass flow rate of hot fluid on heat transfer rate. The conditions for the observation are T_{hi} as 60 °C, and m_{hi} ranges from 40 to 120 LPH and varying inclination angles from 0°, 45° and 90°. In the first test, methanol is used as WF. The heat transfer rate (Q) shows increment for 40–100 LPH at the range of 186.04–606.97 W for ψ as 0°. Similarly, for 120 LPH, the Q value is 552.31 W for ψ as 0°. This plot reveals that 100 LPH shows highest heat transfer rate in the system; in the second test with acetone as WF. The similar type of trends is achieved. The heat transfer rate (Q) shows increment for 40 to 100 LPH at the range of 337.20–947.66 W for ψ as 0°. Similarly, for 120 LPH, the Q value is 941.85 W for ψ as 0°. This observation also reveals that the same results at 100 LPH maximum (Q) values are achieved than 120 LPH. Both the investigations at 100 LPH and ψ as 0° have the highest results, but by comparing both the working fluids, acetone shows superior heat transfer rate than methanol. This increase in heat transfer rate is achieved by the maximum absorption of heat energy by the cold fluid at the condenser zone. When the inlet temperature of the hot fluid is at 60 °C, there is maximum heat transfer which occurs between HCF and working fluids at the evaporator zone. This decrease in trends shows lack of heat transfer at the 45° with a minimum release of sensible heat and effect of friction occurs due to gravitational effect at condenser region of the heat pipe. Hence, the heat transfer rate is calculated using below formulas (2) and (3)

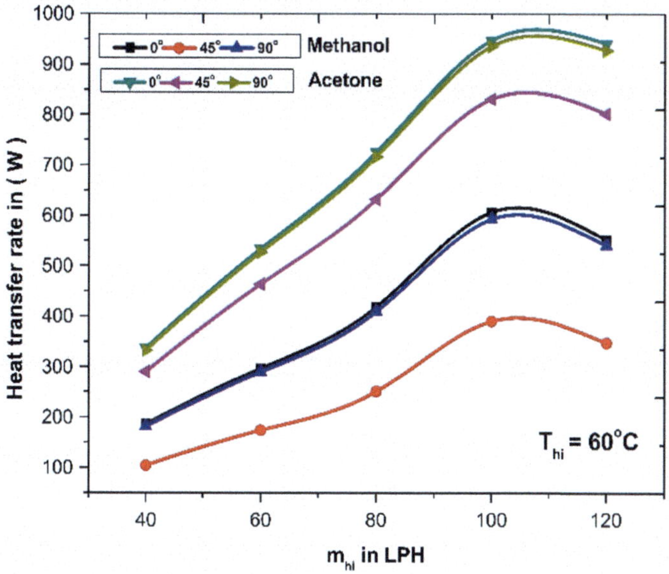

Fig. 4 Mass flow rate on heat transfer rate

$$Q_h = \dot{m}_h c_{\text{ph}}(T_{\text{hi}} - T_{\text{ho}}) \tag{2}$$

$$Q_c = \dot{m}_c c_{\text{pc}}(T_{\text{co}} - T_{\text{ci}}) \tag{3}$$

Figure 5 inferred that mass flow rate of hot fluid on heat transfer coefficient with water as HCF. The conditions for observation are T_{hi} as 60 °C, m_{hi} from 40 to 120 LPH and varying inclination angle from 0°, 45° and 90°. The observed heat transfer coefficient (h) at 40–100 LPH for ψ as 0° is 238.30–797.98 W/m^2 °C. At 120 LPH, ψ as 0° the observed value is 746.08 W/m^2 °C. In acetone as WF, the similar value of trends is observed. At 40–100 LPH, the observed (h) value ranges as 546.46–1664.36 W/m^2 °C for ψ as 0°. Similarly for 120 LPH, the observed (h) ranges as 1498.34 W/m^2 °C for ψ as 0°. Both working fluids show similar trends in the investigation at 100 LPH highest value of (h) which is achieved than 120 LPH. While considering the tilt angles, the horizontal axis ψ as 0° possesses maximum heat transfer coefficient than ψ as 45° and 90°. Using the below equation, the convective heat transfer coefficient of cold fluid on GICTSAHPHE can be calculated by Eq. (4)

$$h = \frac{Q_c}{A * (\Delta T)_{\text{lm}}} \tag{4}$$

In comparing the 0° and 90°, the similar range of values is achieved than at 45°. This predicts the higher heat transfer coefficient at hot fluid for 100 LPH and 60 °C with the 0° inclination angle for acetone as WF. This is due to the higher release of

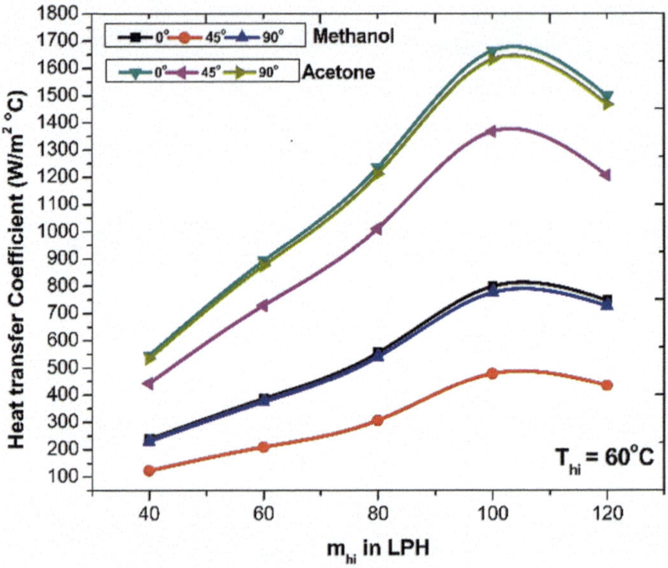

Fig. 5 Mass flow rate on heat transfer coefficient

sensible heat to the cold fluid, by the working fluids which lead to maximum heat transfer along with the system.

4 Conclusion

- It is inferred that optimum conditions observed from the investigation are hot fluid mass flow rates (m_{hi}) as 100 LPH and cold fluid (m_{ci}) as 50 LPH, temperature of hot fluid at an inlet (T_{hi}) as 60 °C and the inclination angle (ψ) as 0° (horizontal axis).
- While comparing both working fluids for above-revealed conditions, acetone shows maximum results when comparing with methanol. There is an increment in results of 72% for effectiveness, 56.13% for heat transfer rate, 108.57% for heat transfer coefficient which are achieved.
- This investigation reveals that the heat pipe induced in concentric-tube heat exchanger with acetone shows superior in results, while comparing with methanol in all the observed conditions of the investigation.

References

1. Khalid Joudi A, Witwit AM (2000) Improvements of gravity assisted wickless heat pipes. Energy Convers Manag 41:2041–2061
2. Alicetin G, Cannistrano C (1991) The inclination effect on the performance of water-filled heat pipes. Renew Energy 1:667–674
3. Lydia Wermer R, Martin Ward J, Justin Simpson D, Robert Zimmerman A, James Stewart A (2018) A high capacity self priming counter gravity heat pipe. Int J Heat Mass Transf 125:1369–1378
4. Zhang D, Li G, Liu Y, Tian X (2018) Simulation and experimental studies of R134a flow condensation in a pump assisted separate heat pipe. Int J Heat Mass Transf 126:1020–1030
5. Esarte J, Domiguez M (2003) Experimental analysis of a flat heat-pipe working against gravity. Appl Therm Eng 23:1619–1627
6. Fonseca LD, Pfotenhauer J, Miller F (2018) Result of a three evaporator in cryogenic helium heat pipe. Int J Heat Mass Transf. 120:1275–1286
7. Shabgard H, Bergman TL, Shariti N, Faghri A (2010) High temperature latent heat thermal energy storage using heat pipes. Int J Heat Mass Transf 53:2979–2988
8. Hossain RA, Chowdhuri MAK, Feroz CM (2010) Design fabrication and experimental study of heat transfer characteristics of a micro heat pipe. Jordan J Mech Ind Eng. 4:531–542
9. Shafai M, Blanoco V, Vafai K, Manco O (2010) An investigation of the thermal performance of cylindrical heat pipes using nanofluid. Int J Heat Mass Transf 53:376–383
10. Kempers R, Ewing D, Ching CY (2006) Effect of number of mesh layers and fluid loading on the performance of screen mesh wicked heat pipes. Appl Therm Eng 26:89–595
11. Pis mennyi EN, Khayrnasov SM, Rassamakin BM (2018) Heat transfer in evaporation zone of aluminium grooved heat pipes. Int J Heat Mass Transf. 127:80–88
12. Said Salem A, Bilal Akash A (1999) Experimental performance of a heat pipe. International Communication of Heat and Mass Transfer. 26(5):679–684
13. Bastakoti D, Zhang H, Cai W, Li F (2018) An experimental investigation of thermal performance of pulsating heat pipe with alcohols and surfactant solutions. Int J Heat Mass Transf 117:1032–1040
14. Vivek Patel K (2018) An efficient optimization and comparative analysis of ammonia and methanol heat pipe for satellite application. Energy Convers Manag 165:382–395
15. Işık HG, Ömür C, Uygur AB, Horuz I (2018) A novel burst test approach for the qualification of heat pipes developed for space applications. Int J Press Vessels Pip 165:241–248

Investigation on Helical Coiled Tube Heat Exchanger for Parallel and Counter Flow Using CFD Analysis

P. Saravana Bhavan and J. Selwin Rajadurai

Abstract Fluid flow design and heat transfer components utilising computational fluid dynamics (CFD) are relatively new because of cheaper and accurate method that becomes popular and reliable today. In this study, the helical coil tube heat exchanger is designed and modelled on solid modelling software. The modelled designs are analysed through ANSYS Fluent. The water is considered as working fluid, and copper as the solid material. The CFD analysis of parallel flow as well as counter flow is compared for the parameters like effective thermal conductivity, velocity and pressure. As an outcome, counter-flow heat exchanger provides effective outcome than the parallel-flow heat exchanger. The study explains about the numerical design and CFD analysis' procedures of the helical coiled tube heat exchanger.

Keywords Heat exchanger · CFD analysis · Helical coil · ANSYS Fluent

Nomenclature

Nu	Nusselt number
L	Tube length (m)
D_h	Hydraulic diameter (m)
u	Fluid velocity (m/s)
A	Area of heat transfer (m^2)
A_f	Cross sectional area of coil (m^2)
B	Outside diameter of inner cylinder (m)
C	Inside diameter of outer cylinder (m)

P. Saravana Bhavan (✉)
Engineering Design, Department of Mechanical Engineering,
Government College of Engineering, Tirunelveli 627007, India
e-mail: bhavan270395@gmail.com

J. Selwin Rajadurai
Department of Mechanical Engineering, Government College of Engineering,
Tirunelveli 627007, India

L.-J. Yang et al. (eds.), *Proceedings of ICDMC 2019*,
Lecture Notes in Mechanical Engineering,
https://doi.org/10.1007/978-981-15-3631-1_56

c_p	Fluid heat capacity (J/kg K)
D	Inside diameter of coil (m)
D_e	Shell-side equivalent dia. of coil (m)
DH	Average dia. of helix (m)
DH1	Inside dia. of helix (m)
DH2	Outside dia. of helix (m)
d_o	Outside dia. of coil (m)
Gs	Mass velocity of fluid (kg/m^2 s)
H	Height of cylinder (m)
h_i	Heat transfer coeff. based on inside dia. (W/m^2 K)
h_{ic}	Heat transfer coeff. inside coiled tube (W/m^2 K)
h_{io}	Heat transfer coeff. inside coiled tube (W/m^2 K)
h_o	Heat transfer coeff. outside coil (W/m^2 K)
j_H	Colburn factor for heat transfer
k	Thermal conductivity of fluid (W/m K)
k_c	Thermal conductivity of coil wall (W/m K)
N	Number of turns
M	Mass flow rate (kg/s)
N_{Pr}	Prandtl number
N_{Re}	Reynolds number
Q	Heat load (W)
q	Volumetric flowrate of fluid (m^3/s)
r	Average radius of helical coil (m)
R_a	Shell-side fouling factor (m K/W)
R_t	Tube-side fouling factor (m K/W)
Δt_c	Corrected log mean temperature difference (K)
Δt_{lm}	Log mean temperature difference (K)
U	Oveall heat transfer coefficient (W/m^2 K)
V_a	Volume of shell (m^3)
V_c	Volume occupied by N turns of coil (m^3)
V_f	Volume available for fluid flow (m^3)
x	Thickness of coil wall (m)
μ	Fluid viscosity (kg/ms)
ρ	Fluid density (kg/m^3)

1 Introduction

Interest to compact heat transfer devices such as heaters or heat exchangers has increased significantly nowadays. The compact heat exchangers are widely used in the process industries, such as chemical, oil and petrochemical, food, pharmaceutical, nuclear, automotive and others. The heat transfer in between the fluids is a significant process to be absorbed upon; thus, heat exchangers play an important role in our daily life. The heat transfer mostly occurs by three principles. i.e., convection, conduction and radiation. The process conduction happens when there is a temperature gradient between the solid walls. The practice of helical tubes is detected in different applications such as processing of foods and the operation of nuclear reactor [1]. The helical coil deals with enormous area of heat transfer within a compact space, in addition to higher coefficient of heat transfer in co-operation with less distribution of residence time [2]. The pressure drops and friction factor have been explored in an inclined helical coil with two-phase flow [3]. A compact double-pipe helical coil provides more fluid interaction and reduces the dead zones [4]. Various correlations between flow rate and pressure drop for helical coil have been established [5]. The simulation of turbulent flow in helical coil was examined [6]. The features of transfer of heat in a double-pipe helical heat exchanger were investigated [7]. The Dean number diverses directly with overall heat transfer coefficient; still the conditions of outer-pipe fluid flow that had a main influence in the overall heat transfer coefficient are carried out [8, 9].

A 3D COMSOL model is to pretend the heat transfer and exothermic chemical reactions in heat exchanger [10]. A study is presented using COMSOL–MATLAB interface to get the optimum helical pitch and radius [11]. The increasing of effectiveness is analysed for different parameters which affects the effectiveness [12]. The performance of heat transfer and the analysis of energy are analyzed in the shell as well as a tube of the heat exchanger through injection air bubbles at various points in the tube [13]. The effect of curved-pipe heat exchanger is compared with the straight-pipe heat exchanger [14]. The result of injection air bubbles at different locations in the entrance of heat exchanger is determined for different percentages of Al_2O_3 nano-particles [15]. The surface of interaction with fluid has to be simulated by applying the initial and boundary conditions, which are done using SolidWorks [16].

Our research work focused on CFD analysis of parallel flow as well as counter flow which are compared for the parameters like effective thermal conductivity, velocity and pressure of the heat exchanger. As a result, the counter-flow heat exchanger provides effective outcome than the parallel-flow heat exchanger. The study explains about the numerical design and CFD analysis' procedures of the helical tube heat exchanger.

2 Helical Coil Heat Exchanger Design

2.1 Design Calculation

To design the heat exchanger, the following data are taken to start the actual design. Thus, following data is taken as input parameters to design the heat exchanger.

Table 1 Properties of fluid

Input data	Units	Shell	Tube
Material		Copper	Copper
Inlet temperature	K	403	313
Outlet temperature	K	383	333
Heat capacity, C_p	J/kg K	4185.5	4185.5
Thermal conductivity, k	W/m K	0.6048	0.6048
Viscosity, μ	kg/m s	0.001	0.001
Density, ρ	kg/m^3	998	998
Velocity, u	m/s	0.1	0.1

Diameter of outer tube $d_o = 0.08$ m; diameter of inner coil $D = 0.07$ m; average diameter of the coil $D_h = 1$ m; average radius of the coil $r = 0.5$ m; pitch of the coil $p = 1.5 \times d_o = 0.12$ m (Table 1).

2.1.1 Length of the Coil

$$L = N\sqrt{(2\pi r)^2 + p^2}$$

$$L = 3.14\,\mathrm{N}$$

2.1.2 Mass Flow Velocity of Fluid

$$\mathrm{Gs} = \frac{M}{\left[\frac{\pi}{4}\left\{\left(C^2 - B^2\right) - \left(\mathrm{DH}_2^2 - \mathrm{DH}_1^2\right)\right\}\right]}$$

$$C = 2r + d_o + (2 \times 0.5d_o),\ \ B = 2r - d_o - (2 \times 0.5d_o),$$
$$\mathrm{DH}_2 = 2r + d_o,\ \ \mathrm{DH}_1 = 2r - d_o$$

$$\text{AREA, } A = \frac{\pi}{4}(d_o)^2 = 0.0038\,\mathrm{m}^2$$

$$M = \rho \times u \times A$$

$$\mathrm{Gs} = \frac{0.379}{\left[\frac{\pi}{4}\left\{\left(1.16^2 - 0.84^2\right) - \left(1.08^2 - 0.92^2\right)\right\}\right]}$$

$$\mathrm{Gs} = 1.5169\,\mathrm{Kg/m^2\,s}$$

2.1.3 Reynolds Number

Volume available in the shell side is

$$V_{\mathrm{f}} = \frac{\pi}{4}\left(C^2 - B^2\right)p\,N - \left(\frac{\pi}{4}\right)d_o^2 L = 0.584\,\mathrm{N}$$

The shell side equivalent diameter is

$$D_{\mathrm{e}} = \frac{4D_{\mathrm{f}}}{\pi d_{\mathrm{o}} L}$$

$$D_{\mathrm{e}} = 0.296\,\mathrm{m}$$

The Reynolds number is

$$N_{\mathrm{Re}} = \frac{\mathrm{Gs} \times D_{\mathrm{e}}}{\mu} = 477.65$$

$$N_{\mathrm{Pr}} = \frac{C_p \times \mu}{K} = 6.94$$

2.1.4 Heat Transfer Coefficient

$$\frac{h_{\mathrm{o}} D_{\mathrm{e}}}{K} = 0.6(N_{\mathrm{Re}})^{0.5}(N_{\mathrm{Pr}})^{0.31}$$

$$h_{\mathrm{o}} = 26.493\,\mathrm{W/m^2K}$$

The heat transfer coefficient inside the coil
The fluid velocity is

$$q = \frac{M}{\rho} = \frac{0.379}{998} = 0.00038\,\mathrm{m^3/s}$$

$$A_t = \frac{\pi D^2}{4} = 3.85 \times 10^{-3}\,\mathrm{m^2}$$

$$u = \frac{q}{A_{\mathrm{f}}} = 0.09\,\mathrm{m/s}$$

The Reynolds number (tube-side)

$$N_{\mathrm{Re}} = \frac{D\,u\,\rho}{\mu} = 6874.2$$

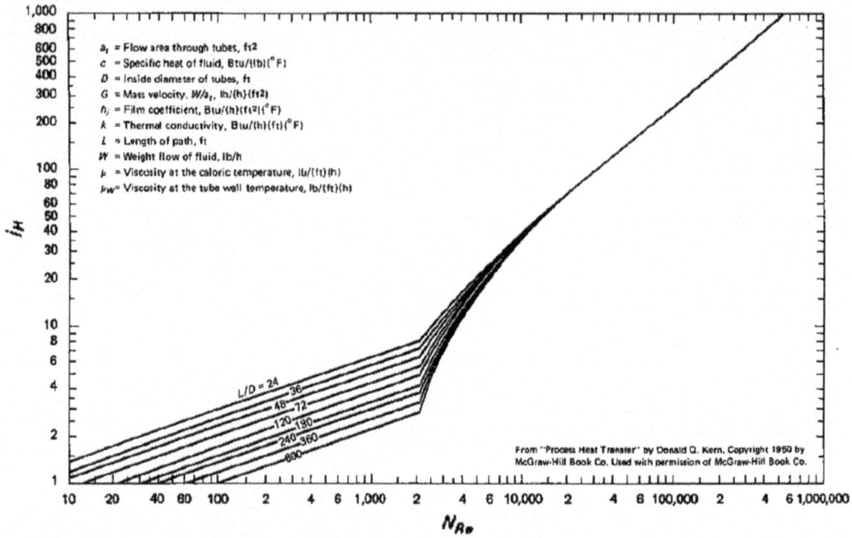

Fig. 1 Graph for determining Colburn factor for heat transfer

From Fig. 1, Graph for determining Colburn factor for heat transfer j_H is 30.

$$h_i = j_H\left(\frac{K}{D}\right)(N_{pr})^{1/3} = 494.4\,\text{W/m}^2\text{K}$$

Corrected for coiled tube

$$h_{ic} = h_i\left(1 + 3.5\left(\frac{D}{D_H}\right)\right) = 615.52\,\text{W/m}^2\text{K}$$

$$h_{io} = h_{ic}\left(\frac{D}{D_o}\right) = 538.59\,\text{W/m}^2\text{K}$$

2.1.5 Overall Heat Transfer Coefficient

Coil wall thickness

$$x = \left(\frac{d_o - D}{2}\right) = 0.005\,\text{m}$$

$$\frac{1}{U} = \frac{1}{h_o} + \frac{1}{h_{io}} + \frac{x}{K_c} + R_t + R_a$$

$$U = 18.57\,\text{W/m}^2\text{K}$$

2.1.6 Determine the Required Area

$$\Delta t_{\text{lm}} = \frac{\left[\left(T_{s,i} - T_{t,i}\right) - \left(T_{s,o} - T_{t,o}\right)\right]}{\ln\left[\left(T_{s,i} - T_{t,i}\right)/\left(T_{s,o} - T_{i,o}\right)\right]}$$

$$\Delta t_{\text{lm}} = \frac{[(403 - 313) - (383 - 333)]}{\ln[(403 - 313)/(383 - 333)]} = 68.05\,K$$

$$\Delta t_{\text{c}} = C_{\text{f}} \times \Delta t_{\text{lm}} = 67.04\,\text{K}$$

The heat loads

$$Q = \text{MC}_p \Delta T = 31{,}722.3\,\text{W}$$

$$A = \frac{Q}{U\Delta T} = 25.36\,\text{m}^2$$

$$N = \frac{A}{\pi d_{\text{o}} L} = 32$$

$$H = (N \times p) + d_{\text{o}} = 3.92\,\text{m}$$

2.2 *Design Model of Heat Exchanger*

See Figs. 2, 3 and 4.

3 Results and Discussion

3.1 *CFD Analysis of Parallel-Flow Heat Exchanger*

3.1.1 Effective Thermal Conductivity Plot

The effective thermal conductivity of the parallel flow is high at the inlet of the helical tube and is gradually decreased till the end of the helical coil as shown in Fig. 5. The value of the helical tube varies from 23.47 to 5062 W/m k. The value obtained at the inlet of the helical tube by placing the probe tool is 4985 W/m k and at the middle of the helical coil is 980 W/m k and at the outlet of the helical coil is 25.135 W/m k.

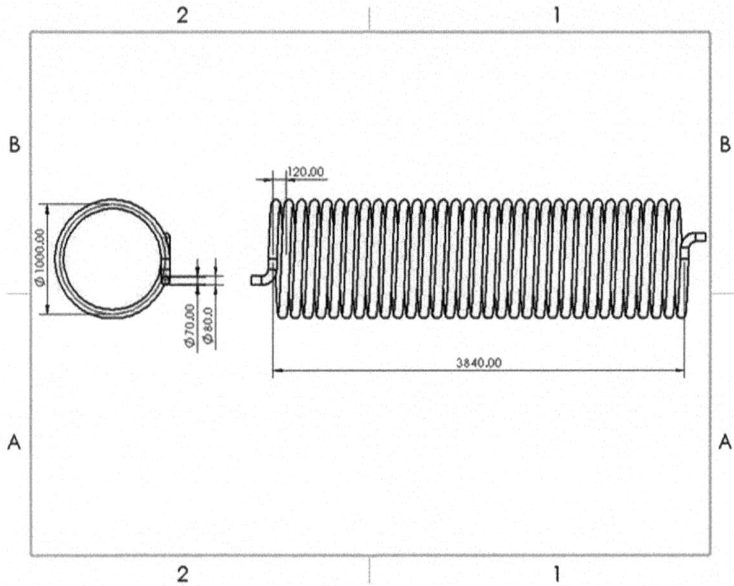

Fig. 2 Front and side view of tube

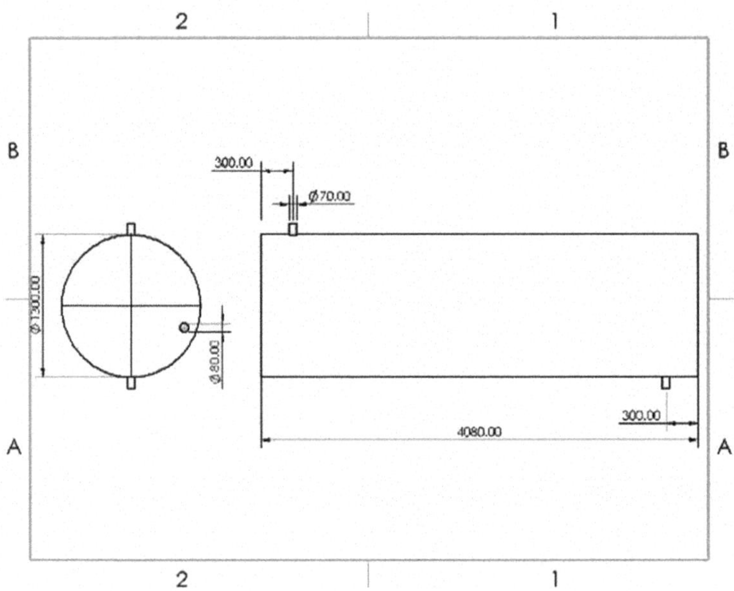

Fig. 3 Projection of heat exchanger

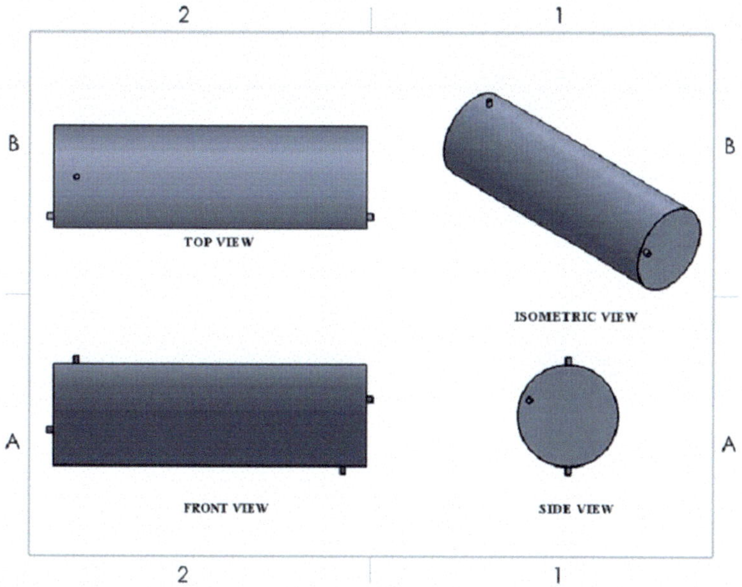

Fig. 4 Front and side view of shell

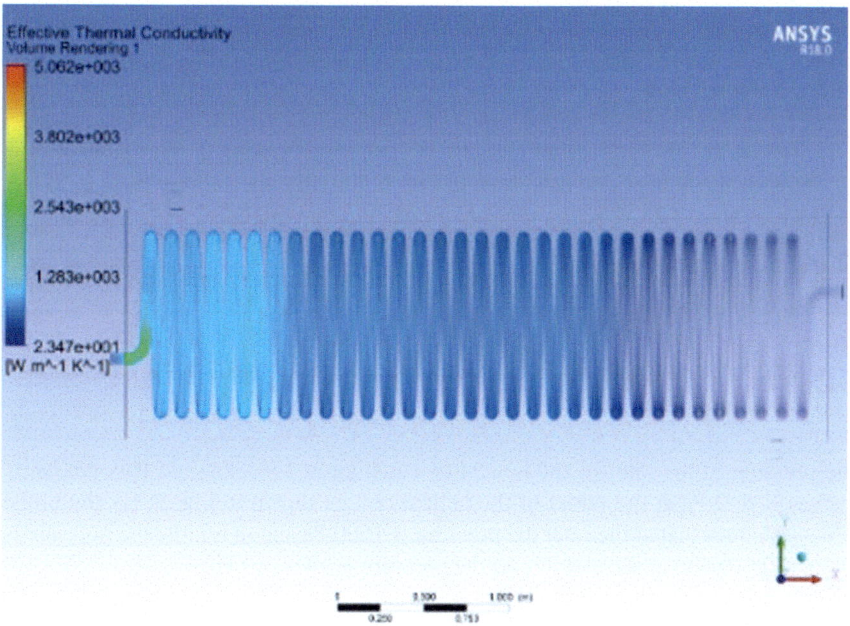

Fig. 5 Effective thermal conductivity plot for parallel-flow

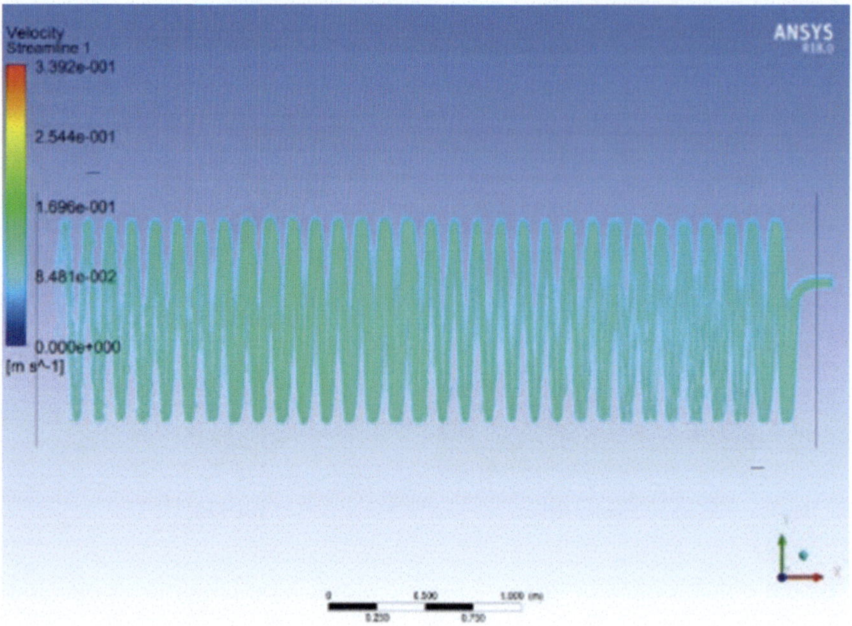

Fig. 6 Velocity plot for parallel-flow

3.1.2 Velocity Plot

It shows that the velocity at the inlet is very low, and it gradually increases towards the outlet of the helical coil as shown in Fig. 6. Initially, the velocity is at 0.08 m/s and increases to 0.23 m/s. By placing the probe, the values show that at inlet of helical tube it is 0.09 m/s and in middle of helical coil it is 1.56 m/s but in between the middle and outlet of coil the velocity gradually decreases and then it increases to 0.2 m/s (Fig. 7).

3.1.3 Pressure Plot

The pressure at the inlet of the helical coil is high at 1505 Pa, and it is gradually decreased to 2 Pa at the outlet of the helical coil as shown in Fig. 9. By the probe value, it defines that at the inlet the pressure is 1500 Pa and in middle the pressure is about 1005 Pa and at the end of coil the pressure is about 50 Pa.

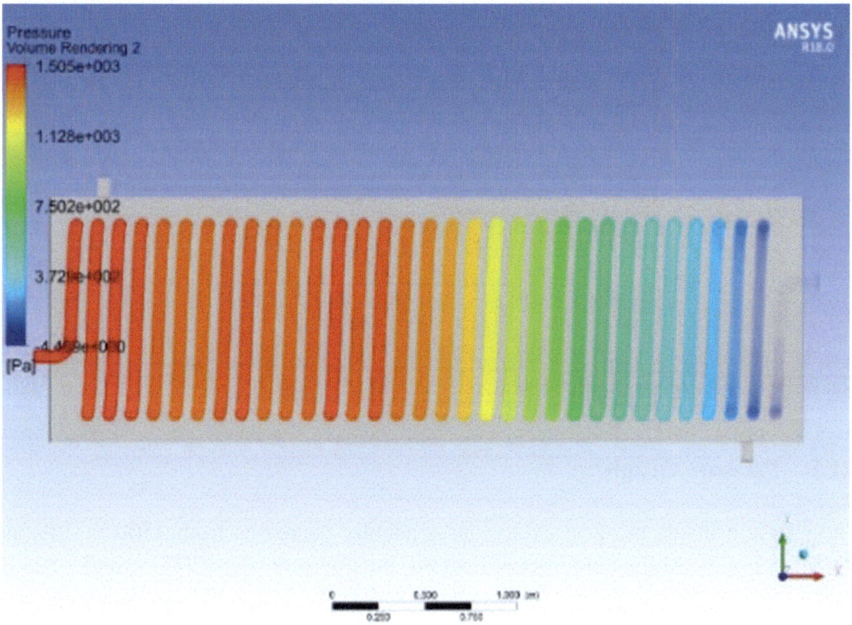

Fig. 7 Pressure plot for parallel-flow

3.1.4 Velocity Graph

At inlet, the velocity is at 0.2 m/s, and it varies alternatively between 0.1 and 0.117 m/s as shown in Fig. 8. At 1 m, the velocity attains 0.115 m/s, at 2 m, the velocity is 0.113 m/s, and at 3 m, the velocity is at 0.112 m/s. The lowest velocity of the heat exchanger is 0.1 m/s.

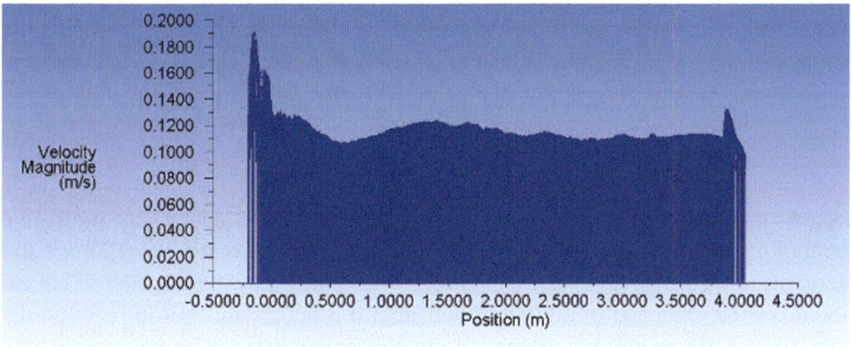

Fig. 8 Velocity graph for parallel-flow

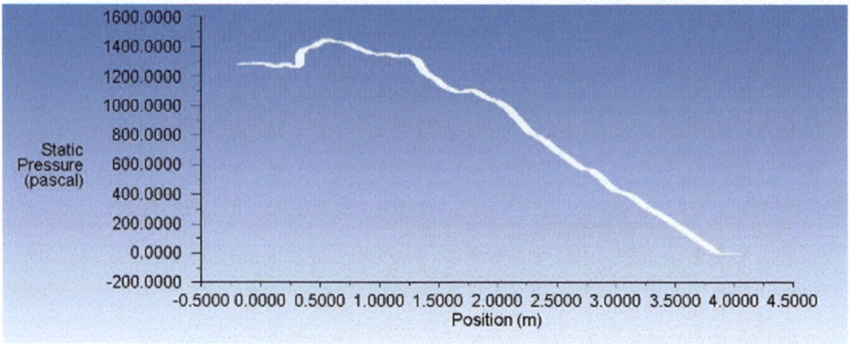

Fig. 9 Pressure graph for parallel-flow

3.1.5 Pressure Graph

The pressure inside the heat exchanger gradually decreases from 1500 to 0 Pa as shown in Fig. 9. At the cold inlet of the coil, the pressure is at 1300 Pa, and when it enters into the shell, the pressure increases to 1400 Pa. At 1 m of heat exchanger, the pressure is 1100 Pa, at 2 m position, the pressure is 900 Pa, and at 3 m, the pressure is about 200 Pa.

3.2 CFD Analysis of Counter-Flow Heat Exchanger

3.2.1 Effective Thermal Conductivity Plot

The effective thermal conductivity of the counter flow is high at the inlet of the helical tube and is gradually decreased till the end of the helical coil as shown in Fig. 10. Values of the helical tube vary from 10.29 to 9976 W/m k. The value obtained at the inlet of the helical tube by placing the probe tool is 9785 W/m k, at the middle of the helical coil, it is 4565 W/m k, and at the outlet of the helical coil, it is 15.85 W/m k.

3.2.2 Velocity Plot

It shows that the velocity at the inlet is high, and it gradually decreases towards the outlet of the helical coil as shown in Fig. 11. Initially, the velocity is at 0.23 m/s and decreases to 0.07 m/s. By placing the probe, the values show that at inlet of helical tube it is 0.195 m/s and in middle of helical coil it is 0.15 m/s but in between the middle and outlet of coil the velocity gradually decreases to 0.079 m/s.

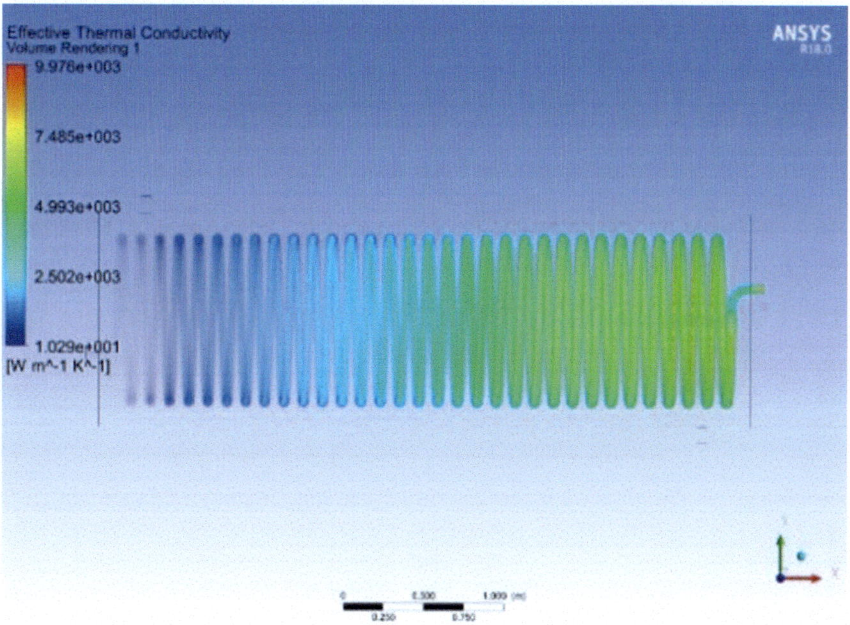

Fig. 10 Effective thermal conductivity plot for counter-flow

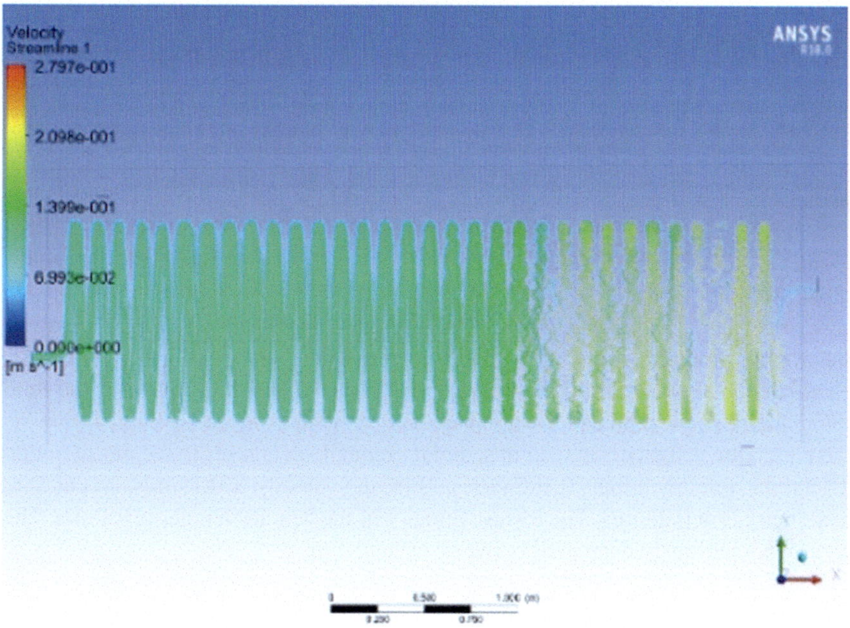

Fig. 11 Velocity plot for counter-flow

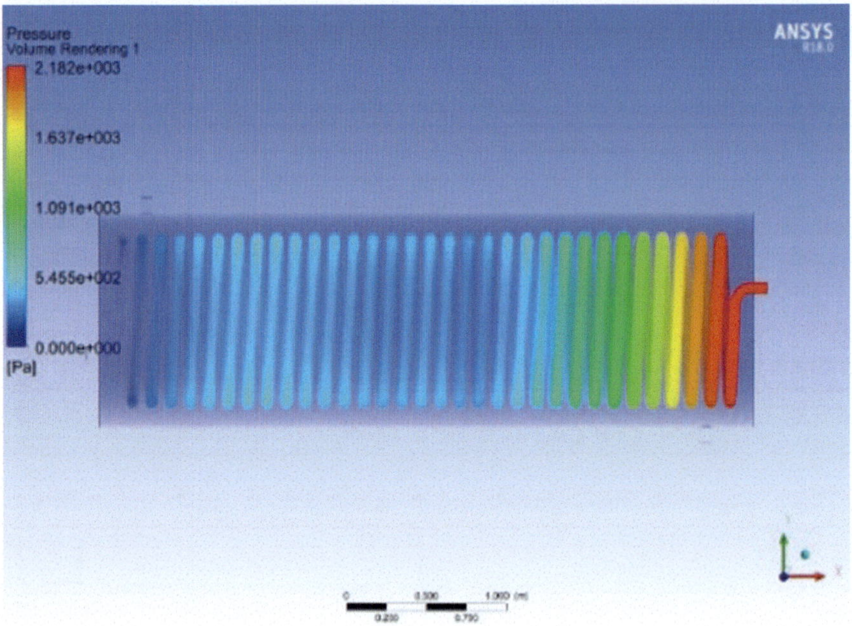

Fig. 12 Pressure plot for counter-flow

3.2.3 Pressure Plot

The pressure at the inlet of the counter-flow helical coil is high at 2182 Pa, and it is gradually decreased to zero pascal at the outlet of the helical coil as shown in Fig. 12. By the probe value, it defines that at the inlet the pressure is 2170 Pa and in middle the pressure it is about 957 Pa and at the end of coil the pressure it is about 20 Pa.

3.2.4 Velocity Graph

The graph is plotted for velocity with the position of the total heat exchanger. *X*-axis defines the position of the coil in metre, and *Y*-axis defines the velocity of the fluid flow in metre per second as shown in Fig. 13.

At inlet, the velocity is at 0.25 m/s, and it gradually decreases to 0.2 m/s at outlet. At 1 m, the velocity attains 0.105 m/s, at 2 m, the velocity is 0.15 m/s, and at 3 m, the velocity is at 0.23 m/s. The lowest velocity of the heat exchanger is 0.1 m/s. The velocity is vigorously decreased between 1.5 and 3 m.

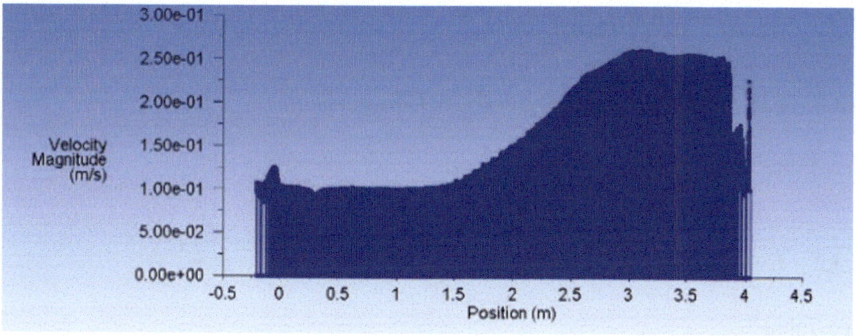

Fig. 13 Velocity graph for counter-flow

3.2.5 Pressure Graph

The pressure inside the heat exchanger gradually decreases from 4000 to 150 Pa as shown in Fig. 14. At the cold inlet of the coil, the pressure is at 3950 Pa, and when it enters into the shell, the pressure decreases to 250 Pa. At 1 m of heat exchanger, the pressure is 750 Pa, at 2 m, position the pressure is 1500 Pa, and at 3 m, the pressure is about 3250 Pa.

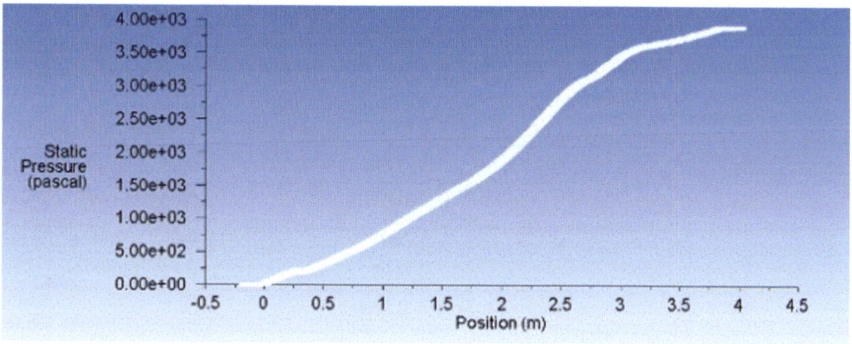

Fig. 14 Pressure graph for counter-flow

3.3 *Result Comparison*

3.3.1 Effective Thermal Conductivity Comparison

See Table 2.

3.3.2 Comparison of Velocity

See Table 3.

3.3.3 Comparison of Pressure

See Table 4.

Table 2 Comparison of effective thermal conductivity

Position (m)	Parallel flow (W/m K)	Counter flow (W/m K)
0	4080	23
1	2013	2308
2	1080	4873
3	320	7209
4	57	9780

Table 3 Comparison of velocity

Position (m)	Parallel flow (m/S)	Counter flow (m/S)
0	0.09	0.05
1	0.07	0.061
2	0.085	0.079
3	0.072	0.12
4	0.063	0.19

Table 4 Comparison of pressure

Position (m)	Parallel flow (Pa)	Counter flow (Pa)
0	1493	25
1	1235	560
2	1080	1028
3	728	1765
4	230	1980

4 Conclusion

In this study, the CFD investigation is conducted on the parallel-flow and counter-flow helical coil tube heat exchanger. The water is taken as the working fluid, and the helical coil has 32 turns. The copper is taken as the shell and tube material. On that, investigation result of the parameters like effective thermal conductivity, velocity and pressure is taken as output. The thermal conductivity of the counter flow has higher rate than the parallel flow. The pressure reduces faster on the parallel flow.

By considering the result parameters, the counter-flow helical coil tube heat exchanger provides the best results for the given input. The more knowledge on the heat exchanger is gathered from various literatures, books and articles. The geometry was created using SolidWorks 18 and CFD analyses in ANSYS Fluent 18.

5 Future Work

On the future research, the cross section of the circular tube is to be changed as elliptical, and fins and turbulator are added so that the performance of the heat exchanger is analysed and improvements are determined.

References

1. Genssle A, Stephan K (2000) Analysis of the process characteristics of an absorption heat transformer with compact heat exchangers and the mixture. Int J Therm Sci 39:30–38
2. Ruthven DM (1971) The residence time distribution for ideal laminar flow in a helical tube. Chem Eng Sci 26(7):1113–1121
3. Guo L, Feng Z, Chen X (2001) An experimental investigation of the frictional pressure drops of steam-water two-phase flow in helical coils. Int J Heat Mass Transf 44:2601–2610
4. Kshirsagar MP, Kansara TJ, Aher SM (2014) Fabrication and analysis of tube-in-tube helical coil heat exchanger. Int J Eng Res Gen Sci 2(3)
5. Ali S (2001) Pressure drop correlations for flow through regular helical coil tubes. Fluid Dyn Res 28:295–310
6. Hüttl TJ, Friedrich R (2000) Influence of curvature and torsion on turbulent flow in helically coiled pipes. Int J Heat Fluid Flow 21(3):345–353
7. Rennie TJ (2004) Numerical and experimental studies of a double pipe heat exchanger, A thesis submitted to McGill University in partial fulfilment of the requirements of the degree of Doctor of Philosophy
8. Rennie TJ, Raghavan VGS (2005) Experimental studies of a double-pipe helical heat exchanger. Exp Thermal Fluid Sci 29:919–924
9. Jakkula S, Sharma GS (2007) Analysis of a cross flow heat exchanger using optimization techniques
10. Rennie TJ, Raghavan VGS (2006) Effect of fluid thermal properties on heat transfer characteristics in a double pipe helical heat exchanger. Int J Thermal Sciences 45:1158–1165
11. Raju M, Kumar S (2010) Modeling of a helical coil heat exchanger for sodium alanate based on-board hydrogen storage system. In: Excerpt from the proceedings of the COMSOL conference Boston

12. Purandarea PS, Leleb MM, Gupta R (2012) Parametric analysis of helical coil heat exchanger. Int J Eng Res Technol 1(8)
13. Ankanna C, Reddy S (2014) Performance analysis of fabricated helical coil heat exchanger. Int J Eng Res 3
14. Nandan A, Singh G (2016) Experimental study of heat transfer rate in a shell and tube heat exchanger with air bubble injection. IJE Trans B: Appl 29(8):1160–1166
15. Al-Dawody MF, Hamzah DA (2018) Comparative study between curved and straight pipe heat exchanger using solid works. J Univ Babylon, Eng Sci 26(2)
16. Thakur G, Singh G, Thakur M, Kajla S (2018) An experimental study of nano fluids operated shell and tube heat exchanger with air bubble injection. IJE Trans A: Basics 31(1):136–143

Design and CFD Analysis of Wickless Heat Pipes Filled with Refrigerants as Working Fluids for HVAC Applications

B. Sivaramakrishnan and J. Selwin Rajadurai

Abstract Heat pipes are heat transfer devices with very good thermal conductivity. Heat pipes with refrigerants as working fluids were investigated. The refrigerants used in this work are R134a and R404a. Gravity-assisted heat pipe was preferred to this application. The various heat transfer limits in the design of heat pipes were found. The thermal performance of the heat pipes was studied by computational fluid dynamics method. ANSYS Fluent 16.0 was used to perform the CFD analysis. VOF method was employed. Very low heat input was applied for this particular application. The temperature distribution along the length of the heat pipe was studied. The CFD results are validated with the experimental values of the reference work. The thermal resistance values are found and compared with the experimental work. Optimization was performed using Taguchi technique.

Keywords Heat pipes · Refrigerants · Computational fluid dynamics · ANSYS Fluent

List of Symbols

Q_{sonic}	Sonic limit
Q_{viscous}	Viscous limit
$Q_{\text{entrainment}}$	Entrainment limit
Q_{boiling}	Boiling limit
A_v	Area of the vapor space
L	Latent heat of evaporation

B. Sivaramakrishnan (✉)
Engineering Design, Department of Mechanical Engineering,
Government College of Engineering, Tirunelveli 627007, India
e-mail: sivarama1196@gmail.com

J. Selwin Rajadurai
Department of Mechanical Engineering, Government College of Engineering,
Tirunelveli 627007, India

L.-J. Yang et al. (eds.), *Proceedings of ICDMC 2019*,
Lecture Notes in Mechanical Engineering,
https://doi.org/10.1007/978-981-15-3631-1_57

ρ_v	Density of the vapor phase
ρ_l	Density of the liquid phase
P_v	Vapor pressure
D_i	Inner diameter of the thermosyphon
r_v	Radius of the vapor space
l_{eff}	Effective length of the heat pipe
l_{ad}	Length of the adiabatic section
l_c	Length of the condenser section
μ_v	Viscosity of the vapor phase
f_1, f_2, f_3	Entrainment factors
g	Acceleration due to gravity
σ_l	Surface tension of the liquid phase
B_0	Bond number
Q	Heat input
T_e	Temperature of the evaporator section
T_c	Temperature of the condenser section
R	Thermal resistance
Seq SS	Sequential sum of squares
Adj SS	Adjusted sum of squares
DF	Degree of freedom
Adj MS	Adjusted mean squares
CFD	Computational fluid dynamics

1 Introduction

Heat pipes are efficient heat transfer devices with very less thermal resistance. Heat pipes are used in heating, ventilation and air-conditioning, electronic circuits, electric motors, air preheating, IC engines, heat exchangers, aerospace, satellites and human body temperature control.

Experimental investigation of the performance of two-phase closed thermosyphon filled with refrigerants R134a, R404a and R407C was reported [1]. The performance of R407C was 50% lower than other two. Thermal performance analysis of nanofluids in a thermosyphon heat pipe using computational fluid dynamics was done. De-ionized water and CuO/water nanofluid were used as working fluids [2]. Rahimi et al. [3] in their experiment reported the performance of thermosyphon at different operating conditions. The experimental measurements were compared with the predicted CFD temperature, and a good agreement was observed. Sozen et al. [4] made an investigation to utilize the Fly ash nanofluids in a two-phase closed thermosyphon for enhancing the heat transfer. The nanofluid filled 33.3% of the volume of the two-phase closed thermosyphon. Three heating power levels (200, 300 and 400 W) were used in the experiments with three different cooling water flow rates. Vafai et al. [5] utilized analytical models to investigate the thermal performance of rectangular-

and disk-shaped heat pipes using nanofluids. Al_2O_3, CuO and TiO_2 with a range of diameters are utilized to form the nanofluid. Khandekar et al. [6] used distilled water and acetone as the working fluid. Experimental investigation of a thermosyphon-based heat exchanger used in energy-efficient air handling units was done [7]. Nine thermosyphons in a modified inline configuration filled with water as working fluid were used. The experimental setup has finned evaporator and condenser sections. The overall effectiveness and thermal resistance of the heat pipe heat exchanger were found. Thermal analysis of heat pipe using Taguchi technique was done. Heat input (W), angle of inclination (deg) and flow rate (kg/min) were taken as the parameters. Efficiency, thermal resistance and overall heat transfer coefficient were found. *L*9 orthogonal array was used [8]. Narayan et al. [9] developed flexible heat pipes. With the straight and bend configurations of the heat pipe, the thermal resistance is reported. This flexibility will prevent damage to the heat pipe against the vibratory motion. Panda et al. [10] developed high-temperature heat pipes and thermosyphons for compact high-temperature reactor. CFD analysis was done. A three-dimensional transient numerical model has been developed to predict the vapor core, wall temperatures, vapor pressure and vapor velocity in the screen mesh wick. Huminic et al. [11] investigated the heat pipes with water–nanoparticle mixtures. Water is used as a base working fluid. Copper oxide was used as the nanoparticles. Volume fraction of 0, 1 and 4% for CuO is considered the analysis. Deshpande et al. [12] theoretically designed radiators using heat pipes. Design calculations for heat pipe were performed. The maximum heat transfer limit was $Q_{\max} = 405.7$ W. Effect on dimensions of the radiator and effect on heat capacity were discussed. As the capacity of the radiator is reduced, the exit temperature of the water is increased. Investigation of the thermal performance of two-phase closed thermosyphon charged with R134a and R404a were done. CFD modeling was done [13].

El-Sharkawy [14] evaluated the performance of heat pipe heat exchanger for automotive applications. Design of experiments was performed. Eight control factors and three levels were taken. Heat pipe design software tool was used to investigate the thermal performance of the heat exchanger design for automotive applications. Umberto Costa Bitencourt [15] explains the CFD procedure for the simulation of pulsating heat pipe. Bhanu Prakash et al. [16] performed thermal analysis of heat pipe with inclined 60° and 90° orientation. Acetone was used as the working fluid. Girish et al. [17] done CFD analysis in the pulsating heat pipe. Pandiraj et al. [18] used MgO nanofluids for thermal performance enhancement in flat plate heat pipes.

Very few CFD works were carried out for heat pipes with refrigerants as working fluids. In this work, theoretical thermal design calculations were performed. Heat pipes with different fill ratios were investigated.

2 Design of Heat Pipes

Design of heat pipes involves the theoretical determination of heat transfer limits using the correlations [12] (Fig. 1; Table 1).

Fig. 1 Geometry of the heat pipe

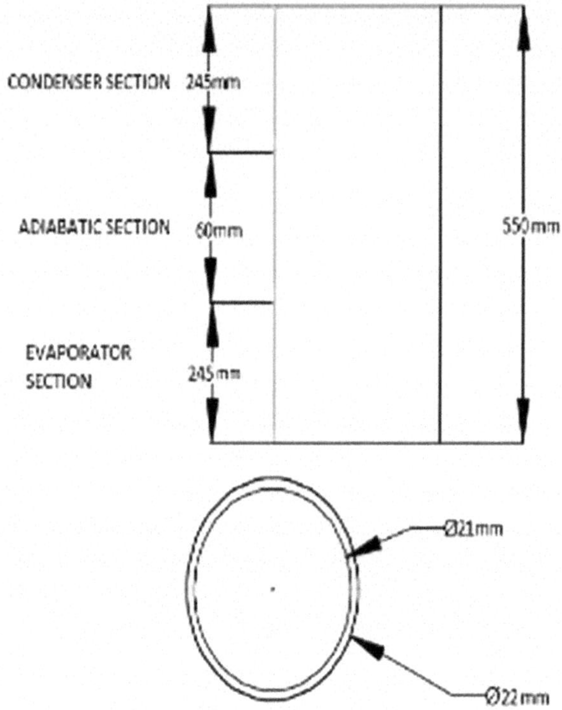

Table 1 Specifications of the heat pipe [1]

Length of the evaporator section (mm)	245
Length of the adiabatic section (mm)	60
Length of the condenser section (mm)	245
Outer diameter of the thermosyphon (mm)	22
Inner diameter of the thermosyphon (mm)	20
Working fluids	R134a, R404a

2.1 *Sonic Limit*

As the temperature of the vapor in the heat pipe is lowered, the vapor pressure drops. The vapor velocity must increase to carry a given amount of heat, which in turn increases the pressure drop from the evaporator to the condenser. Compressible flow effects become important at these low vapor pressures.

$$Q_{\text{sonic}} = 0.474 \times A_{\text{v}} \times L \times (\rho_{\text{v}} \times P_{\text{v}})^{0.5}$$

where $A_{\text{v}} = \frac{\pi}{4} D_{\text{i}}^2$.

2.2 Viscous Limit

The saturation vapor pressure is of the same magnitude as the pressure drop required to drive the vapor flow. There is an insufficient pressure available to drive the vapor. The viscous limit occurs at low operating temperature.

$$Q_{\text{viscous}} = \frac{A_v \times r_v \times L \times \rho_v \times P_v}{16 \times \mu_v \times l_{\text{eff}}}$$

2.3 Entrainment Limit

Dry out of the evaporator section may occur. This limit refers to the case of high shear forces developed as the vapor passes in the counterflow direction over the liquid-saturated wick, where the liquid may be entrained by the vapor and returned to the condenser.

$$Q_{\text{entrainment}} = A_v \times L \times f_1 \times f_2 \times f_3 \times \rho_v^{0.5} \times (g \times \sigma_l \times (\rho_l - \rho_v))^{0.25}$$

f_1 is a function of bond number (Figs. 2 and 3)

Bond Number, $B_o = D_i \times \left(\frac{g \times \rho_l - \rho_v}{\sigma_l}\right)^{0.5}$

f_2 is a function of dimensionless parameter

$$K_p = \frac{P_v}{(g \times (\rho_l \times \rho_v)^{0.25})}$$

$f_3 = 1$ (since the thermosyphon is vertical).

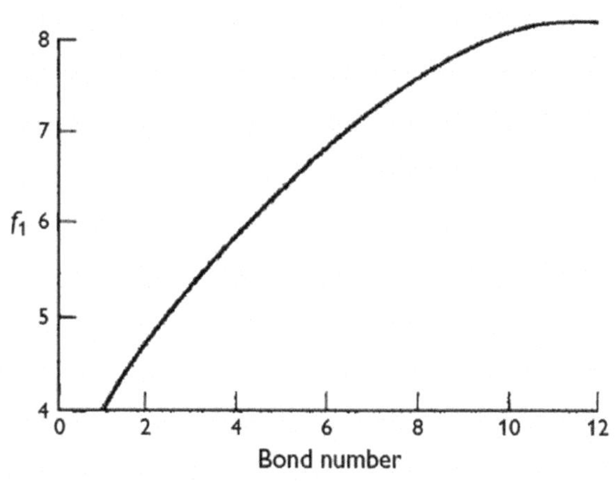

Fig. 2 Effect of bond number versus entrainment factor f_1 [14]

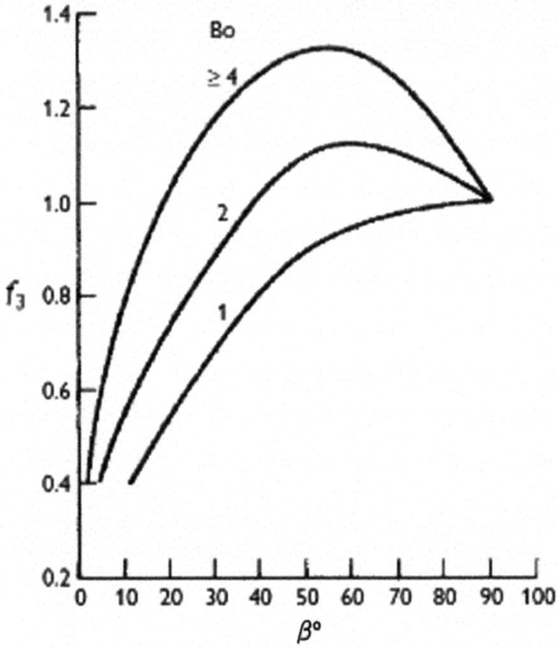

Fig. 3 Inclination angle versus entrainment factor f_3 [14]

2.4 *Boiling Limit*

The boiling limit occurs when the applied evaporator heat flux is sufficient to cause the nucleate boiling in the evaporator. This creates vapor bubbles that partially block the liquid return and can lead to evaporator dryout. Heat flux limit is the other name for boiling limit (Tables 2, 3 and 4).

Table 2 Properties of the working fluids

Properties/working fluids	R134a	R404a
Latent heat of evaporation (kJ/kg)	177.79	140.25
Density of vapor phase (kg/m^3)	32.35	1044.2
Density of liquid phase (kg/m^3)	1207.6	65.247
Specific heat of liquid phase (kJ/kg K)	1.4246	1.5423
Specific heat of vapor phase (kJ/kg K)	1.0316	1.2214
Thermal conductivity of liquid phase (W/m K)	0.081134	0.063625
Thermal conductivity of vapor phase (W/m K)	0.013825	0.015905
Viscosity of liquid phase (kg/m s)	1.9489×10^{-4}	1.2827×10^{-4}
Viscosity of vapor phase (kg/m s)	1.1693×10^{-5}	1.2152×10^{-5}
Molecular weight (kg/k mol)	102.03	97.604

Table 3 Properties of container material (copper)

Density of copper (kg/m^3)	8960
Specific heat (J/kg K)	376.812
Thermal conductivity (W/m k)	401

Table 4 Heat transfer limits

Working fluid	Sonic limit (W)	Viscous limit (W)	Entrainment limit (W)	Boiling limit (W)
R134A	122.72×10^{-3}	210.46	2.05×10^{-3}	118.9
R404A	122.47×10^{-3}	264.7	450.9	158.91

$$Q_{\text{boiling}} = 0.12 \times A_{\text{v}} \times L \times \rho_{\text{v}}^{0.5} \times ((g \times \sigma_{\text{l}} \times (\rho_{\text{l}} - \rho_{\text{v}}))^{0.25})$$

Thermal design calculations were performed. Various heat transfer limits in the design of heat pipes were found and are tabulated.

3 CFD Analysis

The two-dimensional models were created by using the line command, and the surfaces are created using the surface from sketch command in the ANSYS design modeler. The fine mesh was applied. There are 3324 nodes and 3036 elements after meshing. The elements are quadrilateral elements. In the general tab, the transient model was chosen. Because there is a dependence on time and there is no steady-state operation. The gravity was set as 9.81 m/s^2 in the negative y-direction. The multiphase model was clicked twice, and the properties were set. The three Eulerian phases are air, R134a liquid and R134a vapor. The reason for including a phase with air is that if only water liquid and vapor are defined as the phases, the calculation starts as water vapor is already present inside the pipe [15].

The implicit body force box was checked. The energy equation was turned on to allow heat transfer between the phases and the flow is K-ϵ model. The models used in the analysis are the multiphase model and viscous model. The number of mass transfer mechanism is 1. The evaporation and condensation frequency is set as 0.1. The saturation temperature is enabled as 297.69 K. Implicit first-order upwind scheme was used for solving the governing equation. The VOF model is used for analysis. The turbulence model applied for the present analysis was the Reynolds stress model. The SIMPLE scheme was selected. The first-order upwind momentum was checked. To solve the Navier–Stoke equation, the widely used numerical procedure was the SIMPLE algorithm. SIMPLE means for semi-implicit method for pressure-linked equations. The time step size is 0.0005. The number of time steps is 100. Maximum iterations per time step were given as 10.

3.1 Volume of Fluid Method

The volume of fluid method was used in this analysis. It is a numerical method of free surface approximations. It is classified as the Euler method. A good method for volume tracking and locating the free surface is the volume of fluid method. With the most suitable jump boundary conditions at the interface, the motion of different phases is solved by a single set of transport equations. Two or more immiscible fluids can be solved by a single set of momentum equations in the VOF model. There are two conditions when a steady-state VOF calculation is practical. They are (i) the inflow boundary conditions of the individual phases are distinct and (ii) the solution is independent of the initial conditions.

The fill ratio is defined as the ratio of the volume filled by the working fluid to the total volume of the heat pipe.

4 Results and Discussions

Figure 4 shows the temperature distribution along the length of the heat pipe. It was clearly identified that the temperature in the evaporator section (306–301 K) is more than the condenser and adiabatic section. This is due to the fact that the heat is supplied at the evaporator section. In the adiabatic section, there is only some heat loss to the surrounding. In the condenser section, the temperature varies from 295 to 291 K. From the plot, it is clear that the working fluid inside the heat pipe evaporates in the evaporator section and condenses in the condenser section. And the condensate returns back to the evaporator by means of gravity. Since the heat pipes were gravity assisted, there is no need of wick structure inside the heat pipe. Figure 5 shows the liquid volume fraction contours of the heat pipe filled with R134A as working fluids.

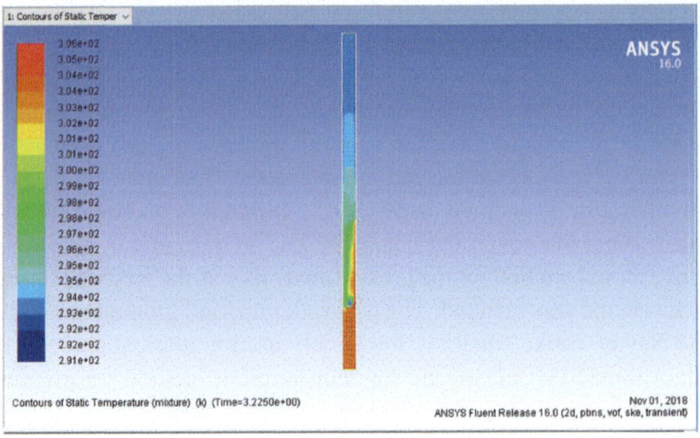

Fig. 4 Temperature distribution of heat pipe with R134A as working fluid with 30% fill ratio

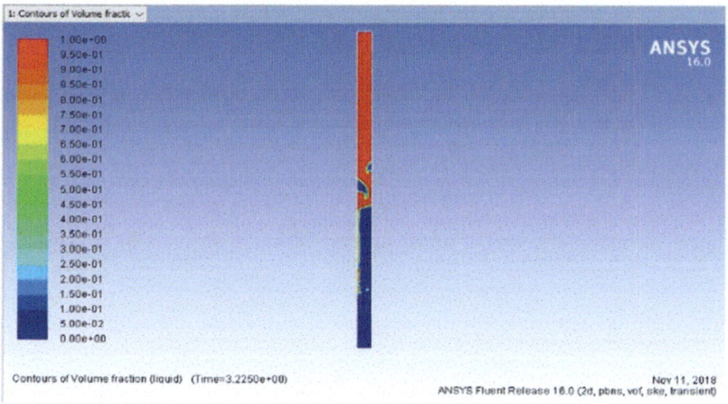

Fig. 5 Volume fraction contours of the heat pipe filled with R134a

The volume fraction contour confirms the process of nucleate pool boiling inside the heat pipe.

The temperature distribution in the heat pipe using R404a is shown in Fig. 6. The temperature distribution in the evaporator section varies from 303 to 300.5 K (Fig. 7). The temperature is maintained constant in the adiabatic section. The temperature varies from 296 to 294 K in the adiabatic section. The plot shows very clearly the working fluid in the heat pipe evaporates at the evaporator section and condenses at the condenser section and also indicates the condensate return across the boundary which is blue in color.

Thermocouple locations from 1 to 3 are in the evaporator section. Adiabatic section has two thermocouples (4 and 5). Thermocouples 6, 7 and 8 are located in the condenser section. The figure shows that there is only little deviation from the experimental and CFD value. And the temperatures found by CFD analysis were

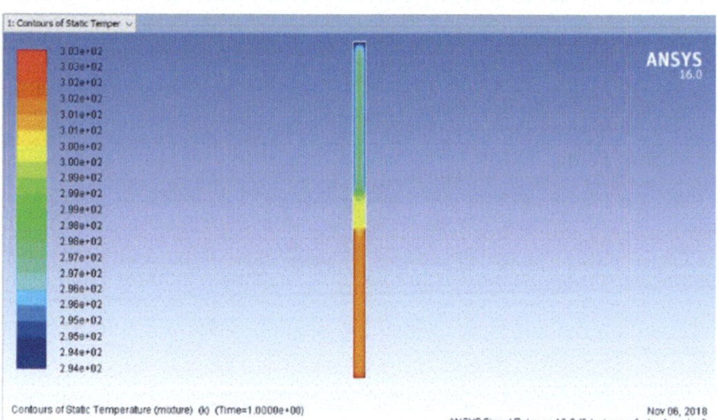

Fig. 6 Temperature distribution of the heat pipe with R404a as working fluid 30% fill ratio

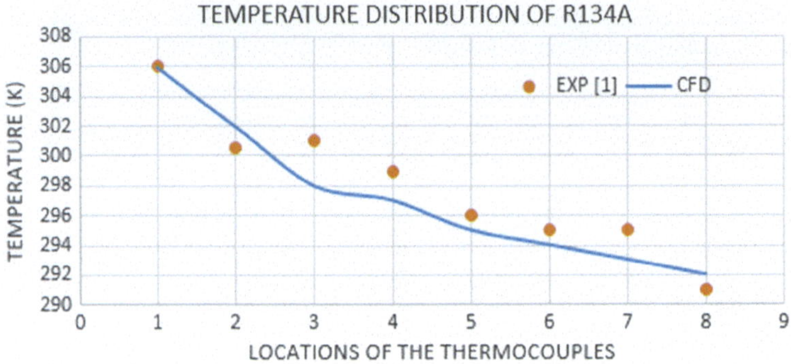

Fig. 7 Temperature distribution with R134a as working fluid along the length of the heat pipe

found to be in agreement with the experimental values. The values obtained from the CFD analysis are validated with the experimental results. The temperature in the evaporator section is more due to the supply of heat. The heat loss takes place at the condenser section. The CFD values are found to be in agreement with the experimental values (Fig. 8).

Figure 9 shows the comparison between the heat pipes with R134a and R404a as the working fluids. For the same heat input, the performance of R134A is almost equal to the thermal performance of the heat pipe with R404a as working fluid. The formula for calculating the thermal resistance is given below (Figs. 10, 11, 12, 13 and 14; Tables 5 and 6):

$$R = \frac{T_e - T_c}{Q}\ \text{K/W}$$

The reason for selecting the fill ratios in increasing order from 10, 20, 30, 40 and 50% is to ensure the evaporation and condensation inside the heat pipe.

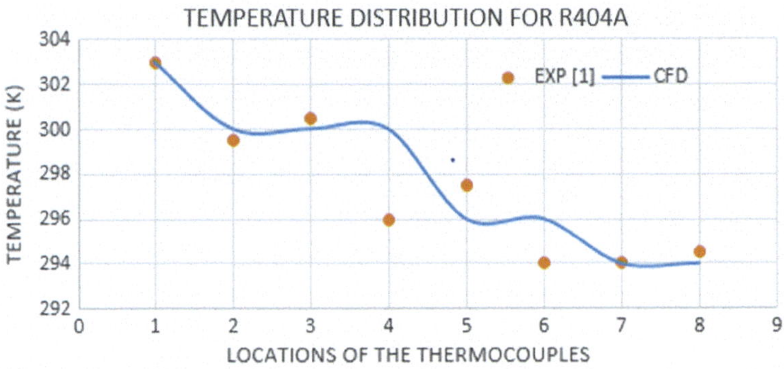

Fig. 8 Temperature distribution of heat pipe with R404a as working fluid with 30% fill ratio

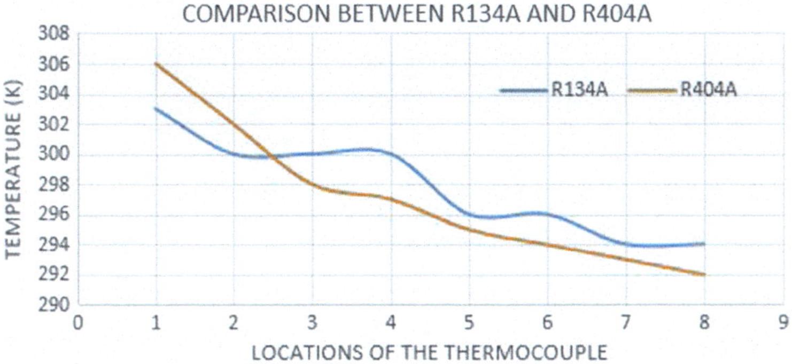

Fig. 9 Temperature distribution of heat pipe with R134A and R404A

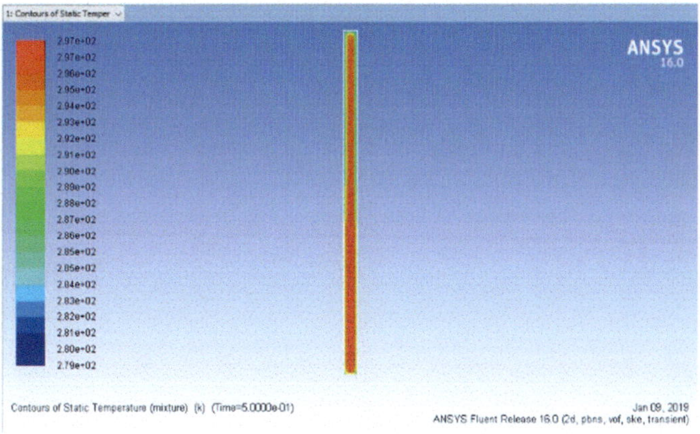

Fig. 10 Temperature distribution of heat pipe with R404a as working fluid with 10% fill ratio

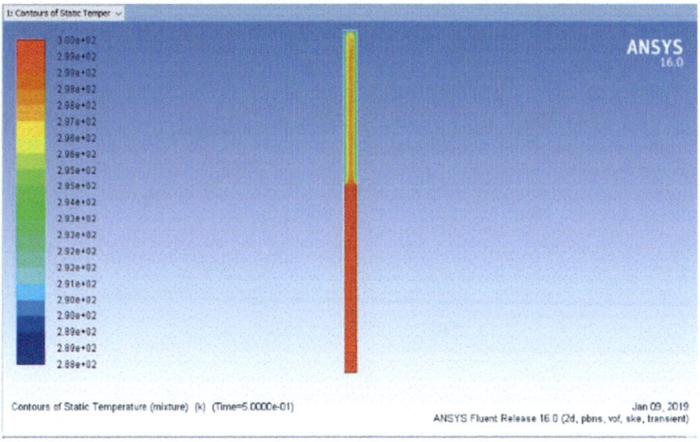

Fig. 11 Temperature distribution of heat pipe with R404a as working fluid with 20% fill ratio

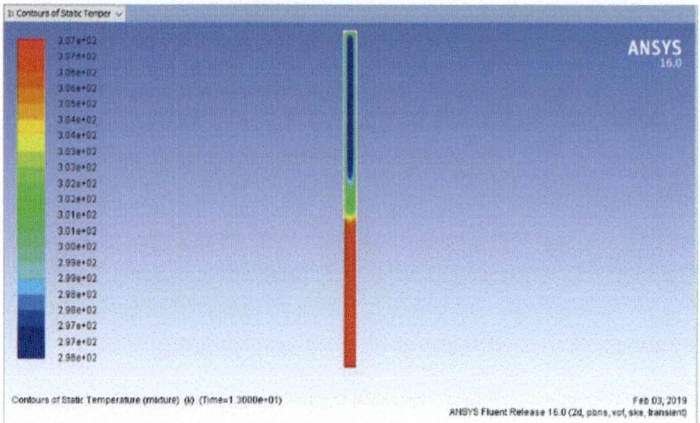

Fig. 12 Temperature distribution of heat pipe with R404a as working fluid with 40% fill ratio

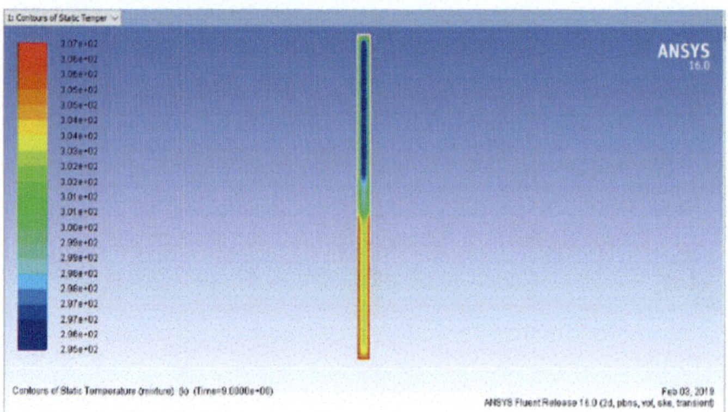

Fig. 13 Temperature distribution of heat pipe with R404a as working fluid with 50% fill ratio

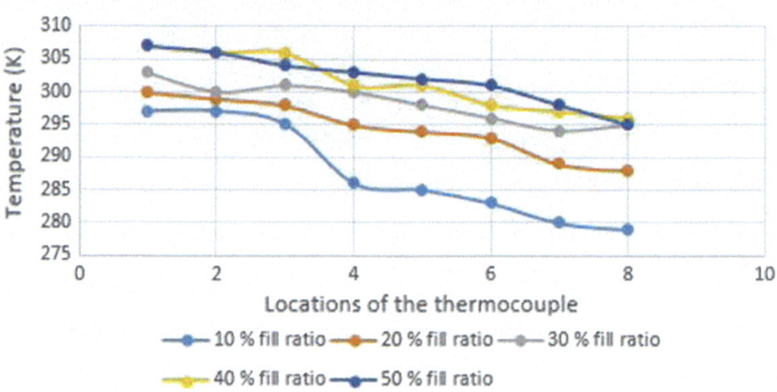

Fig. 14 Temperature distribution of heat pipes with R404a as working fluid with different fill ratios

Table 5 Comparison between CFD and experimental value [1]

S. No.	Working fluid	Thermal resistance experimental value[a] (K/W)	Thermal resistance CFD value (K/W)	Error (%)
1	R134A	0.375	0.35	6.67
2	R404A	0.2125	0.2	5.88

[a]The experimental thermal resistance values are calculated from the temperature distribution plot of the Grezgorz Gorecki [1]

Table 6 Fill ratio and thermal resistance

Fill ratio (%)	Thermal resistance (K/W)
10	0.45
20	0.30
30	0.225
40	0.275
50	0.3

Effect of fill ratio on thermal resistance

The value of the thermal resistance increases with increase in fill ratio up to 40%, and it suddenly decreases at 50% fill ratio.

5 Optimization Using Taguchi Technique

The number of levels taken for this analysis was five and the control parameters were two. Therefore, the *L*25 orthogonal array was selected. Twenty-five runs were conducted with R404a as working fluid. This is because the thermal resistance value is low compared to R134a (Table 7).

5.1 Orthogonal Array for L25 *Design*

See Table 8.

Table 7 Control parameters and number of levels

Control parameters	Level 1	Level 2	Level 3	Level 4	Level 5
Heat input (W)	10	20	30	40	50
Fill ratio (%)	10	20	30	40	50

Table 8 *L*25 design

*L*25	Level 1	Level 2
1	1	1
2	1	2
3	1	3
4	1	4
5	1	5
6	2	1
7	2	2
8	2	3
9	2	4
10	2	5
11	3	1
12	3	2
13	3	3
14	3	4
15	3	5
16	4	1
17	4	2
18	4	3
19	4	4
20	4	5
21	5	1
22	5	2
23	5	3
24	5	4
25	5	5

The formula for calculating S/N ratio is $-10\log\left[\frac{1}{y_i^2}\right]$ (Table 9).

Taguchi analysis: Thermal resistance (K/W) versus heat input (W) and fill ratio (%) response table for signal-to-noise ratios.

Smaller is better (Fig. 15; Table 10).

Taguchi analysis: Heat transfer coefficient (W/m^2 K) versus heat input fill ratio (%) response table for signal-to-noise ratios.

Larger is better (Fig. 16; Table 11).

Table 9 S/N ratios for thermal resistance and heat transfer coefficient

Heat input (W)	Fill ratio (%)	Thermal resistance (K/W)	Heat transfer coefficient (W/m^{2}K)	S/N ratio 1	S/N ratio 2
10	10	1.3	83.32	−2.27887	38.41499
10	20	0.9	120.35	0.91515	41.60892
10	30	0.8	135.4	1.9382	42.63237
10	40	0.9	120.359	0.91515	41.60957
10	50	1	108.3235	0	40.69445
20	10	0.75	144.43	2.498775	43.19315
20	20	0.6	180.53	4.436975	45.13099
20	30	0.4	270.8	7.9588	48.65297
20	40	0.5	216.64	6.0206	46.71477
20	50	0.65	166.64	3.741733	44.43559
30	10	0.67	162.49	3.478504	44.21653
30	20	0.43	249.98	0.43	47.95811
30	30	0.3	361.07	0.3	51.15183
30	40	0.36	295.43	0.36	49.40909
30	50	0.4	270.81	0.4	48.65329
40	10	0.45	240.72	0.45	47.63024
40	20	0.3	361.08	0.3	51.15207
40	30	0.225	481.44	0.225	53.65084
40	40	0.275	393.404	0.275	51.89678
40	50	0.3	361.08	0.3	51.15207
50	10	0.44	246.19	0.44	47.82541
50	20	0.28	386.87	0.28	51.7513
50	30	0.24	451.3457	0.24	53.09019
50	40	0.3	361.07	0.3	51.15183
50	50	0.34	318.6	0.34	50.06492

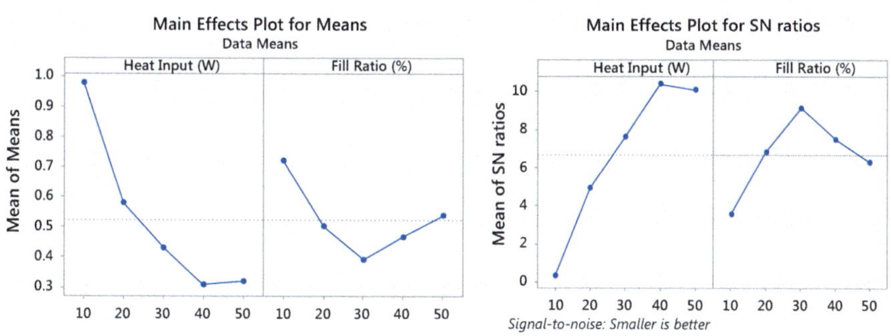

Fig. 15 Effect on thermal resistance

Table 10 Response table for thermal resistance

Level	Heat input (W)	Fill ratio (%)
1	0.2979	3.5530
2	4.9314	6.8394
3	7.6199	9.1413
4	10.4041	7.4961
5	10.0823	6.3057
Delta	10.1062	5.5883
Rank	1	2

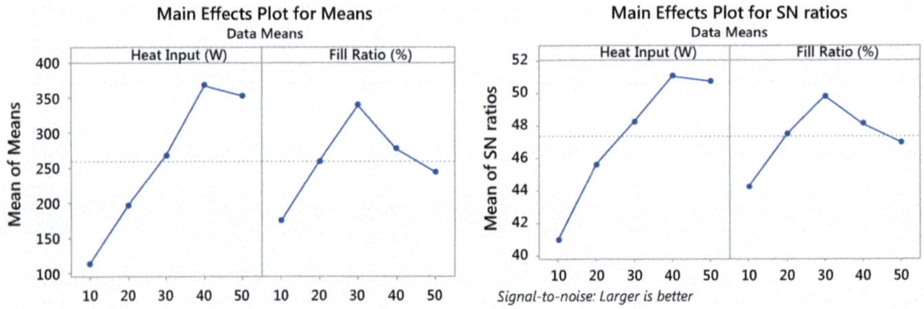

Fig. 16 Effect on heat transfer coefficient

Table 11 Response table for heat transfer coefficient

Level	Heat input (W)	Fill ratio (%)
1	40.99	44.26
2	45.63	47.52
3	48.28	49.84
4	51.10	48.16
5	50.78	47.00
Delta	10.10	5.58
Rank	1	2

5.2 *Analysis of Variance*

See Table 12.

The ANOVA table for thermal resistance shows that the contribution of heat input is 81.31%, and the contribution of fill ratio is 15.97%.

The ANOVA table for heat transfer coefficient shows that the contribution of heat input is 72.83% and the fill ratio is 22.38% (Table 13).

Table 12 ANOVA for thermal resistance

Source	DF	Seq SS	Contribution (%)	Adj SS	Adj MS	*F*-value	*P*-value
Heat input (W)	4	1.53474	81.31	1.53474	0.383684	119.75	0.000
Fill ratio (%)	4	0.30147	15.97	0.30147	0.075367	23.52	0.000
Error	16	0.05126	2.72	0.05126	0.003204		
Total	24	1.88747	100.00				

Table 13 ANOVA for heat transfer coefficient

Source	DF	Seq SS	Contribution (%)	Adj SS	Adj MS	*F*-value	*P*-value
Heat input (W)	4	229,054	72.83	229,054	572,63.5	60.79	0.000
Fill ratio (%)	4	70,386	22.38	70,386	175,96.5	18.68	0.000
Error	16	15,071	4.79	15,071	941.9		
Total	24	314,511	100.00				

6 Conclusion

Theoretical thermal design calculations were performed in the heat pipes with R134a and R404a as working fluid. It was found that the maximum heat transfer limit of the heat pipe with R134a is 118.9 W and R404a is 158.9 W. Temperature distribution plots for both heat pipes were obtained and are validated with the experimental values. The thermal resistance for the heat pipe using R134A as the working fluid is 0.35 K/W and R404a as the working fluid is 0.225 K/W. The average relative error in the thermal resistance is found to be 6.115%. The thermal resistance value is low for 30% fill ratio. Employing this type of heat pipes results in less thermal resistance. Optimization was done using Taguchi technique.

References

1. Gorecki G (2018) Investigation of two-phase thermosyphon performance filled with modern HFC refrigerants. Heat Mass Transf 2018(54):2131–2143
2. Haghshenasfard M, Asmaie A, Mehrabani-Zeinabad A, Nasr Eshahany M (2013) Thermal performance analysis of nanofluids in a thermosyphon heat pipe using CFD modeling. Springer. http://dx.doi.org/10.1007/s00231-013-1110-6
3. Rahimi M, Alizadehdakhel A, Alsairafi AA (2009) CFD modeling of flow and heat transfer in a thermosyphon. Int Commun Heat Mass Transf
4. Sozen A, Menlik T, Guru M, Irmark AF, Kilic F, Akta M (2014) Utilization of fly ash nanofluids in two-phase closed thermosyphon for enhancing heat transfer. Exp Heat Transf

5. Vafai K, Shafahi M, Bianco V, Manca O (2010) Thermal performance of flat-shaped heat pipes using nanofluids. Int J Heat Mass Transf
6. Khandekar S, Battacharya B, Paralikar S, Jaipurkar T (2017) Thermo-mechanical design and characterization of flexible heat pipes. Appl Therm Eng. http://dx.doi.org/10.1016/j.applthermaleng.2017.01.036
7. Jouhara H, Merchant H (2011) Experimental investigation of a thermosyphon based heat exchanger used in energy efficient air handling units. Energy. http://dx.doi.org/10.1016/j.energy.2011.08.054
8. Senthilkumar R, Vaidyanathan S, Sivaraman B (2010) Thermal analysis of heat pipe using Taguchi method. Int J Eng Sci Technol 2(4):564–569
9. Narayan A, Jaipurkar T, Khandekar S, Paralikar S (2015) Development and testing of flexible heat pipes. In: Proceedings of the 23rd national heat and mass transfer conference and 1st international ISHMT-ASTFE heat and mass transfer conference
10. Panda KK, Basak A, Dulera IV (2015) Design and development of high temperature heat pipes and thermosiphons for passive heat removal from compact high temperature reactor. In: Thorium energy conference
11. Huminic G, Huminic A (2009) CFD study of the heat pipes with water-nanoparticles mixture. In: Fourth European automotive simulation conference
12. Deshpande A, Patil V, Patil R (2016) Theoretical design of radiator using heat pipes. Int J Eng Res Technol 5(11)
13. Wrobel LC, Jouhara H, Fadhl B (2015) CFD modelling of a two-phase closed thermosyphon charged with R134a and R404a. Appl Therm Eng
14. El-Sharkawy A, Uddin A (2016) Evaluation of heat pipe heat exchanger for automotive applications. SAE Int
15. Umberto Costa Bitencourt J (2016) CFD simulation of a Pulsating heat pipe using ANSYS Fluent. Res Gate
16. Bhanu Prakash S, Raghuram J, Phani Kumar KVNK, Khiran GV, Snehith K (2017) Thermal Performance of a selected heat pipe at different tilt angles. IOP Conf Ser: Mater Sci Eng
17. Girish S, Lavanya K, Geeta Krishna P (2017) CFD analysis of Pulsating heat pipe using different fluids. Int J Mech Eng Technol
18. Pandiraj P, Gnanavelbabu A, Saravanan P (2018) Experimental and statistical analysis of MgO nanofluids for thermal enhancement in a novel flat plate heat pipes. Int J Nanosci

A Review on the Enhancement of Heat Exchanging Process Using TiO_2 Nanofluids

M. Armstrong, M. Siva Subramanian, and N. Selvapalam

Abstract Because of the decrement of fossil fuels, the energy industries are pushed to consume lesser rate to produce the power with the higher demand, as per the world energy outlook report 2018, the demand for natural gas, coal for power plants and fuel for automobiles increases 1.6% per year upto 2040 and 23 million barrels per day, respectively. Thus, this issue has to be resolved only by optimizing the energy transformation which is achieved by the optimum utilization and enhancement in the performance of the heat exchanging devices since heat exchanging process is very much significant in many industrial applications. Many types of research are undergoing for many years to find a technique with unceasing heat transfer rate. This is insignificantly achieved with the introduction of the nano-sized particles mixed with the base fluid which is having anomalous thermal behavior. In this review paper, the complete investigation on the utilization of TiO_2nanofluid with its thermal properties and volume concentration over the various base fluids in addition to the influence over the Reynolds and Nusselt number during heat exchanging process has been discussed.

Keywords TiO_2 nanoparticles · Base fluid · Heat exchanger · Heat transfer enhancement

---

M. Armstrong (✉)
Department of Mechanical Engineering, KARE, Krishnankoil 626126, India
e-mail: armstrong@klu.ac.in

M. Siva Subramanian
Department of Automobile Engineering, KARE, Krishnankoil 626126, India
e-mail: m.sivasubramanian@klu.ac.in

N. Selvapalam
Department of Chemistry, KARE, Krishnankoil 626126, India
e-mail: n.selvapalam@klu.ac.in

L.-J. Yang et al. (eds.), *Proceedings of ICDMC 2019*,
Lecture Notes in Mechanical Engineering,
https://doi.org/10.1007/978-981-15-3631-1_58

1 Introduction

Heat transfer is one of the major processes involved in many industrial and domestic applications which includes thermal power plants, nuclear power plants, solar water heating, etc. For many years, efforts have been undergoing to enrich the performance of the heat exchangers. Similarly, improvement in heat transfer is always in demand, so the need for more efficient heat transfer systems which increases the rate of heat transfer in less time duration [1]. Heat exchangers are the devices which are used to transfer heat from hot fluid to the cold fluid with or without the separating solid walls. It is of many types like double pipe heat exchangers, shell and tube heat exchangers, plate type heat exchangers, microchannel heat exchangers, etc. These heat exchangers are used according to their varied design and its applications with respect to the heat transfer enhancement competency. Heat transfer enhancement is crucial in the new design of the heat exchangers [2] because of their varied heat transfer coefficient, since many designs have been introduced to enrich their performance. In addition to the design considerations, nowadays the introduction of nano-sized solid particles with high thermal conductivity in the heat exchanging fluid without getting sediment in the pipe is being studied to enrich the performance of the heat exchanging process. Fluid with a suspended nano-sized particle called nanofluid is going to be the next generation heat exchanging fluid since they consist of superior thermal physical properties to enhance the convective heat transfer compared to normal liquids [3]. Because of that in many heat transfer augmentation research, nanofluids play a major part nowadays. The nanoparticles which involved usually of nano-meter size and very fine, henceforth it might behave more like a single-phase fluid than a solid–liquid mixture. The normally used materials for nanoparticles are metals like Ag, Au, Fe, Aland Cuand non-metals like graphite, carbon nanotubes, and oxides ceramics like Al_2O_3, CuO, TiO_2, SiO_2, carbides, and nitrides. The base fluid is usually a conductive fluid, such as water, oil, ethylene glycol and other common fluids, such as paraffin. Investigations from various research papers stated that nanofluids possess enhanced properties such as thermal diffusivity, thermal conductivity, viscosity, and convective heat transfer coefficients compared to those of base fluids like oil or water [3–7]. Thus, its usage is vast in many applications, especially in heat exchangers.

This paper presents a detailed review on the usage of TiO_2nanoparticles and its influence over the varied heat exchangers with the impact of volume concentration and particle size of nanoparticle which gives a clear idea in the selection of TiO_2 nanoparticle performance with respect to the heat exchangers.

2 Titanium Dioxide Nanoparticle (TiO_2)

Titanium dioxide has the chemical entity of titanium (IV) oxide, which is the natural abundant oxide form of titanium metal. However, TiO_2 has good chemical stability, high heat or mass transport potential, fluidity, dispersion stability for energy system

Table 1 Properties of TiO_2 salt

Molecular weight	Color	Stability	Density (g/cm^3)	Boiling point (°C)	Thermal conductivity (W/m.K)
79.866	White solid	Non-flammable, good stability	4.23 (rutile), 3.78 (anatase)	2972	4.8–11.8

applications, etc. [8] and it has been explored as nanoparticle in heat exchanging process. Many researchers have studied TiO_2 nanoparticle preparation and its combination in a base fluid with different concentration over varied heat exchangers and received many interesting results. Table 1 provides the properties of TiO_2 salt.

2.1 *Research Works of TiO_2 Nanofluid in Varied Heat Exchangers*

Eiamsa-ard et al. [9] investigated the double pipe heat exchanger incorporated with double twisted tape with the varied twisted ratio. In that, the smallest twist ratio (y/w = 1.5) results in the increase of heat transfer rates and friction factor up to 224.8% and 8.98 times. He also observed that the TiO_2/water nanofluid induces the behavior of low-pressure loss. Khanlari et al. [10] state that TiO_2 with water as a base fluid in corrugated plate heat exchanger provides an improved overall heat transfer coefficient of 6% both experimentally and computationally using CFD software. Maddah et al. [11] demonstrated the dispersion and random motions of $A12O_3/TiO_2$ nanoparticle significantly near the wall of the tube, leads to the intensification exchange rate between nanoparticles in double pipe heat exchanger incorporated with twisted tape. The experimental results show that the increasing volume concentration of nanofluids and reduction in twist ratio of twisted tapes boost up the exergy efficiency by increasing the Reynolds number, Nusselt number with increased heat transfer coefficient. Yang et al. [8] made a wide study on the TiO_2 preparation and its properties; he concluded that the TiO_2 nanoparticles have good dispersion stability in both hydrophilic and lipophilic liquids, chemical stability, good appearance, non-toxic, non-corrosive with increase in heat transfer coefficient with increase in particle loading, temperature made the nanofluid more inclusive applications in heat transfer or other energy fields. Revenkar et al. [12] discussed that the heat transfer rate and overall heat transfer coefficient was increased and it also attains increases in the pressure drop of 36% with inner pipe grooved double pipe heat exchanger compared to the plain tube. Sacithra et al. [13] experimentally investigated the performance of solar collector with TiO_2 nanofluid, in which it increases the efficiency by 7.02% of the collector and increases the overall coefficient by 22% with increment in mass flow rate. Kilica et al. [14] stated that TiO_2/water nanofluid provided better performance than pure water while using in solar flat plate collector due to the increased working fluid contact surface and high specific heat capacity of TiO_2/water nanofluid. An instantaneous efficiency was obtained with 48.67% for TiO_2/water nanofluid, whereas 36.20% for pure water. Naveen Kumar Gupta et al. discussed the experimental study of TiO_2 nanoparticle as a coating material in the pipe and as a nanofluid with water as the base fluid. Then it is compared with the performance of normal water in the heat pipe. The results showed that there is a high thermal performance in the heat pipe when nanofluid is used comparatively to nano coating and the normal water charge in the heat pipe. Zing et al. [15] experimentally investigated the new porous heat exchanger with 5% of TiO_2, 1% Al_2O_3, 0.03% MWCNT nanoparticle with water as a base fluid and 1% of diamond with 40:60 ethylene glycol/water is used as a nanofluid in the cooling application. Here, TiO_2 plays an improved cooling efficiency with rectangular and square cross-sections inlets than the other nanofluids utilized. Hamid et al. [16] explained the thermal conductivity of TiO_2; it enhanced when there is an increase in temperature and concentration. And also increase in concentration increases the viscosity but decreases with increase in temperature. This leads to the improved heat exchanger performance at 50 and 70 °C and it attained less performance at 30 °C.

Qumars et al. [18] discussed that the effects of fluid and heat transfer parameters in the microchannel aspect ratio 10, 15, 20 and 25 with Re = 300 have the highest heat transfer rate and Re = 10 has the lowest. Increase in aspect ratio has the effect of decreasing the fluid momentum. Heyadri et al. [19] stated that an improved heat transfer performance is attained when the TiO_2 volume fraction is 4% and Re = 400 in the 3D microchannel heat exchanger. Arulprakasajothi et al. [22] in their experiment gave a clear view of the increment of the Nusselt number with the increase in the volume concentration. He observed that Nusselt number enhancement was 13.2% over when used TiO_2 nanoparticle compared to the base fluid with the maximum volume concentration of 0.75%. He also concludes that the enhancement in thermal conductivity of TiO_2/water nanofluids not only depends on the particle volume fraction, it also depends on the pH value, Brownian motion, specific surface area, nanolayer, nanocluster, thermophoresis along with size and shape of the suspended particles. Eiamsa-ard et al. [23] discussed that Nusselt number, friction factor, and thermal performance were increased with decreasing overlapped twist ratio and increasing TiO_2 volume concentration. The thermal performance factor of 1.18 was obtained with the maximum volume concentration of 0.21% and the thermal performance factor of 1.7–4.5% was obtained with (TiO_2/water) nanofluid compared to the (water) base fluid in the tube heat exchanger. Nishant et al. [20] state that there is an enhancement in the heat transfer coefficient in CuO and TiO_2 with the volume concentration of 0.02, 0.04 and 0.06% (at 3 LPM) is 12%, 17% and 27% and 10%, 14% and 17%, respectively, at 50 °C and Nusselt number increased by 13%, 18% and 29% for CuO and 11%, 15% and 19% for TiO_2, respectively, at 80 °C. From this, he concludes that CuO having better heat transfer performance comparing to TiO_2 because of the higher thermal conductivity of CuO nanoparticle.

2.2 *Thermo-Physical Properties of TiO_2 Nanofluid in Different Research Works*

Arulprakasajothi et al. [24] discussed that during the preparation of TiO_2/water nanofluid, the sonication process in the ultrasonic bath for six hours produces good dispersion and stability, and he has included that nanofluid was stable because of the repulsive forces between the nanoparticles. Gupta et al. [17] investigated the deterioration in thermal performance of the heat pipe for every two months which works twelve hours a day. The initial four months result showed that the thermal efficiency was higher in the heat pipe with nanofluid as compared to the TiO_2 coated heat pipe but after aging to six months, the thermal efficiency was higher in the coated heat pipe compared to the heat pipe with TiO_2 nanofluid. Thus, he concluded that to avoid stability issue due to deterioration of nanofluid, nano coating might be the good substitute, since the thermo-physical properties of nano fluid after some period provide agglomeration and sedimentation which shows poor thermal performance.

3 Challenges and Future Work

During the preparation of TiO_2nanoparticle, two-step methods with post-treatment have been recommended comparing to one-step method with some drawbacks. Properties like viscosity, the surface tension of TiO_2 nanoparticle have to be paid attention while mixing with the base fluid. Stability, viscosity and agglomeration properties of the nanofluids are the major challenges without compromising the heat transfer property of the nanoparticles. In most of the studies, stability of the nanofluid varied with the aging of the nanoparticle in the base fluid which caused agglomeration. On the other hand, viscosity of the nanofluid increased with the particle concentration which would also create problem in controlling the fluid mass flow rate. If temperature increases, viscosity of the nanofluid also decreases. The addition of surfactant to the fluid also decreases the agglomeration rate of the system, and thus, the solubility of the particle tend to increase in the nanofluid. The surface tension of TiO_2 nanofluid is more sensitive to surfactant than nanoparticles. Hence, surfactant with low concentration is suggested during experimentation stated by Yang et al. [8]. Although TiO_2 nanofluids have shown greater heat transfer enhancement in many heat transfer applications, its performance was poor when compared to Au, Ag, CuO, and CNT which was reviewed by Yang et al. [8]. Due to its good stability, chemical properties, and economy, its usage in various applications is better compared to thermal applications. But [15] TiO_2 nanoparticle gave better cooling efficiency in the porous medium compared to other nanofluids used such as Al_2O_3, MWCNT, diamond with definite proportions in porous filled heat exchangers. Undoubtedly, the thermal performance is not only proportional to the properties of nanoparticles but also according to the design of heat exchangers. Hence, incorporation of varied and unique heat exchanger designs also enhances the heat transfer coefficient while using low thermal conductivity nanoparticles such as TiO_2. Thus, the future works could include a detailed study on nanoparticle influence over the heat transfer performance in addition to its varied heat exchanger design which is at minimal level.

4 Summary

A comprehensive review on the influence of titanium dioxide (TiO_2) as a nanofluid in different heat exchangers has been studied for the recent five years. From this, we could understand that TiO_2 nanoparticle is one of auspicious material for heat transfer enhancement purpose due to its excellent chemical, physical properties and less expensive. From Table 2, we could be able to appreciate that TiO_2 nanoparticle has been mixed with various low thermal conductivity base fluids like water, ethylene glycol in varied heat exchangers to enhance its thermal behavior. It has been utilized in many heat exchangers to analyze numerically and experimentally through heat transfer coefficient, Nusselt number and Reynolds number with respect to its volume concentration, dispersion over the base fluids. From Table 3, we could realize that stability, viscosity and agglomeration of the nanoparticle in the base fluid also plays a major role in enhancing the performance of the heat exchangers. It is mostly stable when nanoparticle is coated in the system than the fluid inside the pipe. Viscosity increases with increase in concentration. From the literature reports, we have realized that TiO_2 can provide a high volume concentration and it offered higher thermal performance in all type of heat exchangers. In combination with water as a base fluid, it provides low-pressure loss. It provides better performance in heat transfer in the pipes comparing to the water. It is predominantly good as a coolant for both copper and annealed pyrolytic graphite porous matrices at low and high porosity structures, and for both square and rectangular inlet cross sections when compared to other nanoparticles. It enhances the friction factor, Nusselt number and Reynolds number which enhances the heat transfer coefficient of the fluid in the heat exchangers. TiO_2 nanofluid provides a better heat transfer performance than TiO_2 nano coating over the heat pipe. Consequently, all these views would provide to choose a better heat exchanger in combination with TiO_2 nanoparticle with required volume concentration while going for future research.

Table 2 Research works of TiO_2 nano fluid in varied heat exchangers

Researchers and year	Nanofluids utilized	Base fluids utilized	Particle size (nm)	Particle volume concentration (vol%)	Type of heat exchanger	Flow regime	Result and discussion
Zing et al. (2019) [15]	TiO_2, Al_2O_3, MWCNT, Diamond	Water	–	5, 1, 1, 0.03, 1	The porous filled heat exchanger	Fully developed flow	TiO_2 nanofluid has better cooling efficiency in the porous medium comparing to other nanofluids used
Eiamsa-ard et al. (2018) [9]	TiO_2	Deionized water	–	0–0.05	Double pipe heat exchanger with double twisted tape	Swirling-flows with Re = 5000–20,000	Smallest twist ratio (y/w = 1.5), increases in heat transfer rates and friction factor up to 224.8% and 8.98 times
Khanlari et al. (2018) [10]	TiO_2	Water	14	0.2	Corrugated plate heat exchangers	Turbulent flow	Improved overall heat transfer coefficient as 6%
Maddah et al. (2018) [11]	Al_2O_3–TiO_2	Water	170	0.2–1.5	Double pipe heat exchanger with Al twisted tape	Turbulent flow with Re = 3000 to 12,000	Increasing the Reynolds number, Nusselt number, (y/w = 2–8)
Revankar et al. (2018) [12]	Al_2O_3–TiO_2	Deionized water		0.1–0.2%	Grooved double pipe heat	Increased turbulent flow with Re = 3000–23,000	Overall heat transfer coefficient and pressure drop increases compared to plain double pipe heat exchanger

(continued)

Table 2 (continued)

Researchers and year	Nanofluids utilized	Base fluids utilized	Particle size (nm)	Particle volume concentration (vol%)	Type of heat exchanger	Flow regime	Result and discussion
Sacithra et al. (2018) [13]	TiO_2	Demineralized water	30–50	0.1	Flat tube solar absorber with a spiral configuration	Spiral flow with different flow rates	Flow rate increases the Reynolds number and TiO_2 increases the overall coefficient to 22%
Kiliç et al. (2018) [14]	TiO_2	Deionized water	30–60	0.2	Flat plate solar collector	–	Efficiency increased 48.67% for TiO_2/H_2o nanofluid, 36.20% for pure water
Gupta et al. (2018) [17]	TiO_2	Water	40–60	0.5, 1.0, 1.5	Heat pipe	–	Higher thermal performance compared to water
Qumars et al. (2017) [18]	TiO_2	Water	–	0.2–2	Microchannel heat exchanger	Laminar flow. Re = 10, 50, 100, 300	The increase of TiO_2 nanoparticles enhance the heat transfer, friction factor and pressure drop
Heydari et al. (2017) [19]	TiO_2	Water	–	0, 2, 4	3D ribs microchannel with a triangular cross-section	Laminar flow	Increased heat transfer coefficient is achieved in 4% of TiO_2 and in the Re = 400

(continued)

Table 2 (continued)

Researchers and year	Nanofluids utilized	Base fluids utilized	Particle size (nm)	Particle volume concentration (vol%)	Type of heat exchanger	Flow regime	Result and discussion
Abdul Hamid et al. (2016) [16]	TiO_2	Mixture of ethylene glycol and water	50	0.5–1.5%	Tube heat exchanger	Reynolds number of 3000–24,000	The thermal conductivity and viscosity of fluid get improved with increase in concentration
Nishant et al. (2016) [20]	Cuo, TiO_2	Water, mono ethylene glycol	26 nm—Cuo, 9 nm—TiO_2	0.02, 0.04, 0.06%	Shell and tube heat exchanger	Laminar and turbulent	Enhancement in heat transfer and Nusselt number were obtained for both nanofluids
Ramtin et al. (2016) [21]	TiO_2	Deionized water	20	0.3, 0.8 and 1.5%	Brazed plate heat exchanger	Turbulent flow	6.6–23.7% enhancement of HTC with Nusselt number 9–24
Arulprakasajothi et al. (2015) [22]	TiO_2	Water	32	0.1, 0.25, 0.5, 0.75%	Tube heat exchanger	–	The Nusselt number increases with an increase of particle volume fraction
S. Eiamsa-ard et al. (2015) [23]	TiO_2	Water	15	0.07, 0.14 and 0.21%	Tube Heat exchanger with overlapped dual twisted tapes	Overlapped swirling flow, Re = 5400–15,200	At 0.21% volume concentration, the Nusselt number increased up to 9.9–11.2% with 1.7–4.5% thermal performance factor

Table 3 Thermo-physical properties of TiO_2 nanofluid in different research works

Researcher/year of publication	Instruments used to measure	Stability	Rheological properties	Agglomeration
Khanlari et al. (2018) [10]	Scanning electron microscope (SEM), X ray diffraction (XRD)	Sonication, high impact ball milling and adding surfactant leads to the nanoparticle to have longer stability during preparation	Nanofluid viscosity is very close to deionized water viscosity	The surfactant triton X-100 added to increase the solubility
Maddah et al. (2018) [11]	KD2 Pro, Rheometer(Anton Paar equipment), scanning electron microscope (SEM), X ray diffraction (XRD), transmission electron microscope(TEM)	Good nanoparticle dispersion which is discussed with the TEM images	Nanoparticle volume concentration increases the viscosity of the nanofluid	This vibrator generates ultrasonic pulses of 100 W at 36 ± 3 kHz for 5 h to prevent nanoparticles agglomeration process before employing as the heat transfer fluid during preparation
Nishant et al. (2016) [20]	KD2 Pro thermal property analyzer, ultrasonicator, Xray diffraction, transmission electron microscope	All the nanoparticles are dispersed very well with very small aggregation	Low viscosity nanofluid increase the Brownian motion	Convective heat transfer rate increases in the dispersed phase in water. Agglomeration creates problem in various base fluids like water, ethylene glycol and oil

(continued)

Table 3 (continued)

Researcher/year of publication	Instruments used to measure	Stability	Rheological properties	Agglomeration
Hamid et al. (2016) [16]	KD2 Pro thermal analyzer and a Brookfield LVDV-III Rheometer, Fisherbrand ultrasonic bath, FESEM and TEM	Fisherbrand ultrasonic bath was used for two hours for each nanoparticle concentration to enhance the stability of the nanofluid during preparation and after seven months aging period, the nanofluids showed little sedimentation but it was stable throughout the experimentation process	Decreases in viscosity of fluid from 2.3 to 2.4 times with respect to the temperature range from 30 to 70 °C. If concentration has been increased, it directly affects the internal shear stress and increased temperature decreases the intermolecular adhesion force	Less agglomeration while aging of nanofluid
Ramtin et al. (2016) [21]	Ultrasonic homogenizer, transmission electron microscope (TEM)	The surfactant used leads to the better stability	Enhancement in viscosity and density of nanoparticle increases the pressure drop in brazed plate heat exchanging material	Cetyltrimethyl ammonium bromide surfactant is used to increase the solubility in distilled water
Arulprakasajothi et al. (2018) [24]	Ultrasonicator, Zeta potential, Brookfield cone and plate viscometer	Good stability with zeta potential range ±40 to ±60	Volume concentration increases the viscosity	–
Sacithra et al. (2018) [13]	Ultrasonicator, Zeta potential, Brookfield cone and plate viscometer	Zeta potential of TiO_2 nanofluid is 48 mV. It shows the good colloidal stability	Viscosity ranging from 5 to 5000 cP is measured in spiral copper tube solar collector	Triton X-100 is used as a surfactant to increase the dispersion and solubility of nanoparticle in the demineralized water base fluid

(continued)

Table 3 (continued)

Researcher/year of publication	Instruments used to measure	Stability	Rheological properties	Agglomeration
Kiliç et al. (2018) [14]	Ultrasonic processor, impact ball milling	Nanofluid maintained its stability throughout the process. The structure of the nanoparticle leads to good stability	Increase in temperature decrease the viscosity of nanofluid with increase in thermal conductivity	Reduction in density of the nanoparticles decrease the agglomeration. Triton X-100 is used as a surfactant to increase the dispersion and solubility. The minimal agglomeration was achieved at a surfactant concentration of 0.2%
Gupta et al. (2018) [17]	Zeta sizer, KD2 pro thermal analyzer, Brookfield viscometer	The obtained zeta potential value after 30 days aging period of TiO_2 nanofluid is +31.5 mV. This shows the good stability of the nanoparticle in the water as base fluid	Decrease in viscosity was observed with the increase in temperature	Agglomeration and sedimentation occurred at low vapor pressure at low power input. Thermal performance decreased in TiO_2 nanofluid in heat pipe due to agglomeration compared to coating in six month duration
Arulprakasajothi et al. (2015) [22]	Zeta sizer, KD2 Pro and Brookfield viscometer, ultrasonicator	The values obtained from the zeta sizer shows the good stability	Presence of TiO_2 nanoparticle increases the viscosity of the base fluid and which increases the thermal conductivity	–

References

1. Gugulothu R, Somanchia NS, Reddy VKK, Akkirajub K (2017) A review on enhancement of heat transfer in heat exchanger with different inserts. In: Materials today: proceedings 4, pp 1045–1050
2. Ilyas M, Aydogan F (2018) A new dimensionless thermal hydraulics parameter for the heat exchangers. Ann Nucl Energy 118:154–164 (Elseveir)
3. Wang X-Q, Mujumdar AS (2008) A review on nanofluids—part I: theoretical and numerical investigations. Braz J Chem Eng 25(4):613–630
4. Yu W, France DM, Routbort JL, Choi S (2008) Review and comparison of nanofluid thermal conductivity and heat transfer enhancements. Heat Transf Eng 29(5):432–460
5. Tyler T, Shenderova O, Cunningham G, Walsh J, Drobnik J, McGuire G (2006) Thermal transport properties of diamond-based nanofluids and nanocomposites. Diamond Relat Mater 15(11–12):2078–2081
6. Makinde OD, Mutuku WN (2014) Hydromagnetic thermal boundary layer of nanofluids over a convectively heated flat plate with viscous dissipation and ohmic heating. UPB Sci Bull Ser A 76(2):181–192
7. Mutuku WN (2016) Ethylene glycol (EG)-based nanofluids as a coolant for automotive radiator. Asia Pac J Comput Eng 3:1
8. Yang L, Hu Y (2017) Toward TiO_2 nanofluids—part 2: applications and challenges. Nanoscale Res Lett 12:446 (Springer)
9. Eiamsa-ard S, Ketrain R, Chuwattanakul V (2018) TiO_2/water nanofluid heat transfer in heat exchanger equipped with double twisted-tape inserts. Series: materials science and engineering, vol 350, pp 012015
10. Khanlari A, Sözen A, Variyenli Hİ (2018) Simulation and experimental analysis of heat transfer characteristics in the plate type heat exchangers using TiO_2/water nanofluid. Int J Numer Methods Heat Fluid Flow. https://doi.org/10.1108/HFF-05-2018-0191
11. Maddah H, Aghayaria R, Mirzaeea M, Ahmadi MH, Sadeghzadehc M, Chamkha AJ (2018) Factorial experimental design for the thermal performance of a double pipe heat exchanger using Al_2O_3-TiO_2 hybrid nanofluid. Int Commun Heat Mass Transf, 92–102
12. Revankar AA, Venkatachalapathy, S (2018) Experimental studies on internally grooved double pipe heat exchanger using Al_2O_3/TiO_2 nanofluids. J Nanofluids 7(2):275–283
13. Sacithra A, Manivannan A (2018) Turbulent flow analysis of a flattened tube in-plane curved solar collector using Titanium oxide nanofluid. Heat Mass Transf (Springer Nature 2019)
14. Kiliç F, Menlik T, Sözen A (2018) Effect of titanium dioxide/water nanofluid use on thermal performance of the flat plate solar collector. Solar Energy 164:101–108
15. Zing C, Mahjoob S, Vafai K (2019) Analysis of porous filled heat exchangers for electronic cooling. Int J Heat Mass Transf 133:268–276
16. Hamid KA, Azmi WH, Mamat R, Sharma KV (2016) Experimental investigation on heat transfer performance of TiO_2 nanofluids in water-ethylene glycol mixture. Int Commun Heat Mass Transf (Elsevier)
17. Gupta NK, Tiwari AK, Ghosh SK (2018) Experimental study of thermal performance of nanofluid-filled and nanoparticles-coated mesh wick heat pipes. J Heat Transf (ASME) 140
18. Qumars G, Akbari OA, Toghraie D, Marzban A, Mashayekhi R, Karimi R, Pourfattah F (2017) The effect of aspect ratios of rib on the heat transfer and laminar water/TiO_2 nanofluid flow in a two-dimensional rectangular microchannel. J Mol Liq. https://doi.org/10.1016/j.molliq.2017.04.030
19. Heydari M, Toghraie D, Akbari OA (2017) The effect of triangular porous medium and non-Newtonian nanofluid on flow and heat transfer properties in a constant-flux microchannel. Therm Sci Eng Prog. https://doi.org/10.1016/j.tsep.2017.05.010
20. Kumar N, Sonawane SS (2016) Experimental study of thermal conductivity and convective heat transfer enhancement using CuO and TiO_2 nanoparticles. Int Commun Heat Mass Transf, 76:98–107

21. Barzegarian R, Moraveji MK, Aloueyan A (2016) Experimental investigation on heat transfer characteristics and pressure drop of BPHE (brazed plate heat exchanger) using TiO_2-water nanofluid. Int J Exp Heat Transf Fluid Mechanics, 894–1777, 00341–6
22. Arulprakasajothi M, Elangovan K, Reddy KHC, Suresh S (2015) Heat transfer study of water-based nanofluids containing titanium oxide nanoparticles. In: Materials today: proceedings 2, pp 3648–3655
23. Eiamsa-ard S, Kiatkittipong K, Jedsadaratanachai W (2015) Heat transfer enhancement of TiO_2/water nanofluid in a heat exchanger tube equipped with overlapped dual twisted-tapes. Eng Sci Technol Int J 18:336–350
24. Arulprakasajothi M, Elangovan K, Chandrasekhar U, Suresh S (2018) Performance study of conical strip inserts in tube heat exchanger using water based titanium oxide nanofluid. Therm Sci 22(1B):477–485
25. Huminic G, Huminic A (2012) Application of nanofluids in heat exchangers: a review. Renew Sustain Energy 16:5625–5638
26. Yang L, Hu Y (2017) Toward TiO_2 nanofluids—part 1: preparation and properties. Nanoscale Res Lett 12:417 (Springer)
27. Diao YH, Liu Y, Wang R, Zhao YH, Guo L, Tang X (2013) Effects of nanofluids and nano coating on the thermal performance of an evaporator with rectangular microchannels. Int J Heat Mass Transf 67:183–193
28. Gupta M, Arora N, Kumar R, Kumar S, Dilbaghi N (2014) A comprehensive review of experimental investigations of forced convective heat transfer characteristics for various nanofluids. Int J Mech Mater Eng
29. Elias MM, Saidur R, Ben-Mansour R, Hepbasli A, Rahim NA, Jesbains K (2018) Heat transfer and pressure drop characteristics of a plate heat exchanger using water-based Al_2O_3 nanofluid for 30° and 60° chevron angles. Heat Mass Transf. https://doi.org/10.1007/s00231-018-2335-1
30. Barber J, Brutin D, Tadrist L (2013) A review on boiling heat transfer enhancement with nanofluids. Nanoscale Res Lett 6:280
31. Sergis A, Hardalupas Y (2014) Anomalous heat transfer modes of nanofluids: a review based on statistical analysis. Nanoscale Res Lett 6:391. https://www.nanoscalereslett.com/content/6/1/391

Modeling of Flow-Induced Vibration Response of Heat Exchanger Tube with Fixed Supports in Cross Flow

Kore Someshwar Vishwanath and S. Balaguru

Abstract The shell and heat exchanger has either straight or U-tube with baffle support at an intermediate point and fixed at their ends. Cross flow-induced vibration has been found to be most destructive in many industrial applications. This work reveals modeling of flow-induced vibration response: vortex shedding, turbulent buffeting and fluid elastic instability of single straight tube with fixed support, which may cause the amplitude of vibration in terms of fretting wear at the tube support. Modal analysis has been performed using ANSYS to find out natural frequencies and mode shapes, and MATLAB program is developed to check flow-induced vibration response. It is found that results are in good agreement with available literature data. This approach can be used to provide conservative designs or troubleshoot an existing heat exchanger.

Keywords Flow-induced vibration · Vortex shedding, turbulent buffeting · Fluid elastic instability

1 Introduction

Flow-induced vibration has been a serious concern in the design and operation of shell and tube heat exchangers, such as steam generators, condensers and coolers. In shell and tube-type heat exchangers, one fluid flows inside the tubes, while another fluid flows through the shell and across the outside of the tubes. Considering performance and economics, larger-scale heat exchangers are being designed with higher mass flow rates and larger thermal gradients. It is thus desirable to use small diameter tubes and large distances between the tubes support to increase the heat transfer area and reduce shell-side pressure drops. These systems have, therefore, become more

K. S. Vishwanath (✉) · S. Balaguru
Mechanical Department, Vel Tech Rangarajan Dr. Sagaunthala R&D Institute of Science and Technology University, Chennai, India
e-mail: someshwarkore125@gmail.com

S. Balaguru
e-mail: iitmguru239@gmail.com

L.-J. Yang et al. (eds.), *Proceedings of ICDMC 2019*,
Lecture Notes in Mechanical Engineering,
https://doi.org/10.1007/978-981-15-3631-1_59

flexible and prone to flow-induced vibration damage in the form of fretting wear and fatigue failure at the tube supports as well as impacting at the tube middle span [1].

In heat exchangers, there are four types of flow configuration generally classified as: cross flow, internal axial flow, external axial flow and annular or leakage flow. Among these flow types, cross flow-induced vibration has been found to be the most destructive in many industrial applications [2].

The fluid flow causes excitation to the tubes. The performance of the tubes varies according to flow velocities and can be classified as

- Vortex shedding or flow periodicity.
- Turbulent buffeting.
- Acoustic resonance.
- Fluid elastic instability.

Vortex shedding and turbulent buffeting in general cause small amplitude vibrations which in turn may cause long-term fretting wear at the tube supports. And fluid elastic instability may cause tube failure due to its large vibration amplitude characteristics.

To render the problem amenable for most analytical studies and experimental investigations, the flow conditions are idealized as:

- The flow is uniform and steady.
- The incident of the flow is either axial or normal to the tubes.
- The tube motion is linearized, and it is assumed that the frequencies are well defined [3].

The baffle supports provide a simply supported condition.

The specific objective of the work is to check flow-induced vibration response of heat exchanger tube: vortex shedding, turbulent buffeting and fluid elastic instability using computational tools like ANSYS and MATLAB and compares results with available literature.

2 Vibration Evaluation Procedure

The vibration evaluation procedure involves an estimation of certain parameters for various flow-induced excitation mechanisms. The estimated parameters are compared with their respective limiting values to check whether or not vibrations from such excitation mechanisms can cause potential damage to the tubes and the shell. Need to individually examine various zones of interest, namely the nozzle inlet zone, U-bend region and baffle window region, since there is a likelihood of high turbulence and high cross flow velocity in these regions and the variations of span lengths between baffle supports compared to the central baffle region [2].

3 Steps of Vibration Evaluation

1. Calculate the effective mass per unit length.
2. Identify the zones of interest (inlet, baffle window, central baffle zones, U-bend, etc.) to calculate the natural frequency.
3. Calculate the natural frequency (f_n) for spans in various regions of interest. Perform modal analysis.
4. Calculate the damping parameter.
5. Calculate cross flow velocity U.

Formulation of flow excitation mechanisms.
Step 6 for liquid flow.

Vortex Shedding

a. Calculate vortex shedding frequency. Check for the acceptance criteria.

Turbulent Buffeting

Calculate tube response due to random excitation and check for the acceptance criterion.

Fluid Elastic Instability

Calculate the critical velocity and compare with cross flow velocity.
Keep the maximum cross flow velocity below the critical velocity.

Acoustic resonance

a. Calculate acoustic resonance frequency.
b. Calculate vortex shedding frequency. Check for various vortex shedding criteria.
c. Calculate turbulent buffeting frequency. Check for various turbulent buffeting criteria.

4 Problem Formulation and Methodology

4.1 Problem Formulation

Many researchers have developed a semiempirical model and mathematical model to predict interaction damping. Computer codes, such as VIBIC, PIPO-FE, ADINA, have been developed to calculate tube wear work-rates to aid in the prediction of tube fretting wear damage. Modal analysis has been performed to find out natural frequencies and mode shapes. MATLAB program has been developed to check flow-induced vibration response: vortex shedding, turbulent buffeting, fluid elastic instability of heat exchanger tube (fixed-fixed straight tube) which may cause long-term fretting wear at the tube supports.

4.2 Methodology

The flow-induced vibration procedure involves an estimation of certain parameters for various flow-induced excitation responses. The estimated parameters are compared with their respective limiting values to check whether or not vibrations from such excitation mechanism can cause potential damage to the tubes and the shell. Formulation of flow excitation response for liquid flow has been performed in this work. The following methodology is being performed to carry out flow-induced vibration response of shell and tube heat exchanger tube in cross flow.

Step 1 Calculate the effective mass per unit length.
Step 2 Identify the zone of interest.
Step 3 Calculate the damping parameter.
Step 4 Calculate cross flow velocity for the (TEMA) shell under consideration.
Step 5 Develop flowchart for flow-induced vibration response: vortex shedding, turbulent buffeting, fluid elastic instability of shell and heat exchanger single straight tube in cross flow.
Step 6 Perform modal analysis by using ANSYS and find out the natural frequencies and respective mode shapes.
Step 7 Formulate the response due to flow excitation in liquid flow.
Step 8 Check the response of tube for the following responses by utilizing MATLAB tool:

(a) Develop program in MATLAB for the vortex shedding response.
(b) Develop program in MATLAB for the tubular buffeting response.
(c) Develop program in MATLAB for the fluid elastic instability response.

Step 9 Compare results with the available literature data.

4.3 Steps for the Newton–Raphson Method

1. Choose initial values of velocity magnitude $|v|_0$.
2. Use the estimated $|v|_0$ and calculate mismatch δp.
3. Take iteration $n - 1$ for velocity $|v|_0$ and calculate mismatch ΔQ.
4. Use the estimated $|v|_0$ to formulate Jacobian matrix J(0).
5. Check if all the mismatches are below small no. Terminate process. If yes, otherwise go back to step 1 to start the next iteration with updates given below equation.

6. Calculate velocity.

$$|v_1| = |v_0| + \left(\frac{1 + |v_0|}{|v_0|}\right)$$

5 Modal Analysis

5.1 Modal Analysis

A modal analysis determines the vibration characteristics (natural frequencies and mode shapes) of a structure or a machine component. The natural frequencies and mode shapes are an important parameter in the design of a structure for dynamic loading conditions. A modal analysis can be performed using ANSYS in which damping is ignored in modal analysis to determine the natural frequency. Any applied loads are ignored since natural frequency information is also helpful for avoiding resonance reduction noise and as an important meshing check.

5.2 Natural Frequency Analysis

The frequency with which any object will vibrate if disturbed and allowed to vibrate on its own without any external force is called as natural frequency. Inherent characteristics property we recognized each other by face, non-living objects are identified by properties like weight, natural frequency, etc. No external force is applied during the analysis as natural frequency being inherent characteristics property of any component/assembly. Word free in the name "free vibration" indicates free from external force. Constraints: Except free-free run, natural frequency analysis is carried out as per actual constraints. Damping is neglected for natural frequency calculations. Analysis demonstrates magnitude of frequency and corresponding mode shape Mode are normalized. And result plot helps us to understand magnitudes of frequency and corresponding mode shape (usually, a very high value) are not meaningful and should be neglected.

6 Mathematical Calculation of Problem

Case no-01: Let's consider the segmental baffle heat exchanger design, few parameter of heat exchanger are given below for study flow-induced vibration [4].

$$
\begin{aligned}
\text{Tube length} &= 3.58\,\text{m}\\
\text{Shell inside diameter} &= 591\,\text{mm}\\
\text{Baffle thickness} &= 9.5\,\text{mm}\\
\text{Baffle cut} &= 147\,\text{mm}\\
\text{Baffle hole diameter} &= 0.768\,\text{in.}/19.5\,\text{mm}\\
\text{Baffle spacing} &= 447\,\text{mm}\\
\text{Tube outside diameter} &= 19.1\,\text{mm}\\
\text{Tube inside diameter} &= 16.56\,\text{mm}\\
\text{Tube material density} &= 8.5\,\text{g/cm}^3\\
\text{Tube modulus of elasticity} &= 110 \times 10^9\\
\text{Tube pattern} &= 30^\circ\\
\text{Pitch diameter ratio} &= 1.25\\
\text{Snell side fluid} &= \text{water}\\
\text{Fluid density} &= 1\,\text{gm/cm}^3\\
\text{Fluid Kinematic, viscosity} &= 10^{-5}\,\text{m}^2/\text{s}.
\end{aligned}
$$

Calculation procedure for shell-side liquids to check flow-induced vibration response

Case no: 01
Step 1:
Calculating tube mass per unit length (m)

(a) Calculating hydrodynamic mass coefficient C_m by TEMA method
$C_m = 1.756$ (triangular pitch ratio diameter = 1.25)
(b) Calculate the tube mass per unit length

$$
\begin{aligned}
m &= \text{structural mass} + \text{added mass} + \text{contained fluid mass}\\
&= \frac{D^2 \pi Cm\rho s}{4} + \frac{Di^2 \pi \rho i}{4} + \frac{\left(D^2 - Di^2\right)\pi m\rho s}{4}
\end{aligned}
$$

$$
\begin{aligned}
\text{Structural mass} &= \frac{D^2 \pi Cm\rho s}{4}\\
&= \pi \times \left(1.9 \times 10^{-3}\right)^2 \times 1.756 \times 1000/4\\
&= 0.503\,\text{m}
\end{aligned}
$$

$$
\begin{aligned}
\text{Added mass} &= \frac{Di^2 \pi \rho i}{4}\\
&= \pi \times \left(16.56 \times 10^{-3}\right)^2 \times 1.1455/4
\end{aligned}
$$

$$= 2.46 \times 10^{-4}\,\text{m}$$

$$\text{Contained fluid mass} = \frac{\left(D^2 - Di^2\right)\pi m\rho s}{4}$$
$$= \pi \times \left(0.0191^2 - 0.0156^2\right)/4$$
$$= 0.579\,\text{m}$$

$$\text{Mass per unit length} = 0.503 + 2.64 \times 10^{-4} + 0.579$$
$$= 1.08\,\text{m.}$$

Step 2:
The natural frequency for the straight tube

(a) Natural frequency of tube = 6.2706 Hz (From modal analysis)

Step 3:
Calculate damping parameter

$$\delta_1 = \frac{3.41D}{mf_n}$$
$$= 3.41 \times 0.0191/1.08 \times 6.2706$$
$$= 9.617 \times 10^{-3}$$
$$\delta_2 = \frac{0.012D}{m}\frac{\rho s\mu}{f_n}^{0.5}$$
$$= 2.122 \times 10^{-4} \times 0.039$$
$$= 8.47404 \times 10^{-6}$$
$$\delta = \max(\delta_1\delta_2)$$
$$= \max\left(9.617 \times 10^{-3}, 8.47404 \times 10^{-6}\right)$$
$$= 9.617 \times 10^{-3}$$

Correlations of Pettigrew, Rogers and Axisa.

$$D/D^* = (1.7P/D)^{-1}$$
$$= (1.7 \times 1.25)^{-1}$$
$$= 0.47$$

$$\varepsilon = \left(\pi/\sqrt{8}\right)\left(\frac{1 + (D/D^*)^3}{\left(1 - (D/D^*)^2\right)^2}\right)\left(\rho s D^2/m\right)\left(2\nu/\pi f_n D^2\right)^{0.5} + \left(\frac{N-1}{N}\right)\left(\frac{22}{f_n}\right)\left(\frac{\rho s D^2}{m}\right)\left(\frac{t_b}{L_i}\right)^{0.6}$$

$$= \left(\pi/\sqrt{8}\right)\left(\frac{1+0.47^3}{\left(1-0.47^2\right)^2}\right)(1000 \times 0.0191^2/1.08)$$

$$\left(2 \times 10^{-5}/\pi \times 33.84 \times 0.019^2\right)^{0.5}$$

$$+\left(\frac{7}{8}\right)\left(\frac{22}{6.276}\right)(1000 \times 0.0191^2/1.08)\left(\frac{6}{0.447}\right)^{0.6}$$

$$= 0.0359$$

$$\delta = 2\pi\varepsilon n$$

$$= 2 \times \pi \times 0.035$$

$$= 0.2258$$

Step 4:
Calculate the cross flow velocity U.

$$U = 0.15\,\text{m/s (TEMA standard)}$$

Step 5:
Check for vortex shading S_u:

(a) Calculate Strouhal number:

$$S_u = 1/1.73 \times 1.25 = 0.469$$

(b) Calculate vortex shedding frequency:

$$f_s = \frac{S_u U}{D}$$

$$= \frac{0.469 \times 0.15}{0.0191} = 3.683$$

(c) Check:

(1) $0.8\, f_s < f_n < 1.2\, f_s$

$2.9464 < 6.2706 < 1.2 \times 3.683$

$2.9464 < 6.2706 < 4.4196$

(Not resonance)

(2) $\frac{Df_n}{U} > 2S_u$

$\frac{6.2706 \times 0.091}{0.15} > 2 \times 0.462$

$0.798 > 0.924$ (Not acceptable)

(3) $\frac{Df_n}{U} < 0.2S_u$

$0.798 < 0.0924$ (Not acceptable)

Au-Yang et al. criteria
Calculating the reducing damping C_n

$$C_n = 4\pi \in M_n/D^2 \rho s$$
$$= 4\pi \times 0.0359 \times 1.08/\left(1000 \times 0.0191^2\right)$$
$$= 1.33555$$
$$C_n < 64$$
$$8.484 < 64$$

$$\text{(a) } C)\frac{U}{Df_n} < 3.3$$
$$\frac{0.15}{6.2706 \times 0.0191} < 3.3$$
$$1.54 < 3.3$$

(Check OK)

Maximum deflection due to vortex shedding at resonance.

$$C_L = 0.091$$
$$Y\text{max} = \frac{\rho s U^2 D C_L}{4\varepsilon f_n^2 m^2 \pi^3}$$
$$= \frac{0.091 \times 1000 \times 0.0191 \times 1}{4 \times 0.0359 \times 1.08^2 \times 6.2706^2 \times \pi^3}$$
$$= 1.915 \times 10^{-4}$$

Check: If $Y_{\max} < 0.02D$

$$1.915 \times 10^{-4} < 0.02 \times 0.0191$$
$$1.915 \times 10^{-4} < 3.82 \times 10^{-4}$$

(Accept OK)

Step 6:
Turbulence-induced excitation

a. Determine $C_R(f) = 2 \times 10^{-3}$
b. The mean square response for a pinned-pinned span is given by

$$
\begin{aligned}
y_{\max}^2 &= \frac{\left[\rho s U^2 D C_R(f)\right]^2}{128 \varepsilon f_n^2 m^2 \pi^3} \\
&= \frac{\left[12 \times 10^{-3} \times 1000 \times 1 \times 0.019\right]^2}{128 \times \pi^3 \times 0.022 \times 33.84^2 \times 1.03^2} \\
&= 1.802 \times 10^{-11}
\end{aligned}
$$

$$
\begin{aligned}
y^{2rms} &= \sqrt{\bar{y}_{\max}^2} \\
&= 4.2460 \times 10^{-6} \\
4.2460 \times 10^{-6} &< 0.254 \times 10^{-4} \quad \text{(O.K.)}
\end{aligned}
$$

Step 7:
Fluid elastic instability

(a) *Calculating reduced velocity parameter*

$$
\begin{aligned}
\chi &= \frac{m\delta}{\rho s D^2} \\
&= \frac{1.08 \times 0.2258}{1000 \times (0.0191)^2} \\
&= 0.6687
\end{aligned}
$$

(b) *Calculating reduced velocity by*

$$
\begin{aligned}
\frac{U_{\text{cr}}}{f_n D} &= 8.36(P/D - 0.9)\chi^{0.34} \\
&= 8.36(1.25 - 0.9)0.6687^{0.34} \\
&= 2.729
\end{aligned}
$$

Check:

(c) $U < U_{\text{cr}}$

$$
\begin{aligned}
0.15 &< 2.729 \text{ (O.K.)} \\
0.054 &< 0.5 \quad \text{(O.K.)}
\end{aligned}
$$

(a) For the particular tube layout

$$
\begin{aligned}
\frac{U_{\text{cr}}}{f_n D} &= K_{\text{mean}} \left(\frac{m\delta}{\rho s D^2}\right)^{0.5} \\
\frac{U_{cr}}{6.2706 \times 0.019} &= 4.5\left(\frac{1.08 \times 0.2258}{1000 \times 0.0109^2}\right)^{0.5} \\
&= 0.4715
\end{aligned}
$$

$$0.15 < 0.4715 \text{ (O.K.)}$$
$$0.3592 < 0.5 \quad \text{(O.K.)}$$

(b) $\frac{U_{cr}}{f_n D} = 4.0\left(\frac{m\delta}{\rho s D^2}\right)^{0.5}$

$$U_{cr} = 6.2706 \times 0.0191 \times 4 \times 0.408^{0.5}$$
$$= 0.3916$$

Check:

$$U < U_{cr}$$
$$0.15 < 0.3916 \quad \text{(O.K.)}$$
$$\frac{U}{U_{cr}} < 0.5$$
$$0.383 < 0.5 \quad \text{(O.K.)}$$

(c)

$$\frac{U_{cr}}{f_n D} = 2.1\left(\frac{m\delta}{\rho s D^2}\right)^{0.5}$$
$$U_{cr} = 6.2706 \times 0.0191 \times 2.1 \times 0.408^{0.5}$$
$$= 4.358$$

Turbulent buffeting:

$$f_{tb} = \frac{U}{DX_1X_t}\left\{3.05\left(1 - \frac{1}{X_t}\right)^2 + 0.28\right\}$$
$$= \frac{0.15}{0.0191 \times 1.25 \times 1.25}\left\{3.05\left(1 - \frac{1}{1.25}\right)^2 + 0.28\right\}$$
$$= 2.5078\,\text{Hz}$$

7 Output of Program

Case no: 01

Input parameter of program

Mode shape: 1
Mass per unit length (kg/m): 1.08
Tube outside diameter (m): 0.0191
Shell-side velocity: 0.15

Angle of inclination (deg): 30
Structural damping: 0.0359
Fluid density (kg/m^3): 1000
Kinematic viscosity of shell-side fluid (m^2/s): 10^{-5}
Distance between tube supports (m): 0.447
Modulus of elasticity: 110×10^9
Moment of inertia for bending (m^4): 2.89×10^{-9}
Value of lambda (frequency constant): 3.1416
Set natural frequency? [Y/N]: 'y'
Natural frequency: 6.2706.

(1) **Vortex shedding response**:

(i) Value of vortex shedding frequency is 3.683 Hz and velocity of shell-side fluid = 0.15 m/s.
(ii) Value of Strouhal number is 0.465 at pitch ratio = 0.125 with respective 30° layout pattern.

Damping Factor: 0.2256 Reduced damping: 1.3356
Natural Frequency: 6.2706 Hz

(2) **Turbulent buffeting response**:

(i) Fig. 2a shows graph which is same as Fig. 1a which shows the graph vortex shedding vs flow velocity.

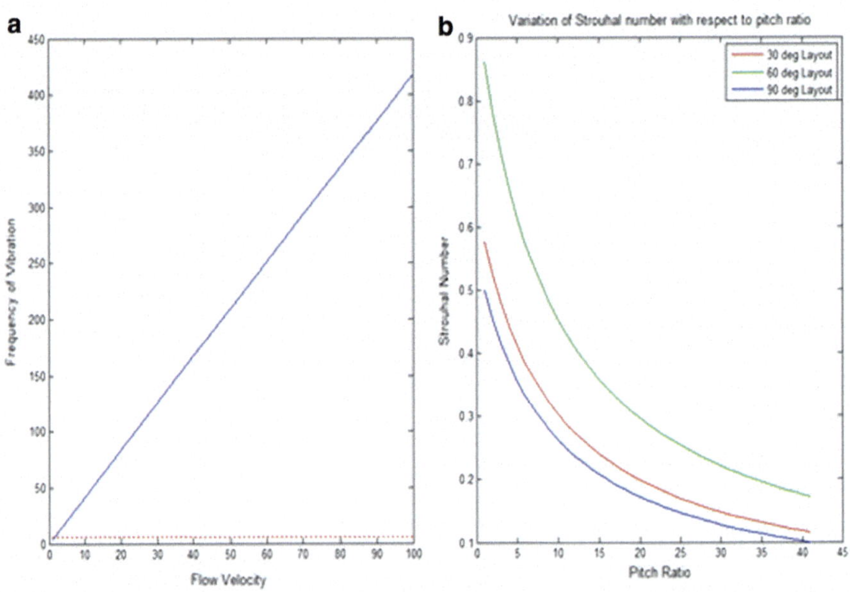

Fig. 1 **a** Frequency of vibration versus cross flow velocity. **b** Strouhal number versus pitch ratio

(ii) Fig. 2a shows the graph of turbulent buffeting frequency on the *Y*-axis and diameter on the *X*-axis. The value of buffeting frequency = 3.8237 Hz at diameter 0.0191 m.

Damping Factor: 0.2256 Buffeting Frequency 3.8237 Hz
Natural Frequency: 6.2706 Hz

(3) **Fluid elastic instability response**:

(i) Fig. 3a shows the graph of the critical velocity vs fluid elastic instability (zeta). The value of critical is equal to 4.649 m/s at the fluid elastic instability parameter $(K) = 4.50$.
(ii) Fig. 3b shows the graph of the critical velocity vs fluid elastic instability (zeta). The value of reduced is equal to 3.569 m/s (near about) at the fluid elastic instability parameter $(K) = 4.50$.

Damping Factor: 0.2256 Through Critical Velocity: 4.6496 m/s
Natural Frequency: 6.2706 Hz

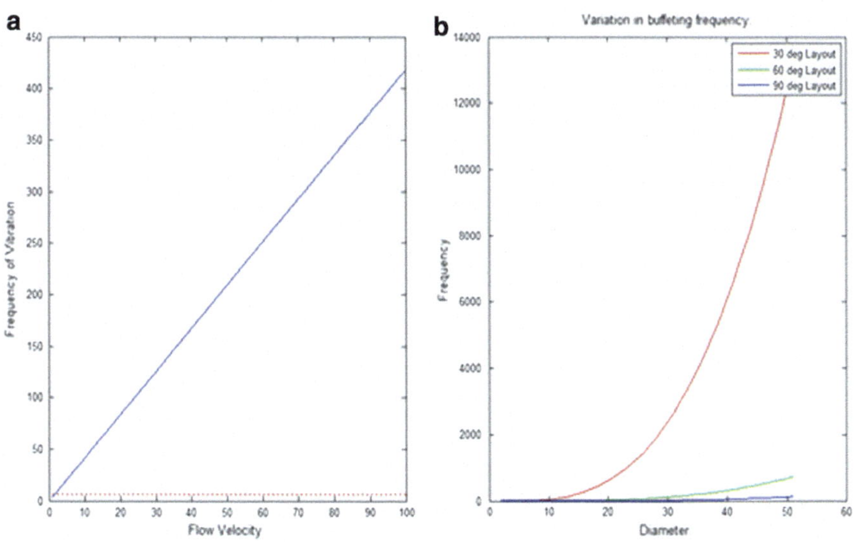

Fig. 2 **a** Frequency of vibration versus flow velocity. **b** Turbulent frequency versus diameter

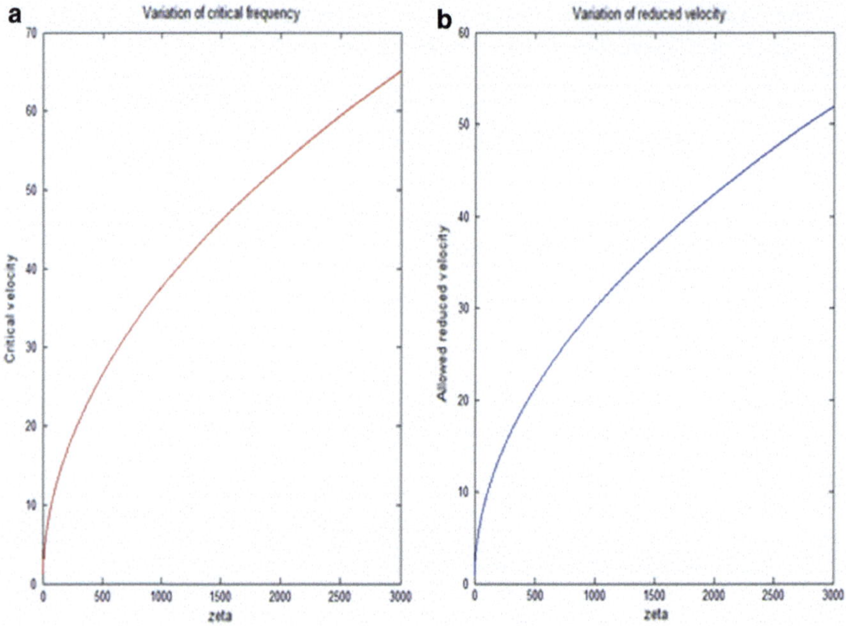

Fig. 3 **a** Critical velocity versus Zeta (K). **b** Reduced velocity versus Zeta (K)

8 Result and Interpretation

Table 1, it is found that the system is free from flow-induced vibration, and results are in good agreement with the available literature data (Table).

Table 1 Result of Case no: 01 (From the available literature)

Parameter	Result of MATLAB program	Result of the available literature
Damping factor	0.2256	0.2256
Reduced damping	1.3356	1.3356
Natural frequency (Hz)	6.2706 (ANSYS)	6.2706 (ANSYS)
Velocity of cross flow (m/s)	0.15	0.15
Through critical velocity (m/s)	4.5 (K = 4.5)	4.358 (K = 4.5)
Buffeting frequency (Hz)	3.8237	2.52078

9 Conclusion

This work outlines flow-induced vibration response such as vortex shedding, turbulent buffeting, and fluid elastic instability of shell and tube heat exchanger single straight tube. Compassion between MATLAB program and the available literature results is listed down:

1. Damping factor, reduced damping and values are same.
2. Critical velocity value for MATLAB program is 4.5 and 4.358 m/s for the available literature when $K = 4.5$; the results are near about same.
3. Buffeting frequency values for MATLAB program and the available literature are 3.8237 Hz, 2.52078 Hz, respectively. Both values of frequency are less than natural frequency value than it is acceptable value for working conditions.

To check flow-induced vibration response based on the vortex shedding, turbulent buffeting, fluid elastic instability by using MATLAB code and results are in accordance with the available literature data. Since the fretting wear of tubes can cause failure of the equipment leading to heavy loss of life and money, this program can be used as an effective design tool at the initial stages to model conservative design of heat exchangers.

References

1. Paidoussis MP (1982) A Review of Flow Induced vibration in reactor and reactor components. Nucl Eng Des 74:31–60
2. Kuppan T (2000) Heat exchanger design hand book. Marcel Dekker, Inc., New York
3. Khushnood S et al (2004) A review of heat exchanger tube bundle vibrations in two-phase cross-flow. Nucl Eng Des 230:233–251
4. Blevins RD (2001) Flow-induced vibration. Krieger Publishing Company Malabar, Florida

Temperature Regulation of CIS Photovoltaic Module Using Eutectic Salt Hydrate as PCM

P. Ramanan, K. Kalidasa Murugavel, D. Hari Kishan, and G. Suriyanarayanan

Abstract The work presents the method to increase the electrical efficiency and power output of photovoltaic (PV) panel with the use of phase change material (PCM). $CaCl_2{\cdot}6H_2O$–$Fe_3Cl_2{\cdot}6H_2O$ eutectic has a suitable melting point and high latent heat for temperature regulation of PV panel. The work has been focused on the experimental setup and simulation heat extraction from the PV panel with the use of ANSYS software. A modification of copper indium diselenide (CIS) PV module from Solar Frontier (SF170-S) was made with a eutectic mixture (70:30) of calcium chloride hexahydrate (70%) and iron (III) chloride hexahydrate phase change material. The cell temperature of the PV panel with and without PCM was measured and compared for two typical days. The simulation of the PV–PCM systems comprising of both PV panels was performed using ANSYS (Fluent) software, followed by the comparison of the results of actual experimental data. The experimental results show that the maximum temperature difference on the surface of PV panel without PCM was 9 °C higher than that on the panel with PCM in a period of one day. Referring to experimental results, the calculation of the maximum and average increase of power gain was made for PV–PCM panel. Final results show that the electricity production of PV–PCM panel was higher for 96.55 Whr in a particular day of experimentation.

Keywords Photovoltaic module · Phase change material · ANSYS · Efficiency · Power gain

1 Introduction

The addition of phase change material (PCM) to a solar cell has been proposed as a method to increase solar PV energy output by keeping the temperature of PV cells close to the ambient. The PCM is a layer of high latent heat capacity which acts as a heat sink, absorbing heat that is transferred from a PV cell. Solar cell efficiency is

P. Ramanan (✉) · K. Kalidasa Murugavel · D. Hari Kishan · G. Suriyanarayanan
Department of Mechanical Engineering, National Engineering College, K.R. Nagar, Kovilpatti, Tamil Nadu 628503, India
e-mail: ramanan_ces@nec.edu.in

L.-J. Yang et al. (eds.), *Proceedings of ICDMC 2019*,
Lecture Notes in Mechanical Engineering,
https://doi.org/10.1007/978-981-15-3631-1_60

dependent on cell temperature, with a drop in efficiency of 0.45% (relative) for every 1 °C rise in cell temperature for crystalline silicon. Therefore, any mechanism which reduces the cell temperature, particularly at times of high irradiance, will increase cell efficiency and PV energy output.

The PV–PCM study was primarily reported by Stultz et al. [1] using eicosane PCM, which increased the electrical efficiency by 1.4% and also revealed that the performance of PCM in PV cooling could be improved by increasing the thermal conductivity of the material. Haung et al. [2] investigated BIPV–PCM system experimentally and validated numerically using paraffin wax as the PCM. The system performance was evaluated by incorporating rectangular aluminum container and also integrating metallic fins with rectangular aluminum container, placed below the module [3]. The result illustrated that PCM with internal fins reduced the temperature rise compared to the single flat aluminum plate. Haung et al. [4] validated the experimental readings with numerically predicted temperature distribution values using 2D and 3D heat transfer models.

Hasan et al. [5] evaluated the thermal performance of five PCMs for the small-scale BIPV system to regulate the surface temperature under indoor environment condition. The results of the study revealed that PV with PCM system brings temperature reduction of 18 °C for pure salt hydrate ($CaCl_2{\cdot}6H_2O$) and a eutectic mixture of capric–lauric acid (C–L), 16 °C for commercial blend (SP22) and a eutectic mixture of palmitic acid (C–P) and 14 °C temperature reduction in paraffin wax (RT 20). Ho et al. [6] performed CFD modeling to analyze the performance of BIPV module integrated with microencapsulated PCM with melting point of 26 °C, which was placed on the rear side of the module to enhance the performance. Only slight increase in electrical efficiency by 0.09% was reported. Also, the temperature reduction from 49 to 47 °C in summer and from 35 to 30.5 °C in winter portrayed that using an MEPCM is not an appropriate solution for cooling BIPV module due to lower thermal conductivity of encapsulation materials. A similar study was performed by Ho et al. [7] using water-saturated MEPCM to enhance the performance of the module, in which electrical efficiency of the module was increased by 2% compared to reference BIPV module.

Laura et al. [8] investigated the BIPV–PCM prototype under real conditions, which was installed on the main façade of Solar XXI office building in Lisbon. The length and the breadth of the prototype under study were 0.73 × 1.75 m, the rear sides of the PV module being placed with PCM gypsum wallboard. The maximum electrical and thermal efficiency reported was 10% and 12%, respectively, on the building façade. Biwole et al. [9] performed a numerical heat and mass transfer studies of PV–PCM and stated that the surface temperature of PV–PCM system attained 40 °C in 80 min, while the reference PV module without PCM reached 40 °C within 5 min.

Stropnik et al. [10] increased the electrical efficiency and power output of the PV module with the help of PCM. The PCM of RT28 HC is encapsulated with the acrylic glass on the backside of the PV module. An average increase of electrical power was 9.2%. The PV cell temperature of the conventional PV panel reached a maximum of 75.2 °C, which was 57.7 °C above the ambient air temperature. The PV cell temperature of the PV–PCM panel reached a maximum of 44 °C, which is

26.7 °C above the ambient air temperature. Sharma et al. [11] studied the integration of RT42 PCM in the building-integrated concentrated photovoltaic module. It was reported that the electrical efficiency of the module increased by 7% and the module surface temperature reduced by 3.8 °C under indoor operating conditions.

Royo et al. [12] developed a mathematical model to determine the behavior of BIPV systems with and without PCM. Moreover, they reported that Glauber salt PCM could reduce the temperature up to 7.5 °C in Zaragoza and 10°C in Seville climatic conditions. Hasan et al. [13] developed PV–PCM system and evaluated with two phase change materials (PCMs): a salt hydrate, $CaCl_2 6H_2O$, and a eutectic mixture of fatty acids with capric–palmitic acid in two different locations at Dublin, Ireland, and Vehari, Pakistan. The peak PV temperature of the module was regulated by 7 °C and 10 °C for capric–palmitic PV–PCM system and salt hydrate system, respectively, compared to the reference PV.

2 Methodology

The methodology adopted in the performance analysis of the PV–PCM system is as follows.

- The aim of the study is to design and fabricate the PV–PCM and regulate the temperature rise of the module using PCM for improved electrical and thermal performance.
- The thermal regulation of the module is provided by an inorganic salt eutectic as PCM ($CaCl_2 \cdot 6H_2O$–$Fe_3Cl_2 \cdot 6H_2O$), and its effect on module's electrical and thermal performance is tested under real outdoor environment.
- The phase change of the PCM has been studied using ANSYS—Fluent.
- Inorganic PCM based on salt has been selected for this study due to their high latent heat of fusion, chemical, non-reactiveness, environment-friendliness, non-toxicity and low cost.

3 Numerical Study

The CIS PV panel was modeled using SolidWorks 16 software. The specifications of the panel are taken from the datasheet of the CIS panel and the thermal parameters obtained from the literature survey. The PV panel contains five layers. Each layer was individually designed and assembled. PV cell was sandwiched between the EVA layer, while the top layer consists of glass and the bottom layer is tedlar sheet. The properties required for the numerical simulation are listed in Table 1. Figure 1a shows the three-dimensional view of the model, and Fig. 1b shows the meshing of the model using ANSYS R18.1.

Table 1 Thermo-physical properties of PV

S. No.	Layer	Thickness (m)	Thermal conductivity (W/m-K)	Density (kg/m^3)	Specific heat capacity(J/kg K)
1	Glass	0.003	1.8	3000	500
2	Cell	225e–6	148	2330	677
3	EVA	500e–6	0.35	960	2090
4	Tedlar	0.0001	0.2	1200	1250
5	Al	–	237	2700	900

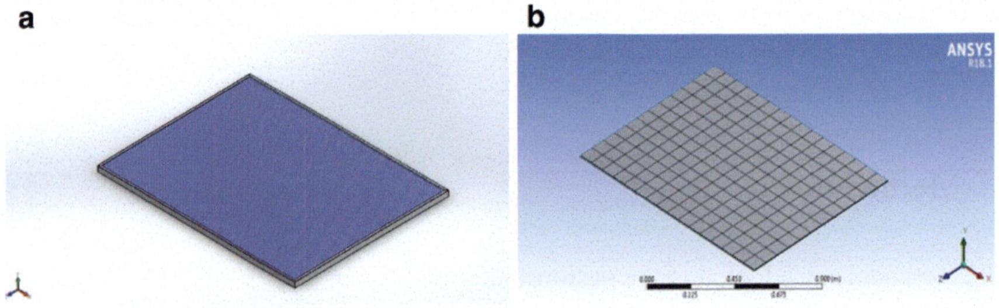

Fig. 1 **a** Complete 3D model of CIS panel and **b** meshing of the PV panel

Convective term in momentum equations and convective energy equations were discretized using second-order upwind interpolation scheme. The coupling between the pressure and velocity is done by SIMPLE algorithm and PRESTO which is suitable for pressure interpolation. Geo-reconstruct is adopted for volume fraction scheme. The iterations are calculated by using a CFD code in ANSYS (Fluent) 18.1.

The initial part of the process is conducted at room temperature, and state of the PCM is solid. Patch the PCM domain for the phase-2, and the volume fraction is set as zero. Auto-save mode is used for every 15 time steps. Standard initialization has been chosen so as to compute from top surface of the panel. The time step is set as 1 s. In post-processing, the continuity equation and the energy equation could be solved to interpret the results. Mass fraction percentage at any instant of time has been be analyzed in the transient model. Using this mass fraction percentage, the influence of heat transfer depending on the phase change is found. The variation on temperature distribution along the panel depending on the averaged irradiance value has been individually simulated for sample days. These results are viewed by creating a contour for PV cell region.

The phase change material would be packed by the tedlar sheet of thickness of 3 mm. It was created using the SolidWorks modeling software and then assembled with the CIS panel. PCM absorbs the energy from the panel to reduce the operating temperature. It will improve the conversion efficiency of the panel. So, the PCM is energy storage model. We are using a mixture of iron II chloride hexahydrate and calcium chloride hexahydrate as a PCM. The specification of the PCM can be obtained from the laboratory.

4 Materials and Methods

It is to experiment with the passive cooling of PV–PCM, and two 170 W CIS PV panels from Solar Frontier (SF170-S) were used. One acts as reference panel, while the other as a PV–PCM system. Photographic view of the experimental setup is shown in Fig. 2. The specifications of the PV panel are shown in Table 2. Two PV panels are mounted on the frame, and the thermocouple was fixed on the back surface of the PV module. To test the electrical power of the PV modules, an electrical load set of rheostat, a voltmeter (150 V) and an ammeter (5 A) were used. The experimentation was carried out for two sample days with almost similar ambient conditions in the month of March 2018.

Fig. 2 Experimental setup with electrical load and data logger

Table 2 Photovoltaic panel technical specification

Sl. No.	Parameter	CIS
1.	Rated power (KWp)	1.36
2.	Module nominal power (W)	170
3.	Module voltage at maximum power V_{mp}	87.5
4.	Module current at maximum power I_{mp}	1.95
5.	Module open-circuit voltage V_{oc}	112.0
6.	Module short-circuit current I_{sc}	2.20
7.	Temp. coeff. of power (%/K)	−0.31
8.	Number of module string	4
9.	Module tilt (°)	10

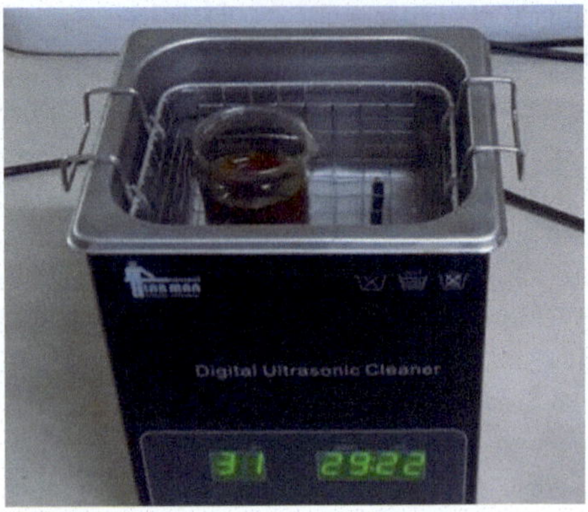

Fig. 3 Preparation of the PCM in ultrasonicator

4.1 Selection of PCM

From the literature survey, we have found only less research works have been reported on cooling by hydrated PCMs that too very few works had been done on eutectic hydrated PCMs; hence, calcium chloride hexahydrate ($CaCl_2{\cdot}6H_2O$)–ferric chloride hexahydrate ($Fe_3Cl_2{\cdot}6H_2O$) was used as a eutectic PCM.

4.2 Characterization of PCM

To find the optimal thermo-physical properties of the eutectic PCM, various proportions have been characterized. Different sample proportions of sample 1 (60% $CaCl_2{\cdot}6H_2O$ and 40% $Fe_3Cl_2{\cdot}6H_2O$), sample 2 (70% $CaCl_2{\cdot}6H_2O$ and 30% $Fe_3Cl_2{\cdot}6H_2O$) and sample 3 (80% $CaCl_2{\cdot}6H_2O$ and 20% $Fe_3Cl_2.6H_2O$) were made, and these proportions were uniformly mixed in ultrasonicator as shown in Fig. 3.

The samples were tested in digital scanning calorimeter (DSC) by which the physical and thermal properties of samples were obtained out of which optimal sample proportion is selected. From the characterization, it was inferred that the sample 2 (70% $CaCl_2{\cdot}6H_2O$ and 30% $Fe_3Cl_2{\cdot}6H_2O$) has optimal physical and thermal properties as stated in Table 3.

4.3 Encapsulation of PCM

In the proposed PV–PCM module, a new method is devised to integrate the PCM. The eutectic salt phase change material is incorporated on the rear side of the PV module. The rear side of module is facilitated to incorporate PCM by amplification with the 3-mm-thick glass of suitable length and breadth. The salt eutectic is packed

Table 3 Properties of the phase change material

S. No.	Property (units)	Value
1.	Melting point (K)	305
2.	Latent heat of fusion (J/kg)	208,000
3.	Density (kg/m^3)	1710 at solid 1500 at liquid
4.	Specific heat capacity (J/kg-K)	1808 at 296 K
5.	Thermal conductivity (W/m-K)	0.54 at 312 K
6.	Viscosity (kg/m-s)	0.001372
7.	Pure solvent melting heat (J/kg)	190,000
8.	Solidus temperature (K)	301
9.	Liquidus temperature (K)	302.5

airtight using tedlar sheet and placed in between the rear side of the PV module. The PCM is encapsulated in the backside of the CIS PV panel, and the encapsulation consists of 11 compartments to minimize the accumulation of PCM salts on one side. Wall of the encapsulation chamber was made of flexible poly glass, and the PCM would be sandwiched between tedlar sheet of solar panel and the tedlar sheet to encapsulate the chamber. As the PCM salt is highly reactive with adhesives, an inert gel of Anabond is used to fix the poly glass and tedlar in the chamber. The thermo-physical properties of the salt eutectic are listed in Table 2. The rear view of the partially PCM-filled solar module is shown in Fig. 4a.

The integration of the PCM on the rear side of the PV cell has more heat transfer rate than the conventional integration in the back surface of the module. After encapsulation of the PCM on the backside of the PV module, it is kept in the sun simulator as shown in Fig. 4b to view the phase change or expansion of PCM from solid to liquid. As sun simulator produces heat on the surface of the PCM, the phase change occurs.

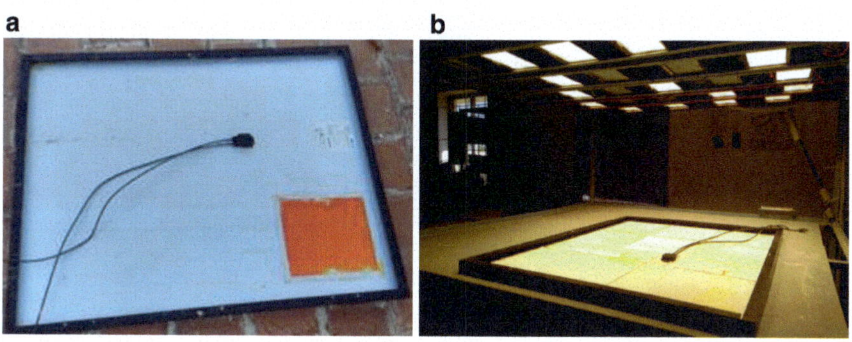

Fig. 4 **a** Encapsulation of PCM on the rear side of the PV module and **b** simulation under sun simulator

5 Results and Discussion

5.1 Contour of Liquid Fraction

Figure 5 represents the variation of liquid fraction with time. We can easily examine the motion of melting interface as the time passes. The red color represents the condition when material is completely liquid ($a_q = 1$), and blue color represents the condition when material is completely solid ($a_q = 0$). It will take 11 min to change their phase completely into the liquid. During the phase change, it will absorb the heat energy from the PV module to reduce the temperature of the panel.

The variation in melting fraction (% of PCM which has been melted) of the PCM increases with increase in time. We can see that $t = 0$, and all the PCM is in solid phase. As the time passes due to heating, the PCM gets melted and the value of liquid fraction increases with time. The value of melting fraction 10%, 20%, 25%, 60%, 80% and 100% is for the melting time of 1, 2, 4, 6, 8 and 10 min. At the end of the melting cycle, i.e., after 11 min, all the PCM has melted (100%). It is noticed that rate of melting is almost same at the starting of the melting process and it increases in middle and end process.

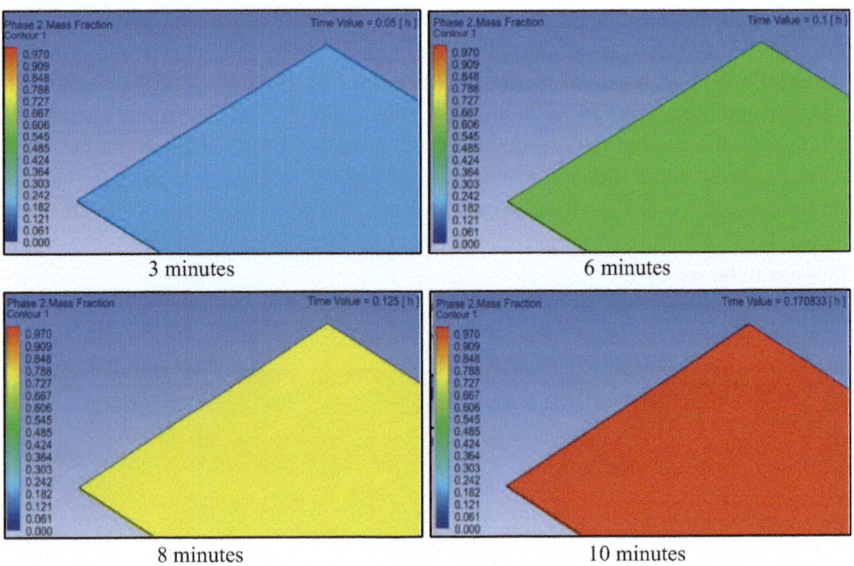

Fig. 5 Contour of liquid fractions of the PCM

5.2 Contour of Thermal Distribution of PV Module

Figure 6a, b represents the comparison between the temperature contour of PV–PCM system and the reference panel. It shows that the maximum temperature can be attained for both panels. The maximum temperature of the reference panel is 338 K while that of PCM-packed panel is about 314 K. So, the PCM can absorb the heat energy that can be derived from the Fourier law of conduction.

5.3 Temperature Regulation of PV Module

The aim of the present experiment is to study of the PV temperature regulations using integrated phase change materials. Experiments were conducted outdoors in the Centre for Energy Studies, National Engineering College, Kovilpatti, Tamil Nadu. The institution is located on the latitude 9.1674° N and longitude 77.8767° E with an average elevation of 106 m above the sea level. The experimentation was done in the period March 2018, on selected certain days of high ambient temperature and solar radiation to assure the rise of the PV module temperature to the required level. The incident solar radiation and ambient temperature at the site are shown in Fig. 7 for the typical day of experimentation. High solar intensity and ambient temperature and low wind speed lead to high surface temperature of the reference PV module and allow the possibility of PV temperature regulation by using PCM which leads to higher output power. The results of the reference panel and panel with PCM were compared, and the temperature distribution is shown in Fig. 8.

From the simulation, it is visible that panel without PCM has reached a peak temperature of about 63 °C where the panel with PCM had reached only a maximum temperature of 54 °C with reduction in temperature of about 9 °C. Initially, the result obtained from the experimentation of panel with PCM and reference panel is compared with simulation. The experimentation is done on two typical days in the month of March 2018 by which the results were compared with the output parameters of the PV module. The ultimate aim of the work is to reduce the back surface temperature of the PV module with the use of PCM; hence, the eutectic PCM material 70% $CaCl_2{\cdot}6H_2O$ and 30% $Fe_3Cl_2{\cdot}6H_2O$ of high heat capacity of 1808 kJ/Kg K is used to reduce the temperature on the backside of the PV module as shown in Fig. 9a, b for two sample days; the optimistic results are obtained by which the eutectic of hydrated inorganic salt absorbs the adequate amount of latent heat.

From the temperature distribution of day 1, the back surface of the PV module is reduced a maximum of 8 °C by which the reference panel reaches the maximum temperature of about 65 °C, whereas the panel with PCM reaches a maximum of 57 °C. Hence, an adequate amount of heat is removed and the same test is conducted on day 2 which has the same trend as on day 1. Also, we infer the variation in simulation and actual result in minimum, and hence the simulation results are met with the experimental result with the maximum variation of only 3 °C.

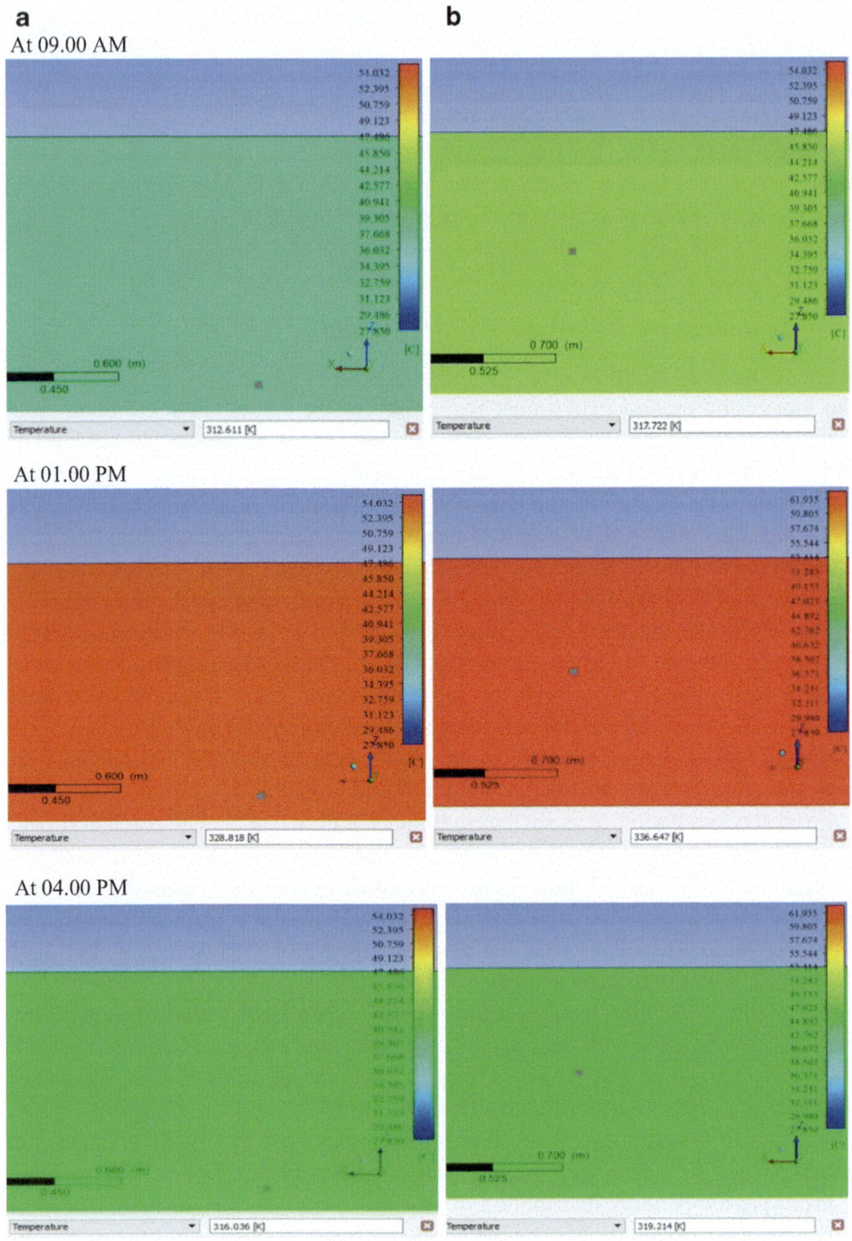

Fig. 6 **a** Contour of temperature for PV–PCM system and **b** contour of temperature for reference PV

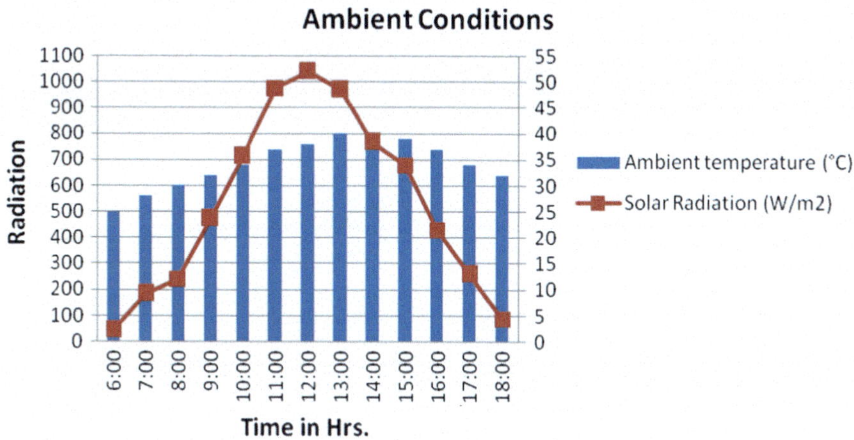

Fig. 7 Average ambient conditions on the typical day of experimentation

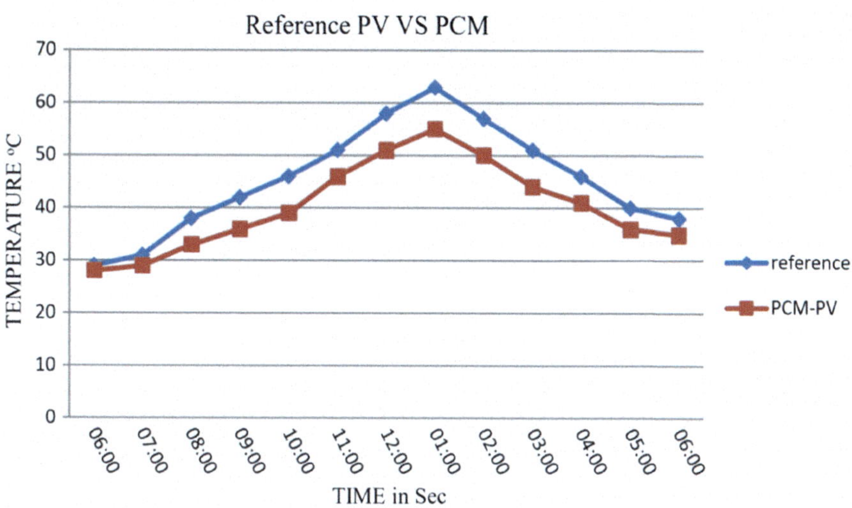

Fig. 8 Comparison of PV surface temperature with PCM and without PCM

From the above figure, it is obvious that reduction in temperature of 9 °C is obtained by which the maximum temperature of reference panel is 63 °C whereas panel with PCM reaches only 54 °C. It inferred the variation in simulation and actual result in minimum, and hence the simulation results are met with the experimental result with the maximum variation of only 2 °C. Also, it is observed that the variation in temperature in different days occurs as the radiation differs with time and day as the radiation on day 1 is high while compared with day 2; hence, the variation in temperature is minimum of about 3 °C. The relative and absolute error for the

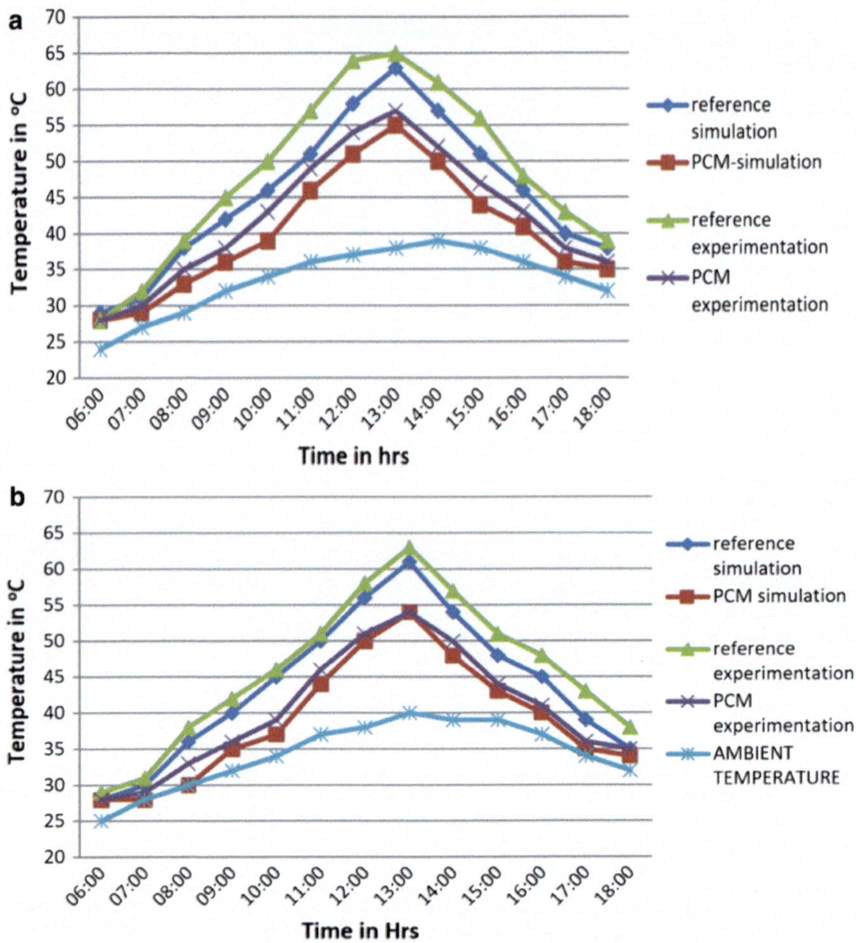

Fig. 9 **a** Comparison of back surface temperature with and without PCM (simulation vs. experimentation)—sample day 1; **b** comparison of back surface temperature with and without PCM (simulation vs. experimentation)—sample day 2

simulation and the experimental reading was calculated for both reference panels, and the panel with PCM has the error obtained so minimum as shown in Table 4a, b.

5.4 *Energy Gain of PV Module*

The reduction in temperature would result in increase in power output as the voltage gain takes place. Even though the power output directly depends on the radiation incident on the front side of PV module, the increase in temperature increases the

Table 4 **a** Uncertainty between predicted and experimental results—sample day 1; **b** uncertainty between predicted and experimental results—sample day 2

Time (H)	Radiation (W/m^2)	Ref. PV temp. (°C)	Predicted temp. (°C)	Uncertainty (%)	PV–PCM temp. (°C)	Predicted temp. (°C)	Uncertainty (%)
a							
6:00	38	28	28	0	28	28	0
7:00	126	32	30	0.06	30	28	0.06
8:00	229	39	37	0.05	35	35	0
9:00	452	45	44	0.02	40	39	0.02
10:00	403	50	50	0	46	44	0.04
11:00	670	57	55	0.03	52	50	0.03
12:00	1002	64	60	0.06	55	53	0.03
13:00	1012	65	63	0.03	56	55	0.01
14:00	809	61	58	0.04	53	50	0.03
15:00	676	56	53	0.05	49	47	0.04
16:00	352	48	46	0.04	45	43	0.04
17:00	256	43	41	0.04	40	39	0.02
18:00	87	39	37	0.05	37	35	0.05
b							
6:00	45	29	28	0.03	28	28	0
7:00	186	31	30	0.03	29	28	0.03
8:00	242	38	36	0.05	33	30	0.09
9:00	474	42	40	0.04	36	35	0.02
10:00	715	46	45	0.02	39	37	0.05
11:00	978	51	50	0.01	46	44	0.04
12:00	1042	58	56	0.03	51	50	0.01
13:00	970	63	61	0.03	55	54	0.01
14:00	769	57	54	0.05	50	48	0.04
15:00	680	51	48	0.05	44	43	0.02
16:00	432	46	45	0.02	41	40	0.02
17:00	261	40	39	0.025	36	35	0.02
18:00	89	38	35	0.07	35	34	0.02

electrical losses and hence the voltage drop takes place. Hence, the adequate heat extraction provides an increase in power gain. The power gain varies with the radiation, and the result obtained shows the increase in power output due to the heat storage capacity of the PCM; the power output is maximum at the peak radiation and decreases gradually when the radiation falls, and this trend follows on both days.

From Fig. 10a, we can find that the panel with PCM has high power gain while compared with the reference panel, from the results of day 1 the power output from the reference panel is 986.75 Whr and the power output from the PCM-packed PV module is 1083.3 Whr. Hence, PCM-packed PV had a power gain of 96.55 Whr than that of reference PV module.

From Fig. 10b, the results of day 2 were obtained even though there is considerable difference in power gain in PCM than reference; its result is less while compared to day 1 as the variation in radiation takes place which affects the power output of the PV module on day 2. From the result of day 2, the power output from the reference panel is 940.6 Whr and the power output of PCM-packed PV module is 1012.95 Whr.

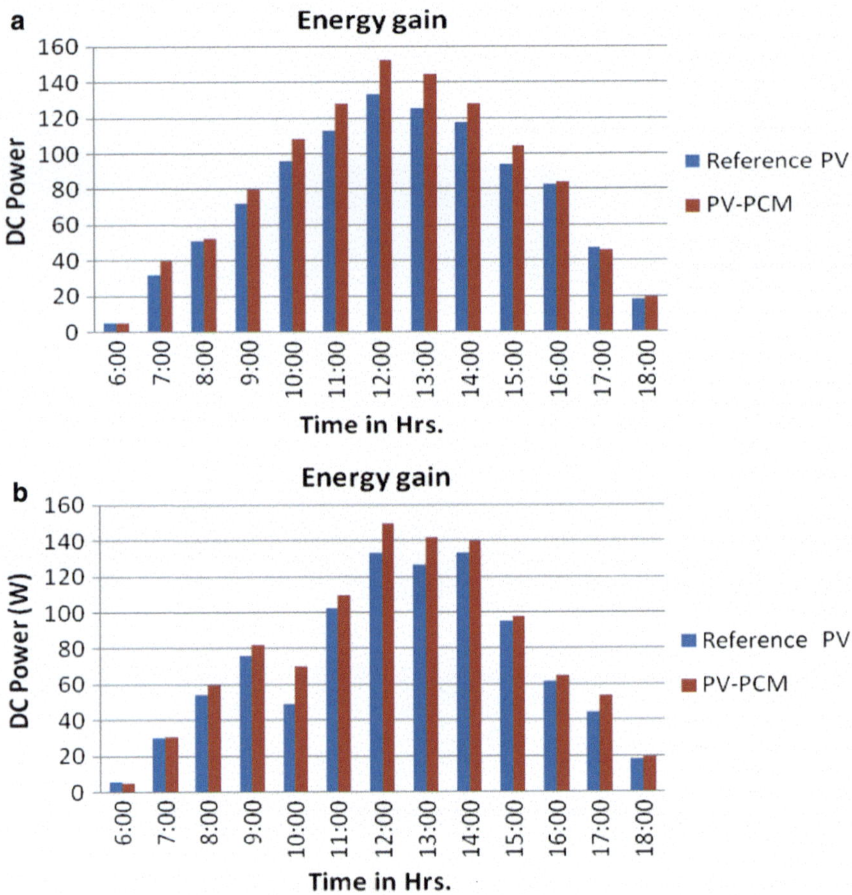

Fig. 10 **a** Energy gain for PV–PCM system (sample day 1) and **b** energy gain for PV–PCM system (sample day 2)

Hence, the PV module with PCM has a power gain of 1345.8 Whr compared to the reference module. But the power gain in day 2 is 940.2 Whr less than that of day 1 due to variation in radiation between days 1 and 2.

6 Conclusion

The simulation studies and experimentation have been done on two CIS PV panels of 170 W power rating in which one panel is encapsulated with PCM on the back surface while other act as reference. The following conclusions were obtained.

- The simulation and experimental result show the incorporation of PCM on the back surface of the PV module decreases the back surface temperature of about 9 °C from the reference panel.
- The PCM–PV module has the power output of 1012.95 Whr on day 1 and 1091.3 Whr on day 2 with the average energy gain of 103.6 Whr.
- A considerable amount of reduction in PV temperature has been noticed, and hence by varying the thickness better performance can be archived by providing optimal thickness of the eutectic PCM.
- Various proportions of eutectic mixtures of a variety of PCMs can be studied by which significant improvement in the performances can be noted.

References

1. Stultz JW, Wen LC (1977) Thermal performance testing and analysis of photovoltaic modules in natural sunlight LSA. Task Report 5101-31. Jet Propulsion Laboratory, Pasadena, California
2. Huang MJ, Eames PC, Norton B (2004) Thermal regulation of building integrated photovoltaics using phase change materials. Int J Heat Mass Trans 47:275–733
3. Huang MJ, Eames PC, Norton B (2006) Phase change materials for limiting temperature rise in building integrated photovoltaics. Sol Energy 80:1121–1130
4. Huang M, Eames P, Norton B, Hewitt N (2011) Natural convection in an internally finned phase change material heat sink for the thermal management of photovoltaics. Sol Energy Mater Sol Cells 95(7):1598–1603
5. Hasan A, McCormack SJ, Huang MJ, Norton B (2010) Evaluation of phase change materials for thermal regulation enhancement of building integrated photovoltaics. Sol Energy 84:1601–1612
6. Ho CJ, Tanuwijava AO, Lai C-M (2012) Thermal and electrical performance of a BIPV integrated with a microencapsulated phase change material layer. Energy Build 50:331–338
7. Ho CJ, Jou B-T, Lai C-M, Huang C-Y (2013) Performance assessment of a BIPV integrated with a layer of water-saturated MEPCM. Energy Build 67:322–333
8. Laura A, Pereira R, Helder G, Andreas A (2014) Thermal performance of a hybrid BIPV–PCM: modeling, design and experimental investigation. Energy Procedia 48:474–483
9. Biwole PH, Eclache P, Kuznik F (2013) Phase-change materials to improve solar panel's performance. Energy Build 62:59–67
10. Stropnik Rok, Stritih Uros (2016) Increasing the efficiency of PV panel with the use of PCM. Renew Energy 97:671–679

11. Sharma S, Tahir A, Reddy KS, Mallick TK (2016) Performance enhancement of a Building-Integrated Concentrating Photovoltaic system using phase change material. Solar Energy Mater Solar Cells 149:29–39
12. Royo P, Ferreira VJ, López-Sabirón AM, Ferreira G (2016) Hybrid diagnosis to characterise the energy and environmental enhancement of photovoltaic modules using smart materials. Energy 101:174–189
13. Hasan A, McCormack SJ, Huang MJ, Sarwar J, Norton B (2015) Increased photovoltaic performance through temperature regulation by phase change materials: materials comparison in different climates. Solar Energy 115:264–276

Influence of Nanofluid and Inclination Angle on the Temperature Distribution of the Thermosyphon

Sidhartha Das, Asis Giri, and S. Samanta

Abstract An experimental investigation on two-phase closed thermosyphon was carried out with deionized water and TiO_2 nanofluid as a working medium. The thermosyphon was set at different inclination angles of 30°, 45°, 60° and 90° to observe the effect of inclination angle on the wall temperature gradient of the thermosyphon. The heat input was varied from 6 to 9 W. Result indicated that inclination angle plays a significant role in wall temperature reduction of the thermosyphon in which better reduction was observed at an inclination angle of 60°. Moreover, wall temperature decreases significantly with the use of TiO_2 nanofluid compared to that of deionized water.

Keywords TiO_2 nanofluid · Wall temperature · Thermosyphon

1 Introduction

A thermosyphon is a passive heat transfer device which is used for its better thermal performance compared to that of a copper pipe. The TPCT works under the action of gravity for which the condensing section is placed above the evaporating section to facilitate the flow of condensate back to the evaporator section. Thermosyphon is a wickless heat pipe [1–5] which can be oriented at any inclination angle except the horizontal position [6–8].

The TPCT has a simple structure and low thermal resistance, higher efficiency and low manufacturing costs. Therefore, it finds its application in many fields such as solar heating systems, cooling of electronic component, heat recovery systems, etc. [9, 10]. A lot of research has been carried out with nanofluid as a medium of heat transfer in thermosyphon. Nanofluid is fluid which consists of nanoparticles suspended in the base fluid which have enhanced thermal properties compared to that of deionized water [11, 12].

S. Das (✉) · A. Giri · S. Samanta
Department of Mechanical Engineering, NERIST, Itanagar, Arunachal Pradesh 791109, India
e-mail: sidhartha714@gmail.com

L.-J. Yang et al. (eds.), *Proceedings of ICDMC 2019*,
Lecture Notes in Mechanical Engineering,
https://doi.org/10.1007/978-981-15-3631-1_61

A lot of research have been made using DI water, ethanol, methanol, refrigerant-based nanoparticles and pure water-based nanofluid, but very few works have been done with ethylene glycol as a surfactant in TiO_2 water-based nanofluid. Moreover, the effect of inclination angle on the wall temperature distribution of the thermosyphon is not discussed in detail. Therefore, an attempt is made to study the influence of inclination angle on the wall temperature distribution of the thermosyphon at different heat inputs and inclination angles.

2 Materials and Methods

2.1 Nanofluid Preparation

TiO_2 nanoparticle was purchased from Srlchem (Mumbai). The average size of the particle is around 120 nm as seen in the scanning electron microscope (SEM) image in Fig. 1. For the preparation of TiO_2 nanofluid, two-step method [13] was followed in which required nanoparticles were initially mixed in the deionized water, and after that, same proportion of ethylene glycol is added as a surfactant, and sonication is done in an ultrasonic bath for four hours. The concentration of nanofluid prepared was 0.001 vol% of TiO_2 nanofluid. The sample looks milky and has a colloidal appearance.

2.2 Thermosyphon and Experimental Setup

To analyze the performance of nanofluid on the wall temperature of the thermosyphon, an experimental setup is prepared. The total length of the thermosyphon

Fig. 1 SEM images of TiO_2 nanoparticles

is 205 mm, and the evaporator and the condensing section is 85 mm and 75 mm, respectively. The outer diameter of the thermosyphon is 6 mm. For measuring the wall temperature on the thermosyphon, K type thermocouple is used. Eight K types thermocouples were attached to the thermosyphon which were used to measure the temperature at different locations. All these thermocouples were individually calculated in an isothermal bath; after that, the thermocouples were attached to the temperature indicator for calculating the temperature at different heat inputs and inclination angles of the thermosyphon.

For applying constant heat, a heating unit is applied to the evaporator section. The heat input is measured with the help of wattmeter. The power is varied with the help of an autotransformer, and no forced convection was applied for cooling in the condenser section. The study was made to measure the wall temperature of the thermosyphon. The setup was allowed to operate for one hour before any reading was taken during which steady state was attained. The heat input was varied at 6 and 9 W to see the effect of nanofluid on the wall temperature of the thermosyphon. The temperature was measured with the help a temperature indicator.

3 Results and Discussions

3.1 *Wall Temperature of the Thermosyphon*

For a medium of heat transfer from the evaporator section, both deionized water and TiO_2 nanofluid as a working medium were used. From Fig. 2a–d, it can be observed that for every heat input of 6 and 9 W, wall temperature for TiO_2 nanofluid is less compared to that of deionized water. Moreover, the wall temperature from the evaporator section to the condensing section shows a decreasing trend, which shows the effect of using thermosyphon as a heat transfer device. The wall temperature is always higher for all combination of heat loads for deionized water. The trend of decrease of temperature is almost linear from the evaporative section to the condenser section. The lower temperature observed in the wall temperature for 0.001 vol% of TiO_2 nanofluid may be due to the higher thermal conductivity of the nanofluid [3], and moreover, due to the formation of a porous layer inside the thermosyphon, surface area increases, which increases the nucleation sites. The higher the nucleation sites, more will be heat transfer rate, thus reducing the wall temperature of the thermosyphon.

3.2 *Effect of Heat Input and Inclination Angles*

The heat transfer rate of the thermosyphon was found to increase with the increase in heat input, application of TiO_2 nanofluid and change of inclination angle of the

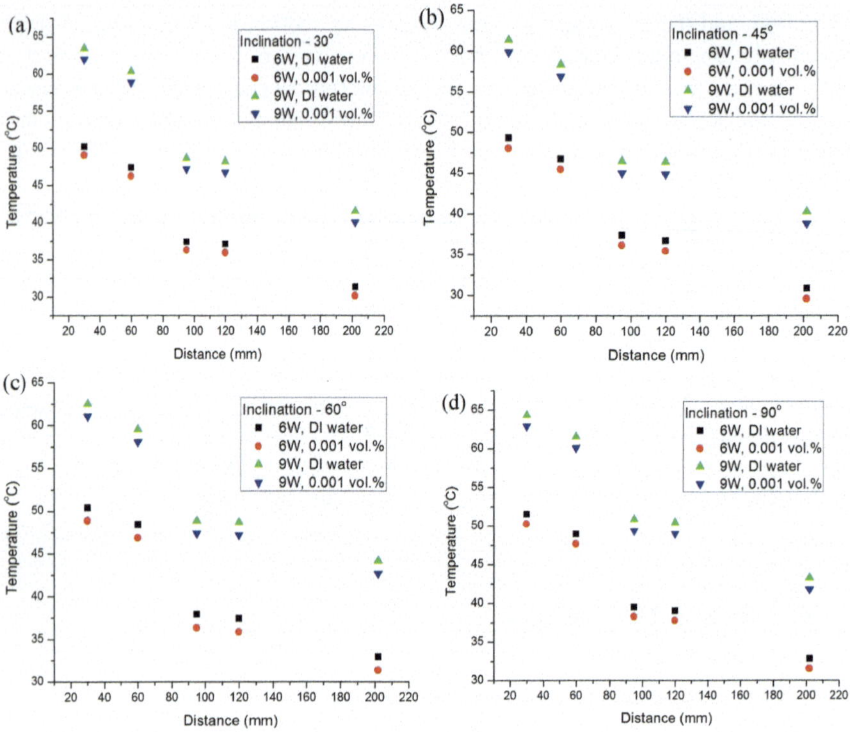

Fig. 2 Wall temperature of the thermosyphon at different heat inputs and inclination angles of the thermosyphon **a** 30° angle of thermosyphon; **b** 45° angle of thermosyphon; **c** 60° angle of thermosyphon; **d** 90° angle of thermosyphon

thermosyphon. The wall temperature of the thermosyphon increases with increasing the heat input at the evaporator section. At higher heat input of 9 W, the amount of heat transfer was higher than the lower heat input of 6 W. At a higher heat input, more vapor pressure is generated which leads to the transfer of heat at a faster rate to the condenser section. Therefore, the efficiency increases. The inclination angle plays a major role in heat transfer. To see the role of inclination on heat transfer on the thermosyphon, different inclination angles of 30°, 45°, 60° and 90° are plotted. From Fig. 2a–d, it can be noticed that at 60° inclination wall temperature of thermosyphon is lower compared to the other inclination angles of 30°, 45° and 90° inclination. This may due to the fact that at an inclination angle of 60°, the vapor flow through the evaporator and condenser section occurs uniformly. The hot liquid after being evaporated in the evaporator section goes to the condenser section where it releases the heat, condenses and flows back to the evaporator section through the adiabatic section. However, the influence of 45° inclination of the thermosyphon is not far off. Similar observation is observed for both heat inputs of 6 and 9 W.

4 Conclusions

The experiment performed showed that with the application of TiO_2 nanofluid and wall temperature of thermosyphon can be decreased compared to that of deionized water. Enhanced performance of thermosyphon was observed at 60° inclination of the thermosyphon for both heat loads of 6 W and 9 W, respectively. Wall temperature decreases from evaporator section to the condenser section of the thermosyphon. With increase in heat input, temperature difference between the evaporator section and condenser section decreases. The wall temperature is always higher for DI water, compared to the other TiO_2 nanofluid.

References

1. Wang X, Xu X, Choi SUS (1999) Thermal conductivity of nanoparticles-fluid mixture. J Thermophys Heat Transfer 13(4):474–480
2. Eastman JA, Phillpot SR, Choi SUS, Keblinski P (2004) Thermal transport in nanofluids. Annu Rev Mater Res 34:219–246
3. Jang SP, Choi SUS (2004) Role of Brownian motion in the enhanced thermal conductivity of nanofluids. Appl Phys Lett 84:4316–4318
4. Keblinski P, Phillpot SR, Choi SUS, Eastman JA (2002) Mechanisms of heat flow in suspensions of nanosized particles (nanofluids). Int J Heat Mass Transfer 45:855–863
5. Han K, Yee SS, Park SH, Lee SH (2002) A study on the improvement of heat transfer performance in low temperature closed thermosyphon. KSME, Int J 16(9):1102–1111
6. Payakaruk T, Terdtoon P, Ritthidech S (2000) Correlation to predict heat transfer characteristics of an inclined closed two phase thermosyphon at normal operating condition. Appl Therm Eng 20:781–790
7. Khandekar S, Joshi YM, Mehta B (2008) Thermal performance of closed two-phase thermosyphon using nanofluids. Appl Therm Eng 47:659–997
8. Pappas CA, Jordan DA, Norris PM (2014) Experimental investigation of the heat transfer performance of a hybrid cooling fin thermosyphon. J Heat Transfer 136:104502–104505
9. Kang SW, Wei WC, Tsai SH, Yang SY (2006) Experimental investigation of silver nanofluid on heat pipe thermal performance. Appl Therm Eng 26:2377–2382
10. Ong KS, Tong WL (2011) Inclination and fill ratio on water filled two phase closed thermosyphon. Int 10th Heat Pipe Symposium, Nov 6–9, Taipei, Taiwan
11. Lee S, Choi SUS, Li S, Eastman JA (1999) Measuring thermal conductivity of fluids containing oxide nanoparticles. J Heat Transfer 121:280–289
12. Choi SUS, Eastman JA (1995) Enhancing thermal conductivity of fluids with nanoparticles. In ASME Int Mech Eng Cong & Exp, Nov 12–17, San Francisco
13. Singh K, Sharma S, Gangacharyulu D (2013) Experimental study of thermophysical properties of Al_2O_3/water nanofluid. Int J Res Mech Eng Tech 3(2):229–233

The manufacturer's authorised representative in the EU is Springer Nature Customer Service Centre GmbH, Europaplatz 3, 69115 Heidelberg, Germany. If you have any concerns regarding our products, please contact ProductSafety@springernature.com

Printed and bound by CPI Group (UK) Ltd, Croydon, CR0 4YY
30/09/2026
02225317-0017